The Modern Coral Reef Aquarium

Svein A. Fosså · Alf Jacob Nilsen

Volume 4

Close up of the sea urchin *Astropyga radiata*
Photo: M. Kirschner

Credits:
All the photos, drawings, diagrams and tables are by the autors, except where otherwise stated in the captions.

Agents:

* Australia - ABS Technologies
P. O. Box 697
Stirling
South Australia 5152
Tel.: (61-8) 83 70 10 66
Fax: (61-8) 83 70 10 68

* Norway - NorZoo A/S
Marken 4
5017 Bergen
Tel.: 55 31 11 21
Fax: 55 96 07 28

* UK - Tropical Marine Centre Ltd.
Solesbridge Lane
Chorleywood
Herforshire
WD3 5SX England
Tel.: 0 19 23 - 28 41 51
Fax: 0 19 23 - 28 58 40

* U.S.A. - CPR Aquatic, Inc.
P.O. Box 1111
3749 West End Road
Arcata, CA 95521
Tel.: (707) 826 - 96 36
Fax: (707) 826 - 96 23
Email: cpr@cprusa.com

- Hollywood Import & Export, Inc.
P.O. Box 220024 Hollywood
FL 33022-0024
Tel.: (954) 922 - 29 70
Fax: (954) 926 - 54 76
Email: aqualogusa@cs.com

The Modern Coral Reef Aquarium, Volume 4
First edition, 2002

Original title: "Korallrevsakvariet", © Svein A. Fosså/Alf Jacob Nilsen

Translation: U. E. Friese, St. Ives, NSW, Australia

Editors and layout: Dr. Dieter Brockmann, Mülheim, Germany, and
Werner Schmettkamp, Bornheim, Germany

Colour separations: Viscan Graphics, Singapore

Printed and bound by J.C.C. Bruns GmbH, Minden, Germany

ISBN 3-928819-22-4

Table of Contents

Chapter 4:
Tunicates and Lancelets 405

Appendix 465

Forskals' Sidegill Sea Slug, *Pleurobranchus forskali*. **Foto: Dr. M. Bergbauer**

A beautiful "Modern Coral Reef Aquarium"

owned by Claude Hug, France.

Preface

If we look back to the seventies, when we started out with marine aquarium keeping, and compare our first reef tanks with what can be achieved to day, there is only one expression that can justify what we see: "enormous progress"! In the seventies, we naturally hoped for progress, and certainly believed in progress, in marine aquarium husbandry, but not in our wildest imagination did we predict what was going to happen. From our point of view, marine aquarium keeping has developed in ways that no one would believe to be possible back in the seventies.

This volume starts out with "perspectives for the future", a most interesting chapter written by Dr. Bruce Carlson, the former director of the Waikiki Aquarium, Honolulu, Hawaii and now Vice President for Life Sciences, Georgia Aquarium, USA. Dr. Carlson really opens our eyes and encourages us to look ahead. He also stands out as a supporter of our work from day one, and we want to extend our most sincere gratitude to him!

We continue with a large chapter on Molluscs, a group represented in almost every reef aquarium in the world. In this chapter Dr. Leslie Newman, Southern Cross University, NSW, Australia, has written two most interesting sections on pelagic molluscs; the Heteropods and the Pteropods. We want to thank Dr. Newman for giving us the opportunity to include this valuable material in our book.

The mollusc chapter also deals with the Giant Clams of the family Tridacnidae, a group of animals that is most familiar to marine aquarist. Daniel Knop, Sinsheim, Germany, has kindly provided information and photographs on the group and edited the section on Tridacna farming. Many thanks to Daniel!

We are also indepted to Thora Whitehead, Honorary Associate in Malacology, Queensland Museum, South Brisbane, Australia, for general identification of molluscs; to Dr. Kathe R. Jensen, Zoological Museum, Copenhagen, Denmark, for valuable information on sea slugs in general; and to Jeff Zimmerman, Omaha Nebraska, USA, for supplying valuable information on the aquarium care and successful raising of the Lettuce Slug. Many thanks to Dr. Richard Pyle, Bishop Museum, Hawaii, for providing information of "deep reefs" and to Daniela and Erich Stettler, Bern, Switzerland, for providing photos and information on their *Tubastraea*-aquarium.

Dr. Lester Cannon of the Queensland Museum, South Brisbane, Australia, has helped with various identifications of Echinoderms, many thanks to him!

The Ascidians or Sea Squirts – presented in the last chapter in this volume – are poorly known by aquarists and usually not at all well covered in aquarium literature. We have been so lucky as to get indispensable help with this group from Dr. Patricia Kott, Queensland Museum, South Brisbane, Australia, a leading expert on Ascidians. Dr. Kott has kindly and most carefully edited the manuscript, a work that has put a kind of quality to the text that we by no means could not have achieved on our own hands. We thank Dr. Kott most sincerely! Many thanks also to Gregory Schiemer, New York, USA, for supplying information on the aquarium care and photos of the ascidian *Nephteis fascicularis*.

Like with the rest of the volume we could not have presented such a diverse photographic material without the help from many photo contributors. We would like to express particular gratitude in this respect towards: Robert Brons at DeJong Marine Life BV, Spijk, The Nederlands; Clay Bryce, Western Australian Museum, Perth, Australia; Dr. Bruce Carlson, Georgia Aquarium, USA; Wendy and Michael Cufer, Oyster Bay, NSW, Australia; Jeff Macaré, Laguna Hills, California, USA; Scott W. Michael and Janine Cairns-Michael, Lincoln, Nebraska, USA; Dr. Leslie Newman and Dr. Andrew Flowers, Brisbane, Australia; Erling Svensen, Ocean Photo, Egersund, Norway - who all in principle has given us unlimited access to complete archives of incredible photos.

Many, many people from all over the world have supplied us with valuable information, photos and given us indispensable advises. It is utterly impossible to mention all of them here, but some do deserve a special mention. Many thanks to: Marj Awai, The Florida Aquarium, Tampa, Florida, USA; Dr. Patrick Baker; Western Australian Maritime Museum, Fremantle, Australia; Dr. Rudiger Bieler, The Field Museum, Chicago, USA; Dr. Guy Cox, University of Sydney, Sydney, Australia; Rolf Hansen, Nor Zoo, Bergen, Norway; Professor Dr. Gerhard Haszprunar, Zoologische Staatssammlung, München, München, Germany; Dr. Gerald Heslinga, Indo Pacific Sea Farms, Hawaii, USA; Dr. I. Illich, Salzburg, Österreich; John Jackson, Odessey Publishing, El Cajon, California, USA; Ken Kates, Mollymook, NSW, Australia; Torsten Luther; Rostock, Germany; Dr. John Norton, Dep. of Primary Industries, Townsville, Australia; Jan Olsen, Aqua Design, Oldenburg, Germany; Kenneth Olsen, Oslo, Norway; P. Schupp, Würzburg, Germany; Dr. Ron Shimek, Wilsall, Montana, USA; Dr. Rob van Soest (†), Rijksmuseum van Natuurlijke Historie, Leiden, Holland; Ann Tanner, Brisbane, Australia; Professor Endre Willassen, Zoologisk Museum, Bergen, Norway and Prof. Dr. Peter Wirtz; Universidade da Madeira, Funchal, Madeira, Portugal.

At last, but certainly not least, we would like to thank our publisher, Birgit and Werner Schmettkamp and Dr. Dieter Brockmann in Birgit Schmettkamp Verlag (BSV), Bornheim, Germany, most sincerely for once again having co-operated with us so trustworthy and given the volume its characteristic feature. Also many thanks to Mr. U. E. Friese, St. Ives, NSW, Australia, for doing the hard work on translation.

Also warm and heartily thanks to our beloved wives, Kristin Hystad Fosså and Lise M. Løhre Nilsen who have – perhaps not without a sight of relief – seen yet another volume coming to an end.

And last, but certainly not least, again many thanks to you, our readers and fellow aquarists!

Grimstad and Hidra, March 2002
Svein A. Fosså
Alf Jacob Nilsen

Chapter 1:

The Future of Reef Aquariums

by Bruce A. Carlson, PhD

Vice President for Life Sciences, Georgia Aquarium (formerly Director, Waikiki Aquarium)

When I began keeping tropical marine fishes in 1962, there was only one authoritative book on how to set up a marine aquarium, and, in the United States, only a handful of people to turn to for advice. While the techniques worked to keep marine fishes alive, the knowledge and advice we followed back then seems almost primitive today, and few could have foreseen the tremendous advances that have taken place during the past forty years, both in the science of aquariology as well as our understanding of corals and coral reef ecosystems. Today there are dozens of excellent books on reef aquariums, hundreds of well-written articles in popular and scientific journals, and thousands of people around the world who have become expert in the husbandry of a wide variety of tropical marine animals and plants. Forty years ago, many people (including many coral reef scientists) scoffed at the suggestion that living corals might someday thrive in closed-system aquariums, but the pessimists were wrong. In fact, if anything the predictions were too conservative in forecasting the state of the reef aquarium hobby, as it now exists. Certainly, much more has taken place since 1962 than even the most optimistic hobbyists could have dreamed of at that time. So as we consider what might take place in home aquariums during the next forty years, the lesson is to be bold and imaginative when thinking about the future.

On the next few pages, I have

May be the reef aquarium of tomorrow will enable us to keep to days most difficult animals alive for a long period of time? To the left is a *Dendronephthhya* soft corals, which presumably needs a lot of phytoplankton to stay alive (photo: P. Lange). Above is the crinoid *Oxycomanthus bennetti* which must be regarded as absolutely impossible to keep alive and which belongs to a group of animals that to day neither should be caught for nor kept in captivity (photo above: Dr. Bruce Carlson).

The Waikiki Aquarium, Honolulu.

written down some thoughts on what we might expect to see during the next few decades in marine aquarium keeping. All of these thoughts are rooted in developments of today, but I hope that some completely unexpected and unpredictable discoveries will also happen.

Culturing ornamental reef fishes and invertebrates

Culture techniques for marine ornamental fishes and invertebrates will become the leading edge of aquarium science during the next twenty years and beyond. As I write this chapter in January 2002, I have on my computer some exciting messages and images concerning the first *Centropyge (C. fisheri)* and *Genicanthus* angelfish to be spawned and reared in captivity – both in Hawaii. These have long been considered among the most difficult tropical marine fishes to spawn and rear. Most interesting is the fact that one of these species, *G. personatus*, has spawned naturally and repeatedly at the Waikiki Aquarium in a relatively small tank (300 gallons/1200 litres), and even more exciting, the hatchlings were raised entirely on cultured food! In the past, most, if not all, marine angelfishes reared in tanks originated from eggs and sperm stripped from ripe adults, with the larvae being fed wild-caught plankton. Only aquarists with access to adult fish on nearby coral reefs, and a source of wild plankton could pursue this kind of research, but now the possibility exists that anyone anywhere might someday be able to spawn and rear these exquisite fishes. Certainly, this will be one of the most exciting new developments to watch over the coming years, but where might it take us as we peer farther into the future? Let us answer that question by first examining the problems that now exist and how they might be solved.

The most significant bottleneck that has stymied the development of culture techniques for marine fishes has been the lack of an appropriate first-food for larval fishes. The same is true for the culture of many invertebrates. I have always been amazed that our arsenal of foods for larval fishes has remained almost the same for decades: rotifers, brine shrimp, copepods, and more recently, trochophores, as well as a variety of phytoplankton species. Success in rearing larval fishes will come when this list is expanded 10, 20 maybe 50 times or more. The ocean is filled with tens of thousands of species of microscopic plants and animals, and many of these are potential food-sources for larval fishes. Our menu of first-food items will someday include protozoa and other very small living prey that microscopic larval fishes will eat. Different species of micro-plankton will be specially cultured for various species of reef fish larvae. Perhaps when demand increases, cultures of these food items will be mass-produced and sold either alive or in dried form to hobbyists world-wide to feed to their larval "navarchus" - angelfish and yellow tangs, or a tank full of microscopic newly-hatched clown-triggerfish!

Many small reef fishes such as mandarin fishes and dart fishes will spawn in small home aquariums, but larger reef fishes such as angelfishes will require much larger aquariums. As previously mentioned, masked angelfish at

The rare Anglefish *Genicanthus personatus* is endemic to Hawaii where it lives in relatively deep water. This fish has recently been raised at the Waikiki Aquarium. The male (left) and female show clear colour dimorphism.

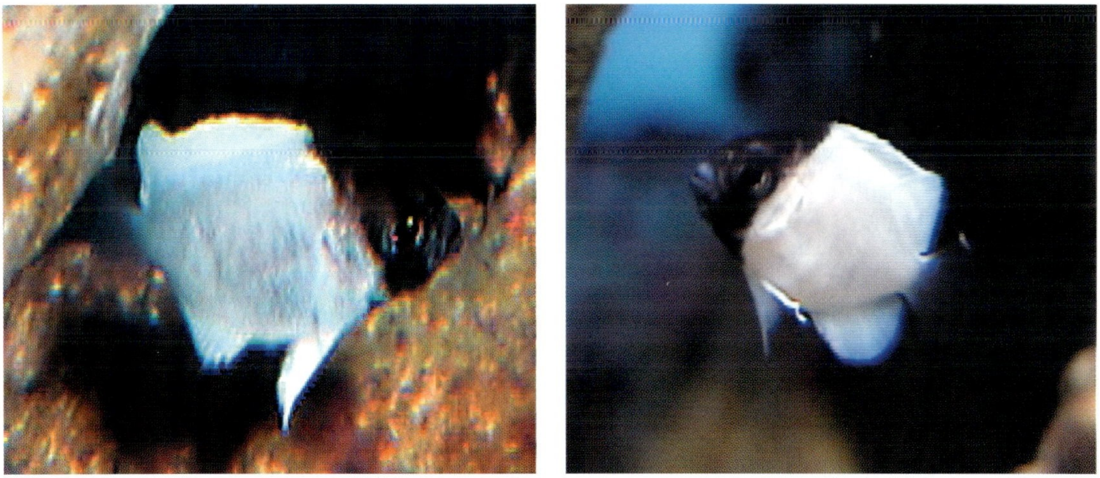

First successful rearing of *Genicanthus personatus*. Here is a section of the Press Release from Waikiki Aquarium, Honolulu, Hawaii (January 17, 2002): "The Waikiki Aquarium today announced the first ever success in rearing the endemic Hawaiian "masked angelfish", *Genicanthus personatus*. ... There are more than 70 species of angelfishes, and aquaculture scientists have struggled for decades to unlock the secrets to rearing these beautiful animals. Waikiki Aquarium biologist Karen Brittain made a breakthrough when she discovered a new, living food source to sustain the small larval angelfishes. The first ever aquarium-bred larval masked angelfish began metarmorphosis over the past two weeks. The juveniles are currently miniatures of their parents and will go on display to the public when they are larger. The parental masked angelfish are on display at the Waikiki Aquarium's Midway Island exhibit." The left photo shows the juvenile at 16 days of age, the right shows the juvenile at 21 days of age. Photos: B. Carlson

the Waikiki Aquarium have spawned repeatedly for years in a relatively small 300-gallon (1200-liter) aquarium. Other angelfishes, including "lemonpeel"-angelfish and bicolor angelfish have also spawned in these aquariums, but other species will require more room. For example, yellow tangs spawn almost every day at the Waikiki Aquarium in an outdoor 1200-gallon (4800-liter) reef tank. Most home aquarists do not have the room or the money to build such large aquariums in their homes, so breeding facilities will be limited to a few dedicated hobbyists, to public aquariums, universities and aquaculture facilities. Commercial facilities will keep their discoveries secret, which is understandable but unfortunate. As we all know, the sharing of information has been one of the primary driving-forces behind the evolution of successful reef aquariums, and likewise the broad sharing of experience and knowledge in the culture of marine fishes will greatly hasten the time when the great majority of reef fishes and invertebrates will all be tank-raised. At first, the secrets will remain proprietary, but as more hobbyists and researchers publish their results the successes will begin to increase exponentially.

Another interesting future possibility will be flash-frozen fish eggs and invertebrate eggs and larvae. At present, larval trochophores are available in this form – simply thaw them out and they come back to life! If home aquarists don't have room to breed large reef fishes such, perhaps future aquaculture facilities can ship frozen eggs and larvae to thaw and rear. That could be cheaper for the aquaculture farms and more enjoyable for the hobbyist - provided he or she also has a variety of "new" microscopic

The pioneer of the German marine aquarium keeping, Mr. Jürgen Grobe was also the first person to successfully breed and raise the Bangaii Cardinal Fish, *Pterapogon kauderni.* This species is endemic to the Bangaii Island in North Sulawesi (Indonesia) and it is thus believed that the species easily can be over exploited. The ease of breeding this particular species in captivity is therefore especially welcomed. The top photos show how the male encircling the female (left). The female lays one egg at the time (right). The large, yellow-orange eggs are picked up by the male, which carries the eggs in his mouth for about 22 days, whereafter the 9 mm long juvenlies are released.

Photos: J. Grobe

foods to feed the larvae.

And what about larval rearing tanks, what will they look like? Conventional tanks for easy-to-rear species, such as anemonefishes, may not change much, but for other species look for design changes to aid the larvae in swimming and locating prey. Will large tanks work better than small tanks? Will dark tanks be superior to clear or light-coloured tanks; will natural sunlight prove better than fluorescent lights? Will multi-species larval culture tanks prove superior to single-species culture tanks? This is more than speculation since a lot of research is already underway on these variables and more, so even if the results are not certain, it is highly likely that present-day aquaculture techniques and facilities will look primitive compared to facilities of the future.

Two 6 months old juvenile Comets, *Calliplesiops altivelis*, raised by the German marine enthusiast Thomas Röwer.

Sadly, a driving force behind the necessity for tank-reared fishes and invertebrates is the rapidly deteriorating condition of

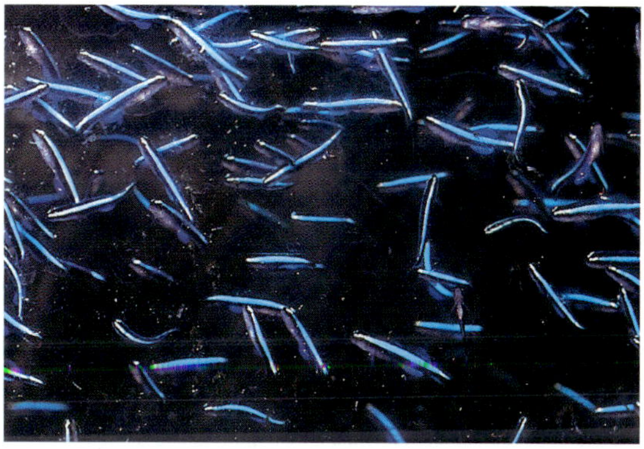

Farm raised aquarium fishes: Arabian Bluelined Dottyback (*Psedudochromis aldabraensis*, left) and Sharknose Goby (*Gobiosoma evelynae*, right).

Farm raised aquarium fishes of Red Sea Fish Farm, Eilat, Israel: Sunrise Dottyback (*Pseudochromis flavivertex*, left) and Orbiculate Batfish (*Platax orbicularis*, right).

coral reefs worldwide. Unless current trends are reversed, in another 40 years a significant percentage of the world's coral reefs will be seriously degraded or dead due to pollution, coastal development, dredging and blasting, sedimentation, and over-harvesting of fishes, corals and other reef animals and plants. During the most recent El Niño and La Niña events (1998 and 2000), millions of corals died on reefs throughout the world. Some island groups in the Pacific and Indian Oceans experienced nearly 100% mortality of corals. These reefs will probably recover but it may take decades to reach their full diversity again – provided ocean warming doesn't recur. The loss of coral reefs will lead to increased conservation measures and restrictions on collecting. As a result, the need for aquaculture of marine ornamentals will grow in importance in the future.

Marine Habitats in Aquariums

Today's aquariums are often conglomerations of fishes and corals from the Pacific, Caribbean, Indian Ocean and Red Sea maintained as colourful aquatic "gardens". These mixed habitat aquariums will probably always be around as they are visually stunning and very enjoyable, but the future should also see the development of a greater variety of specialized habitat aquariums. A partial list is discussed on the following pages:

- **Large-scale surge zone habitats**

Public aquariums will probably be the first to truly create exhibits representing dynamic outer-reef environments containing a wide variety of large table acroporids, massive brain corals, and pocilloporid antler corals. The first attempts to create such exhibits have already appeared in Australia but undoubtedly others will be created in the future complete with crashing waves and surge, shoals of planktivorous reef fishes, reef sharks, and possibly a few coral-feeding butterfly fishes. These aquariums will be too large for most hobbyists to consider in their homes, but new surge devices will certainly be developed for home aquariums too.

The first attempt to create a large-scale surge-zone aquarium was done in the Great Barrier Reef Aquarium, Townsville Australia. This photo is from 1991, but the facility still exists.

- **Reef-pass habitats**

A wide variety of filter feeding organisms occur in reef passages that are characterized by strong tidal water flows alternating with periods of slack flow. Crinoids, sponges, tunicates, sea fans, and soft corals are commonly found in these areas and many are adapted to specific flow rates that facilitate the capture of both phytoplankton and zooplankton and "marine snow". Few of these organisms have ever been kept alive in aquariums despite many years of trial and error, e.g., dendronephthiid soft corals. True, some aquarists have reported limited success, but nothing compared to what has been accomplished with stony corals. However, new studies are beginning to reveal the specialized requirements of these filter-feeding organisms and someday in the future we will see them successfully maintained and propagated in home aquariums.

- **Mangrove and seagrass habitats**

Mangrove trees are already being grown by hobbyists and public aquariums, and to a lesser extent so are seagrasses. These plants form important habitats for a variety of unusual animals such as mudskippers, archerfish, sea horses and pipefishes, fiddler crabs, and even for some corals such as *Catalaphyllia elegans* and *Montipora digitata*. In the future the water systems of these exhibits might be coupled with reef aquariums, and also may be used as refugia or perhaps as nurseries for juvenile reef fishes in much the same way as they function in nature.

Every two minutes a strong surge flows through this exhibit at the Waikiki Aquarium which helps to maintain vigorous grows of a variety of corals. This is a closed-system aquarium using a calcium reactor, an ETS protein skimmer, two 400-watt metal halide lamps (10.000 K), and natural sunlight. The surge is created by two overhead automatic siphon tanks.
Photo: Dr. B. Carlson

Most of the private reef tanks of today contain a mixture of invertebrates and fishes that in the nature show certain preferences to the selection of habitats. Only rarely do whe see tanks that display special habitats or designed for selected animals. We hope that this will change. The modern coral reef aquarium certainly has the potential for studying the biology of reef creatures in close up. The Elegance Coral, *Catalaphyllia jardinei*, for instant is almost always found in the rubble zone or among sea grasses on the innermost reef flat, and only rarely living among other corals on the outer reef flat or in the reef slope. The Elegance Coral should therefore be kept in a special display like like shown in the photo above where the species is kept at the Waikiki Aquarium on sandy bottom and among sea grass. Here it grows together with the branching *Montipora digitata*, which often occur in the same type of habitat.
Photo: M. Awai

We are at a ledge outcropping long a deep reet drop-off in Fiji. The depth is 100 m, and the diver in the photo is using a closed circiut, mixed-gas rebreather. The reefs at these depths are dominated by gorgonian corals, and are home to many new species of fishes and invertebrates. Photo: R. L. Pyle

Richard Pyle tells the exciting story about the discovery of a new *Cirrhilabrus* on a deep-reef dive: *"I first saw this fish at a depth of 100 m while diving on a coral-reef drop-off in Fiji in March, 2001. I was with Howard Hall and his team, working on an IMAX film called "Coral Reef Adventure". On this particular dive, the camera had malfunctioned, so we decided to just explore the reef. As soon as I saw this large, beautiful male* Cirrhilabrus, *I knew instantly from the dramatic tail that it was a new species. As I started chasing the fish to collect it, I heard the other divers making a commotion, because a large Tresher shark was swimming past. I was nuch more interested in this little fish than the shark, however, so I chased it with my net until I was able to collect it. Later in the trip I collected several more of them - both females and males, at depths ranging from about 90 to 125 meters. One moth late, while diving inside Fagatele Bay on Tutuila, American Samoa, I saw a whole harem of this distinctive new species at a septh of 114 meters."*

Photo: R. L. Pyle

• **Deep reef habitats**

Through the pioneering deep-diving efforts of scientists like Richard Pyle in Hawaii, an entirely new fauna is being revealed. Colourful new reef fishes are being discovered on nearly every dive below 100 meters and those that have been kept in aquariums seem to survive very well. Undoubtedly in the future we will see many new fishes, corals, and other invertebrates maintained in specially illuminated aquariums representing the "twilight zone" from about 100 meters to 200 meters depth.

Dr. Ricard Pyle and Charles "Chip" Boyle discovered these three beautiful fish species during deep reef dives in Rarotonga, Cook Islands, in 1988-1990. *Centropyge boylei* (below left) lives in harems in rubble and drop-offs usually deeper than 75 m, down to 100 m, but is very difficult to keep alive in reef tanks due to a special diet. *Centropyge narcosis* (below right) lives in harems under ledges starting at 100 m and going down to at least 120 m and is - like many of the other members of the genus - easily kept and fed in captivity. *Belanoperca pylei* (right) lives solitary, from about 75 m down to at least 120 m and is like Dr. Pyle expresses it: *"... a fantastic and really charismatic aquarium fish"*. All three species prefer temperatures below 25 °C and a rather low light intensity in their tanks, which is directly related to their natural habitat.
Photos: R. L. Pyle

In addition to new exhibits simulating natural habitats, I foresee significant advances in our understanding of the culture methods for tropical marine organisms:

- **Coral farms**

The 1990's will be remembered by aquarists as the decade when coral-culture first became a viable industry. Small-scale coral farms and research programs, such as the Tropicorium in Romulus Michigan, and at the Waikiki Aquarium and Monaco Aquarium, are already producing significant numbers of coral fragments for hobbyists, public aquariums and for research. The future will see more facilities in operation, both large and small. Will these coral farms be financially viable? Perhaps, especially in the early years, but in the future it may be so easy to grow corals in home aquariums that hobbyists will gladly give away excess corals to fellow hobbyists and public aquariums. The commercial coral farms (and giant clam farms) will therefore have to diversify and devote more and more resources to research to remain competitive.

One of the tanks of Tropicorium used for propagating corals is studied by a group of aquarists.

- **Coral breeding**

Researchers working at the larger coral farms will eventually unlock the secrets to successfully spawning many species of corals in captivity. This will be especially important when the day arrives that collection of corals from the wild is completely banned worldwide. Coral farms located in the tropics will provide coral spat to villagers to grow in lagoon environments for eventual sale to hobbyists, but possibly also for reef restoration projects and research.

Left: Dick Perrin holding a magnificent *Acropora* colony grown at his facilities. Right: One of Tropicorium's most popular items is the "true Stüber's *Acropora*", offsprings grown from mother colonies directly descending from Dietrich Stüber's original 1980-ies colony in Berlin.

Coral Farming in the Philippines (Photos: Prof. Dr. T. Heeger)

Left: Women of the coral farm team are tying the coral fragments to customized cut fossil limestone substrates. Exposure time is not exceeding 10 minutes and has proven to be harmless for coral survival. Right: Two divers deploy fragments into Coral Nursery Units (CNU's) using the surface air supplying system. The baskets with the fragments result in negative buoyancy and the divers remove their fins to avoid stirring up sediment.

Left: A coral farm diver shows two fragments, which have overgrown the substrate with alive coral tissue. Those fragments are ready for reef rehabilitation. At the rehabilitation site the fragments are carefully deployed on suitable reef substrate at an average density of two fragments per square meter (right). It is important to space the farm-grown colonies to resident corals to avoid space competition.

"According to REEF CHECK data there are no more pristine reefs in the world (Hodgson, 1999). In the Philippines, host to one of the most diverse reefs on earth, surveys on the status of the coral reefs at 14 sites revealed that only 2.4 % could be considered in excellent conditions with 75 to 100 % live coral cover. While three fourth were in the fair to poor category at 0 to 49 % coral cover (Gomez et al., 1994). Given the direct relation of coral cover and productivity, the reefs have become less productive. In support of this, the catch per unit effort of small pelagic fishes in the Philippines decreased from almost 11 tons/hp in 1948 to less than 2 tons/hp in 1985 (Dalzell et al., 1987). Should this trend of marine degradation be stopped or even reversed, new strategies in managing coastal resources by integrating livelihood options with strong entrepreneurial components (Moffat et al., 1988) and limiting future resource users have to be implemented."

Heeger in Fosså & Nilsen (2001)

Fish-eye's view of the coral farm. Up to 300 Coral Nursery Units can be stocked with a total of approx. 25.000 fragments.

The reproduction of animals naturally leads to more animals. To benefit from this process in aquaristic is equal to avoid harvesting the reef, but instead "make" ones own animals, in other words: our goal as aquarists must be to be as independent of the natural habitat as possible and to keep more and more animals that come from farms and hatcheries. Propagation of giants clams is one example of an industry that has evolved to the best for marine aquatics. The photo shows the facilities of Palau Mariculture Demonstration Centre, Palau. Photo: Dr. B. Carlson

• Flash-frozen planulae

Coral researchers today are hampered in their experimental research on coral planulae by the fact that they are available only one or two days each year. One possibility to extend this period is to develop methods for flash-freezing planulae in liquid nitrogen (and fish eggs, as previously discussed). Hobbyists might also benefit from such technology. Even if the collection of adult colonies from the wild is eventually banned, it might be possible to harvest a small fraction of the annual spawn produced on the Great Barrier Reef and other localities. After flash-freezing, it can be distributed to hobbyists and researchers anywhere to thaw at a later time and rear.

A collection of small *Tridacna* clams with the most beautiful colours, raised at Palau Mariculture Demonstration Centre.

• Designer clams and corals

As our culture methods for clams, corals and other marine organisms improve so will our ability to produce desirable characteristics, particularly colouration. I have no doubt that someday we will be able to produce cultures of 100% morpho-blue giant clams, as well as pink, purple, green and blue corals. Perhaps this will be as simple as providing the right environmental stimulus such as light intensity and wavelength, but it might also involve genetics or the right strain of symbiotic algae.

• New technologies

Devices to simulate tidal changes, laminar water flow, crashing waves and turbulence are already available but frequently unreliable or ineffective, especially for large exhibit aquariums. Undoubtedly a wider range of equipment will emerge in the future that is both effective and reliable - and computer controlled. Likewise, we can expect continued improvements in lighting possibly with radically new (and cheaper!) light sources becoming available. Aquarium systems overall will become more standardized and computerized to the point where a "turn-key" marine aquarium system for keeping living corals will be available. This will be a boon to beginners and also to educators and scientists wishing to teach students or conduct research on marine animals.

• Ultraviolet light

Ultraviolet (UV) light bathes natural coral reefs and corals and many other marine animals and plants have pigments and other adaptations to protect themselves from this otherwise harmful light. But UV light has beneficial effects too, by destroying bacteria in the water, and breaking down large organic molecules. Other beneficial effects may also exist but are unknown at this time. For example, do fishes require some ultraviolet light for behavioral purposes? Current research indicates that some reef fishes do have colour patterns that are only visible in UV light. Public aquariums are also noticing that fishes maintained outdoors in full sunlight seem to have brighter colours and fewer diseases than fishes maintained indoors under artificial light. Does UV light play a role here? If so, perhaps we will see more UV light fixtures as standard equipment on reef aquariums in the future.

- **Encyclopedia of marine life**

Many well-illustrated guides to Indo-Pacific marine life began to appear in the mid-1990's, but a single comprehensive guide to the fauna and flora of this region is presently beyond our ability to produce, both due to our limited knowledge of the organisms of this region as well as the expense of producing such a massive publication on paper. However, in the coming decades the growing body of scientific data and nomenclature as well as photographs of marine life will all be combined on computer disc (or some other technology) and will be widely available at reasonable cost to anyone anywhere with a handheld computer. Better yet, video clips showing the natural behaviors of marine animals will also be accessible, as well as real-time underwater views from coral reefs around the world.

- **Aquariums for endangered species**

At the start of the new millennium, no coral reef fishes or corals have been designated "endangered species", but all of the evidence today indicates that this will change in the future. Concern is growing for the survival of species with low reproductive potential, such as seahorses, which are highly sought after for medicinal purposes as well as the aquarium trade. Endemic species with low population sizes may also become threatened in the future, e.g., some of the valuable endemic angelfishes of remote island groups. Corals too may someday fall in this category due to extreme habitat destruction. Home aquariums and public aquariums may become the last refuges for these species providing both a possible source for reintroduction

Tubastraea **are non-photosynthetic stony corals that lack symbiotic algae and are completely depending upon plankton and other sources of nutrient being carried to their large polyps. Generally they have been regarded as difficult to keep in captivity, and indeed this is true! However, this does not mean that they cannot be kept! The example here, which shows the aquarium of Daniela and Erich Stettler, Bern, Switzerland, really proves the opposite. An extensive feeding with brine shrimps etc. is a must for obtaining such growth.**
Photo: D. and E. Stettler

Offsprings of *Tubastraea* sp. in the aquarium of Daniela and Erich Stettler. In this tank Cup-Corals grow very well, and after one year - usually during full moon - they began to reproduce and release planula larvae, the general larva type of corals. The planula larvae, particulary like to hold on to shady, spots where the current is strong. There they develop during one single year from the size of a single primary polyp to coral colonies in the size of a tennis ball!
Photo: D. and E. Stettler

The Angelfish *Chaetodontoplus conspillatus* is endemic to the Lord Howe Island, which has the southernmost coral reefs in the world. The fish feeds exclusively on coral and sponges in the wild.

of animals back into the wild, as well as diverse genetic material required to maintain healthy populations. But one drawback for the reintroduction of species to the wild is knowledge of the original source of the organisms, as well as concern for the accidental release of pathogenic organisms. Hobbyists and researchers today should begin tracking the strains of corals and other organisms cultured in their aquariums noting the source of the original animals and the date when they were obtained.

Left: *Chaetodon declivis* is endemic to the Marquesas Island and found in the lower reef slope where it lives mainly from small invertebrates like worms and crustaceans. Unless such a diet can be provided, or the fish can be successfully trained to take frozen or freezed-dried food, the species should not be kept. On the other hand, the reef aquarium can even in this case learn us about the biology of the species.

• **Methuselah corals**

Scientists have discovered that some corals have survived for centuries and may continue to survive for many more centuries. These long-lived corals have been dubbed Methuselah corals. I have no doubt that 20, 50 or even 100 years from now some of the corals now living in our aquariums will still be thriving - perhaps not in their original aquarium homes, but rather as daughter colonies passed on from one generation of hobbyists to the next. One very large coral colony at the Waikiki Aquarium (*Goniopora* sp.) has been on display since 1979 and shows no signs of aging. And Dietrich Stuber's corals (especially "Stüber's Acropora", see Volume 2, page 359) have been passed on to hundreds of hobbyists, as have corals from the Waikiki Aquarium, and they may live on for decades hence. I certainly hope that aquarists will keep good records of the origins of their living collections - like vintage wine, perhaps their value will increase with age!

"Stüber's Acropora" was the first *Acropora* sp. to be grown in a home aquarium. The species identity of the coral remains uncertain. Offsprings from the parent colony, which we see here in Dietrich Stüber's aquarium in Berlin in 1985, have been spread to hundreds if not thousands of enthusiasts all over the world.

The one event that I do not foresee happening in the future is the eventual demise of the marine aquarium hobby. There will be tough times ahead with many new regulations restricting the collection and importation of wild-caught marine ornamentals. However, I believe the hobby will prevail and that marine aquariums will continue to flourish providing immense enjoyment for aquarists, as well as being outstanding educational and scientific tools. The tremendous advances made during the 1980's and 1990's will continue in the new millennium, but the pace will be slower as the challenges become more difficult

The reef aquarium hobby can have a powerful influence on children, and thus on the future. The coral reef scientists and aquarists of tomorrow who will make great discoveries may well be children peering into a reef aquarium at this very moment. Perhaps they are watching the behavior of colourful reef fishes and studying the intricate designs of beautiful corals while they dream of far off magical places called coral reefs. Maybe this encounter will inspire them to become marine biologists devoting their careers to the study and conservation of reef plants and animals. I hope that someday, perhaps 20 or 30 years hence, a coral reef scientist will find this book and read this chapter and recall his (or her) first encounter with a marine aquarium and how it influenced his career. Perhaps he will frown at our naiveté and mistakes. Or perhaps her reaction will be one of amazement at how much we have accomplished in so short a time. But most of all I hope they will feel gratitude to all the aquarists of the past few decades whose accomplishments have helped to inspire a new generation who will ensure the survival of coral reef ecosystems forever.

The Ascension Pygmy Angel, *Centropyge resplendens,* is endemic to the Ascencion Island in the tropical Atlantic. American military personnel used to catch the fish and offered it to the trade. Nowadays the fish is no longer caught and the last specimens that died in 1997 in our tank, was probably the very last fish of this species at that time living in a home reef aquarium.

Chapter 2:

Mollusca

The phylum Mollusca is extremely diverse, and encompasses seven classes with a total of more than 100,000 species. Therefore, in this volume we can cover only a fraction of these within an overview and so we will confine ourselves to those species that are often kept in aquariums and/or which play a very specific role in their natural habitat.

Molluos- or soft-bodied animals are divided into the following classes:

❶ Primitive molluscs – Class Monoplacophora (5 species);

❷ Worm-like molluscs – Class Apalocophora (approximately 200 species);

❸ Chitons – Class Polyplacophora (approximately 600 species);

❹ Snails, Clams, Mussels, Sea Slugs and Sea Hares – Class Gastropoda (approximately 60,000 to 75,000 species);

The phylum Mollusca (soft-bodied animals) encompasses a great number of species, contained in seven diverse classes of animals. These include not only the (among coral reef aquarists to popular) giant clams - left: a *Tridacna maxima* on a reef flat in the Red Sea, at a depth of only 2.5 m (photo: T. Luther), but also the primitive-appearing pearl boats or chambered nautiluses - right: *Nautilus belauensis*, at Palau (photo: P. Schupp).

Left: Nudibranchs are among the most colourful molluscs, where a shell is absent and which have gills located outside the body on the posterior dorsal region. Unfortunately, nudibranchs are distinct food specialists. Therefore, it is nearly impossible to keep these molluscs in an aquarium. The species *Chromodoris bullocki*, shown above on the left, feeds exclusively on sponges. Right: Next to snails, clams and mussels make about the second largest class of animals in phylum Mollusca. Some species, like this *Spondylus* species show an attractively coloured mantle inside their protective shell.
Photo: S. W. Michael

❺ Double-shelled or Bivalve molluscs – Class Bivalvia (approximately 15,000 to 20,000 species);

❻ Tusk- or tooth shells – Class Scaphopoda (approximately 350 species);

❼ "Head-footed" molluscs, nautiluses, octopus, squid, cuttlefish – Class Cephalopoda (approximately 650 species)

From an aquarist's perspective only the snails and the two-shelled (bivalve) molluscs are of particular significance. These two classes cover more than 97 % of the known mollusc fauna.

Evolution

Although there are many variations within the phylum, this animal group is generally characterised by the fact that its members have become adapted to almost all sorts of habitats on earth. In essence then, this high degree of adaptability has lastly been the reason for the successful evolution of the molluscs.

There are various theories about the origin of molluscs; however, today they are generally considered to be a side branch of a genealogical lineage that commenced with flatworms of phylum Platyhelminthes (see Vol. 3, Chapter 4, pages 69 – 86) and ended with joint-footed animals of phylum Arthropoda. This theory is also supported by the fact the respective larval stage is of the same type.

The first molluscs lived in the sea. Fossil molluscs have been found dating back to the Cambrian Period, which makes them about 530 million years old. Fossil finds from many other locations suggest that all currently living classes of molluscs had already become established 500 million years ago. The first terrestrial molluscs appeared 200 mil-

Ammonites predominated in ancient seas, about 100 million years ago. They became extinct at the end of the Cretaceous Period and can nowadays be found as fossils in many locations (photo right: P. Hoffmann).

lion years ago, but they were rare. It was not until the Tertiary Period (65 million to 2 million years ago) that molluscs apparently became common on land. Some of the many fossil molluscs that have been found have relicts living to day that still show ancient characters, as for instance the Chambered- or Pearly Nautilus of genus *Nautilus*. Currently, only six species of the unusual cephalopod are still alive. These animals have developed a spirally coiled shell with internal air (gas) chambers that give them buoyancy. Fossil finds, however, indicate that there were once more than 3,000 species that have become extinct a long time ago. Chambered- or Pearly Nautiluses were quite numerous during the Palaeozoic Period (570 to 225 million years ago), but then they gradually disappeared during the Mesozoic Period (225 to 65 million years ago).

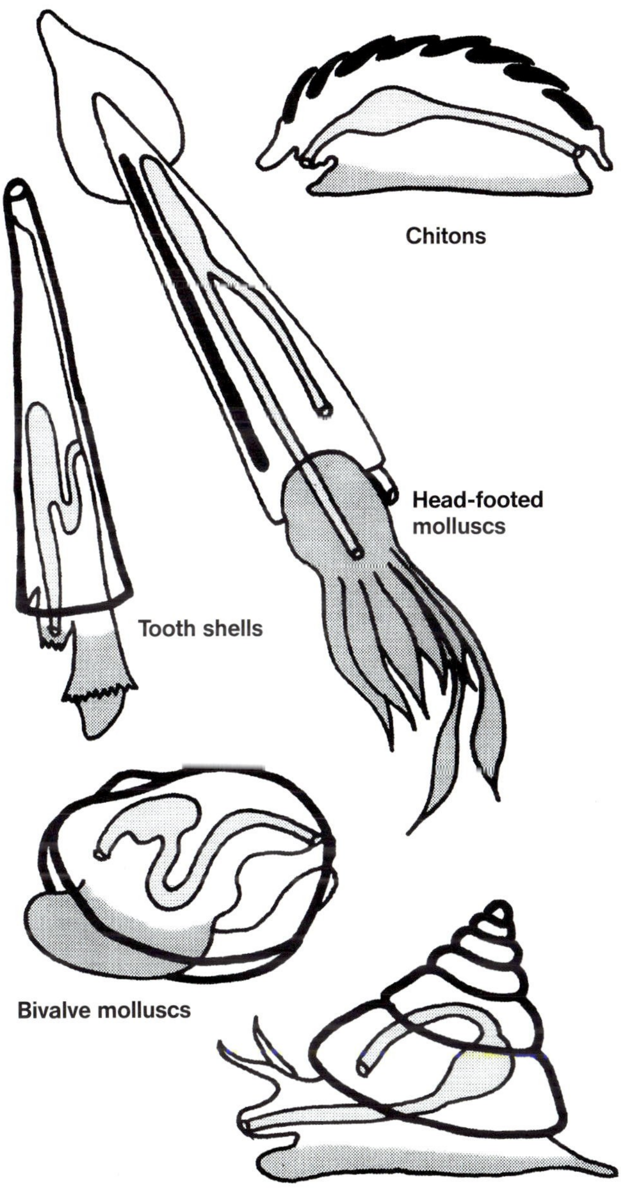

Body plan of the five largest mollusc classes: the shells, also referred to as valves in some groups, are indicated as bold printed lines, the digestive tract is drawn in light grey and the feet are represented by grey-shaded areas.
Drawing after BUCHSBAUM (1967) and ANDREWS (1994)

After the decline of nautiluses, the ammonites (subclass Ammonoidea) became dominant in ancient seas. They too had spirally-shaped, wound air (gas) chambers that had considerable resemblance to those of their predecessors. Today we can find large numbers of fossil ammonites in many locations, such as those along the beaches of southern England. At the end of the Cretaceous Period (about 65 million years ago) ammonites became extinct.

General Biology

Although soft-bodied animals can externally be rather variable, they still share characteristics; they have a soft body that consists of a head, a muscular foot and an abdominal cavity for the digestive system covered by a mantle. Many molluscs have a heart, a simple vascular system, a basically-structure kidney, a digestive gland and a primitive nervous system. There are also additional, well-developed internal organs, but no internal skeleton that supports these. Although there are "giants" among the molluscs, such as the fearsome deep sea squid, yet, most are relatively

This snail (*Cymbiola* sp., found along a muddy coastal zone at Bunaken, Indonesia) clearly shows the typical mollusc characteristics: the fleshy foot, the shell (or valve) and the long siphon.

small, probably due to the absence of an internal skeleton.

The **mantle** consists of the external cell layer (epidermis) and it encloses the mollusc animal within a skin fold. Therefore, most internal organs are located in a cavity (referred to as **mantle cavity**) that has various functions. In most species the mantle excretes a **shell** (chonch[a]), which serves as external skeleton (exoskeleton), and its most important task is the protection of the animal. This shell consists of calcium carbonate, sodium phosphate and magnesium phosphate. These three substances are embedded in proteins and polysaccharides. In addition, many marine species can also bond salts, which are then incorporated in the shell. This then makes the shells walls particular thick and strong, which renders greater resistance against excessive wear due to wave action, for instance. Growth of the shells keeps pace with any increase in body size, and is accompanied by the development or more less distinct "ribs" and other external structures that are important species characteristics.

In some species the mantle gives off mucus or colour substances that serve as defence against predators. In many clams and mussels parts of the mantle have been modified in to a respiratory tube. This facilitates gas exchange even when the animal is buried in sand or mud, or lives embedded in a coral head.

The tissue below the epidermis is often developed in to a powerful muscle colloquially referred to as **foot** which enables these molluscs to actively burrow or simply move about. Many molluscs, especially some clams (scallops), octopuses, squids and relatives utilise the principle of "jet propulsion" for locomotion. The "jet effect" is being created by forcing water under high pressure through a narrow opening. This type of locomotion enables a quick and so often successful escape from enemies.

The primitive molluscs have internal gills, but sea slugs and related forms (sea hares, nudibranchs) have gills that are located on the outside of the body, so that they always stay in open water, which makes them easily recognizable. As in most marine invertebrates, ammonia is also the most important metabolic waste product in marine molluscs. It is being discharged directly in to the surrounding water.

Molluscs also possess **sensory organs**, which range from small, light-sensitive cells located in the mantle of clams up to highly developed eyes in octopuses, squids and related forms. Special equilibrium organs, the statocysts, enable molluscs to maintain their balance. These are organs that also occur in many other invertebrates.

In such a species-diverse phylum **reproduction** is, of course, highly variable. Most species are heterosexual, however, hermaphrodites (both sexes in same animals) also occur, especially in

Molluscs can be extremely venomous. One should not copy what Dr. Johnson Seeto (University of South Pacific) does. Here he is seen demonstrating a *Conus geographus* specimen caught on a reef off Suva (Fiji Islands). These snails must be treated with extreme caution.

Some fishes, like this Red-lipped Wrasse, *Cheilinus fasciatus*, feed on molluscs. Fundamentally, these animals are a significant food source for coral reef animals.

Photo: S. W. Michael

snails. On the other hand, asexual reproduction is rare among molluscs. Chitons, clams and primitive snails release their gametes (eggs and sperm) through an opening in the mantle, which is then followed by fertilization in open water. In higher developed snails, fertilization takes place in the mantle cavity. Many snails deposit their eggs in large clusters or strips. In nudibranchs and sea hares, kept in aquariums, this type of reproduction is easy to observe. A few species are livebearers.

The fertilised eggs go through one or two larval stages. Initially, there is a **trochphore larva**, which often continues to develop in to a **veliger larva** while still inside the egg. The trochophore larva occurs in many invertebrates, while the veliger larva is characteristic for molluscs. The latter feeds on phytoplankton and remains planktonic until it loses its swimming organs, descends to the bottom and there commence the normal life as a mollusc. The duration of the planktonic stage is species-dependent and can range from one day to several months.

The biology among the various classes of phylum Mollusca can be highly variable, and so further details are being provided within the respective discussion on a particular class.

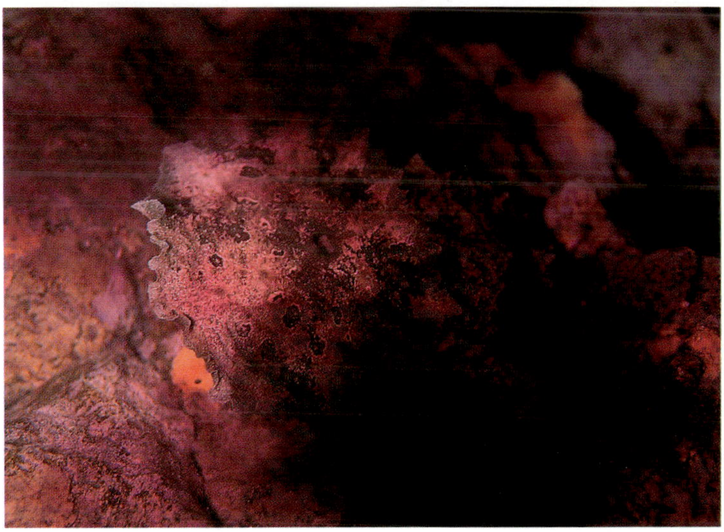

Molluscs have various means of protection against predators. Clams and mussels can close their shells (valves), and snails can bury in sand or withdraw in to their shell. However, sea slugs and related forms do not have these possibilities. We found this sea hare (class Hastropoda, family Aplysidae) in the vicinity of a beach at the Maldives. When we touched the animal it excreted a purple substance (upper photo). Within a short period of time the water all around us was purple-coloured (bottom photo).

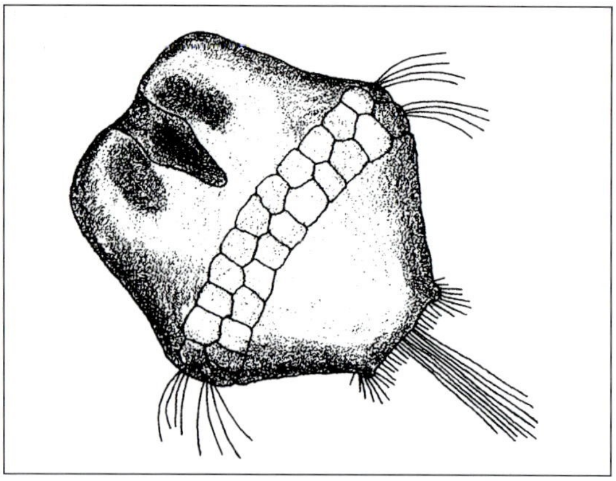

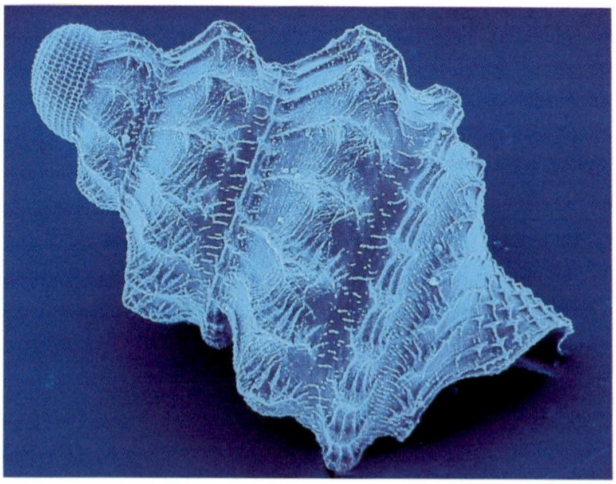

Trochophore larva (top left: drawing after BARNES, 1980) and three different veliger larvae.
Photos: top right – R. Brons; below – L. Newman and A. Flowers

Molluscs on Coral Reefs

Molluscs are very common on coral reefs. They occupy virtually all ecological niches, from the inter-tidal zone down to greatest depths. There are tiny species that drill in to rocks or burrow in to sand or mud, up to giant clams with a length of more than 1 metre that accommodate symbiotic algae, and live among corals in shallow waters with a strong irradiance. In addition, we can also find fascinating octopuses and cuttlefish as well as deadly venomous snails. This enormous diversity is amply documented in zonal studies by MERGNER (1979) of a reef in the Red Sea. A total of 1,147 mollusc specimens, encompassing 113 species, were found in an area covering 25 m². Normally, a shallow water mollusc fauna is horizontally distributed and is dependent upon sediment conditions and exposure to current, wave action and light penetration.

The table on page xxx from TAYLOR (1971), gives an indication of how the number of snails and mussels are distributed over the various habitats on corals reefs around the Seychelles (Indian Ocean). This particular publication by TAYLOR also contains many details about the composition of the mollusc fauna in the Indian Ocean, for instance, the distribution of snails and clams – by genus – in individual habitats, as well as much general information on tropical molluscs.

When SHEPPARD (1984) investigated the fauna around the Chagos Islands (Indian Ocean), he found 384 mollusc species, with 220 of these being new to the Chagos Islands. The mollusc fauna increased with depth and reached its greatest species diversity between 10 and 20 m depth. Below that depth range species diversity declined again. This pattern also conforms to the depth distribution of corals. Along reef slopes, where species diversity is greatest, the number of individuals per species is nearly always low.

Molluscs make up an important ecological component of a coral reef. They are of significance as algae grazers and detritus feeders. Numerous species burrow in the substrate, turning over the upper layers and expo-

Left: Small, inter-tidal snails (here a *Littorina* sp. at Puerto Rico) are often significant algae eaters and can be quite numerous. Many of these are suitable as algae eaters for a coral reef aquarium. Right: Many of the larger clams live in deeper areas of the reef, such as this *Pteria* species at the Maldives, seen here hanging in a black coral. Other molluscs can become established on its shells, such as the clam *Lopha* sp., easily recognisable by its zigzag-shaped shell opening, and also some worm snails.

sing them to fresh water and with that to oxygen. Many molluscs are predators, feeding on other animals in or on the bottom. For instance, the large Triton's Trumpet, *Charonia tritonis*, feeds on echinoderms and among these also on the dreaded coral predator, the Crown-of-Thorns Sea Star, *Acanthaster planci*. But there are also snail species that even feed on stony corals. For instance, in recent years the populations of coral-feeding *Drupella* species in Western Australia have significantly increased and so have caused considerable damage to the coral fauna (BLACK & JOHNSON, 1994; STODDARD, 1989). Most clams are sessile and play an important role as filter feeders.

A considerable number of molluscs live in association with stony corals. Several molluscs have become established on this *Porites* colony: giant clams, *Tridacna maxima*, various boring clams of genus *Arca* and worm snails, *Dendropoma maxima* (photo taken at the Maldives).

The number of snails and mussels in coral reef habitats in the Seychelles. (After TAYLOR, 1971)			
Habitat	Snails	Mussels	Total number
Mangrove	5	5	10
Coral sand	0	3	10
Mud and sand	19	14	33
Sand flats with marine grass	49	26	75
Sandy substratum without vegetation	41	21	62
Sandy substratum in deeper water	49	60	109
Hard bottom in the tidal zone	34	5	39
Zone with small, rounded rocks	43	7	50
Zone with large, rounded rocks	64	13	77
Among corals	27	24	51

These molluscs can turn over large volumes of water within a single day.

Even after the death of a mollusc, its shell can still be useful. Empty shells are often taken over by hermit crabs, to afford them protection for their soft abdomen. Weathering mollusc shells, whether these are from snails, clams or others, contribute to the production of coral sand.

Clams and snail shells have been collectors' items for centuries. Hobby malacologists, as they are often called, are actually as interested in nature as are aquarists. Some have made significant contributions to science with their collections, which must not be overlooked or under estimated. Yet, it cannot be denied that certain populations of very popular collecting objects have

The Triton Trumpet, *Charonia tritonis* (seen here at Madang, Papua-New Guinea), is a very important snail, since it feeds on the Crown-of-Thorns Sea Star, *Acanthaster planci.* Photo: L. Newman and A. Flowers

The small snail *Drupella cornus* feeds exclusively on stony coral polyps. In recent years its population has substantially increased, especially on western Australian reefs. This photo shows a stony coral grazed over by this snail (seen here on Morning Reef, at Houtman Abrolhos, Western Australia).
Photo: P. Baker

been become severely over-exploited. It is simply unacceptable that rare "sea shells" be exhibited as mere trinkets. Although it has never been proven that particular species have become extinct due to (over-) enthusiasm of shell collectors, yet, the populations of certain species have become noticeably smaller. Around the world several museums display shell collections to the public. If you ever travel to Phuket, Thailand do not miss the island's "Sea Shell Museum" as it has an extraordinary collection of snails, mussels and fossils displayed in a very elegant and informative way.

Collecting aquarium animals must be viewed as being just as damaging as gathering sea shells, as long as this is being done in an uncontrolled manner. We are all consumers of natural resources, even if these are renewable. Often they provide essential food sources and an economic livelihood for native populations. For those reasons too these natural resources must be managed with diligence and on a scientific basis.

Many molluscs are considered to be culinary items, such as our local blue-mussels or oysters. But for tropical species too (e.g. giant clams of family Tridacnidae) there is a considerable demand, not only from the local population but also from the worldwide gastronomy. For years, giant clam populations have been over-exploited, so that these molluscs have become rare in some areas. On the other hand, the increased demand has also brought benefits: the artificial rearing of giant clams on breeding farms. Today, giant clams are being bred on a large scale on Hawaii, several islands of the Pacific and along the east coast of Australia. Some of the breeding farms sell juvenile specimens for aquarium purposes, so that collecting these animals in the wild is no longer necessary.

Many molluscs, like the snails seen here in the Caribbean Region, are not only being used by indigenous populations as food, but there is also a worldwide demand by the gastronomy for these molluscs as highly prized culinary items.
Photo: Dr. A. Spreinat

Molluscs in the Coral Reef Aquarium

It goes without saying that molluscs have their place in coral reef aquariums; however, species selection must be made cautiously. Some, such as nudibranchs are almost impossible to keep since most species require a specific type of food. Others, such as many of the cowries, are predatory, and damage corals and other invertebrates. Some snails, including some cone shells (*Conus* spp.) are also venomous to humans. Although their biology is rather interesting, they cannot be recommended as aquarium animals. Octopuses and cuttlefishes require special aquariums and are suitable only for advanced and very experienced aquarists.

Giant clams of family Tridacnidae – like this *Tridacna maxima* specimen – can be very colourful. They are among the most popular invertebrates in coral reef aquariums. Rearing these molluscs in special breeding farms contributes to their continuous availability to the aquarium hobby, without having to remove them from natural populations in the wild.

Fortunately, however, there are also quite a few molluscs that are very well suited for a coral reef aquarium, when kept in combination with corals and live rocks. Here we can include, for instance, the algae-grazing snails of genera *Astrea, Turbo* and *Nerita*. Giant clams of family Tridacnidae have been kept successfully for many years. They have become a standard exhibit component in modern coral reef aquariums.

The diversity of phylum Mollusca is further emphasised by the fact that some species can be kept in virtually every aquarium, while others hardly ever survive even under the care of experienced marine aquarists. This then underlines how important it is to have in depth knowledge about this animal group.

Yet it is inevitable that the treatment of molluscs in this book

Snails can be marked very attractively – like this *Cypraea eglantina*; however, only a few *Cypraea* species are algae eaters. Most species feed on corals and other invertebrates, which makes them unsuitable for maintenance in a coral reef aquarium.

must remain incomplete, but fortunately there is now – in part due to contributions made by hobby malacologists – a comprehensive, specialised literature. For further reading we recommend ABBOT & DANCE (1982), ANDREWS (1994), CERNOHORSKY (1967, 1972, 1978), DANCE (1977), HINTON (1972, 1978), KNOP (1994), VAUGHT (1989), WELLS & BRYCE (1988) and WILSON (1993, 1994). Furthermore, there are several journals on snail- and clamshells. Relevant addresses and useful information about molluscs can be found in WILSON (1993, 1994).

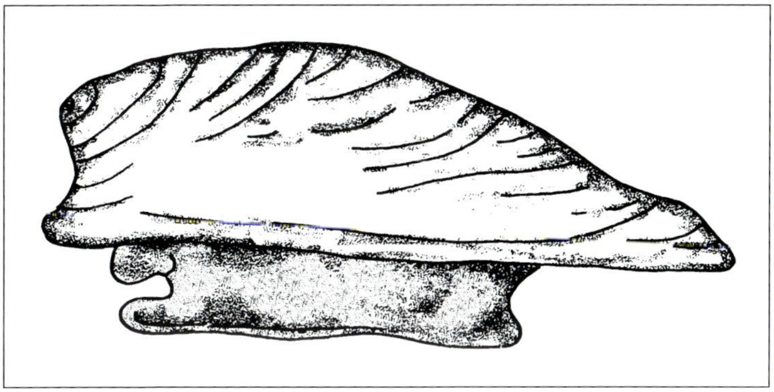

Neopilina **sp. is a typical representative of primitive molluscs.**
Drawing after MARGULIS & SCHWARTZ (1995)

Class Monoplacophora – Primitive Molluscs

The molluscs in this class are primitive soft-bodied animals that have been known for a long time only as fossils, dating back to the Cambrian- and Devonian Periods. However, in 1952, in deep water off Costa Rica, ten live *Neopilina* specimens were brought up in sediment scrapings. Since then further finds have been made in deep water (2,000 to 7,000 metres) at several other locations around the world. So far five species have been described. *Neopilina* species are very small, only a few millimetres long. Externally they look much like a cross between snails and sea slugs.

The discovery of ancient soft-bodied animals was very important for the understanding of the evolution of molluscs. Monoplacophorans have only one shell (monos, Greek = alone, single; plax, Greek, Genetive plakos = area, plate; phoros, Greek = carrier). The internal organs are arranged in pairs. These primitive molluscs have, e.g. eight pairs of muscles, six pairs of gills and a nervous system that is structured like a ladder with ten cross-connections. Many scientists believe that this supports the theory that molluscs have originated from a segmented worm-arthropod genealogy, and that molluscs have lost their segmentation during an early developmental stage.

From an aquarist's perspective these ancient molluscs are without any significance. But anyone interested in learning more about the natural history of these animals is advised to refer to publications by LEMCHE (1959) and LEMCHE & WINGSTRAND (1959).

Class Aplacophora – Worm-like Molluscs

Molluscs in this class are of an elongated, worm-like shape. A shell is absent, but instead they carry calcareous spicules (sclerites) in their mantle. Sometimes these spicules are so numerous that one gets the impression as if the animals are wearing a "fur coat". Like most other molluscs, the Aplacophora have a rasping tongue radula, used for scraping off food. The internal organs are located in a mantle cavity that is enclosed by a mantle. Locomotion is accomplished by means of a muscular foot. The approximately 200 species live in bottom mud, often far offshore and in deep water. Consequently, they are not important for aquarium purposes.

Chaetoderma argenteum **is found in shallow muddy environments along the West Coast of North America, where it eats foraminiferans.**
Photo: Dr. R. L. Shimek

Class Polyplacophora - Chitons

Although chitons are usually not among frequently imported molluscs, they are still quite frequently seen in coral reef aquariums. Usually they have been introduced as "stowaways" together with live rocks. Chitons are rather common in the sea, especially in the surf zone. So far about 600 species have been described that are all marine forms.

Chitons have a peculiar oval shell that consists of eight, transverse and partially overlapping plates (poly Greek = many, numerous; placophora = refer to Monoplacophora for a translation). Since the shell consists of several plates, it is flexible. This facilitates locomotion on the bottom, but also makes it possible to roll up the body when the animal is frightened or has been disturbed. Fossil chiton finds dating back to the Cambrian Period (more than 500 million years ago) are barely distinguishable from current species. A certain body structure and adaptation favourable for the survival of an organism is often being retained unchanged. The systematic summary (on the left below) provides an overview over the most common families and genera encountered on coral reefs of the Indo-Pacific region, but under no circumstances is that to be taken as the total picture of Class Polyplacophora.

Chitons are adapted perfectly to a life in habitats exposed to strong wave action. Underneath their shell they have a muscular, oval-shaped foot, with which they can attach themselves so firmly that the aid of a knife or similar appliance is required to release them from their substrate. The mantle is large and easily recognisable since it generally protrudes from underneath the shell. Microscopic structures along the outer rim of the mantle are important species characteristics. They are often the only certain feature enabling correct species identification, which makes identification rather difficult to amateurs. Therefore it can often be difficult to determine particular species. Chitons have separate sexes and fertilisation takes place outside the body.

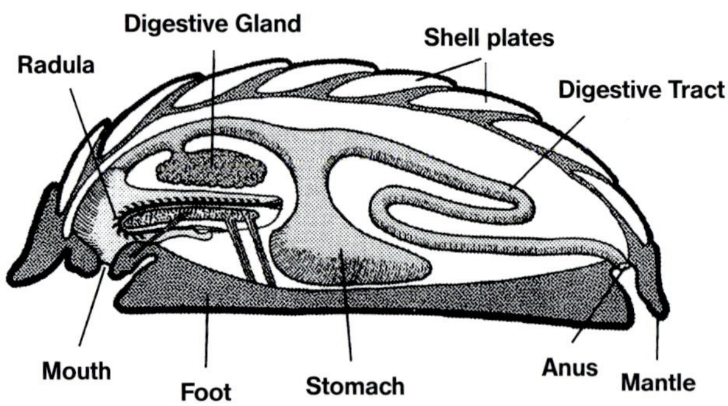

Cross-Section through a chiton. **Drawing after BUCHSBAUM (1967)**

Class Polyplacophora
- Order Neoloricata
  - Family Lepidopleuridae
    - *Leptochiton*
  - Family Chitonidae
    - Subfamily Chitoninae
      - *Tegulaplax, Rhyssoplax*
    - Subfamily Acanthopleurinae
      - *Acanthopleura, Stenoplax*
    - Subfamily Toniciinae
      - *Onithochiton, Tonicia*
  - Family Ischnochitonidae
    - Subfamily Ischnochitoninae
      - *Ischnochiton*
    - Subfamily Chaetopleurinae
      - *Chaetopleura*
    - Subfamily Callochitoninae
      - *Callochiton*
    - Subfamily Callistoplacinae
      - *Callistochiton*
  - Family Mopaliidae
    - *Plaxiphora, Placiphorella*
  - Family Schizochitonidae
    - *Schizochiton*
  - Family Acanthochitonidae
    - *Acanthochitona, Craspedochiton*

Systematic overview showing the most important families and genera of class Polyplacophora from the coral reefs of the Indo-Pacific.
After MATHER & BENNET (1993)

Many chitons are sensitive to light and therefore they remain in dark or shady areas, such as underneath rocks, under overhanging cliffs or in small hollow spaces. They have a powerful rasping tongue (radula) covered with tiny, hard teeth, used for scraping algae off the substrate.

There are also predatory chitons, such as *Placiphorella velata* (family Mopaliidae), which occurs along the west coast of the United States and in the Gulf of California. In addition to algae, this species also feeds on small crustaceans and worms. Other chiton species feed on sponges, bryozoans and tunicates.

Predatory chitons, *Placiphorella velata* in a display tank at the Monterey Bay Aquarium (Monterey, California, USA).

Family Chitonidae

This family encompasses species occurring in tropical-, subtropical as well as in temperate and cold seas. It also contains those species that are of interest for the coral reef aquarium. Typical for all species is a convex shape of shell plates. Some species live in the inter-tidal zone and so can survive out of water for several hours.

***Stenoplax* sp., here at a depth of 3 m, off Puerto Rico.**

Subfamily Chitoninae

This taxonomic group contains small species with a total length of about 4 cm and smooth, straight shell. *Chiton olivaceus* occurs on reef flats in the Red Sea. It is often attached to dead corals (VINE, 1986). *Tegulaplax hululensis* and *Rhysoplax venusta* are two species from the Indo-Pacific region.

***Acanthopleura gemmata* – seen here in southeastern Queensland (Australia) – is one of the largest chitons. It is common throughout the central Pacific.**
Photo: L. Newman and A. Flowers

Subfamily Acanthopleurinae

The best-known group of this subfamily is genus *Acanthopleura*. Its species can reach a length of up to 13 cm. They pos-

Acanthopleura grannulata
Caribbean Chiton

Photo:
Specimen seen here off Puerto Rico.

Distribution: Widely distributed throughout the Caribbean Region.
Size: 3.5 to 10 cm.
Reproduction: Sexual; no information available about captive breeding.
Food: Feeds on microscopic algae.
Description: Mantle grey with black bands, covered by short, coarse hair-like spines. Plates brownish, but often eroded and overgrown with calcareous algae. Lives in the inter-tidal zone where it is normally firmly attached to rocks. Can occur in larger numbers in certain localities.
Aquarium requirements: Like all chitons, this species is difficult to keep. Under aquarium conditions it requires strong water movement, ideally with cyclic tidal conditions. In addition, microscopic algae must be available. Rarely ever commercially traded.

GA: +/-; **SE:** -; **CO:** +

sess, strong, often substantially eroded shells. *Acanthopleura grannulata* is very common in the Caribbean Sea (see Fact Sheet above). We found this species by the hundreds in the inter-tidal zone along the coast of Puerto Rico. *A. gemmata* and *A. spinosa* are two species commonly found in the Indo-Pacific region. With a total length of up to 10 cm both species reach a substantial size. *A. spinosa* is easy to recognise by its dense cover of spines protruding from the mantle. (spinosus, Latin = spiny). The inter-tidal zone in the Red Sea is widely inhabited by *A. haddoni*, where it grazes on filamentous algae.

***Tonicia* sp. (?) - seen here at Heron Island, Great Barrier Reef - belongs to the more colourful chitons. Photo: L. Newman and A. Flowers**

Subfamily Toniciinae

This taxonomic group contains many very beautifully coloured species, with a maximum length of about 7 cm. The most frequently occurring genus is *Tonicia*, with *T. lamellosa* and *T. fortilirata* two common species from the Indo-Pacific. *T. schrammeri* lives in the western Atlantic, but according to HUMANN (1994) it is not very common there. This species is found in shallow reef areas and within a reef it attaches itself to rocks or other solid substrates.

Family Ischnochitonidae

The representatives of this family are rather common in subtropical seas, but there are also a few tropical species. With a maximum length of 3 cm, these chitons re-

Two *Ischnochiton* species at Eilat, Israel. Photo: R. Brons

main rather small. Their preferred habitats are shaded areas, for instance, underneath rocks or below overhanging reef structures. If they are pulled out in to bright light, they will quickly crawl back in to dark areas.

A rather typical tropical species is *Ischnochiton verburi* that occurs wide-spread in the Red Sea, where it can be found on reef flats hiding underneath coral blocks. Other Indo-Pacific species are *I. luticolens, I. adlaeidensis* and *I. arbutum*. Along the coast of Florida, from Tampa to the Florida Keys, as well as throughout the Caribbean Sea, we find *I. papillosus*. This species has a whitish base colouration and is covered by olive-green spots.

Chitons in the Coral Reef Aquarium

Chitons are generally algae eaters, so that it should be possible to supply them with sufficient food in an aquarium. Nevertheless, most species are apparently difficult to keep in a coral reef aquarium. This applies particularly to species from the inter-tidal zone. When they are introduced (by chance) in to an aquarium, while attached to live rocks, they usually survive only for a brief period. We are uncertain why this is so; however, we assume that chitons have become adapted to wave action in the inter-tidal zone. Consequently, they require corresponding conditions in an aquarium. The same also appears to apply to other animal groups from such habitats, and this may be related to respiration, feeding and temperature.

If chitons are to be kept successfully they must be given an aquarium with ample algae growth. It is important to offer very strong water movement, because chitons have become adapted specifically to wave action.

Family Acanthochitonidae

With a maximum length of 15 cm, this family contains the largest chitons. *Acanthochitona penicillatus* occurs below coral blocks on reef flats in the Red Sea. *A. complatana* and *A. shirleyi* are species that live in the Indo-Pacific region.

Some species in this family are conspicuously elongated, almost worm-like, such as *Cryptoplax larvaeformis*. This is a chiton found rather widely along the east coast of Australia.

Cryptoplax larvaeformis at Heron Island, Great Barrier Reef.
Photo: L. Newman and A. Flowers

This is the largest and most diverse class of phylum Mollusca. More than 65,000 species have been described, which amounts to about three-quarter of all currently known mollusc species. Gastropods occur in the sea and in brackish water, as well as in freshwater. Moreover, they are the only molluscs that have managed to make a successful transition to a terrestrial life. Constantly, there are new, interesting mollusc discoveries. Recently, the so-called "hot-vent" limpets have been rediscovered. These animals live in the deep sea close to warm submarine volcanic springs. Up to now it was assumed that these limpets had become extinct already millions of years ago. This discovery and further, important fossils finds, as well as new information derived from the comparative anatomy have substantially modified our understanding of gastropod systematics in recent years.

Classification

Subclass Heterobranchia was established in 1985. In the species included are several parasitic forms, such as *Heliacus* spp., which are also of interest to the aquarium hobby (HASZPRUNAR, 1985); however, the validity of subclass Heterobranchia is being disputed. Nowadays not all scientists accept this particular taxon within the current systematics of Gastropoda (e.g. see BIELER, 1990). Similarly, even within the classification of subclass Prosobranchia that encompasses those gastropods of greatest interest to the aquarium hobby (snails, whelks, periwinkles, abalones), there have been significant changes.

The systematics of Gastropoda is under constant discussion, so that we can expect further changes in the years to come. For further details on the systematics of Gastropoda see BIELER (1990 and 1992), HASZPRUNAR (1985 and 1988), HICKMANN (1988), MORE (1960), PONDER & WAREN (1988), WILSON (1993, 1994) and VAUGHT (1989). The systematics of Class Gastropoda given on page 45 (down to the level of orders) is based on the work by these authors. Those groups marked with (*) are not being discussed in detail.

The adaptability of the more than 65,000 snail species is enormously large. The have conquered most habitats. We encountered this *Chicoreus* species (family Muricidae) in a cave at Bunaken (northern Sulawesi). In reality it is not orange it's colouration is due to a sponge that covered the snail. This affords excellent camouflage and protection against enemies. The snail itself is a predator that feeds on other invertebrates.

Many cowries - like this *Cypraea cribaria* at Rosemary Island, Western Australia - have a beautifully coloured mantle used for polishing their shell. Cowries are rather frequent in shallow water of reef flats and along reef slopes. They feed on algae, detritus and various invertebrate animals.

Photo: C. Bryce

General Biology

Gastropods belong to the more advanced invertebrates; however, within this class there are significant variations in body structure. Many species are specialised, for instance, in respect to body structure and food preferences. Land-living (terrestrial) snails have a primitive lung, which can take up atmospheric air via moist skin, while gastropods that live in the sea and breathe via gills. In nudibranchs of order Nudibranchia (subclass Opisthobranchia) a shell is absent and the gills have regressed. Instead, secondary gills have developed, which are often well visible as bushy outgrowths along the abdomen. In other orders within this subclass there is a shell, which is partially or totally covered by a mantle. In species from subclass Prosobranchia the shell is located externally and there forms the well-known snail shell, a character that is an important reason for the evolutionary survival success of prosobranch snails. In marine gastropods the shell is thick and strong, while in land-living forms it is usually very thin and fragile, but it is easier to carry around.

A very strong, muscular foot used for locomotion and burying, is characteristic for gastropods. The rasping tongue (radula) is

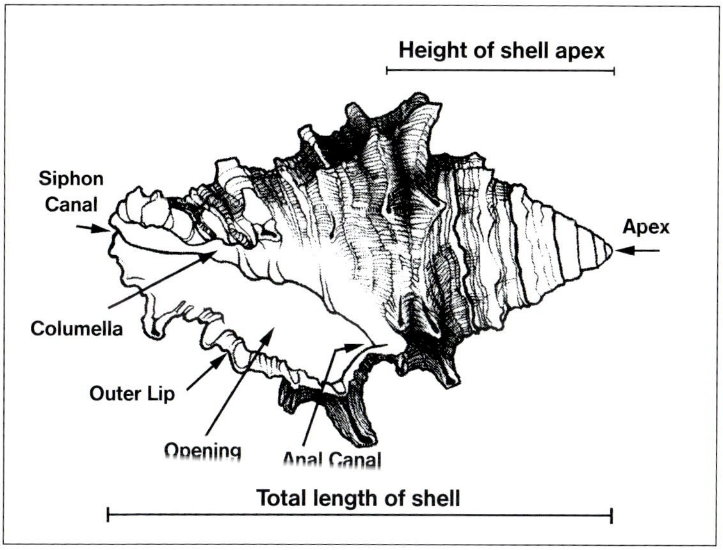

Simplified shell structure of snails. **Drawing after WILSON (1993, 1994)**

Class Gastropoda · Snails

- **Subclass Prosobranchia**
  - Order Docoglossa
  - Order Cocculiniformia
  - Order Neritimorpha
  - Order Archaeogastropoda
  - Superorder Caenogastropoda
    - Order Architaenioglossa
      - Suborder Discopoda
      - Suborder Heteropoda
      - Suborder Ptenoglossa
    - Order Neogastropoda
- **Subclass Heterobranchia**
  - Order Heterostropha
- **Subclass Ophistrobranchia**
  - Order Cephalaspidiea
  - Order Acochlidiacea *
  - Order Rhodopemorpha *
  - Order Saccoglossa
  - Order Anaspidea
  - Order Notaspidea
  - Order Theocosomata
  - Order Gymnostomata
  - Order Nudibranchia
- **Subclass Pulmonata**
  - Order Systellommatophora
  - Order Basommatophora *
  - Order Stylommatophora *

The tiny *Littoraria philippiana* (family Littorinidae) has become adapted to a terrestrial life. It occurs on mangrove trees, and feeds on algae.
Photo: L. Newman and A. Flowers

 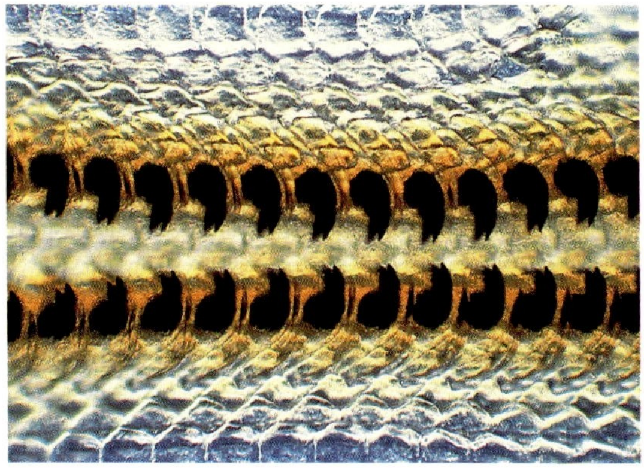

Many snails can scrape algae off the substrate with their modified tongue, referred as radula. The mouth is often in the shape of a tunnel, as seen here (left photo) on the front glass of an aquarium. In addition, we can also see part of the foot, the edge of the mantle below the shell and the eye-carrying tentacles. On the right (photo: R. Brons) is a photomicrograph of a radula.

particularly well developed. It is very effective for picking up small organisms and microscopic algae. Man gastropods are algae eaters, but there are also predatory species, detritus feeders and even several parasitic species.

Gastropods on Coral Reefs

Gastropods are numerous on coral reefs. They occur from the inter-tidal zone, where some species live partially on land, down to the deepest parts of a reef. Gastropods bury in sand, live under rocks and among corals, or they live as parasites well camouflaged on animal hosts. There are tiny species that can only be found with a magnifying glass, but also large and colourful, easily distinguishable species. There are harmless algae eaters as well as venomous predatory snails, which – in extreme situations – can even kill fish. The gastropod fauna of a reef is so diverse, that even a limited area requires detailed studies in order to gain an overview.

McCLANAHAN (1990) has investigated shallow areas of coral reefs off Kenia. He found 135 different gastropod species from 25 families. This is a number that corresponds well to respective studies in the Indian Ocean.

Gastropods in the Coral Reef Aquarium

Gastropods are almost always introduced in to coral reef aquariums together with live rocks or corals, as a sort of "complimentary" addition. The majority of these gastropods are small and difficult to identify. Our experiences have shown that most are harmless and feed on algae and detritus. But occasionally species can appear that feed on corals and other invertebrates. Such parasites can cause considerable damage, and – moreover – some can reproduce quite rapidly. Unless an aquarist has particular interest in the natural history of parasitic snails, these gastropods should be removed from the aquarium as quickly as possible.

Large gastropods are periodically available from the aquarium trade. In many cases these are predatory species (order Neogastropoda). Occasionally there are also delicate and very colourful nudibranchs. These, with very few exceptions, belong to a group of marine invertebrates, which – according to our current knowledge and aquarium-technological state of the art – are definitely not recommended for the aquarium trade. Virtually all nudibranchs are food specialists that will not be able to find suitable food in an aquarium and therefore they usually die within a few days.

General Information

Over the following pages there are photo plates assembled from several genera and/or species of the (from an aquarist's perspective) important families of class Gastropoda. Not all of these species are, however, necessarily being discussed in the subsequent text. Instead, the plates are merely intended to depict gastropod families in general. We are kindly indebted to Clayton W. Bryce from the Western Australian Museum (Perth, Australia) for making these illustrations available.

Those groups and species that are important to the aquarium hobby are being presented in the form of Fact Sheets. Moreover, various groups of gastropods are being shown in their natural habitat. Dr. Leslie Newman and Dr. A. Flowers, Lismore, (NSW, Australia) took most of these photos.

All "common" marine snails are included in Prosobranchia. The shell, which varies significantly in shape, plays a significant role in this group. While the more primitive prosobranchs have (largely) flat shells (reminiscent of those found in their evolutionary precursors), the more highly developed species often have very artistic and spirally coiled shells. The prosobranchs are considered to be more primitive snails. They have a long evolutionary history, dating back to the Cambrian Period. Today it seems likely that the other subclasses have evolved from prosobranchian ancestors.

The Prosobranchia have conquered most habitats. They are found in the sea, in brackish- and freshwater, as well as on land. All species possess a mantle cavity that accommodates the gills where gas exchange takes place. Some species, for instance, beach snails of genus *Littorina* live in the inter-tidal zone. These molluscs must always return to the sea to dispose of their metabolic waste products, which otherwise would accumulate in the small amounts of water retained in their shells and so lead to their death. Species that live on land have developed different systems for their excretions, which have freed them from frequent contact with seawater. Many prosobranchs have developed bodyadaptations with which they are well equipped for their survival. Their feeding behaviour ranges from detritus feeders and algae grazers to carnivorous forms and even parasitic species with highly specific food requirements. One of the example of this is the small Caribbean Flamingo Tongue, *Cyphoma gibbosum*, which feeds exclusively on gorgonians.

Most prosobranchs are characterised by a shell opening that is protected by lid (operculum) created by glands located in the foot. In species living in the inter-tidal zone or in terrestrial forms the lid also prevents the loss of water. An exception to this rule is the limpets of family Patellidae, where the operculum is absent. In contrast to other subclasses of snails, which are all hermaphrodites, the prosobranchs have separate sexes. Their larval stage is relatively short, whereby veliger larvae or trochophore larvae assure the species distribution.

Subsequently we will discuss the various orders and also provide an overview down to genus and/or species level, depending upon our assessment of their respective significance to the aquarium hobby. Many of the details provided here are based on WILSON (1993, 1994).

Many snails can close off the opening of their shell with a thick, protective lid, the operculum as seen here in *Turbo perspeciosus*.
Photo: L. Newman and A. Flowers

Order Docoglossa - Limpets

Limpets live in the inter-tidal zone, often at sites exposed to strong wave action. With their fleshy foot they are attach themselves so firmly to a substrate, that it is difficult to dislodge them even with a knife. The shell is usually adapted to a specific site on a particular rock, where it then forms a tight fitting "house" that covers the snail. The shape of the shell is peculiarly conical and bilaterally symmetrical, being reminiscent of an "elbow" (the colloquial name for limpets in many countries is "elbow snails"). Currently, the order Docoglossa encompasses five families:

- Acmaeidae
- Patellidae
- Lepetidae
- Nacellidae
- Lottidae

Family Acmaeidae

This family contains mostly small species, which have one true gill, which – when fully extended – becomes visible around the edge of the shell. It is possible that the two most tropical genera – *Colli-*

sella and *Patelloidea* – are sometimes imported in conjunction with live rocks.

Family Patellidae

Species with true gills are absent from this family. Instead, these limpets possess a series of filaments referred to as secondary gills, which are located in a depression between mantle and foot. Many species are relatively large, such as those of genus *Patella*. Even though this genus is widely distributed throughout the Indo-Pacific, it is however relatively uncommon in the western Atlantic and in the Caribbean Sea.

The Red Sea Limpet, *Cellana eucosmia* is endemic to the Red Sea and lives within the inter-tidal zone. It reaches a diameter of about 4.5 cm and feed on filamentous algae as well as on microscopic algae, which are scraped off the substrate with the radula. This species would probably be quite suitable for a coral reef aquarium. Photo: R. Brons

Snail shells of families Acmaeidae and Patellidae: 1 - *Cellana radiata*, 4 cm, Indian Ocean and western Pacific Ocean. 2 - *Patella laticostata*, 12 cm, South and Western Australia. 3 - *Patella peronii*, 3 cm, New South Wales to Western Australia. 4 - *Patella flexuosa*, 5 cm, Indian Ocean and western Pacific Ocean. 5 - *Patella chapmanni*, 3 cm, New South Wales to Western Australia. 6 - *Patelloidea nigrosulcata*, 3 cm, Western Australia. 7 - *Collisella onychitis*, 2 cm, South- to Western Australia. 8 - *Patelloidea saccharina*, 5 cm, Indian Ocean and western Pacific Ocean. 9 - *Patelloidea alticostata*, 5 cm, New South Wales to Western Australia. Photo: C. Bryce

The genus *Cellana* is very reminiscent of genus *Patella*, however, the apex of its shell is slightly displaced forward. While the shell of *Patella* species looks like a perfect cone that of *Cellana* species has the shape of a leaning cone. *Cellana eucosmia* is endemic to the Red Sea, while *C. tramoserica* is rather common along the Australian coastline.

Order Neritimorpha – Nerite Snails

This order contains snails with a solidly built shell. Usually they live in the inter-tidal zone. They are characterised by a thick, calcareous operculum and a distinctly basal tooth. According to Wilson (1993, 1994) this order contains three families:

- Neritopsidae
- Neritidae
- Phenacolepadidae

Of these three only family Neritidae is of interest to the aquarium hobby.

Two different forms of the Australian Limpet, *Cellana tramoserica*, seen here in southeastern Queensland, where it lives in the inter-tidal zone. It reaches a diameter of about 6 cm and feeds on filamentous algae as well as on microscopic algae, which are scraped off the substrate with the radula. This species would probably be quite suitable for a coral reef aquarium.
Photo: L. Newman and A. Flowers

Family Neritidae

Species of this family have been imported for several years and used as algae grazers in the reef aquarium. The most common genera are *Nerita* and *Puperita*, which live in the inter-tidal zone and so they are often found above the water level. Similarly, in an aquarium they often crawl out of the water or stay exactly at the edge of the water level.

In their natural habitat neritid snails feed on algae, but under aquarium conditions it can hardly be said that they are perfect algae eaters. These snails tend to stay too much at or above the water surface and move only rare over the aquarium decoration, where – from our perspective – algae removal would be required. Nerite snails are exceedingly well adapted to life in the inter-tidal zone. The shell is thick and the operculum closes the opening tightly so that the animals can retain water, even when they are crawling about on nearly dry land.

The genus *Nerita* has a cosmopolitan distribution along all tropical and subtropical coastlines. Apart from the typical family characteristics, the column (columella) usually has several small spines and teeth.

According to WILSON (1993, 1994) at least 12 *Nerita* species are known to occur in Australia. One of the most common species of the Indo-Pacific is *N. albicilla*, which according to VINE (1986) is also distributed in Red Sea. During the day *N. albicilla* hides under rocks. Upon onset of darkness it re-appears to feed on algae. The colouration of this snail is highly variable; often it is black-white or sprinkled with grey dots or covered with bands and spots. This species is regularly imported.

The black *Nerita picea* occurs in large aggregations on lava rocks in the surf zones around Hawaii and the Johnson Islands, but it is rare in other parts of the Pacific.

Nerita polita looks like *N. picea* but with a diameter of 3 to 4 cm it

***Littorina* sp. on rocks along the coast of tropical Queensland, Australia.**

A group of *Nerita picea* along the coast of Hawaii (Hawaiian Islands, USA).

Shells of Nerita snails (family Neritidae) from the Indo-Pacific: 1 - *Nerita radula*, 1 cm. 2 - *Nerita albicilla*, 2 cm. 3 - *Nerita squamulata*, 1 cm. 4 - *Nerita plicata*, 2 cm. 5 - *Nerita maxima*, 3 cm. 6 - *Nerita polita*, 2 cm. 7 - *Nerita costata*, 3 cm.
Photo: C. Bryce

is significantly larger. It carries a grey and white shading against a black background colouration. Its area of distribution is the Indo-Pacific and the western Pacific, however, it never occurs in such large aggregations as *N. picea*.

The Caribbean Sea also accommodates several *Nerita* species. *N. peloronata* has characteristic red and black zigzag bands against a yellowish background colour. We have found this species to be quite numerous along the coastline of Puerto Rico (see Fact Sheet page 51).

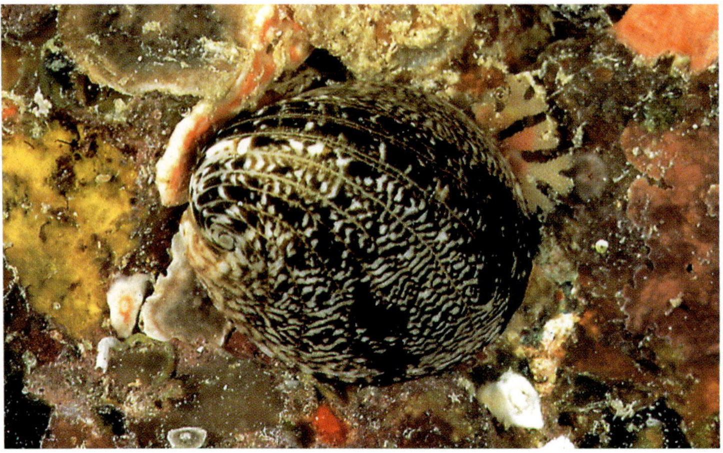

Nerita albicilla **at Heron Island, Great Barrier Reef.**
Photo: L. Newman and A. Flowers

Nerita versicolor is dirty white with red and white spots. It also has four strong spines along the columella. *N. tesselata* has a strongly shaded black-white colour pattern and poorly developed teeth along the columella. *N. fulgurans* has a particular shell structure with distinct, longitudinal grooves. It is uniformly dark-grey or black.

The beautiful species of genus *Puperita* – which resemble *Nerita* species – also live in inter-tidal habitats, but they have a thinner shell. In many species the shell has ornamental colour markings and structures. This genus is distributed in tropical and subtropical seas; however, little is known about it. The Fact Sheet on page 52 presents the Zebra Snail, *Puperita pupa*, from the Bahamas.

Nerita albicilla

Distribution: Widely distributed throughout the Indo-Pacific, where it lives in the inter-tidal zone; often out of water and exposed to air for several hours.
Size: Up to 2.5 cm.
Reproduction: The female lays the eggs in white capsules, and deposits these on rocks or the shells of other snails. Such capsules are commonly seen in aquariums, but unfortunately these clutches rarely ever develop.
Food: Feeds principally on filamentous- and microscopic - algae.
Description: The small shell is covered with a few spiral bands. Colouration grey with tiny, irregular spots. The margin of the lip is often brown; mantle greenish-yellow or white; operculum with granular surface.
Aquarium requirements: Frequently available in the aquarium trade. Has been kept extensively during the 1980's as algae grazer; at that time imported to Germany mainly from East Africa. However, as an inhabitant of the inter-tidal zone this snail prefers to stay at or above the water surface. Easy to keep, but in conformance with its natural life cycle, it has a longevity of only one or two years.

GA: +; **SE:** +; **CO:** +

Photo:
Nerita albicella.

Nerita peloronata

Distribution: Lives in the Caribbean Sea, where it occurs in the inter-tidal zone, commonly under rocks; in Florida often found together with *Nerita versicolor*.
Size: Up to 2.5 cm.
Reproduction: The female deposits egg clutches in capsules attached to rocks or to the shells of other snails.
Food: Feeds on filamentous- and microscopic-algae.
Description: Half-moon shaped, firm shell with a flat apex. Shell with traces of coarse spirals. Shell opening large and round. Colouration of shell yellowish with red and black zigzag bands. Parietal area (connection of body to the shell) with blood-red spots and with two or three white teeth.
Aquarium requirements: Rarely ever traded available in the aquarium trade. Since it is an inter-tidal organism, this snail prefers to stay at or above the water surface.

GA: +; **SE:** 0; **CO:** +

Photo:
Some *Nerita peloronata* specimens at Puerto Rico.

Puperita pupa
Zebra Snail

Distribution: Bermuda and southern Florida to the Caribbean Islands. Found intertidal in rock pools and splash pools just above the high-tide mark.
Size: Most species up to 1 or 2 cm.
Reproduction: Largely unknown, but possibly similar to *Nerita* spp.
Food: Feeds on filamentous- and microscopic algae.
Description: Little is known about this species. The species resembles *Nerita* species, but they are smaller and have a thinner shell. Operculum with a barely visible hook. Often with very attractive markings. Earlier placed in the genus *Neretina*.
Aquarium requirements: Only few details known. These snails must have an abundant supply of microscopic algae available.

GA: 0; **SE:** 0; **CO:** +

Photos:
Top: Habitat of *Puperita pupa* at the Bahamas.
Bottom left: *Puperita pupa* in a tide pool at the Bahamas.
Bottom right: A group of *Puperita pupa* in an aquarium on the shell of a young Giant Clam, *Tridacna maxima*, together with one *Littorina* specimen from the same habitat.

Order Archaeogastropoda

This order contains many marine snails that are highly suitable for a coral reef aquarium; however, we will discuss only the following families:

- Haliotidae
- Fissurellidae
- Trochidae
- Turbinidae

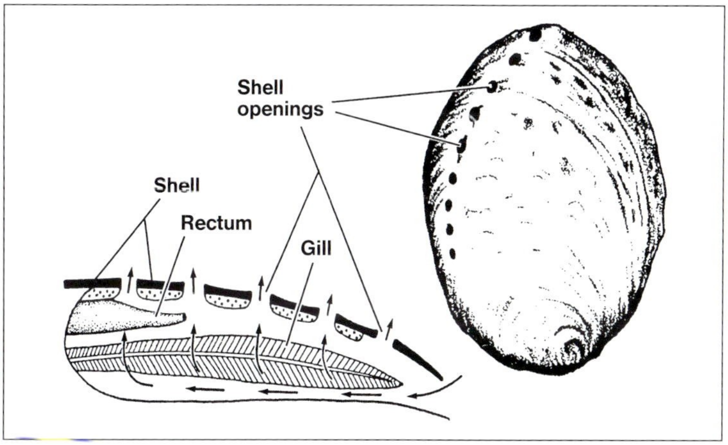

Structure of a *Haliotis* species. Left: Cross-section through mantle cavity; arrows indicate the direction of water flow. Right: Dorsal view of the shell.
Drawing after BARNES (1980)

Family Haliotidae – Abalones, Ear Shells

Abalones are easily recognised by a row of small holes along the edge of the (single) shell. Tentacle-like outgrowths, which have a sensory function, protrude from these holes, and give abalones an impressive appearance. This family of primitively developed gastropods contains about fifty species, most of which are tropical, but a few Abalones do also occur in temperate waters. Although some of the literature refers to an alleged existence of several genera, only *Haliotis* is recognised as a scientifically valid genus.

In the wild, abalones occur in the lower inter-tidal zone, often at sites where there is strong water movement. Certain species live in deeper water. *Haliotes asinine* (8 - 12 cm long) is rather common on coral reef throughout the Indo-Pacific. This species has a relatively thin and more elongated shell than that of other species. Its colouration is an intense olive-green. Probably the most frequently occurring species in the Indo-Pacific is *H. varia* (see Fact Sheet page 54). It is relatively small, with a maximum diameter of 4 cm, and has been described repeatedly under different names, such as *H. concinna, H. semistriatus* and *H. viridis*.

It is not unusual to have *Haliotis* species introduced in to an aquarium attached to imported live rocks. The suction foot is very strong, and any attempt to remove the snail from its substrate leads to the demise of the animal. Abalones feed on algae and are active only at night, and so one has to look for them with an electric torch. It is easy to confuse abalones with members of the subfamily Stomatellinae (see page 57), but in the latter the (for *Haliotes* so typical) holes along the margin of the shell are absent.

***Haliotis asinine* at Heron Island, Great Barrier Reef. Well visible is the expanded mantle and the respiratory holes in the shells that are characteristic for abalones (sea ears). Photo: L. Newman and A. Flowers**

Haliotidae - Abalones, Sea Ears

Haliotis varia
Variable Abalone

Distribution: Widely distributed throughout the Indian Ocean and western Pacific Ocean, where it occurs under rocks in shallow reef water or in rocky coast habitats.
Size: Up to 5 cm.
Reproduction: Separate sexes; males and females reach sexual maturity at the same time; eggs and sperms are being released in to open water; very large number of eggs. No details available about any captive breeding.
Food: Normally feeds on filamentous- and microscopic algae, but can also bit off small pieces from leafy algae.
Description: Most commonly occurring abalone in the Indo-Pacific. Elongated, oval shell, rounded off towards the top. Apex located at the posterior end. Colour of shell brown or greenish, normally with white or cream-coloured spots. The respiratory holes form a row close along the left shell margin. The fleshy mantle and the two feelers are well visible.
Aquarium requirements: Occasionally imported, attached to live rocks. Easy to keep, usually nocturnal. Does not damage any other inhabitants of a coral reef aquarium.

GA: +; **SE:** +; **CO:** +

Photos:
Top: *Haliotis varia* in the Red Sea.
Bottom: Abalone shells; the largest has a diameter of 4 cm.
Photos: R. Brons

Family Fissurellidae – Slit Limpets, Keyhole Limpets, Elephant Snails

The shells of these (false) limpets are conically shaped and are reminiscent of order Docoglossida (see page 47). Slit limpets have an anal opening or slit in their shells, through which they discharge their metabolic waste products. This feature also gave rise to the family name: fissum (Latin) = slit, cut. It is also analogous to the holes in the shell of abalones and is considered to be a primitive (evolutionary) development characteristic. In general, split limpets have a series of primitive development features. Often the body is so large that there is not enough space for it under the shell (*Scutus*, Elephant Snails).

An operculum and a mantle are absent in split limpets. Frequently the foot- and head parts can be seen outside the shell. This then totally covers up the shell and creates an impression as if it was a nudibranch. On occasion even this has fooled us. For instance by *Scutus unguis*, which can frequently be found in aquariums (see page 55).

Generally-speaking, slip limpets are algae eaters and are therefore well suited for a coral reef aquarium. Yet, a few species are specialist feeders, using sponges and tunicates as their principal diet. Overall, little research appears to have been done so far on the biology of split limpets.

The genus *Diodora* (keyhole limpets, subfamily Diodorinae) has a worldwide distribution. Its species are being distinguished on the basis of a hole (instead of a slit) in the shell. In contrast to the other genera, *Diodora* species

Scutus unguis
Shield Limpet

Distribution: Widely distributed throughout the Indian Ocean and western Pacific Ocean; inter-tidal species, often under rocks or in narrow crevasses.
Size: Shell up to 4.5 cm; mantle and foot are large and fleshy, maximally 6 to 7 cm when fully expanded in adult specimens; but usually smaller (2 to 5 cm) in aquarium specimens.
Reproduction: Reproduces sexually under aquarium conditions. Juvenile specimens have occasionally been found in aquariums.
Food: Fundamentally herbivorous; grazes algae off substrate with its fan-shaped radula. May occasionally "nibble" on sponges and dead animals.
Description: Originally described by LINNEUS (1758), sometimes listed as *S. granulata* in the literature. Shell developed as a fingernail-shaped shield (scutus, Latin = shield, and unguis, Latin = fingernail), with the apex about one-third in front of posterior end; concentric ripples with small bubbles; anal notch shallow, but wide and v-shaped; colouration whitish or shades of tan. Mantle and foot are black and fleshy, covering the shells, which is visible only as a small, white slot along the dorsal side. Two internal gills (same as in all members of family Fissurellidae). Can easily be mistaken for a nudibranch or sea slug.
Aquarium requirements: Often introduced in to aquariums together with live rocks. Principally nocturnal, but often visible during darkness while grazing algae of aquarium windows. Several specimens can be kept at the same time. Hardy and easy to keep in a coral reef aquarium; can reach a longevity of several years.

GA: +; **SE:** +; **CO:** +

Photos:
Top: Shield Limpet among algae in an aquarium.
Centre: Juvenile in an aquarium.
Bottom: Exposed shell.

The foot is fully expanded in this *Diodora singaporensis.* The small shell cannot completely cover the body of the snail. Photo: C. Bryce

cannot cover the shells with their mantle. These snails live under rocks or in small crevasses in the inter-tidal zone. WILSON (1993, 1994) lists 12 species for Australia, but this genus has so far been insufficiently researched. *Diodora junkesii* (max. size to 6 cm) is among the most frequent species in this region.

Diodora cayensis is rather common in the western Atlantic. It occurs in the inter-tidal zone from Bermuda in the north to Brazil in the south. The colouration of its shell varies from lilac, white to dark grey, but often these snails are covered with algae, which camouflages them rather well.

***Diodora junkesii* at Dunwich, Stradbroke Island (Queensland, Australia). Photo: L. Newman and A. Flowers**

The species in genus *Macroschima* (subfamily Fissurellinae) have long, saddle-like shells with an elongated hole. The shell covers only the anterior part of the animal. The posterior part of the foot extends beyond the shell.

In subfamily Emaginulinae the hole-like opening is replaced by a cleft. Genus *Emaginula* has a cosmopolitan distribution with many tropical species. The apex points backward, so that one gains the impression as if the cone-shaped shell was askew. Other genera in this subfamily are *Hemitoma, Montfortula, Opuncturella* and *Rimula*, which are all occasionally being imported attached to live rocks.

While the slit limpets mentioned so far appear only sporadically in aquariums, genus *Scutus* with the black *S. unguis* is relatively frequent. This genus has representatives in European as well as tropical seas. These species have a large shell where the hole-like opening is absent or in a v-shape only weakly indicated.

Scutus antipodes grows to a size of 11 cm and is distributed from the southern region of the Great Barrier Reef and the south Australian coast to the Houtman Albrolhos Island Group, off Western Australia. Because of this geographical distribution *S. antipodes* is rarely ever seen in the aquariums of European aquarists. Yet, *S. unguis* occurs very frequently in the aquarium trade. It has a black mantle that encloses the shell almost completely. Its appearance is reminiscent of that of a sea hare. We too have also mistaken this species and referred to it as a sea hare of family Aplysidae in volume 1 of the German edition of this book series. In reality, however, it is a split limpet with such a large mantle and foot that enclose and virtually cover the entire shell.

S. unguis is frequently imported in conjunction with live rocks (see Fact Sheet on page 55). A small, special aquarium for this and other limpets would surely be an attraction.

Superfamily Trochoidea
Spiral-shelled Gastropods

According to WILSON (1993, 1994) this superfamily contains three families, thirteen subfamilies, and it is of considerable generic and species diversity. Below we are discussing merely a selection of species of interest to the aquarium hobby, and those are only from families Trochidae and Turbidae.

In previously discussed families (Haliotidae and Fissurellidae) the respective shells contained either slots or holes for the discharge of water and together with accumulated metabolic waste products. The features are totally absent in the Trochoidea. These gastropods have a long evolutionary history. Fossil records date back to the Permian Period (290 to 245 million years ago).

Family Trochidae
Top Shells

Snails of family Trochidae usually have a conical shell. The opening of the shell is relatively large and can be closed off with a thin, round or oval-shaped operculum. These snails are strongly reminiscent of turban snails (family Turbinidae, discussed below), and they are distinguishable from them merely by a differently shaped radula. Trochid snails have variable feeding strategies. There are herbivores (plant eaters), carnivores (meat eaters) as well as detritus feeders. External sex organs are absent in the Trochidae. They reproduce by releasing eggs and sperm directly in to the open water, where fertilisation takes place. Juvenile snails develop after a planktonic larval stage has been gone through.

Most trochid species are found on a hard substrate, and many live among the corals on a coral reef. Occasionally specimens are being imported attached to live rocks or corals, and here it is important to remember that some trochids can feed on corals. According to WILSON (1993, 1994) the systematics of the Trochidae have – in general – been insufficiently researched only, so that subsequent revisions can be expected for this family.

The species of **subfamily Caliostomatinae** are carnivorous and feed on coral polyps. Genus *Calliostoma* has a worldwide distribution and contains a few species, which are sporadically introduced in to coral reef aquariums in conjunction with corals and live rocks. Sometimes they are also specifically imported and are then available from the aquarium trade.

Most *Calliostomatinae*-species have a typical pyramid shape, and an open "umbilicus" is absent in all species. That means internal spiral walls are being developed inside the shell without a simultaneous development of a central opening. *Calliostoma similare* is a frequently occurring species in the central Indo-Pacific region. The Caribbean Sea is inhabited by *C. euglyptum* and *C. jujubinum*.

Genus *Astele* resembles genus *Calliostoma*. It too has a cosmopolitan distribution with several tropical species periodically being imported for the coral reef aquarium trade.

In **subfamily Stomatellinae** the shells have a large opening and are usually ear-shaped. The foot is large and cannot be totally retracted into the shell. It carries three or more tentacles, which is a distinctive feature of this subfamily. The Stomatellinae have an unusual way of defending themselves against predators. When they are being attacked they sever a small part of their foot. This part then lies on the bottom and by moving it, the attention of the predator is attracted, while the snail as such flees. The (lost) part of the foot is later regenerated. This type of behaviour is most conspicuous in *Stomatella auricula*, but it occurs generally in the entire group. Most *Stomatella* species are algae eaters and are widely distributed throughout the Indo-Pacific. *Stomatella* sp. "KA5-GAS-02" is being imported attached to live rocks and sometimes it reproduces rapidly in a coral reef aquarium (see Fact Sheet on page 58).

Snail shells of family Trochidae: 1 - *Tectus niloticus*, 15 cm, Indian Ocean and western Pacific Ocean. 2 - *Tectus fenestratus*, 4 cm, Indian Ocean and western Pacific Ocean. 3 - *Trochus hanleyanus*, 5 cm, Indian Ocean and western Pacific Ocean. 4 - *Tectus pyramis*, 8 cm, Indian Ocean and western Pacific Ocean. 5 - *Trochus histrio*, 4 cm, Indian Ocean and western Pacific Ocean. 6 - *Calliostoma monile*, 2 cm, Western Australia. 7 - *Calliostoma ciliaris*, 3 cm, Western Australia. 8 - *Angaria tyria*, 7 cm, Western Australia. 9 - *Angaria delphinus*, 7 cm, Indian Ocean and western Pacific Ocean. Photo: C. Bryce

Subfamily Stomatellinae has so far only been insufficiently investigated, but traditionally, one includes genera *Microtes, Pseudostomatella* and *Stomatella* in it. Nevertheless, the need for a revision is clearly indicated.

Subfamily **Trochinae** contains a number of genera and species that are small and cone-shaped and have a thick shell. This is the largest subfamily in family Trochidae. The largest genera are *Clanculus* (represented worldwide in tropical and temperate seas), as well as *Tectus* and *Trochus*, both

***Stomatella* spp.**

Distribution: Indian Ocean, western Pacific Ocean and southeastern Austral (Tasman) Sea.
Size: Most species are only a few millimetres big; maximum size 2.5 cm.
Reproduction: Eggs and sperms are released in to the open sea.
Food: These snails feed mainly on filamentous- and microscopic algae.
Description: Body not completely covered by the shell. This genus and subfamily Stomatellinae (traditionally including genera *Microtes* and *Pseudostomatella*) have only marginally been researched and are insufficiently known. The largest species is *Stomatella impertusa* (syn. *S. auricula*). The shell is thin, compressed, oval and generally ear-shaped.
Aquarium requirements: *Stomatella* sp. "KA5-GAS-02" was introduced in an aquarium together with live rocks. This specimen thrived and kept moving rapidly over aquarium décor. The increasing number of specimens seen suggested that this animal reproduced in the aquarium.

GA: +; **SE:** +; **CO:** +

Photos:
Top: *Stomatella* sp. "KA5-GAS-02" in the aquarium of K. Nagy, Flekkefjord, Norway.
Left: Close-up photo of a *Stomatella* sp. "KA5-GAS-02" specimen, about 1 cm long.

Left: *Pseudostomatella decolorata* at Heron Island, Great Barrier Reef (photo: L. Newman and A. Flowers). Right: *Clanculus pharaonis* from the Red Sea (photo: R. Brons).

distributed throughout the Indian Ocean and western Pacific Ocean. The best-known species is *Trochus niloticus* (max. size 15 cm, bottom to top), which is collected for human consumption. It inhabits the inter-tidal zone as well as shallow water. *Trochus* species are sporadically imported for the aquarium trade.

Family Turbinidae
Turban Snails

The snails of this family have derived their popular name from the shape of their shell that is reminiscent of a turban. Some of the species belong to those snails most frequently introduced in to coral reef aquariums together with live rocks. Turban Snails occur in tropical and temperate seas, either in the inter-tidal zone or in relatively shallow water.

The shape of the radula clearly distinguishes family Turbinidae from family Trochidae. Turbinid snails are herbivorous, the reason why they are also quite suitable for aquarium maintenance. Although the species of this family commonly occur in shallow water, the family as such has only been insufficiently researched and little is known about their biology. Most species release their eggs and sperm in to open water, which has also been repeatedly observed under aquarium conditions. Some species also deposit sex cells on to substrates, in the form of egg masses.

Most important for the aquarium hobby is **subfamily Turbiniae**. Several *Turbo* species appear regularly in marine aquariums, often in association with corals or live rocks. Frequently, this involves *Turbo brunneus* (see Fact Sheet on page 60), which has repeatedly been misidentified as *T. marmoratus*. *T. marmoratus* is, however, a very large species with a maximum size of 20 cm, which is not likely to be imported. *T. perspeciosus* (see photo on page 47) is another typical species from the Indian Ocean and the western Pacific Ocean. It grows to about 7 cm.

Snail shells of family Turbinidae: 1 - *Astralium tentorium*, 5 cm, Western Australia. 2 - *Astralium stellare*, 4 cm, western Pacific Ocean. 3 - *Astralium rotularia*, 4 cm, Western and Northern Australia. 4 - *Astralium squamifera*, 3 cm, New South Wales and Western Australia. 5 - *Astralium pileola*, 6 cm, Western Australia. 6 - *Turbo intercostalia*, 9 cm, Western Australia. 7 - *Turbo jourdani*, 20 cm, South Australia to Houtman Abrolhos, Western Australia. 8 - *Turbo undulates*, 4 cm, New South Wales to Western Australia. 9 - *Turbo lamiferus*, 5 cm, Indian Ocean and western Pacific Ocean. 10 - *Turbo torquatus*, 9 cm, New South Wales to Western Australia. 11 - *Turbo petholatus*, 5 cm, Indian Ocean and western Pacific Ocean. 12 - *Turbo cinerues*, 4 cm, Indian Ocean and western Pacific Ocean. 13 - *Turbo angyrostomus*, 8 cm, Indian Ocean and western Pacific Ocean.
Photo: C. Bryce

Caribbean members of the family include other interesting algae grazers in the genus *Astraea*, of which *A. tectum* is the most frequent species. This is an excellent algae remover in a coral reef aquarium (see Fact Sheet on page 80). Formerly, genus *Astraea* was also considered to be common for the Indo-Pacific region, but nowadays most species are incorporated in genus

Astraea tectum
American- or Red Turban Shell

Distribution: Common throughout the Caribbean Region. Lives in shallow water zones in variable habitats; sea grass areas, under rocks and in reef crevasses.
Size: Up to 5 cm.
Reproduction: Eggs and sperm are released in to the open sea.

Food: Feed mainly on filamentous- and microscopic algae.
Description: According to Colin (1978) there are two subspecies: *Astraea tectum tectum* occurs around the West-Indian Islands, while *A. t. americanum* can be found along the Florida coastline. Humann (1994) describes these snails as being two different species of genus *Lithopoma*: *L. tectum* and *L. americanum*. So far we have been unable to resolve what the correct taxonomic status is. Shell with well-defined apex, often over-grown, e.g. with calcareous algae.
Aquarium conditions: Excellent algae eater. Should be one of the first animals when conditioning a newly set up aquarium, but preferably as a large number of specimens. Strangely enough though, most specimens live only for about a year, while a few can survive for many years.

GA: +; **SE:** +/-; **CO:** +

Photo:
American- or Red Turban Shell in an aquarium.

Turbo brunneus
Green Turban Shell

Distribution: Indian Ocean and western Pacific Ocean.
Size: Up to 5 cm.
Reproduction: Eggs and sperm are released in to the open sea.

Food: Feeds and filamentous- and microscopic algae.
Description: Shell higher than diameter; shades of tan or pale green with dark brown axial bands and spots on the ripples. Operculum with purple-brown centre, and a tan-coloured margin, remainder is pale.
Aquarium requirements: Often imported together with live rocks (see Volume 1, page 188). Very useful as an algae remover; easy to keep. Does not injure any other aquarium inhabitants.

GA: +; **SE:** +; **CO:** +

Photo:
Green Turban shell.

Astralium, i.e. species that had once been placed in one of several subgenera of *Astraea* (LINDNER, 1978).

Order Neotaenioglossa

This order contains the following suborders:

- Discopoda,
- Heteropoda,
- Ptenoglossa.

Suborder Discopoda

From this suborder only those families listed below will be discussed:

- Batillariidae
- Cerithidae
- Planaxidae
- Potamididae
- Turitellidae
- Littorinidae
- Strombidae
- Vermetidae
- Cypraeidae
- Ovulidae
- Lamellariidae
- Naticidae
- Bursidae
- Cassidae
- Ranellidae
- Tonnidae

Rhinoclavis aspera - seen here at Heron Island, Great Barrier Reef - is a typical needle snail. Photo: L. Newman and A. Flowers

Family Batillariidae
Mud Snails, Horn Snails

This family belongs to superorder Cerithioidae, a large group of snails that contains at least 14 families. As is indicated by the popular name, snails of family Batillaridae live on or in muddy bottoms. They are often inconspicuously coloured and they can occur in large aggregations distributed over a small area. Formerly, this family was considered to be a subfamily belonging to family Potamididae (see page 63); however, they are distinct from the latter family on the basis of a differently structured radula (HOUBRICK, 1988). Included in family Batillaridae are four Indo-Pacific genera, including *Batillariella, Pyrazus* and *Velacumantus*.

Family Cerithidae
Cerith- or Needle Snails

As indicated by the popular name, the snails in this family have an approximately needle-shaped shell, usually with distinct spirals. Structure and colouration of the shell can be highly variable within a particular species, which tends to make species identification difficult. This variability may be due to local conditions, but even within the same population there can be substantial deviations in shape and colouration (WILSON, 1993, 1994). Moreover, it has also been shown that species within a particular family can interbreed (hybridise). Needle snails are most commonly found in shallow water. Most species are detri-

Bottom: Mud snails at Stradbroke Island, southern Queensland (Australia): *Velacumantus australis* (left) and *Pyrazus ebeninus* (right). Photos: L. Newman and A. Flowers

Snail shells of family Cerithidae from the Indo-Pacific: 1 - *Rhinoclavis sinensis*, 3 cm. 2 - *Rhinoclavis articulatus*, 3 cm. 3 - *Cerithium echinatum*, 5 cm. 4 - *Rhinoclavis aspera*, 6 cm. 5 - *Rhinoclavis diadema*, 3 cm. 6 - *Cerithium citrinum*, 3 cm. 7 - *Cerithium columna*, 2 cm. Photo: C. Bryce

tus feeders, which make them ideally suited for a coral reef aquarium.

The dominant genus is *Cerithium*, with many species distributed throughout the Indo-Pacific; how many there are exactly, is not known. But there are also *Cerithium* species in the Caribbean Sea, such as *C. atratum*, which is often found on hard substrate, and *C. lutosum*, that is typically a mud dweller, occurring among marine grasses. Genus *Rhinoclavis* incorporates a rather large number of species, which are also rather widely distributed in the Indo-Pacific. Most of these – such as the 6 cm long *R. aspera* – live in sandy areas of the inter-tidal zone.

Genus *Clypeomorus* includes at least 12 species of which some are highly variable. All inhabit the

Needle snails often occur in large aggregations: *Cerithium* sp. "KA5-GAS-03" on a muddy bottom in the proximity of a mangrove zone at Key Biscane, Florida (left), and *Clypeomorus bifasciata* at Heron Island, Great Barrier Reef (right; photo: L. Newman and A. Flowers).

inter-tidal zone, normally on a rocky bottom. There they can be very numerous, occurring in dense aggregations that can readily endure exposure to air for prolonged periods of time. One common species is *C. bifasciata*. It is widely distributed throughout the Indo-Pacific and reaches a maximum size of 3 cm.

Family Planaxidae

The Planaxidae are a small family that lives on hard substrate in the inter-tidal zone. It encompasses about 20 species in 6 genera. On first sight these snails are strongly reminiscent of family Littorinidae (see page 64). Like most snails of the inter-tidal zone, the Planaxidae are algae eaters. In fact, they feed on microscopically small algae that are "rasped" off the substrate.

The reproductive biology of this family is rather interesting. Males do not possess a penis but instead produce packages containing sperm cells (spermatophores). Females protect the larvae inside special chambers and in most cases release the larvae only if they have reached the veliger stage. In some species the trochophore larvae develop directly into fully-grown individuals, i.e. not via the pathway of the planktonic veliger larval stage.

The best-known genus is *Planaxis*, which includes – among other species – *P. sulcatus*, a species common in the inter-tidal zone. HOUBRICK (1987) believes that the (embryonic) development of this species is variable, depending on its geographical distribution. Populations in the northwestern Indian Ocean are viviparous, where the larvae develop inside the maternal animal, which then gives birth to fully developed snails. On the other hand, populations from the western Pacific retain their larvae up to the early veliger stage. Other genera within this family are *Angiola, Fissilabia* and *Hinea.*

Family Potamididae
Mud- or Club Snails

Like species of family Batillaridae (see page 61), the small family Potamididae also lives on muddy substrate, where these snails are sometimes seen in large numbers. But they can also occur in mangroves or in areas with terrestrial plants, which are covered at high tide by seawater. They feed principally on microscopic algae and detritus.

The best-known genus in this family is *Telescopium*. Its species have a thick, conical shell with distinct spiral structures. The only species that has survived to this day is *Telescopium telescopium*, which is often encountered in brackish water. The distribution of these 11 cm long snails extends from Madagascar and East Africa to the Great Barrier Reef and the Philippines. Beyond, this genus is widely represented in fossil finds from the Tertiary Period. But genera *Cerithidea* and *Terebralia* contain several (current) species, however, these are not of interest to the aquarium hobby.

Family Turitellidae
Turret Snails

Among turitellid snails there are some species that have the longest and most slender shells of all snails. The shells have many spiral coils, terminating in a needle-sharp apex. In appearance they are somewhat reminiscent of shells from family Cerithidae (see page 61), but the shell opening is generally smaller and is covered by a strong, corneous operculum.

Turret snails live nearly always buried in sand- or mud bottom; where they are searching for food that consists of planktonic organisms, plant remains and detritus. The food is filtered from the water with the aid of bristles, and is coated with mucus before it is swallowed. Many species occur in the mud of deep-water region, however, a few can also be found in shallow water. In some areas turret snails occur in large aggregations. Since this family has so far

Planaxis sulcatus **is a common snail along the interface between land and sea – seen here at Caloundra, southeastern Queensland (Australia).**
Photo: L. Newman and A. Flowers

Telescopium telescopium **at Cape Domett, Western Australia.**
Photo: C. Bryce

Terebralia semistriata **at Cape Domett, Western Australia.**
Photo: C. Bryce

Cerithidea largillierti **at Exmouth, Western Australia. Photos: C. Bryce**

received little attention from systematists, its genus *Turitella* had long been considered as a repository for all species. Yet, more recent research has revealed, that this family contains several genera.

Archimediella species are beautifully coloured. They are widely distributed throughout the Indian Ocean and the western Pacific Ocean. Genus *Colpospira* contains a number of species common in the Pacific Ocean. All species are characterised by an upright knot (sinus) at the outer shell edge, around the opening of the shell (i.e. along the outer lip). Other recognised genera of the Turitellidae are *Gazameda, Haustator, Maoriclopus, Turitella, Turitellopsis, Zaria* and *Zeacolpus*. The aquarium trade carries turret snails only sporadically. Because of their mode of life these snails are considered to be part of the interstitial fauna of an aquarium, i.e. organisms that perform a useful function but are rarely ever seen. For further details refer to GARRAD (1972, 1974 and 1982).

Family Littorinidae
True Periwinkles, or Beach Snails

The periwinkles of genus *Littorina* are well known from local, temperate waters. They live in shallow water and in the inter-tidal zone, and can occur in such number that more than a hundred specimens may be distributed over a few square metres. Often they can be found at water's edge (high tide mark) or just above the water level.

Various species occurring at the same locality often show a distinct zonation within the inter-tidal zone. For example, in cold and temperate waters of the Northern Hemisphere we find the "Rough Periwinkle", *Littorina saxatilis*, and the "Small Periwinkle", *L. neretoides*, inhabit the highest position in the inter-tidal zone (i.e. furthest inland position); the "Common Periwinkle", *L. littorea*, is generally at the edge of the high-tide mark and slightly above,

***Archimediella* sp. on a fore reef at Kuredu, Maldives.**

while the "Yellow Periwinkle", *L. littoralis*, occurs among algae of genus *Fucus* in the lowest belt of the inter-tidal zone, which dries out.

Life permanently above the water surface is problematic for these snails, since metabolic waste products tend to accumulate inside the shell that can cause the death of the animal. Therefore, the snails have to return to sea at regular intervals in order to exchange the water inside the shell. Species that live farthest inland have solved this problem by excreting solid urea and uric acid as metabolic waste products, instead of liquid ammonia compounds.

The distribution of periwinkle snails extends from cold seas at high (northern and southern) latitudes, via temperate- up to tropical seas. They are cosmopolitan and occur in all oceans. The tropical species live just like our native beach snails, usually just above the water surface and they are often found (e.g. *Littoraria scabra*) on the roots of mangroves and on tree trunks (see page 45). The shell is small but strong, in order to withstand the severe exposure to the elements that can occur in such habitats.

Periwinkles have a thin, corneous operculum. Within a particular species the shells can show distinct signs of sexual dimorphism, whereby those of females are often significantly larger than those of males.

The reproduction is particularly interesting in these snails. Males have a penis and there is internal fertilization. For some species it has been shown that males produces two different types of sperm cells. The "normal" type fertilizes the eggs, the other, atypical sperm cells develop as "auxiliary cells", for which it is assumed that they aid normal sperm cells to reach the eggs.

Periwinkles feed exclusively on microscopic algae, which are scraped off substrate using the radula. One should be able to assume that the "Common Periwinkle", *Littorina littorea,* which also occurs in temperate seas, could be introduced in to coral reef aquariums as algae eaters. That has indeed been tried, but without great success. We have observed ourselves that this species was unable to endure the high temperatures of a coral reef aquarium for a prolonged period of time. Moreover, the Common Periwinkle – like most other members of the genus – tends to crawl out of the water and indeed out of the aquarium. Therefore, we advise against introducing *L. littorea* in to a coral reef aquarium.

BANDEL & KADOLSKY (1982) have published a revision of family Littorinidae, whereby five subfamilies are being accepted. Of these, subfamily Littorininae, with genera *Littoraria, Littorina, Nodlittorina* and *Paesiella* is of significance to aquarists. Genus *Littoraria* encompasses at least 36 species, of which at least 20 species occur in the Indian Ocean and western Pacific Ocean. The structure of the shell exhibits spirally coiled ribs and grooves, often with striped colour patterns.

Littoraria occinea and *L. undulata* are two species frequently occurring in the Indo-Pacific, often in same type of habitat in the

Littorina pintado at Hawaii, USA).
Photo: Dr. B. Carlson, Waikiki Aquarium, Honolulu (Hawaii)

Littorina unifasciata at Caloundra, southeastern Queensland (Australia).
Photo: L. Newman and A. Flowers

Nodlittorina pyramidalis at Caloundra, southeastern Queensland (Australia).
Photo: L. Newman and A. Flowers

Bembicium nanum, seen here at Caloundra - is endemic in southern Queensland and South Australia.
Photo: L. Newman and A. Flowers

upper inter-tidal zone and on coral rocks. *L. undulata* has variable coloured markings, but it is usually yellowish brown or brown with brown lines or rows of brown dots. *L. coccinea* is nearly uniformly white. *L. nebulosa*, native to the Caribbean Sea, grows to a maximum size of 15 mm. It occurs only in seawater and avoids brackish water- and mangrove areas. The shell has a yellow white to bluish base colour with white and red-brown spots.

Genus Littorina is further divided in to two subgenera: subgenus *Littorina*, which occurs in cold and temperate seas of the Northern Hemisphere, and subgenus *Austrolittorina*, which is distributed in tropical seas. *Littorina (Austrolittorina) acutispira* occurs on the Great Barrier Reef. *L. (A.) unifasciata* also occurs in Australian waters, but further south. It reaches a size of 2.5 cm, is white or grey and occurs mostly on rocky shores.

Littorina (Austrolittorina) pintado is being found on isolated Pacific Ocean, such as the Ryukyu Islands, in Micronesia and Hawaii. The Caribbean Sea contains at least five species. The small *Littorina (Littorina) ziczac* (max. size 1.3 cm) is white with black or blackish brown wave-like stripes. This species is typically found in depressions among rocks in the inter-tidal zone. It can easily be mistaken for *Littorina (Austrolittorina) lineolata*, however, that species is smaller and has darker colour markings.

Nodlittorina has a stronger shell, which – in contrast to that of other species – has a knobby structure. This genus is widely distributed in the Indo-Pacific, and it also occurs in the western Atlantic. *N. pyramidalis* can serve as an example here.

Paesiella species have a flattened, depressed shell and they are often very small, sometimes mere millimetres. They occur in the Indo-Pacific as well as in the eastern Pacific, where they live on the roots of mangroves or on rocks in the upper inter-tidal zone. *Paesiella roepstroffiana* and *P. tantilla* are two Australian species.

Another genus is *Bembecium*. Most of its species are endemic in South Australia, with a northernmost distribution at Lord Howe Island.

Family Strombidae – Stromb-, Spider- or Conch Snails

The majority of Conchs have a large shell, are sometime equipped with strong spines and other outgrowing projections, that serve as defensive mechanism. All strombid shells have U-shaped indentation at the edge of the opening, that is an excellent identification feature. These snails live on sand or mud bottom, where they feed on algae and detritus. On certain sites, where food conditions are highly favourable, these snails can be very numerous. In such surroundings they often lay their eggs in the form of long, gelatinous-like and coiled strips, which become covered in sand and coral fragments. These strips contain a great man eggs that develop in to veliger larvae upon hatching (WILSON, 1993, 1994).

Another peculiarity of Strombidae is the eyes that are located on long stalks and which can be extended outward from underneath the edge of the shell. The shell has an elongated opening and has a mother-of-pearl colouration underneath. The shell opening is covered by a longish, cor-

Spider snails - shown here is a *Lambis* species in the Coral Sea - can extend their eyes from beneath their shell.
Photo: E. Svensen

Conch-, vase- and harp shells of family Strombidae from the Indo-Pacific: 1 - *Lambis chiragra.* 2 - *Lambis crocata.* 3 - *Strombus epidromis.* 4 - *Strombus aurisdiane.* 5 - *Strombus canarium.* 6 - *Lambis millepeda.* 7 - *Strombus sinuatus.* 8 - *Strombus luhuanus.* 9 - *Strombus bulla.* 10 - *Vasum ceraminucs.* 11 - *Lambis scorpius.* 12 - *Strombus lentiginosus.* 13 - *Lambis lambis.* 14 - *Harpa articularis.* 15 - *Harpa amouretta.* Photo: G. Allen

neous operculum, which also has defensive purposes.

Family Strombidae is distributed in tropical and subtropical seas and contains five recent genera. Best known of these are *Lambis* (spider shells) and *Strombus*. The former encompasses nine species that all have large, serrated projections on their shell, a characteristic that sets them apart from those of genus *Strombus*. The large *Lambis lambis* is common throughout the entire Indo-Pacific. It represents the best-known species and is sporadically being imported for the aquarium trade.

Larger yet is *Lambis truncates*. It can reach a maximum size of 35 cm. The shell of this species has usually 7 spines including the most anterior apex spine; however, in the two population of this species in the western Pacific, Indian Ocean and Red Sea the shells are blunt. *L. crocata* is similar to *L. lambis*, but it is smaller and has seven bent or sharp-pointed spines. The most anterior spines are bent hook-like. *L. millepeda* has ten relatively short spines along the outer shell edge. This species reaches a size of up to 14 cm and occurs in the central Indo-Pacific, but it is absent from Australia. *L. scorpius* has seven spines, whereby the apex spines are shaped hook-like. Two of the spines in *L. scorpius* are long, but in comparison to those of other species they are depressed and knobby. This species is widely distributed in the western Pacific Ocean and Indian Ocean. *L. chiragra* has a robust shell, with six large spines, which – in contrast to those of the other species – all point in to different directions. When the shell is held with its rear side upward, the first spine points to the left, the second forward and to the right, the third one directly to the right, the fourth and fifth spine point backwards, while the sixth is pointing in a forward direction but slightly to the left. All spines are more or less bent. This species can reach a size of up to 25 cm, whereby male specimens are the largest. The shell is white with brown spots, and the shell opening is pink to orange.

All *Lambis* species can appear in the aquarium trade sporadically. Our own experiences are limited to *L. lambis*. When these snails are being provided with an ample food supply of algae and detritus they can be kept in a coral reef aquarium for a long time. However, we have observed that spider shells can get stocked in between coral and rocks in the reef aquarium because of the shape of the shell. A sand-zone tank with algae or sea grass and only a few live rocks is the perfect aquarium habitat for these beautiful and interesting snails. For further details on genus *Lambis* refer to ABBOT (1961).

While genus *Lambis* is not particularly species diverse, making it relatively easy to gain an over-

Strombus gibberulus - seen here at Heron Island, Great Barrier Reef - is one of the most commonly occurring *Strombus* species on Indo-Pacific reefs.
Photo: L. Newman and A. Flowers

view over this group, the opposite is true for genus *Strombus*. The shell of larger *Strombus* species, such as *S. latissimus* and *S. sinuatus*, are reminiscent of those in *Lambis* species, however, the spines are absent. In small species, such as *S. mutabilis* and *S. labiatus*, the shell has a distinct tip with spiral coils.

Very much like *Lambis* species, *Strombus* species usually live on soft bottom substrate with epiphytic algal growth. There are at least 51 species, which are placed in ten subgenera, which we are not going to discuss in any detail here. Anyone who is interested in the systematics of genus *Strombus* is advised to consult the publications by Wilson (1993, 1994) and Abbot (1960).

Many *Strombus* species are large and colourful. *S. aurisdianae* and *S. dendatus* are the most common species. They are distributed throughout the entire Indo-Pacific, where they are usually found in algal belts and in coral sand. Some species are associated with coral reefs, where they can be found in areas with lush coral growth. Others are relatively rare, which places them in high demand with seashell collector.

Strombus mutabilis is a very widely distributed species in the Indo-Pacific and in subtropical seas. Its colour markings are highly variable. Normally, this snail is crème-coloured with yellow, red-brown, orange or black bands and spots. This species live in shallow water among coral fragments and on sandy substrate, where it can sometimes be quite numerous. *S. canarium* is often encountered on muddy sand bottom. It is widely distributed in the central Indian Ocean and in the western Pacific Ocean. In East Asia this conch is commonly used in local commerce.

Well known from the Caribbean Sea is the Queen Conch, *Strombus gigas*. Unfortunately, it has become threatened with extinction in many areas due to overexploitation (i.e. shell collecting). The decorative shell of this species makes it a highly desirable collector's item. Normally, the queen conch is found on sand bottom or in sea grass meadows on the inner reef. *S. costatus* and *S. raninus* are two other species from the same geographical area. For further details about this genus please refer to specialist literature.

Strombus gigas - seen here at Key Biscayne, Florida - is among the largest marine snails, and reach a size of 30 cm. Their shells are in great demand as souvenirs, so that nowadays many populations are seriously endangered.

Family Vermetidae
Vermetid- or Worm Snails

Vermetid snails are somewhat of a curiosity. In contrast to all other snails they are unable to move about, because they live permanently inside a tube. This tube is often more or less spirally coiled, attached to a substrate and can be hidden in corals or over-grown with other organisms. Worm snails can occur in colonies or live as solitary specimens. The scientific family name has been derived from vermis = (Lat.)

The tubular structures seemingly enclosing the dead branches of a black coral – seen here on Latheef Reef at Kuredu (Maldives) – are the tubes of worm snails (*Petaloconchus*) totally overgrown by colourful sponges.

worm, and refers to the similarity of the tube to that of calcareous tubeworms.

The intake of food by worm snails is variable, and can be observed in detail in a coral reef aquarium. Some species excrete a net of mucus from a gland at the tip of the foot, which is then extended from opening of the tube. Such mucus net easily captures detritus and small organisms, which then serve as food for the snail. For that the net is retracted and eaten together with the "catch". Other species filter the water and capture food particles with the aid of gills (as in mussels), but this only rarely ever seen in snails.

Worm snails occur in all seas. They are difficult to identify and can easily be mistaken for bristle worms; however, if the tube is removed the snail shape becomes visible. Relatively little research has been done on the systematics of this group.

Genus *Dendropoma* is distinguished from other worm snails by the presence of a strong operculum used by these snails to close off the tube. The large *Dendropoma maximum* lives partially embedded in corals and is commonly found on coral reefs. Another species associated with corals is *D. platypus*, which lives totally hidden in corals.

Genus *Petalochonchus* contains a species that often becomes established in coral reef aquariums by the hundreds. This is a cosmopolitan genus, and it assumed that feeding is accomplished with the aid of methods described above. A Fact Sheet on page 69 provides further detailed

Worm snails live partially encased in corals and extrude from their tubes mucus net for collecting food. Both photos show the Large Worm Snail, *Dendropoma maximum*, in the Coral Sea. This species is, however, widely distributed throughout the entire Indo-Pacific region, including the Red Sea. Unfortunately, it is rarely ever available in the aquarium trade, because it would indeed be a suitable animal for a coral reef aquarium.

Petalochonchus sp. "KA5-GAS-04"
Small Worm Snail

Distribution: Cosmopolitan; common in the Indo-Pacific region; no further detail available.
Size: Depends on the presence of predators and water movement. In an aquarium, in the absence of predators and kept in a relatively weak water movement, there is an irregular size pattern, and the snail's tube can reach a length of 10 to 20 cm. It stands erect from a coiled base. In the sea with a prevalence of predators, the tubes usually remain smaller.
Reproduction: Separate sexes; the eggs are incubated by the female in a mantle cavity or in capsules attached top the tube's opening.
Food: Suspension feeder, that acquires food by two different mechanism: 1) A mucus net is being excreted by glands located along the edge of the foot, and is suspended in the open water. When organic particles or small organisms become entrapped in the net, it is pulled in and swallowed. 2) Food particles swept together with water in to the interior of the shell are being captured with the gills, coated with mucus and then passed in to the mouth.
Description: *Petaloconchus* species are the only worm snails that have an operculum and internal shell lamellae. The latter are very difficult to see. They are located inside the first spirals of the tube and are connected to a muscle that attaches the snail's body to the shell. There are about 40 to 50 *Petalochoncus* species with a worldwide distribution (BIELER, pers. communication).
Aquarium requirements: Very hardy aquarium inhabitant. Live rocks can provide the base for such large populations that need to be culled regularly so that corals and other sessile invertebrates are not damaged by the development of the snails. Can also become established in filter chambers, where the snails are provided with a steady water flow and an abundance of food particles.

GA: +; **SE:** +; **CO:** +

Photos:
Top: *Petaloconchus* sp. "KA5-GAS-04" colony in an aquarium.
Centre: A number of worm snails that established themselves in the filter chamber of our coral reef aquarium.
Bottom: Worm snails removed from their tubes.

 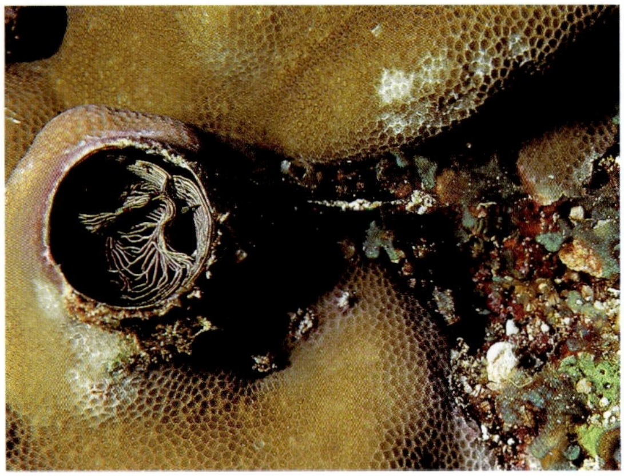

Worm snails of genus *Serpulorbis* do not posses a corneous operculum, as is typical for the *Dendropoma* species (see page 68). The left photo shows *Serpulorbis* sp. "KA5-GAS-05" at Cod Hole, Great Barrier Reef, and the photo on the right depicts *Serpulorbis* sp. "KA5-GAS-06", at the Maldives.

information on this genus. In snails of genus *Serpulorbis* the operculum is absent, which is a distinctive feature of this genus. Occasionally, very large, decorative species are being imported, however, these do not appear to be particularly hardy in coral reef aquariums. The attractive *Serpulorbis sipho* occurs around Australia. This is a large species, which we have seen being kept successfully at the Wonderland (Great Barrier Reef) Aquarium in Townsville (Queensland, Australia). It belongs to a species complex, which has not yet been fully delineated (BIELER, personal communication).

***Serpulorbis* sp. "KA5-GAS-07" at the Great Barrier Reef (Wonderland) Aquarium in Townsville (Queensland, Australia).**

Genus *Vermetus* is cosmopolitan. The tubes of these species are only weakly coiled and exhibit a rather irregular structure. These snails capture their food with a mucus net; their operculum is small and they resemble *Serpulorbis* species.

Small worm snails that occur on live rocks are easy to keep in an aquarium. In fact, sometimes it is rather difficult to keep their population in check. Among these, genus *Thylaeodus* is rather common. One reason as to why these snails thrive in coral reef aquariums is the absence of predators. Consequently, the tubes can develop unimpeded and so grow in to long structures, while those in the sea are often being kept shorter by predators.

Family Cypraeidae
Cowries

Cowries (cowry snails) of family Cypraeidae are probably the best known of all tropical marine snails. It is being said, that the ancestral populations in Asia and Africa used cowry shells as monetary currency. The name "cowry" is of Hindustani origin. Cowries can be recognised by their

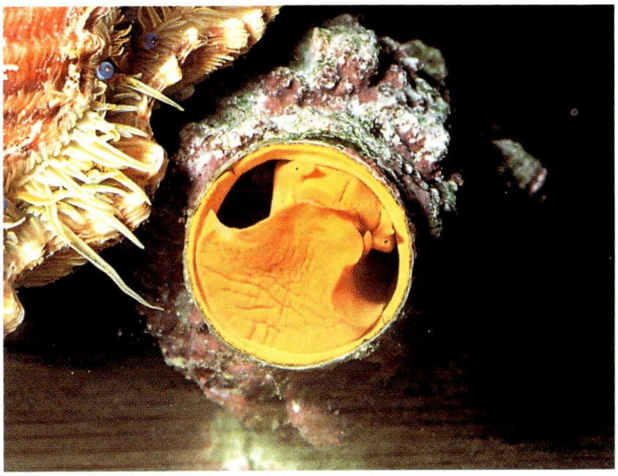

Left: This attractive worm snail *Serpulorbis* sp. "KA5-GAS-08" is sometimes being imported. Right: An unidentified worm snail "KA5-GAS-09" in the Red Sea (photo: R. Brons).

smooth, porcelain-like surface, which is covered by a layer of shiny glaze. This glaze is excreted by the snail's mantle, which is periodically pulled over the shells. Cowry shells have decorative markings that make them valuable collectors' items. There are species that are being traded for thousands of dollars per shell!

The shell's opening on the ventral side is only small and equipped with small teeth along each side. The shape of these teeth is an important species characteristic. Colouration too can be the basis for identification; however, colour patterns can vary substantially within a species, which makes this feature relatively unreliable.

The origin of cowry snails dates back 50 million years ago. They amply represented in fossil finds. Today there are about 200 living species, but the systematics of this family Is still under di-

The shells of cowry snails (family Cypraeidae) from the Indo-Pacific: 1 - *Cypraea testudinaria*, 11 cm. 2 - *Cypraea tigris*, 13 cm. 3 - *Cypraea histrio*, 7 cm. 4 - *Cypraea mauritiana*, 9 cm. Photo: C. Bryce

Cowry snails of genus *Cypraea* from the Indo-Pacific and the Red Sea

Cypraea stolida — Photos: L. Newman and A. Flowers — *Cypraea asellus*

Cypraea cribaria — Photos: L. Newman and A. Flowers — *Cypraea nucleus*

Cypraea nebrites — Photos: R. Brons — *Cypraea turdus*

scussion among malacologists. Family Cypraidae is divided in to four subfamilies with a total of 25 genera, whereby genus *Cypraea* is the dominating form.

Cowries are numerous on coral reefs. Usually they live on a hard substrate, but they are not always well suited for maintenance in a coral reef aquarium. Most species appear to be omnivores: they feed on algae, sponges, bryozoans, microorganisms, coral polyps and detritus. Very few species, such as *Cypraea moneta* (see Fact Sheet on page 74) and *C. caputserpentis* – both typical representatives of the inter tidal zone – are pure algae eaters and so they are well-suited for a coral reef aquarium. A rather similar species is *C. annulus* (see Fact Sheet on page 74), which lives in the same habitat. However, there is no reliable information available on their feeding behaviour. With other species of the genus one has to be cautious, because it is conceivable under aquarium conditions that they could attack corals and other invertebrates. But, here we must hasten to add that different aquarists have had variable experiences with these snails.

For *Cypraea tigris* (see Fact Sheet on page 75) and *C. eglantina* it has been observed that in an aquarium these cowries feed on fish- and mussel flesh however, it is not known whether fish is part of the natural diet of these snails. Overall, it appears that the food requirements of cowry snails are rather variable. Therefore, it is difficult to predict whether a particular cowry snail is suitable for a coral reef aquarium or not. Fortunately, it is not difficult to remove a predatory cowry snail from an aquarium before it can totally decimate the entire invertebrate population.

One of the most attractive *Cypraea*-species is *C. asellus*. It has a black mantle and wide, dark-brown or black cross bands on its shell. This species is widely distributed in the Indo-Pacific; nevertheless it is not among the most common species one can find. Another conspicuous species is *C. nucleus*, with long, branched filaments on its mantle. It is principally tropical, however, it can also be found in subtropical seas. The same distribution pattern applies to *C. cribraria*, where the tip of the shell is red-brown and has sharply delineated, white spots.

Family Ovulidae – False Cowry- or Egg Snails

Egg snails resemble cowries and indeed they are closely related to them. The teeth along the shell opening – typical for cowries – are generally absent in ovulid snails. The mantle can be spread over the shell and so polish its surface. All egg snails live on cnidarian colonies, such as soft corals, gorgonians, hydroids and stony corals They feed on coral polyps, and often they are quite specific in the selection of their host. Reproduction also takes place on the host, whereby the egg masses are deposited among corals. The best-known example is probably is the Caribbean Flamingo Tongue, *Cyphoma gibbosum*, which lives exclusively on Gorgonians.

Many egg snails have developed camouflage colours that make it quite difficult to recognise them on their host and so protect

Cypraea **sp. "KA5-GAS-12" from Western Australia.** **Photo: C. Bryce**

Cypraea annulus
Yellow-ringed Cowry

Distribution: Indo-Pacific; from East Africa to eastern Polynesia. Lives in reef habitats within the inter-tidal zone; often found in tide pools.
Size: Up to 3 cm.

Reproduction: Separate sexes. The female deposits 300 to 400 oval egg capsules, which can contain many individual eggs, and then covers these capsules for one to two weeks with the foot, until the larvae hatch. The veliger larvae feed on the smallest of plankton.
Food: Uncertain. Based on the occurrence of this snail it feed probably most on algae.
Description: Shell margin without nodules. Shell without spots, white, yellow or grey; bluish on top (WILSON, 1993, 1994). A characteristic feature is a yellow ring that distinguishes this species from the Money Cowry, *Cypraea moneta* (see below).
Aquarium requirements: Probably ideally suited for a coral reef aquarium; however, caution is advised since the feeding behaviour of this species has not yet been determined yet. It is import to observe whether this cowry also takes food other than algae.

GA: +/-; **SE:** +/-; **CO:** 0

Photo:
Yellow-ringed Cowry at Heron Island, Great Barrier Reef.
Photo: L. Newman and A. Flowers

Cypraea moneta
Money Cowry

Distribution: Indo-Pacific; from the Red Sea and East Africa to Polynesia.
Size: Up to 3 cm.
Reproduction: Separate sexes. The female deposits 300 to 400 oval egg capsules, which can contain many individual eggs, and then covers these capsules for one to two weeks with the foot, until the larvae hatch. The veliger larvae feed on the smallest of plankton.

Food: One of the few cowry snails that feeds exclusively on algae.
Description: One of the most common cowries in the intertidal zone, and which can often been outside water. Top of the shell is yellow, orange or greenish-grey with three faint transverse bands. A faint ring-shaped band is sometimes visible. Colours pale towards the edge of the shell (WILSON, 1993, 1994). Can be mistaken for the Yellow-ringed Cowry (see above).
Aquarium requirements: Occasionally imported. Ideally suited for a coral reef aquarium. Caution: must not be mistaken for other, predatory cowries!

GA: +; **SE:** +; **CO:** +

Photo:
Money Cowries in their natural habitat, at Heron Island, Great Barrier Reef. Photo: L. Newman and A. Flowers

Cypraea tigris
Tiger Cowry

Distribution: Widely distributed throughout the Indo-Pacific, including the Red Sea.
Size: Up to 13 cm.
Reproduction: Separate sexes. The female deposits 300 to 400 oval egg capsules, which can contain many individual eggs, and then covers these capsules for one to two weeks with the foot, until the larvae hatch. The veliger larvae feed on the smallest of plankton.
Food: Uncertain. Under aquarium conditions it has been observed to feed on algae, but also on fish- and mussel/clam meat.
Description: Found in shallow reef areas down to a depth of 30 m. Shell base is white, shell can be white or pale brown with large brown spots. Mantle with tentacular filaments.
Aquarium requirements: Often available in the aquarium trade. Can be kept in coral reef aquariums, but its feeding behaviour must be monitored as the Tiger Cowry has been observed to harm corals on several occasions. Looks rather spectacular when the filaments of the mantle are fully expanded.

Photo:
Tiger Cowry with expanded mantle filaments.
Photo: C. Bryce

GA: +/-; **SE:** +/-, **CO:** +/-

Cypraea arabica
Arabian Cowry

Distribution: Indo-Pacific; from the Red Sea to Hawaii (although the presence in Hawaii might be due to accidental introduction; GOSLINER et al., 1995). Often found under coral boulders.
Size: Up to 10 cm.
Reproduction: Separate sexes. Eggs capsules are deposited on the substratum and covered.
Food: Uncertain. Might feed on algae but might also take in animal food.
Description: Variable species. Base usually creamy grey. A hieroglyphic pattern of brown markings that might form longitudinal lines interrupted by clear spaces. Brown spots and blotches present. Can be confused with *C. elegantina*, but lacks dark spire blotch.
Aquarium requirements: Occasionally seen in the trade. Can be kept, but its feeding habits must be carefully observed. We have kept this species ourselves and observed it to eat algae for a long time. Suddenly it shifted diet and attacked corals.

Photo:
C. arabica feeds on algae on the edge of a run-off chamber in our aquarium.

GA: +/-; **SE:** +/-, **CO:** +/-

Egg snails live in association with gorgonian corals and often have an excellent camouflage colours.
Photo: L. Newman and A. Flowers

them. Their feeding behaviour makes them relatively uninteresting for the coral reef aquarium. It is conceivable, that some corals are attacked by egg snails, the snails reproduce and then become a thread to the entire coral population, provided they are not being removed in time. Therefore, it is important to watch for egg snails, because they are annoying coral parasites that can cause harm to such colonies in the reef aquarium.

According to WILSON (1993, 1994), there are more than 100 species in 16 genera, however, a revision of this family is clearly indicated and therefore it must currently considered to be a family

Left: This Egg Snail *Dimiovula* sp. shows here how the mantle can enclose the shell, and the extent of the foot protruding behind the shell (above). In smaller specimens (below) the shell is visible without being covered by the mantle.
Photo: L. Newman and A. Flowers

The Flamingo Tongue, *Cyphoma gibbosum* is distributed throughout the Caribbean Sea and along the southeastern coastline of North America. This species lives principally on *Gorgonia flabellum* and *G. ventalina*, and it also feeds of the tissue of these corals. Because of its decorative appearance it is often kept in aquariums, but it is unsuitable for coral reef aquariums.
Photos: E. Svensen

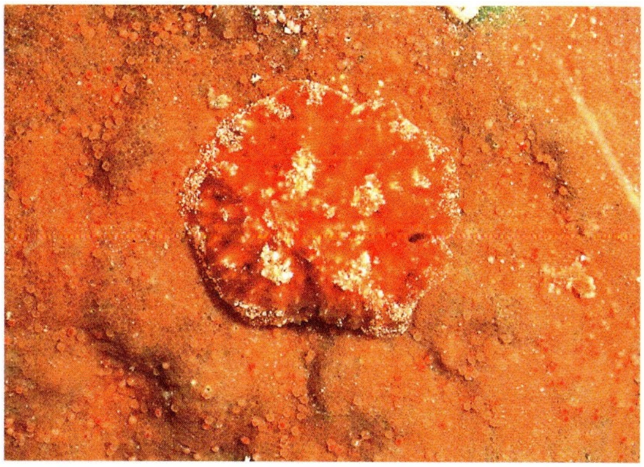

Lamellaria sp. at Heron Island, Great Barrier Reef. Photos: L. Newman and A. Flowers

that has been insufficiently researched. The same author also provides details as to what species live on various groups of cnidarian animals. The family contains genera, such as *Margovula, Ovula, Phenacovolva*, *Prinovolva* and *Procalpurnus* from the Indo-Pacific region. The best-known species is *Ovula ovum* that lives on soft corals of genus *Sarcophyton.*

Family Lamellariidae

This small family could perhaps be ignored here, however, its natural history makes it possible that representatives may be inadvertently introduced in to our aquariums, as "stowaways" on corals and live rocks. These snails live under rocks and rarely ever venture in to broad daylight. The shell is substantially reduced and always covered by a mantle that has large appendages and are beautifully coloured. Little research has been done on this family, and current research is related only to genera *Chelynotus, Lamellaria* and *Mysticoncha.*

Family Naticidae
Moon Snails

As far as is known, all moon snails of this family feed on mussels and related gastropods (Kabat, 1990). During the day moon snails lie motionless on the sand; only at night do they hunt for prey. Clams buried in sand are being preyed upon and restrained with the snail's foot, which is specifically developed for that purpose. Following that the radula, in the shape of a trunk, bores through the shell and feed on the clam. It is common to find clamshells with small, round holes – a clear sign of the work of moon snails.

Family Naticidae is divided in to three subfamilies. Subfamily Naticinae is characterized by ball-like, coloured shell. The operculum is flat and shaped like a semicircle; moreover, it often has a granular texture. The foot is very large, and yet the animal can withdraw quite rapidly in to the protective confines of the shells.

Subfamily Naticinae is dominated by genus *Natica*, which is distributed worldwide. It has one to

The small Moon Snail, *Polinices conica* (left), and its clutch of eggs in a sandy-muddy habitat at Stradbroke Island, southern Queensland (Australia). Photos: L. Newman and A. Flowers

Polinices mammilla in a reef habitat at Heron Island, Great Barrier Reef. Photos: L. Newman and A. Flowers

three spiral-shaped depressions along the edge of the operculum. *Natica gualtieriana* and *N. seychellium* are two common species in the Indo-Pacific region, while *N. canrena* and *N. pusilla* occurs in the Caribbean Sea.

In subfamily Polinicinae the shells are thick, oval or flat. The operculum is corneous and thin. The foot appears larger and more dominating than in subfamily Naticinae. In fact, occasionally it is so larger that it can no longer be withdrawn quickly back in to the shell. *Polinices* is a frequently occurring genus, of which *P. sordidus* is common on muddy bottom along Australian coasts. There are three other species, widely distributed in the Indo-Pacific, are *P. melanostomus*, *P. mellosus* and *P. conicus*.

In subfamily Sininae, the various species have thin, often round or flat, depressed shells. Where such a shell occurs, the operculum is granular. The predominant genus is *Sinum*, whereby *S. maculatum* and *S. perspectivum* are common in the Caribbean Sea.

Bursa granularis **at Heron Island, Great Barrier Reef.**
Photos: L. Newman and A. Flowers

Family Bursidae
Frog Snails

The popular name of these snails has been derived from the many knot-like outgrowths from their shells. The shell surface is covered with irregular, spiny structures that are somewhat reminiscent of similar features in Family Muricidae (see page 88), yet, they are related closest to family Ranellidae, and they have also many characteristics in common with triton snails (see page 79).

Frog snails have a corneous operculum. These are predatory snails that feed primarily on bristle worms. The genera *Bursa* and *Tutufa* occur in the shallow water of tropical and subtropical seas, where they can be found among live coral and coral blocks. *Bufonaria* species can be found on soft bottom, often in somewhat deeper water. These snails are imported only sporadically.

Family Cassidae
Helmet Snails

These snails have a white, helmet-shaped shell, and are very decorative and popular collectors' items. *Cypraecassidus rufa* is probably the best-known species, which has been used for hundreds of years in the production of cameo jewellery.

Helmet snails live on a sand bottom and feed on echinoderms, mainly sea urchins (WILSON, 1993, 1994). The long, tubular mouth can be raised from below the substrate. The shell has a typically folded anterior section, which enables the respiratory tube (siphon) to be pushed upward and so to take in fresh water, while the snail remains buried in the substrate. This adaptation is commonly found in snails with such burying behaviour. Part of the radula has become modified in to an effective venom apparatus, which paralyses the prey with potent nerve venom. Simultaneously, the animal releases acids so that the skeleton of a sea urchin becomes soft and can be eaten. This type of feeding behaviour of helmet snails has been utilised in the fight against the coral-feeding Crown-of-Thorns Sea Star, *Acanthaster planci*. *Cassis cornuta* (maximum size 35 cm), is widely distributed throughout the Indo-Pacific and feeds exclusively on *Acanthaster*. This helmet snail is now a protective species in the area of the Great Barrier Reef in Australia.

Helmet snails exhibit sexual dimorphism, and males are usually smaller. Females deposit egg capsules, which can contain a larger number of eggs. These can develop directly in to fully functional snails or give rise to a planktonic stage via veliger larvae.

Cassius flammeus is well known from the Caribbean Sea. With its decorative shells it is – like *Strombus gigas* – a popular collectors' item and a trading commodity. In many locations this has lead to over-exploitation and a threat of extinction.

Other genera of family Cassidae are *Galeodea*, *Casmaria, Echinophoria, Phalium* and *Semicassis.*

Family Ranellidae
Triton Snails

Previously, this family was carried under the family name Cymathiidae. BEU CERNOHORSKY (1986), however, came to the conclusion that triton snails belong in Ranellidae. In this family – widely distributed in the Indo-Pacific region – we find many large predatory snails, which feed on tunicates, echinoderms, clams and other snails. HOUBRICK FRETTER (1969) and LAXTON (1971) have published details about the feeding behaviour of triton snails.

The best-known species is the Triton Trumpet, *Charonia tritonis* (see Fact Sheet on page 80). The shell opening is equipped with sharp teeth, and on inside it is orange coloured. The Triton Trumpet feeds on echinoderms, especially on the Crown-of-Thorns Sea Star, *Acanthaster planci*, that is threatening corals in many areas since it feeds unimpeded on stony corals. It has been considered whether an increase in Triton Trumpets could make an impact on the large Crown-of-Thorns sea star populations. However, for the time being the Triton Trumpet population is so small that this is not feasible. With a shell size of up to 45 cm, *Charonia tritonis* is one of the largest snails in the world. For many years it has been a highly desirable collectors' item for malacologists. Moreover, it was also popular as a souvenir. That has caused the Triton Trumpet to become almost extinct in many areas of its former range.

The genus *Charonia* also includes *C. lampas*. This species is part of a complicated species complex with several subspecies and variable geographic distribution patterns (WILSON, 1993, 1994).

Shells of families Tonnidae (1), Cassidae (2) and Ranellidae (3-10) from the Indo-Pacific: 1- *Tonna perdix,* 2 cm. 2 - *Casmaria erinaceus,* 5 cm. 3 - *Cymatium lotorium,* 9 cm. 4 - *Charonia tritonis,* 45 cm. 5 - *Cymatium hepaticum,* 5 cm. 6 - *Cymatium nicobaricum,* 6 cm. 7 - *Cymatium rubeculum,* 4 cm. 8 - *Linatella succincta,* 3 cm. 9 - *Distorsia reticulata,* 8 cm. 10 - *Distorsio anus,* 4 cm. Photo: C. Bryce

Charonia tritonis
Trumpet Shell

Distribution: Indo-Pacific; from the Red Sea and East Africa to Polynesia and Hawaii, southern Japan and New Zealand.
Size: Up to 45 cm.
Reproduction: Separate sexes. The female deposits the eggs in egg capsules, and bundles of these are then attached to a substrate. Free-swimming veliger larvae hatch from these eggs, which go through a planktonic stage.
Food: Predatory snail. Feeds on sea stars (on the Crown-of-Thorns Sea Star, *Acanthaster planci*, among others), and has been used in attempts to control the growing numbers of coral-feeding *A. planci*. Therefore, the Triton Trumpet is protected in Queensland.
Description: Lives on sandy bottom among corals, often at medium depths. It is the largest species in family Ranellidae. Shell with spiral ribs, attractively structured. Colouration from shades of tan to light brown, with sickle-shaped, purple-brown markings.
Aquarium requirements: Because of its specialised food requirements, not suited for aquarium maintenance. Should not be collected or imported by the aquarium trade!

GA: -; **SE:** -; **CO:** –

Photo:
Triton Trumpet in natural habitat. Photo: C. Bryce.

Genus *Cymatium* incorporates many species and has a worldwide distribution. The shell is thick and solidly built, with a long shell opening that is pointing slightly to the right. An example is *Cymatium aquatile,* commonly known as the oyster drill because of its unusual behaviour to bore holes in oyster shells. On the other hand, *C. nicobarium* is a very small species that is frequently found in the inter-tidal zone.

Oyster Drill, *Cymatium aquatile*, at Stradbroke Island, southeastern Queensland (Australia). Photo: L. Newman and A. Flowers

Family Tonnidae

The snails in this family have a very thin shell, probably the reason why they can move so rapidly along the bottom. Foot and mantle are large and have beautiful coloured markings. The head is equipped with long tentacles. The shell is large, round or oval with a wide opening; however, the apex is only moderately or very little developed These snails are distributed in tropical and subtropical seas, where they occur in shallow water. During the day they remain buried in sand, to appear only at night in search of prey. They feed on sea cucumbers, which are being swallowed in one piece!

The dominating genus is *Tonna* with three frequently occurring species: *Tonna allium, T. chinensis* and *T. perdix.* Sporadic imports are conceivable.

Suborder Heteropoda

Dr. Leslie Newman
School of Environmental Science & Management (EnvSM)
Southern Cross University, Lismore, NSW, Australia

Heteropods (sometimes known as side-keeled snails) are one of the only two groups of prosobranch molluscs to have successfully adapted to living within the water column. The other group, the janthinids, are restricted to floating at the surface with the aid of bubble rafts. Heteropods, however, are highly modified to their planktonic existence and use a large single muscular swimming fin and highly developed eyes for seeking out and catching prey. They are mainly epipelagic being found in tropical and subtropical seas (VAN DER SPOEL, 1976; VAN DER SPOEL & HEYMAN, 1983; VAN DER SPOEL et al., 1996). The heteropods are easily separated into three distinct families with radically different morphologies.

Taxonomic outline
(including ALL genera, * discussed or illustrated in this chapter)

- Suborder Heteropoda
  - Family Atlantidae
    - *Atlanta*, Oxygyrus*, Protatlanta*
  - Family Carinariidae
    - *Cardiapoda, Carinaria*, Pterosoma*
  - Family Pterotracheidae
    - *Firoloida*, Pterotrachea**

Family Atlantdae

The smallest and least specialised of the heteropods are the shelled Atlantidae. One example being *Atlanta gaudichaudi* (see Fact Sheet on page 82). They are relatively small, ranging in size from a few to 11 millimetres in shell diameter. The dextrally coiled shell is compressed, extremely fragile, completely transparent and the outermost whorl is surrounded by a thin keel. In two genera the keel and/or the body whorl is not calcified but cartilaginous. The animal can completely retract into its shell which is closed by the operculum. Animals possess a long snout or proboscis, highly developed and mobile eyes with a clear lens and pigmented retina, a single large swimming fin with a sucker and a muscular foot with an operculum. The apical whorls of the shell and the body show some pigmentation but these molluscs are transparent and all the internal organs can be seen through the shell. These snails are extremely difficult to identify due to shell simila-

Atlanta gaudichaudi

Distribution: Occurs worldwide; in tropical and subtropical seas.
Size: Maximum shell diameter 5 mm.
Reproduction: Separate sexes (dimorphic). Copulation and spermatophore transfer have not yet been observed. Females deposit their eggs in drifting egg masses that develop in to veliger larvae, with the velum used for capturing food (see drawing on page 85)
Food: Feeds on smaller zooplankton, especially on pteropods.
Description: Shell with four-and-a-half whorls, laterally compressed, transparent and very fragile. The body whorl, surrounded by a flat, thin keel, is substantially larger than the other whorls. The body is transparent and colourless.
Aquarium requirements: Can probably not survive for long under aquarium conditions, since these snails tend to become entangled in their own mucus and are then unable to feed.

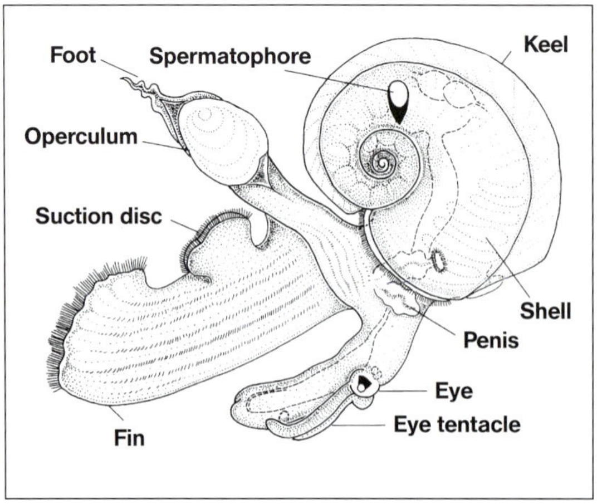

Drawing and photos:
Top left: Body features of atlantids, using *Atlanta gaudichaudi* as an example.
Drawing from Jamieson & Newman (1989)
Top right: Live *A. gaudichaudi*.
Photo: L. Newman and A. Flowers
Left: *A. gaudichaudi* feeding on pteropod *Creseis vurgula*. Photo: L. Newman and A. Flowers

Top: Shells of anlantids: *Atlanta fuca* (left), *A. turriculata* (centre) and *A. gaudichaudi* (right).
SEM-Photo: L. Newman and A. Flowers

rities and they are rarely collected intact or even studied alive. The other two families of heteropods **Carinariidae** and **Pterotracheidae** are well adapted to their planktonic lifestyle and are larger, highly transparent and streamlined. The body is clearly divided into a head, trunk and tail region. The shell is greatly reduced and covers only the viscera or is absent in the adults. The only darkly pigmented areas are their large eyes and visceral mass although the body maybe covered with bright scattered pigment spots.

The carinariids possess a conical shell with a spiral protoconch (or larval shell) and thin keel. However, these molluscs cannot retract into their shell since it only covers the stalked visceral nucleus. This visceral mass or nucleus includes the digestive and reproductive organs and the gills which are covered by the mantle. The body is generally deep and cylindrical with a well developed, muscular proboscis. The digestive, reproductive and nervous systems can be clearly seen through the body wall. Carinariids are the largest of the holoplanktonic molluscs and can reach 50 cm in length. They are generally considered rare and their fragile shell, if found intact, is greatly prized by shell collectors.

The pterotracheids are considered the most advanced of the holoplanktonic molluscs due to the loss of the shell in the adults and further streamlining and pronounced elongation of the body. These animals are bizarre in shape and some resemble chaetognaths more than molluscs, al-

Live *Atlanta plana* in swimming position. Photo: L. Newman and A. Flowers

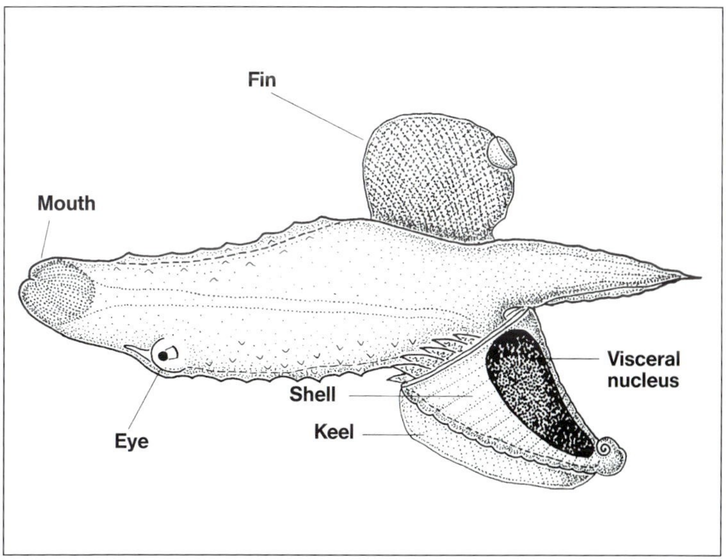

Features of a female *Curinaria cithara*.
Drawing: L. Newman in Beesley & Ross (1996)

though they can reach 30 cm in length. The proboscis is long, slender and highly mobile. The visceral nucleus is embedded in the body and the gills project dorsally. Concealment is enhanced by their long and slender visceral nucleus which is covered by a reflective layer of iridophores eliminating the ventral silhouette to upward searching predators (SEAPY & YOUNG, 1986). They are also neutrally buoyant due to ion exchange in their gelatinous tissue and their transparency aids in their concealment (LALLI & GILMER, 1989).

All heteropods generally swim upside-down and undulate their single large fin in a sculling motion. However, they do not swim all the time. Large carinariids and pterotracheids have been observed to grasp their tail with their radula and hang suspended while digesting their last meal (HAMNER et al., 1975). The gelatinous nature of these animals also aids in buoyancy much like a jellyfish. In contrast, atlantid heteropods swing their shell when they swim. Their slender keel is thought to act as a stabiliser and cuts the water on descent but it must impede swimming to some extent. Atlantids have also been observed to remain motionless and deploy a mucus parachute to hang within the water column while they wait for prey (NEWMAN, 1990).

Heteropods are voracious carnivores and feeding opportunists as they will take a variety of prey including salps, chaetognaths, other pelagic molluscs and even small fish (LALLI & GILMER, 1989). Their well developed eyes are used to scan for prey from below or within the water column. The radula is well equipped with hook shaped lateral teeth for grabbing prey which they swallow whole or tear apart. Although small in size, atlantids are well adapted to feed on shelled pteropods which they grasp with their fin sucker while they pull their prey's body from its shell. Heteropod predators are known to include fish and sea turtles.

Unlike the holoplanktonic pteropods (see page 121), the sexes are separate in the heteropods and they display sexual dimorphism. Male atlantids have a large lobed penis on the right side of their neck and often a small oval black and white spermatophore which can be seen through the transparent shell. Carinariids and pterotracheids also have a large lobed penis and may also show other morphological differences between sexes such as unequal

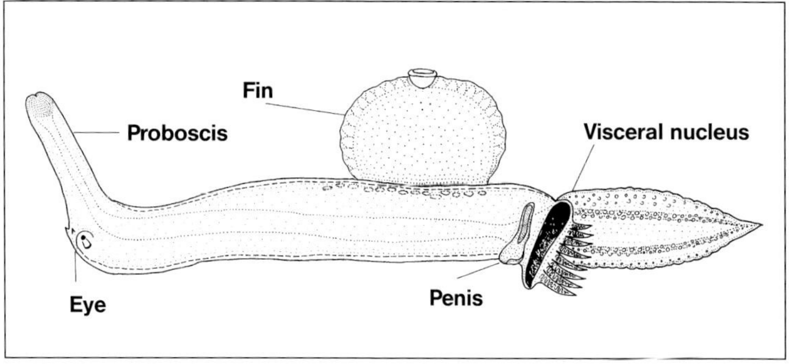

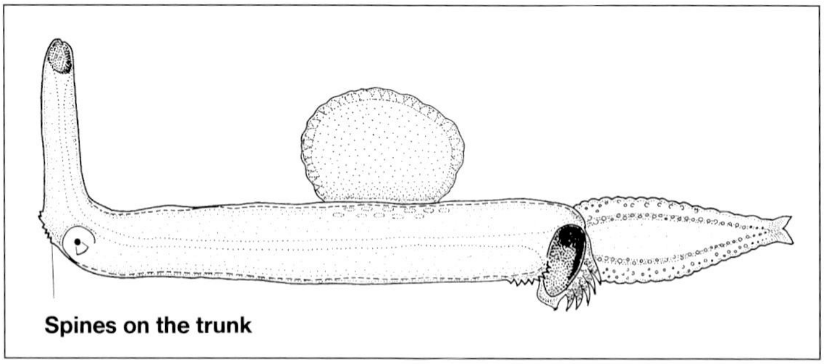

Features of a male of *Pterotrachea coronata* (top) and the female of *P. hippocampus* (below). **Drawings: L. Newman in BEESLEY & ROSS (1996)**

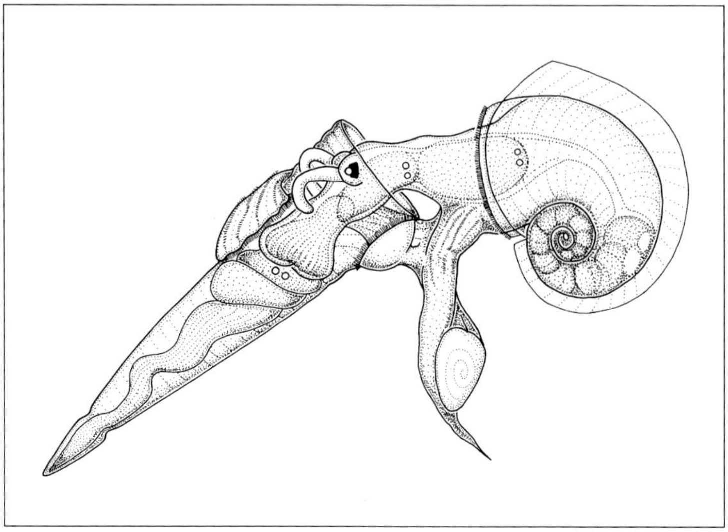

***Atlanta gaudichaudi* feeding on the pteropod *Creseis chierchiae*. The prey is being restrained with the suction disc, and its tissue is removed from the shell by the radula.** **Drawing: L. Newman (1990)**

eye tentacles and tubercules on their proboscis. Copulation and spermatophore transfer have not been oserved in these molluscs (LALLI & GILMER, 1989).

Gravid females deposit eggs in floating mucous strings. Veligers hatch with a bilobed velum which is used for both swimming and collecting food. The velum has two to six lobes which are ingested at metamorphosis. The shell-less pterotracheids cast off their larval shell at metamorphosis. One pterotracheid, *Firoloida desmaresti,* is known to carry a permanent egg string with continually developing embryos, the most distal ones being the most developed.

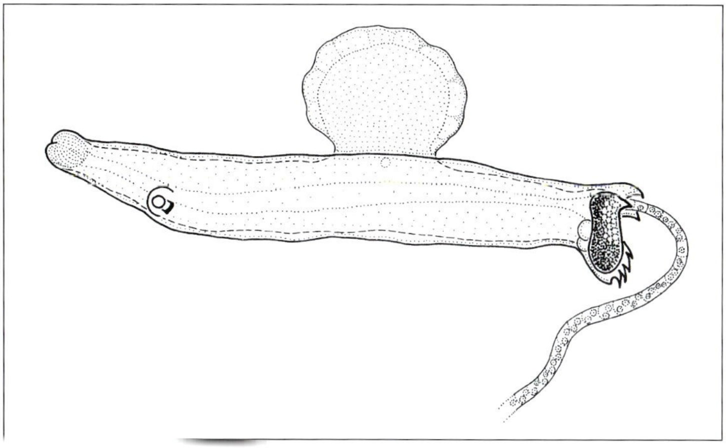

***Firoloida desmaresti*; female with permanent egg string.**
Drawing: L. Newman in BEESELY & ROSS (1996)

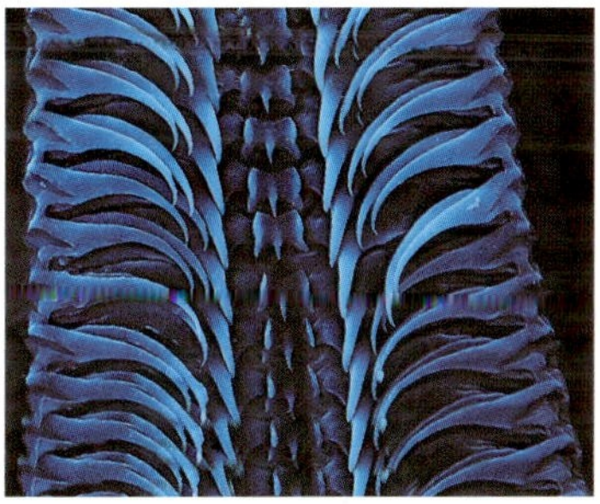

Radula of a carinariid. Zooplankton prey is captured with a hook shaped lateral teeth.
SEM-Photo: L. Newman and A. Flowers

Atlantid heteropods have occasionally been found with yellow egg-like objects attached to their shells. TESCH (1949) surmised that these were large spermatophores. SEAPY (1996) however, determined that these objects were egg cases of the ocean strider or sea skater, *Halobates sericeus*, which are known to deposit their eggs on male atlantid shells. How the atlantid can swim with this added weight is unclear. A possible explanation of this is that male atlantids might swarm at the ocean surface, not unlike mating swarms of some insects in which females enter the swarm, become captured by a male and then drop down to mate. This behaviour has been suggested in insects for bringing together mating adults in low density populations where individuals are normally widely dispersed. The relatively large heteropod, *Firoloida desmaresti*, has been found in such large swarms off southern Australian waters as to cause conic scattering layers (BLACKBURN, 1956).

***Atlanta turriculata* with the egg sac of an ocean strider attached to its shell**
Photo: R. Seapy

There are about 34 species of heteropods within worldwide tropical and subtropical waters (RICHTER & SEAPY, 1999; SEAPY & SKOGLUND, 2001). Their distribution is generally restricted to between 40° N to 40° S and they are epipelagic, within the upper 200 m of the water column (VAN DER SPOEL, 1976; VAN DER SPOEL et al., 1996). Numerically, heteropods do not represent a large proportion of the zooplankton (SEAPY, 1974; LALLI & GILMER, 1989). However, atlantids are known to contribute empty shells to pteropod oozes in some ocean basins.

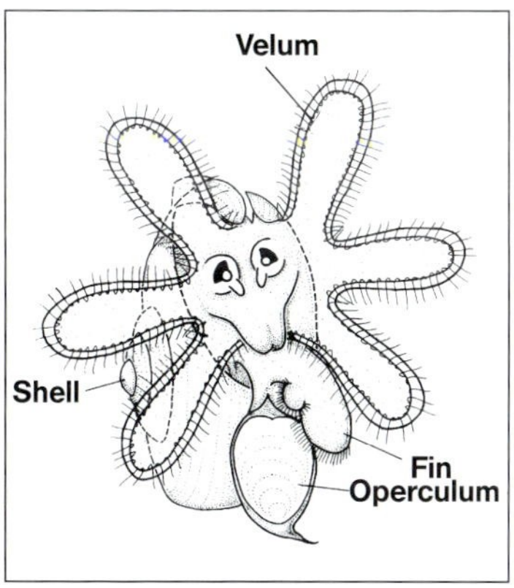

Atlantid veliger larva: *Atlanta gaudichaudi* (left; drawing: L. Newman, 1990) and *A. helicinoides* (right; photo: L. Newman and A. Flowers).

Their rarity may be a consequence of the difficulties encountered in collecting and studying these elusive molluscs. Due to their excellent eyesight they can avoid slow towed plankton nets (Seapy, 1990) and if caught tend to get distorted making identification difficult. Atlantid shells are so thin and fragile and will dissolve in formalin. Carinariids are rarely collected with their shells intact.

Suborder Ptenoglossa

The majority of species of suborder Ptenoglossa live parasitically on other organisms. Only four from among the many families in this suborder are going to be discussed here as representative groups.

The species of families Cerithiopsidae and Triphoridae live on sponges and they also feed on them. In fact, some accept only on a particular species of sponge as food. Family Epitoniidae is very large and its species occur in tropical as well as subtropical seas. Their shells are often spirally coiled and can have distinct structures. This family is of interest to aquarists because many species live and feed on sea anemones or other cnidarians. One such example is *Epitonium bullatum* that lives on *Tubastraea* spp. and eats their polyps (see photo on the right). It is conceivable that species of family Epitoniidae are being introduced in to aquariums together with corals or other cnidarians. *Epitonium* is the largest genus of this family, but there are also a number of other genera, such as *Cirsotrema, Dannevigena, Eglisia, Filiscala, Mazescala* and *Opalia*, which are all distributed throughout the Indo-Pacific.

Snails of the similarly large family Eulimidae live as parasites on echinoderms. Here too many species are host-specific. In many instances, these snails live alternately in the bottom substrate and on a host, and it is assumed, that the snails suck fluid from the body cavity of echinoderms. Sea stars, sea urchins

Above: *Epitonium bullatum* (suborder Ptenoglossa), seen here at Heron Island, Great Barrier Reef, uses its long proboscis to feed on polyps of a stony coral *Tubastraea* sp. Photo: L. Newman and A. Flowers

To the right: Murex shells of genera *Murex* (left), *Haustellum* (centre) and *Chicoreus* (right) from family Muricidae (order Neogastropoda).

and related forms have a body cavity that is unique in the Animal Kingdom. This is probably one of the reasons they are so vulnerable to infestations by eulimid snails (WILSON, 1983). Yet, sometimes damage inflicted by the snails is being repaired by the echinoderms, because of their very large regenerative capabilities.

Order Neogastropoda

While the preceding orders of class Gastropoda contained snails that eat algae, detritus, meat or they are omnivores, the Neogastropoda are exclusively predatory. They feed on worms, crustaceans, other invertebrates and even on fishes! Anatomically they are ideally adapted for such feeding behaviour.

The radula has fewer teeth than in other snails and is adapted more to feeding on animals than to grazing on algae. Generally, there are two different types of radula in this order: the rachioglossa-type has three rows of teeth. In this case the radula is used in conjunction with corneous jaws for biting, and occurs in superfamily Muricoidea. The taxoglossa-type has only one row of teeth, but these are long and pointed, and also connected to venom glands. This type of radula occurs in superfamily Conoidea.

Mouth and radula are located in the centre of a long, protrusible proboscis, used by neogastropods for capturing prey. The proboscis can be inserted into prey organisms, which is the principal reason as to why neogastropods are so successful in capturing prey. In some groups, e.g. family Conidae (see page 95), proboscis and radula are connected to a well-developed venom apparatus.

Apart from the proboscis and a specifically adapted radula, neogastropods also have a siphon, which is actually a tube-shaped fold at the margin of the mantle. Originally, the siphon served only as a snorkel that was extended from underneath the bottom, in order to supply the gills with respiratory water. Yet, in the course of evolution the siphon has developed further. It also contains an effective smelling organ that is capable of recognising whether the inhaled water is fresh and oxygenated. Moreover, with the aid of chemoreceptors is can also detect prey organisms. For that purpose the proboscis is extended forward, so that potential prey can be localised over a relative long distance. The shell has an indentation to accommodate the siphon – a feature that has

Murex shells (family Muricidae) from the Indo-Pacific: 1 – *Chicoreus saulii*, 10 cm. 2 – *Naquetia triquetra*, 8 cm. 3 – *Marchia martinetana*, 4 cm. 4 – *Pteronotus tripterus*, 6 cm. 5 – *Thais armigera*, 10 cm. 6 – *Morula coronata*, 3 cm. 7 – *Thais intermedia*, 3 cm. 8 – *Nassa serta*, 4 cm. **Photo: C. Bryce**

The Violet Coral Snail, *Coralliophila radula* – seen here at Madang, Papua-New Guinea - feeds on corals (left). This species is easy to recognise by its violet colouration. Photos: L. Newman and A. Flowers

undergone extreme development in some species, for instance in *Murex* spp.

The foot can often be totally withdrawn in to the shell that can be closed off with a strong operculum. In addition, the foot – in some families – has a "drill" that give off an acid, which can dissolve the shell of prey organisms. This is a useful device for these snails, when they encounter prey with a strong shell, e.g. clams or acanthocephalans. The foot is also used for burying, and many neogastropods spend a large part of their life buried in sand- or mud bottom.

Exactly as in many other snails, there have been significant changes in the classification of order Neogastropoda in recent years. In total, more than 1,100 genera and subgenera are known. Since Neogastropoda are not of the greatest significance from an aquarist's point of view, we confine ourselves to an overview of the following families only:

- Muricidae
- Turbinellidae
- Buccinidae
- Volutidae
- Olividae
- Mitridae
- Turridae
- Conidae

The family Conidae is especially interesting and contains a few very poisonous species.

Family Muricidae
Murex Shells

Murex shells are generally aggressive predators. The family contains several hundred species, which occur in tropical and subtropical seas. Some species, especially those of genera *Murex* and *Chicoreus* have long, needle-sharp spines and strong shell structures, which probably stabilise the shell and protect the animal. In other species, for instance in those of genus *Drupella*, long spines are absent. Instead, they have a knobby and very compact shell. Feeding behaviour is highly variable within this family, and indeed it can be quite specific for a particular species.

Many murex snails drill holes in to the shells of other molluscs and others feed on soft-bodied invertebrates, such as bristle worms.

The snails of the **subfamily Coralliophilinae** are generally called "coral snails", since they are usually found on corals, especially among the polyps of stony corals. There are several genera, but the most common one is *Corallophila*. The snails feed by sucking the coral polyps out from corallites. Some species, such as *Coralliophila violacea*, live in groups, and the same group can remain for months on the same spot on a particular coral colony. When the snails finally move they leave only the dead skeleton behind. All living tissue has been eaten away. One species, i.e. *Rapa rapa*, is so specifically tied to corals, that it actually lives hidden in the coral skeleton.

Subfamily Muricinae contains the most bizarre species among the Muricidae. According to WILSON (1993), genus *Chicoreus* encompasses about 30 species, which makes it the largest within the family. The sturdy shell is equipped with coarse, more or less bent spines and other outgrowths. It also has rough, irregular structures. *Chicoreus microphyllus* (a synonym for *Murex torrefactus*) is a rather common species in the Indo-Pacific, where it grows to a size of 11 cm. *C. microphyllus* is relatively often encountered during reef dives, especially along the reef edge and on the upper section of a reef slope. However, the shells are frequently difficult to see since they are generally overgrown with algae or sponges. *C. dilerctus* is a common species in the Caribbean Sea.

The most artistically sculptured Murex shells are those of genus *Murex* where the shell is equipped with sharp spines. The most conspicuous and best-known species is *Murex pectin* that can reach a size of up to 15 cm. It has a long shell apex and an unusually long siphon canal, as well as many needle sharp spines. This genus is distributed widely throughout the Indo-Pacific. *M. cabritii* which grows to about 5 cm is distributed in the Caribbean Sea, yet it is not often seen. Much more common in this region are *M. fulvescens* and *M. porum*.

In *Haustellum* sharp spines are absent from the shell; however, the siphon canal is longer than the shell and is club-shaped. *H. haustellum* can reach a maximum size of 14 cm and is a frequent species in the Indo-Pacific.

Drupella species of **subfamily Thaidinae** feed on stony corals and have come in focus in recent times because their populations in the wild have grown substantially and with that there has been an increase in damage to the corals (see photo to the right). There are three widespread species: *D. cornus, D. rugosa* and *D. fragum*. All species vary a lot in size, shape, and colour (JOHNSON & CUMMING, 1995). *D. cornus* grows to a size of up to 3.8 cm and occurs widely in the Indo-Pacific, where it lives among corals in shallow water. It was principally this species that, during the mid 1980's, attacked Ningaloo Reef off Western Australia, known as one of the most beautiful of all coral reefs. In the course of this infestation, the number of snails increased from normally 100 – 200 per kilometre of reef, to 1 – 2 million! Between 1987 and 1989 live coral cover decreased by up to 75 % over a 280 kilometres stretch of reef, which means that the total number of *Drupella* specimens involved amounted to several hundred million (AYLING & AYLING, 1987; STODDART, 1989). BAIRD (1999) reports on a dramatic increase in *Drupella* population at Pelorus Island in the Townsville region of the Great Barrier Reef. The site was one of the worst affected during the 1998 mass coral bleaching, and it is possible that there is a link between *Drupella* aggregations and coral mortality. A similar phenomenon was seen after the hurricane "Allen" where predation by corallivorous snails in the months following the hurricane killed many more corals than were killed during the hurricane itself (KNOWLTON et al., 1981).

***Drupella* species feed on corals, as seen here at Heron Island, Great Barrier Reef. It is important to watch for these snails, because they can cause considerable damage to corals in a coral reef aquarium.**
Photo: L. Newman and A. Flowers

***Thais bufo* is occasionally imported. In an aquarium this snails has been seen burying in sand in search of food. Caution is advised, since it feeds on other invertebrates, and therefore it is fundamentally unsuitable for a coral reef aquarium.**

***Morula marginalba* at Caloundra, southeastern Queensland (Australia).**
Photo: L. Newman and A. Flowers

The population explosions of *Drupella* snails do remain unexplained. All ecosystems are in a continual state of flux. Predator-prey systems often displays a cycle where the number of predators increases because of an increase in the numbers of preys, then decline dramatically through malnutrition, allowing prey numbers to build up and start the cycle again (STODDART, 1989). The *Drupella*-"explosions" might therefore be part of a natural cycle, or it might be caused by changes in the environment – nobody know for sure, but the effects on coral growth is for sure dramatic!

Genus *Morula* also contains a large number of species with a cosmopolitan distribution in tropical and subtropical seas. Many are only a few centimetres large and live in corals. *Morula uva* is particularly common in inter-tidal zones and shallow water areas along rocky coasts, but it also occurs in reef habitats.

Genus *Nassa* has only two species. *N. francolina* is endemic to the Indian Ocean and *N. serta* lives in the western Pacific Ocean, along eastern Australia and also – rather strangely – off the Cocos Island, but nowhere else in the Indian-Ocean (WILSON, 1994).

The species of genus *Thais* are – like the other Murex snails – predators. They live along rocky shorelines on sandy substratum searching for food. Some species, for instance *Thais bufo*, are occasionally encountered in the aquarium trade. WILSON (1994) lists eight species.

Family Turbinellidae
Vasum- or Vase Shells

Vasum snails of family Turbellidae are large and heavy. They live in shallow water and feed on bristle worm- and peanut (sipunculid) worms. They catch their prey using a long, thin proboscis. In subfamily Vasinae the shells are often substantially structured, with an operculum that is strong and distinctly bent. *Vasum ceramicum* and *V. turbinellum* are two species that have a wide distribution in the Indo-Pacific.

Family Buccinidae
Whelks

Whelks occur in deep- as well as in shallow water, with a distribution in tropical, subtropical and cold seas. For us, as aquarists, the small species of genus *Nassarius* **subfamily Nassarinae** are the most important whelks. They live in habitats from where they are easily imported together with live rocks. Whelks are particularly numerous on muddy or sandy bottoms in the inter-tidal zone, but they also occur very frequently in mangrove areas. This is a very large genus, and many species feed on decaying organic material. These snails are small, oval with a medium- or high apex. The

Dogwhelk shells of genus *Nassarius* (family Buccinidae) from the Indo-Pacific: 1 – *Nassarius papillosus*, 4 cm. 2 – *N. splendiddulus*, 1 cm. 3 – *N. granifer*, 2 cm. 4 – *N. pauperus*, 1 cm. 5 – *N. reeveanus*, 2 cm. Photo: C. Bryce

Cantharus undosus at Heron Island, Great Barrier Reef. Photos: L. Newman and A. Flowers

shell is often attractively sculptured, and there is a corneous operculum. The most common species is *Nassarius dorsatus* (3 cm), but *N. granifer, N. albescens* and *N. bicallosus* are also frequently encountered in the Indo-Pacific. The most commonly occurring species in the Caribbean Sea are *N. acutus* and *N. vibex*, which both only grow to a size of 12 mm. Genus *Cantharus* occurs worldwide. *Canthanus undosus* grows up to 4 cm and has a stable shell.

Subfamily Fasciolarinae contains the so-called tulip- or spindle snails of genus *Pleuroploca*, which are regularly imported. *P. filamentosa* is the most common species on coral reefs of the Indo-Pacific. Our own experience with these snails is limited, however, with their robust shell that contains a long siphon canal they are so attractive that their aquarium maintenance should be

Volute shells (family Volutidae) from the Indo-Pacific: 1 - *Amoria jamrachi*, 7 cm. 2 - *Amoria ellioti*, 11 cm. 3 - *Amoria macandrewi*, 9 cm. 4 - *Volutoconus bednalli*, 13 cm. 5 - *Amoria praetexta*, 5 cm. 6 - *Lyria mitraeformis*, 5 cm. 7 - *Volutoconus hargreavesi*, 8 cm. 8 - *Paramoria guntheri*, 4 cm. 9 - *Amoria dampieria*, 3 cm. 10 - *Volutoconus capricorneus*, 4 cm. 11 - *Amoria grayi*, 10 cm. 12 - *Amoria damoni*, 10 cm. 13 - *Aulicina nivosa*, 9 cm. Photo: C. Bryce

Melo georginae has an extraordinarily wide shell aperture.
Photo: L. Newman and A. Flowers

attempted. This subfamily also includes genus *Latrius*, with a cosmopolitan distribution in tropical and subtropical seas. It includes medium-size species.

Family Volutidae
Volutes

The snails in this family have an elongated, decoratively coloured shell with a very large, oblong aperture and a short apex. An operculum is absent in most species, but there is a long siphon and a strong, colourful foot. The shells of many species are so ornamental, that they have been highly prized collectors' items for a long time. There are about 200 known species that live in shallow water, buried in sand bottom, often in the inter-tidal zone or on reef flats. The majority occurs in seas around Australia and in the central Pacific Ocean. At night they leave their hiding place in order to hunt for other snails or dead animals. The Volutidae deposit their eggs on the bottom, from which fully developed juvenile snails will hatch, and so a pelagic larval stage does not occur in volutes.

Subfamily Zidoninae contains the genera *Cymbiola, Ericusa* and *Melo*. *Cymbiola magnifica* is frequently imported. Subfamily Armorinae contains genus *Armoria* with some of the most beautiful snail species, as well as genera *Cymbiolista, Nannamoria* and *Notovoluta*. Subfamily Odontocymbiolinae includes genus *Volutoconus* with four rare species. Far more common are genera *Lyreneta, Lyria* and *Teramachia* of subfamily Volutinae.

Family Olividae

Because of their shiny appea-

To the left: Olive shells (family Olividae) from the Indo-Pacific: 1 - *Oliva caerulea*, 5 cm. 2 - *Oliva lignaria*, 6 cm. 3 - *Oliva australis*, 3 cm. 4 - *Oliva caldania*, 2 cm. 5 - *Oliva oliva*, 4 cm. 6 - *Ancillista cingulata*, 10 cm. 7 - *Oliva annulata*, 5 cm. 8 - *Amalda elongata*, 4 cm. Photo: C. Bryce

Olive snails in their natural habitat: An *Oliva* species (left; photo: L. Newman and A. Flowers) and probably an *Ancillista* species (right; photo: C. Bryce).

rance and size, olive shells are very popular among shell collectors. Their colouration varies from one species to the next. So far more than 60 species have been described from tropical- and subtropical seas. Olive snails are predators that prey on other invertebrates; usually molluscs, or they are scavengers that feed on dead organisms or organic material. During the day they remain buried in sand, but at night they become active and commence searching for food. Their foot is large and has an enlargement at its anterior end, referred to as propodium. The propodium is being used as a "plough" while moving through the sand in search of food.

This family encompasses four subfamilies. True olive snails belong to subfamily Olivinae, with genus *Oliva* as the dominating

Mitre shells (family Mitridae) from the Indo-Pacific: 1 – *Mitra retusa,* 2 cm. 2 – *Mitra litterata,* 3 cm. 3 – *Mitra turgida,* 1 cm. 4 – *Mitra coronata,* 1 cm. 5 – *Cancilla filaris,* 3 cm. 6 – *Mitra multiplicata,* 3 cm. 7 – *Imbricaria conovula,* 2 cm. 8 – *Imbricaria punctata,* 2 cm. 9 – *Imbricaria olivaeformis,* 1 cm. 10 – *Mitra edentula,* 2 cm. 11 – *Scabricola bicolour,* 2 cm.
Photo: C. Bryce

group. The best known and most widely distributed species is *Oliva oliva.* It has a highly variable colouration, and is commonly encountered in sandy areas of reefs. PEUTSCH & SARGENT (1986) have compiled an atlas of genus *Olivia.*

Olive snails are occasionally imported. They are fundamentally difficult to keep and because of their specific feeding behaviour they are not really suited for a coral reef aquarium.

Family Mitridae
Mitres or Mitre Shells

About 375 current species are included in family Mitridae. This then makes it one of the largest marine snail families. Unfortunately, many of the scientific publications dedicated to this family over the years have lead to a confusing disarray in terms of its systematic and nomenclature. Consequently, there are now more than 2,600 names within family Mitridae (WELLS & BRYCE, 1988).

Most of the mitre species occur on coral reefs, where they live on sand bottom among corals or on and in live rocks. The shell is spirally coiled and often very attractively coloured. The shell aperture is oval-shaped. Many species are small and occur within the intertidal zone; however, the best known mitre shell, i.e. *Mitra mitra* grows to a size of 11 cm. It has decorative colours, consisting of a white base colouration, densely covered by large, red spots. However, in some of the geographical distribution areas in the Indo-Pacific this snail is not very numerous.

Mitre snails are predators. A large venom gland has been found in some species, and it is assumed that this is used for killing prey organisms. Several species of this family can be found in an aquarium on live rocks. Our experience has shown that they cause little, if any, damage to aquarium animals. The genera *Imbricaria, Mitra, Neocancilla, Pterygia, Pusia* and *Scabricola* are among the most commonly encountered tropical Mitre Snails.

Family Turridae

This is the largest family within class Gastropoda. Its systematic structure is confusing and under much debate by malacologists (POWELL, 1964, 1967 and 1969). Members of this family are found in all seas, from the Arctic to the Antarctic.

Turrids are normally encountered on or in soft bottom substrate, and a few species can also occur in rocky areas and among corals on reefs. The shells are variable in shape, but usually oblong and with a high apex and a long siphon canal. A specific feature of these snails is the presence of a so-called sinus, i.e. a cleft in the outer lip, between connective tissue and the periphery of the body

Shells of turrid snails (family Turridae) from the Indo-Pacific: 1 – *Xenoturris cingulifera*, 3 cm. 2 – *Xenoturris kingae*, 2 cm. 3 – *Turridrupa cerithina*, 2 cm. 4 – *Turridrupa astricta*, 1 cm. 5 – *Turridrupa bijubata*, 1 cm. 6 – *Philbertia nexa*, 1 cm. 7 – *Philbertia lata*, 1 cm. 8 – *Clavus pica*, 1 cm. 9 – *Clavus lateus*, 2 cm. 10 – *Leinardia roseotincta*, 3 cm.

Photo: C. Bryce

Cone snails in their natural habitat: *Conus marmoreus* (left) and *C. miles* (right) at Lizard Island, Great Barrier Reef. Photos: L. Newman and A. Flowers

coils. A corneous operculum is also present.

Turrids are carnivores that feed on a great variety of organisms, and utilise various methods of prey capture. The radula differs from species to species and in some it is connected to a venom gland. The radula teeth are shaped harpoon-like, and in their structure they are reminiscent of those found in cone snails (the next family to be discussed).

Family Conidae
Cone Shells

No other snails have achieved such notoriety as the cone snails. Humans seem to be particularly fascinated by dangerous animals, a characteristic that has also lead to misconceptions about coral reefs. Many people believe, that sharks and other fearsome creatures dominate in the ocean, and that diving on coral reefs means exposure life-threatening dangers. Consequently, the irrational

To the right: Cone shells of genus *Conus* (family Conidae) from the Indo-Pacific: 1 - *Conus rufimaculatus.* 2 - *C. achatinus.* 3 - *C. retifer.* 4 - *C. striatellus.* 5 - *C. planorbis.* 6 - *C. sugillatus.* 7 - *C. musicus.* 8 - *C. sponsalis.* 9 - *C. miliaris.* 10 - *C. ranunculus.* 11 - *C. ebraeus.* 12 - *C. chaldaeus.* 13 - *C. frigidus.* 14 - *C. catus.* 15 - *C. coronatus.* 16 - *C. stramineus.* 17 - *C. caracteristicus.* Photo: G. Allen

fear of cone snails is hardly surprising. Of course, there are some dangerous animals on coral reefs, a fact that cannot be denied, and these include a few species of cone snails, but it would certainly be false to think that you are generally in great danger when you visit a coral reef. Another reason for the attention enjoyed by cone snails, is the fact, that this family also contains some very rare species, e.g. *Conus glorimaris* and *Conus milneedwardsii*.

Conidae is monotypic, containing only a single genus, *Conus*. However, with its over 300 species, the genus contains the largest number of species in the entire Animal Kingdom. With few exceptions, *Conus* species occur in tropical seas only. Many species display a very attractive colour pattern on their shell, which in most species has a weak apex. In a few species, e.g. *Conus excelsus*, the apex is more distinct, while it completely absent in *C. litteratus*, where the top of the shell is totally flat.

The shell aperture is narrow and oblong, and usually there is only a very small operculum. Normally, each individual *Conus* species has a characteristic appearance, so that the most commonly occurring species are relatively easy to identify, although closely related species can be very similar. The colours and patterns of the shells in this group can be useful species characters. A few species (especially *C. anemone)* are rather variable in their colour markings, however, ranging from blue-grey to white, brown, lilac, orange or yellow.

Cone shells of genus *Conus* (family Conidae) from the Indo-Pacific: 1 - *Conus litteratus*. 2 - *C. leopardus*. 3 - *C. betulinus*. 4 - *C. virgo*. 5 - *C. imperialis*. 6 - *C. generalis*. 7 - *C. vitulinus*. 8 - *C. striatus*. 9 - *C. figulinus*. 10 - *C. litoglyphus*. 11 - *C. thalassiarchus*. 12 - *C. tessulatus*. 13 - *C. tulipa*. 14 - *C. textiles*. 15 - *C. episcopus*. 16 - *C. aulicus*. 17 - *C. geographus*. 18 - *C. omaria*.

Photo: G. Allen

Cone snails are predators that live buried in sand- or mud bottom, and only the siphon protrudes above the substrate. This supplies the snail with water rich in oxygen and at the same time also removes its metabolic waste products.

Cone snails are among the highly advanced snails. They have developed a very effective venom apparatus for capturing prey and as a defensive mechanism. The radula has become modified in to needle-sharp, harpoon-like teeth. Each of the chitinous teeth is shaped in to a rolled-up cord that is located freely in a cavity at the front of the radula. Tooth after tooth is being used individually and is transported through the tubular mouth. On its way to the outside, the tooth is coated with venom that is produced in a long, spirally shaped tract, which leads into the pharynx. A powerful muscle is located at the other end of this spiral tract, which probably serves to force the venom through and out of the tract. On the basis of its mode of feeding, the family can be divided in to three groups: 1. bristle worm feeders, 2. mollusc feeders, and 3. fish feeders.

Most *Conus* species feed on bristle worms. A few species display an extremely narrow food selectivity, such as *C. zonatus* and *C. imperialis;* each feeding on only one particular bristle worm species. In this group, the venom apparatus is only marginally developed. Species that feed on other molluscs (principally on other snails) such as *C. marmoreus* or the highly poisenous *C. textilis* that both feed on other cone snails, however, actively use their venom apparatus. The venom paralyses the victim, which then loses the ability to withdraw in to its protective shell, and so becomes easy prey for the cone snail.

The most advanced venom apparatus together with the best-developed capturing methods are found in cone snails that feed on small fishes. In these particular snails the venom tooth is much larger than in the preceding groups. Using their sense of smell the snails locate suitable prey. Once the prey has been detected, the mouth with the venom tooth is pushed forward. Then the snail searches along the fish's body and injects the venom tooth at a suitable site, where there is no bone. The fish is helplessly restrained by the venom cord and is then swallowed with the aid of the proboscis. This is followed by a slow digestive process, which often lasts for several days. During this period the prey organism takes up so much room inside the stomach cavity that the snail is unable to retract its foot back in to the shell. The best known of all cone snails, the extremely toxic *C. geographus* is an example of a fish-feeder.

Conus species also use their venom apparatus for their own defence. In some cases that has had fatal consequences for humans. This involved mainly to *Conus geographicus*, which has caused 12 of the known 16 deaths (WILSON, 1994). Two deaths are attributable to *C. textilis*, which is otherwise a mollusc feeder. The two remaining fatalities were caused by stings from unidentified *Conus* species. Research on the venom efficacy has shown that the venom is very effective on the muscular system of vertebrates. Initially there is a burning pain at the point of penetration by the

In cone snails the tip of the radula has been modified in to a harpoon. The venom of some species can be fatal to humans. Photo: C. Bryce

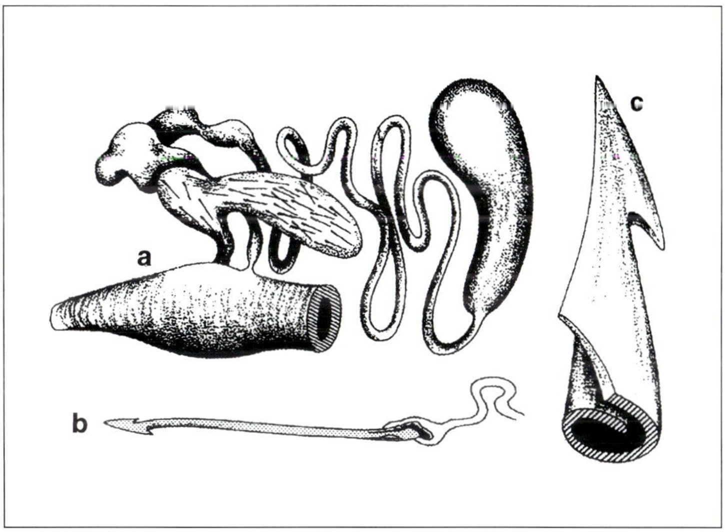

Venom apparatus of a cone snail of genus *Conus*; the venom is produced in a long, tubular venom canal and then stored in a venom sac. Arrow-shaped radula teeth are produced in a radula sac (a) and are then stored. When the venom sac is being compressed, the venom is pushed in to the oesophagus, where it comes in to contact with a single radula tooth (b). The radula teeth are hollow, made up of rolled chitinous plates, and are fitted with a barb at the tip (c).

Drawing after MEBS (1989)

Cone snail swallowing a damselfish after the snail has paralysed it. **Drawing made up from various sources**

snail's venom tooth, followed by loss of tactile feeling. This is followed very quickly by muscular paralysis. Death occurs due to heart- and lung failure.

Serious stings that have not always been fatal, are known to have been inflicted by *Conus geographus, C. textilis, C. tulipa* and *C. omaria*. Less serious reactions in humans have been caused by stings from *C. obscurus, C. catus, C. imperalis, C. lividus, C. quercinus, C. sponsalis, C. pulicarius, C. litteratus* and *C. anemone* (WILSON, 1994).

It appears that stings from worm-feeding cone snails usually heal relatively quickly, while stings from mollusc-feeding species require a long convalescence period.

Relatively few details are known about the distribution of cone snails, including the frequency of occurrence of species that feed on fish. They prefer sandy beaches, muddy bottom substrates and shallow sections of coral reefs. During the day they remain buried or seek refuge under rocks, but at night they become active. Once a cone snail has been discovered, it should be left alone. Although these snails have an interesting natural history, they are organism that in principle should not be kept in aquariums. If, in spite of these cautionary remarks, an aquarist wants to keep cone snails, it is imperative to know what species it is and what its food requirements are. Moreover, they must be kept in a special aquarium, where they can bury and be comfortable. If this is successful and the cone snails can be given sufficient food, an aquarium is an excellent venue for observing the fascinating behaviour of cone snails. But it is very important not to forget that: **some cone snails are very dangerous!**

Conus geographus **(left; photo: C. Bryce) and *C. textile*, – seen here at the Philippines at a depth of 4 m (right; photo: P. Schupp), – are the two most dangerous and venomous cone snails.**

The taxonomy of this subclass is contentious. In this book we follow HASZPRUNAR (1985), who views the Heterobranchia as being an independent subclass. Yet, it must be noted, that not all malacologists recognize this status, but consider this group as part of subclass Prosobranchia.

Heterobranchia includes several parasitic snails. *Heliaceus* spp. and *Architectonia* spp. (family Architectonicidae) are parasites on encrusting anemones and so they appear regularly in coral reef aquariums. A long, planktonic larval stage is known to occur in many, widely distributed species. *Architectonica nobilis* (syn. *A. grannulata*) occurs in the Caribbean Sea, along the coastline of southeastern North America, Texas, Surinam and Brazil, as well as in the eastern Atlantic.

Architectonica maxima **(top) and** ***A. perspectiva*** **(bottom) are often introduced in to an aquarium together with live rocks. Their shells are extremely depressed. Photo: C. Bryce**

It bores itself in to the sand in shallow water areas. At night it feeds on soft-bodied cnidarian animals. *A. maxima* comes from the Indo-Pacific, where it occurs from Sri Lanka to Japan, as well as at the Philippines and around Australia; however, it is never very frequent. *Heliacus variegates* has a distribution extending from East Africa to Japan, Polynesia and Australia, where it usually lives in reef habitats, often on coral polyps.

Other species of parasitic snails that appear in aquariums belong to family Pyramidellidae. These snails have a slender, turret- to cone-shaped shell with thick spindle folds. According to LASERON (1959) there are more than 1,000 species of this family in the Indo-Pacific region. With few exceptions, these snails are

Shells of family Pyramidellidae from the Indo-Pacific: 1 – *Pyramidella acus.* 2 – *Pyramidella sulcata.* 3 – *Pyramidella dolabrata.* 4 – *Otopleura auriscata.* Photo: C. Bryce

***Heliacus variegatus* sp. "KA5-GAS-13" is a parasite of encrusting anemones.** **Photo: Scott W. Michael**

only a few millimetres large. They live as ectoparasites on bristle worms, sponges, tunicates and on molluscs, including giant clams of family Tridacnidae, and others. These snails have a long proboscis with a stiletto-like point, which is capable of penetrating the shell of prey organisms; however, a radula is absent.

Pyrimidella crenulata occurs in the Caribbean Sea. With a maximum size of 10 to 12 mm it belongs to the larger species in this family. From an aquarist's perspective the genera of interest are *Tathrella* and *Turbonilla*, since their species often occur as parasites on *Tridacna* clams (see Fact Sheet page 101).

***Heliacus* sp. "KA5-GAS-14" (left) has a conical operculum. Sometimes this species is introduced in to an aquarium together with corals and live rocks. *Heliacus* sp. "KA5-GAS-15" (right) on the glass bottom of an aquarium.**

Turbonilla spp. and *Tathrella* spp.

Distribution: Widely distributed throughout the Indo-Pacific.

Size: Very small; most species only a few millimetres; the majority of species reach a length of 2 to 4 mm.

Reproduction: *Turbonilla* spp. deposit their gelatinous egg clumps on the mantle or at the base of giant clams. Other types of substrate are rarely used. The eggs are transparent and have an irregular shape. All egg clumps deposited on a giant clam are positioned close to each other. The veliger larvae hatch in about ten days. Attempts to control these snails on clam farms, by deploying the crab *Thalamita sima* have only been partially successful (CUMMING, 1988).

Food: These snails feed on the blood or tissue fluid of their host. Species that are parasitic on giant clams, can – unless they are promptly removed – lead to the demise of clams on breeding farms and in the aquarium. In an aquarium, *Turbonilla* spp. display variable behaviour in the course of a day. During the day, these tiny snails remain in shaded areas along the bottom or underneath giant clamshells. At night they crawl up to the edge of clamshells in order to feed on the margin of the mantle. With pumping movements of their rather long proboscis they are able to suck body fluid from the mantle of a giant clam.

Description: The genus *Turbonilla* contains several species and was earlier placed in the genus *Pyrgiscus*, which now has the status as one out of several subgenera in *Turbonilla*. All species of family Pyramidellidae live as ectoparasites on invertebrates, usually on bristle worms and sessile molluscs. They do not possess gills, and a radula is absent. Instead, there is a very long proboscis, modified to suck out the blood or body fluid from its host. Aquarists can encounter these snails as ectoparasites on giant clams of genus *Tridacna*.

Aquarium requirements: Basically, all snails of family Pyramidellidae can be considered as hardy animals that can live in coral reef aquariums for a long time and readily reproduce there. The tiny snails of genera *Turbonilla* and *Tathrella* have been observed as parasites on giant clams of genus *Tridacna*, kept in aquariums, aquarium shops and on clam breeding farms. However, it is also possible that other genera in this family exhibit the same behaviour. Consequently, it is important to detect and remove these snails as quickly as possible. Because of their small size, this is often rather difficult.

GA: -; **SE:** +; **CO:** –

Photos:
Top: Two small snails of family Pyramidellidae, seen here using their proboscis to suck body fluid from a giant clam. Also visible close to the snails is a tiny calcareous sponge.
Photo: S. W. Michael
Bottom: Snail shells of family Pyramidellidae, found on a giant clam.

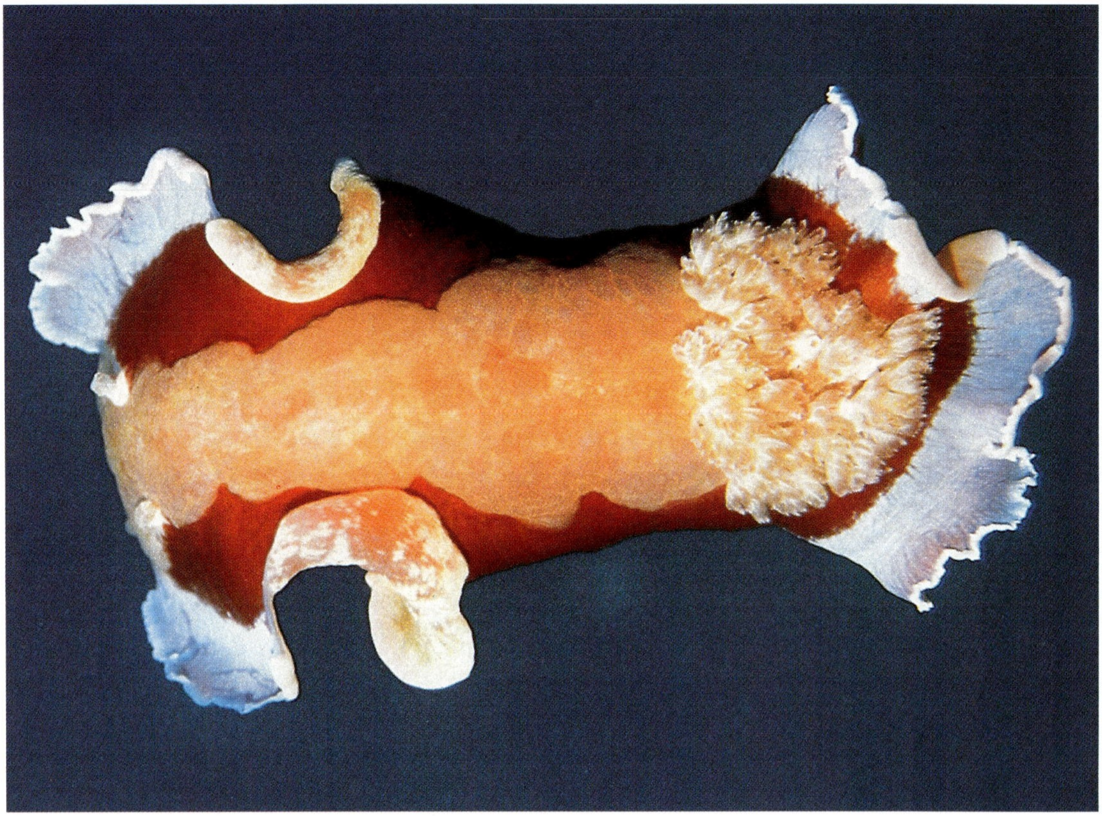

The Spanish Dancer, *Hexabranchus sanguineus*, seen here off Guam swimming in open water at a depth of 7 m, is a nudibranch that can grow to a size of up to 30 cm. **Photo: P. Schupp**

Subclass Opisthobranchia - Sea Slugs, Sea Hares, Nudibranchs

Subclass Opisthobranchia contains snails with a substantially reduced shell, or a shell can also be totally absent. The Opisthobranchia – usually referred to as sea slugs – have probably evolved from prosobranch-like ancestors that possessed only the left gill (RUPPERT & BARNES, 1994).

Since a protective shell is largely absent in opisthobranchs, they had to develop other means of defence against predators. These animals succeeded by developing the capability to produce specific, chemical substances and to discharge them in to the surrounding water, e.g. bad-tasting acids. Some opisthobranchs have developed camouflage colours that blends them perfectly in to their surroundings. Others have colours that are used to imitate poisonous organism. And yet others again have taken on extremely conspicuous colours to signal predators: "Caution, I am poisonous!" Many sea slugs simply hide during the day and are only active at night. Beyond that, there are those that live on animals that can defend themselves, such as cnidarians, but they themselves are immune against the weapons or venoms of their host.

There are several thousand known species, however, this subclass has so far been only insufficiently investigated. Major systematic changes have been made in recent years (for a systematic structure of taxonomic levels down of orders, see page 45). In the subsequent discussion we will follow the pattern of previous chapters, whereby only a limited selection of species and their respective major taxonomic groups are presented here.

The about 2000 highly diverse species of Opisthobranch molluscs have several anatomical characteristics. The species are largely marine, and many have a particularly beautiful colouration that attracts aquarists and divers. Only the species in order Cephalaspidiea have a normal-appearing shell. The body is in the majority of species bilaterally symmetrical. The mantle cavity and its organs are located on the right side of the body. The gills are modified, and some opisthobranchs have lost them completely or developed secondary gills. When the gills are missing, the sea slugs then have the capability to absorb oxygen across the outer skin surface (LINDTNER, 1976). When present the gills are positioned behind the heart and are usually pinnate (ctenidium; feather-like), but sometimes they are folded or re-formed from folds in the mantle wall. The gills can also be openly exposed in those species where the mantle cavity has regressed, as for instance in order Notaspidea. In nudibranchs of order Nudibranchia there is an outer circle of gills located on the posterior part of the body.

The head is normally not distinctly offset from the body. There is a tactile organ at the mouth used for orientation. Located behind the tactile organ are the rhinophores, complex sensory organs for the perception of currents and their directions. The foot has a "creeping sole", often with attached parapodia (lateral lobes), often modified into swimming organs. In these one can see a general tendency toward regression of the shell. In those species that still have a shell, it is thin, and the species of the most primitive groups are still able to withdraw into the shell.

Nearly all opisthobranch molluscs live in the sea; only very few are found in freshwater. Their life

A typical bubble shell of genus *Hydatina*, with a large mantle and shell reduced to a "bubble". Photo: M. Cufer

Close-up photograph of the feather-like external gills of the Spanish Dancer, *Hexabranchus sanguineus*. Photo: P. Schupp

Nudibranchs and Bubble Shells are hermaphrodites that act as male as well as female during copulation. These photos depict two *Ceratosoma tribolata* specimens starting copulation, at a depth of 12 m, off Papua-New Guinea. Photos: P. Schupp

Egg masses of the Spanish Dancer, *Hexabranchus sanguineus*. The snail meticulously transfers venomous substances, extracted from sponges, to its eggs to make them unpalatable to predators. Photo: Prof. Dr. P. Wirtz

does not last long, often less than a year and sometimes only one or a couple of months. The life cycle is often directly linked to that of the food they consume (WELLS & BRYCE, 1993). The slugs feed on an available food source, such as a colony of hydroids or a sponge, and when the food is finished the slugs are gone. During the short and dramatic life, the molluscs must reproduce in order to have their genes spread. The reproductive behaviour is varied within the group, but all opistobranchs are hermaphrodites and act as male and female either simultaneously or in sequence. Spermatozoa are generally transferred with a penis and self-fertilisation is usually prevented. Two animals copulate with each other and then act simultaneously as male as well as female. Sea hares of family Aplysidae (page 118) are an exception to this rule. They are known to form animal chains during the breeding season. Each specimen in the chain – except for the first and last ones – act as male for the specimen in front, and as female for the specimen behind (WELLS & BRYCE, 1993). The eggs, which are usually deposited on the substratum (see photo above), hatch to free swimming veliger larvae that settle and grow rapidly to an adult sea slug. This life-cycle has even been observed in the modern coral reef aquarium (see page 114).

Order Cephalaspidiea

This is the most primitive order in subclass Opisthobranchia, but it still contains the largest number of species. Most families in this order include predatory species. They feed on other molluscs, corals or small invertebrates, as well as on single-celled organisms that live buried in the bot-

Hydatina physis (left) and *H. zonata* (right) at Rottnest Island, Western Australia. Photos: C. Bryce

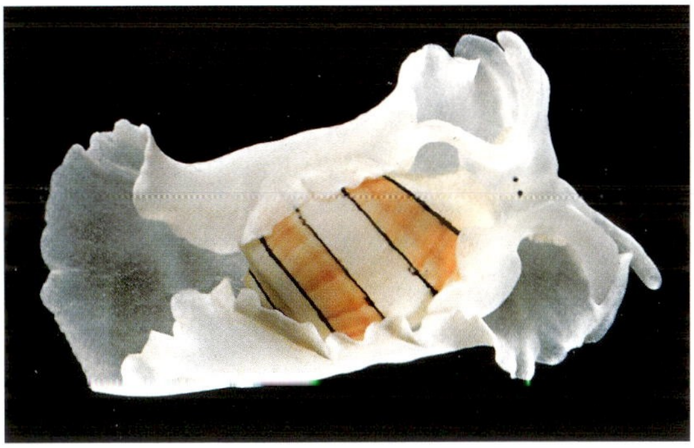

Hydatina amplustris (left; photo: C. Bryce) at Rottnest island, Western Australia, and *Micromelo undulatus* (right; photo: L. Newman and A. Flowers) at Madang, Papua-New Guinea.

tom. This order is structured in to eight superfamilies, which contain a total of 18 families.

Snails of family Acteonidae still have the capability to withdraw in to their shell and then close it with an operculum. In genus *Acteon* the coiled nervous system is only marginally developed. In genus *Bulla* the shell is bubble-shaped, the mouth is just as long as the shell and the external lip protrudes substantially. The species of family Cylindrobullidae have small cylindrical shells. The following families are being discussed in detail below:

- Amplustridae
- Aglajidae
- Haminoeidae
- Bullidae

Family Amplustridae
Bubble Shells

This family contains genus *Hydatina*, where the species have a medium-size, roundish to egg-shaped, thin-walled shell. Mantle

Philinopsis tarouga (left) and *P. lineolata* (right). Photos: C. Bryce

Aglaja orientalis **at Rottnest Island, Western Australia (left; photo: C. Bryce) and *Philinopsis* sp. "KA5-GAS-16" in Papua-New Guinea (right; photo: L. Newman and A. Flowers).**

Chelidonura inornata*, here at Madang (Papua-New Guinea), is seen feeding on a soft coral of genus *Dendronephthya.

Photo: L. Newman and A. Flowers

Two *Chelidonura inornata* specimens, seen here off Guam at a depth of 11 m, while spawning. **Photo: P. Schupp**

and foot usually protrude substantially beyond the shell. *Hydatina* species often live in tide pools. *Hydatina physis* is common throughout the Indo-Pacific. Much less frequent is the closely related *H. amplustris*. Another bubble shell found frequently in tide pools throughout the entire Indo-Pacific is *Micromelo undulatus*, which feeds on bristle worms. For the time being we are unable to advise whether these species are suitable for a coral reef aquarium.

Family Aglajidae
Lobetail Sea Slugs

Most species in this family are rather small, with a thin body equipped with parapodia located along the sides. The parapodia folds over the central body and extend as "tails" at the back of the body, usually one "tail" longer than the other. This is probably the most distinguishing characteristic of this family. The head has a series of sensory bristles at the front and a radula is absent. Lobetail Sea Slugs have a reduced shell, which is completely covered by the parapodia, making the group look similar to nudibranchs. Lobetail Sea Slugs are occasionally available from the aquarium trade, however, basically they are very difficult to maintain in an aquarium because of their mode of feeding.

Genus *Philinopsis* contains many very beautiful species, for instance *P. cyanea, P. lineolata, P.*

Chelidonura amoena (left; photo: L. Newman and A. Flowers), – seen here off Western Australia, – feeding on turbellarian (flat) worms (dark spots on the coral) and *C. hirudina* (right; photo: C. Bryce).

Chelidonura pallida (left) and *C. tsurugensis* (right). Photos: C. Bryce

Chelidonura sp. (left) at a depth of 2 m in the Philippines, and *C. varians* at a depth of 18 m in Papua-New Guinea. Photos: P. Schupp

Haminoea cymbalum **at Heron Island, Great Barrier Reef.** **Photo: L. Newman and A. Flowers**

tarouga and *P.* sp. "KA5-GAS-61", which occur widely throughout the Indo-Pacific. They are of medium size (1 to 7 cm) and are likely to be encountered in the inter-tidal zone as well as on reefs. They move quite rapidly along the substrate in search of food and then bury their prey (mainly other large opisthobranchs) beneath them.

Genus *Chelidonura* contains equally beautiful species. These can be distinguished from those in genus *Philinopsis* by the two posterior lobes, which are longer in *Chelidonura* species. Species of the latter are distributed widely in the Indo-Pacific and the Caribbean region, as well as in subtropical seas off Japan and southeastern Australia (RUDMAN, 1974). They occur inter-tidally as well as on reefs. Their size range extends from less than 1 cm to 7 cm. These slugs release many tiny eggs, which give rise to veliger larvae that undergo a planktonic stage. *Chelidonura* species are active hunters that feed on other invertebrates, e.g. on corals, worms and other molluscs. Some species feed on the tiny, commensal flat-worms from the Order "Acoela", that can multiply to form uncontrollable populations in the coral reef aquarium. *Chelidonura* spp. do therefore have the potential to be introduced to the reef aquarium to control such populations.

According to RUDMAN (1974) genus *Chelidonura* contains 20 described species as well as a large number of species that have not yet been scientifically described.

For a long time, genera *Aglaja* and *Melanochlamys* were considered to be synonyms, but according to RUDMAN (1973) they both are valid genera. *Melanochlamys* species have a cylindrical body. The parapodia are small and normally pressed firmly against the sides of the body. The partially coiled shell is strong and totally calcified. In genus *Aglaja* the parapodia are not pressed quite as firmly against the sides of the body, and the shell (like that *Philinopsis* and *Chelidonura*) is more flattened and only partially calcified. It must also be noted here, that the frequently referred to genus *Navanax* is a more recent synonym of *Aglaja* (RUDMAN, 1974).

Family Haminoeidae
Bubble Shells

Bubble shells are up to 2 cm long molluscs with smooth, whitish shells with an elongated aperture. The animals prefer sandy and soft substratum and are often found in tide pools. There are, according to GOSLINER et al. (1995), at least 65 described species from the Indo-Pacific. The Green Bubble Shell, *Haminoea cymbalum*, is the best known species and one of only a few that is brightly coloured. This tiny, but most beautiful species occurs widely throughout the Indo-Pacific, where it often occurs in the tidal zone. The Green Bubble Shell is herbivorous and feeds mainly on small algae. The shell is transparent and very fragile. Regrettably, this snail has not yet been imported for the aquarium hobby, although it would probably be quite suitable for a coral reef aquarium.

Family Bullidae
Bubble Shells

The snails of this family, which encompasses only the genus *Bulla* are again herbivores, which make them potential aquarium inhabitants. The shell is thick and heavy and comes in many sizes. It has a swollen body whorl and a spire sunken into the top of the shell. An operculum is lacking. Some species reproduce rather prolifically at certain times of the year and can occur in huge numbers in tide pools and in shallow water of the inter-tidal zone. The Bubble Shells are nocturnal and hide during the day. The Botany Bay Bubble Shell, *Bulla quoyii* is common off Western Australia, *Bulla ampulla* lives in the Red Sea, while *B. striata* is distributed widely in the Caribbean Sea.

Order Sacoglossa
Sacoglossans

This order contains many opisthobranch molluscs with a wide range of shell structures, although many species lack shells completely. These animals are among the most attractive of all molluscs. But apart from being spectacular they also show a most interesting mode of feeding and fully demonstrate what "food specialists" are all about! The radula is highly specialized with a single row of knife-like teeth that are used – one at a time – to pierce the cell wall of the algae on which the molluscs feed. The used radular teeth are stored in a sac ("the ascus"), which gives name to the order (glossa, Greek = tongue; saccus, Latin = sac). A number of sacoglossans do also keep the chloroplasts from the algae inside their cell, where the plastids for a while continue to carry out photosynthesis providing the slugs with an additional source of energy. This process is known as "kleptoplasty", but more on this later.

A characteristic of sacoglossan snails is the presence of two long rhinophores, seen here is *Thuridilla moebii*. Photo: C. Bryce

Some Sacoglossans are distinctly rare. In fact, in some localities they are already considered to be among those species **threatened with extinction**, which – and that must be unequivocally admitted – is in part due to careless collecting for the aquarium hobby (Clark, 1994; Jensen, pers. comm.). Therefore, we believe that it is most important to present relevant information about the sacoglossans. According to Jensen (1981) there are now more than 200 described species, however, only a few of these have been studied thoroughly.

Many sacoglossans do not have a shell. The body is flattened with large parapodia that among others captures solar energy for the eaten plastids. There are two rhinophores, often cylindrical or grated, on the head. They are important organs for detecting food. The eyes are located laterally on the head behind the rhinophores, sometimes on the top of the head between the rhinophores. The animals can distinguish between light of different intensities and wavelengths, (Jensen, 1994). Two lobes, projecting from the foot, are clearly visible in many species.

As we mentioned a while ago, the feeding habits of sacoglossans are especially interesting and rather unique. A few species feed on the eggs of other molluscs; however, the vast majority feeds on cytoplasm (= cell fluid without the nucleus) of green algae (Jensen, 1994). Apparently structure and chemical composition of the cell wall is the most important factor determining diet within sacoglossans, and very few species include algae with more than one type of cell wall structure in their diet, (Jensen, 1993). Unlike most other snails, sacoglossans do not rasp on the algae, but instead suck cytoplasm out of the cells by the use of their radular teeth. Most species are specialized to feed on certain algae genera, sometimes even on particular algae species, and most are further more stenophagous, meaning that they are completely unable to switch to another source of food. Many species prefer the growth section and/or thallus region that has the highest photosynthetic rate.

Elysia viridis, which is distributed along the coasts of Europe and in the Mediterranean, is in fact one of a very few species that have been observed to have a diet consisting of different species of green algae and a species that can change its selection of food from the fleshy *Codium* algae to the filamentous *Chaetoromorpha* algae (Jensen, 1994).

A rather specific, well-known

***Oxynoe viridis* in its natural habitat (left) and a close-up photo (right).** **Photos: C. Bryce**

and much discussed characteristic of the sacoglossans is their capability to remove functional chloroplasts from their algal diet in the cells of their digestive gland and to process them further. The mechanism that facilitates this capability is not totally known; however, chloroplasts are consumed (phagozytosed) by cells of the digestive gland. After a certain period of time the consumed cells dissolve their membrane so that the chloroplasts are exposed to the cytoplasm of the digestive gland. At that stage the chloroplasts (which at stage have lost their ability to photosynthesise and are only externally reminiscent of true chloroplasts) are called cleptoplastids. Now the snail produces a substance, which stimulates the cleptoplastids to release their previously absorbed photosynthetic products. It is not fully understood yet how important this energy source is for the sacoglossans. The following hypotheses have been proposed:

- Supply with a specific nutrient, that only occurs in *Caulerpa* algae
- Provision of energy for the production of mucus
- Provision of an energy reserve for a period of starvation and/or during the reproductive phase.

Beyond that it is also assumed that this type of energy uptake occurs only in adult specimens (JENSEN, 1993b).

This type of feeding has led to a modification of radula teeth. It appears that the evolution of members of order Sacoglossa is closely tied to the energy release of food consumed (JENSEN, 1993c). The cell wall of algae consumed contains either xylan, mannan or cellulose as structural polysaccharide.

Radula with blade-like teeth in *Elysia ornata*. **SEM-Photo: C. Bryce**

The radula teeth of sacoglossans are divided in to three groups:

1. tricuspidate (three-pointed)
2. sandal-shaped
3. blade-shaped

Tricuspidate teeth occur in those species that feed on calcareous- or *Caulerpa* algae. Those are algae that contain xylan in their cell walls. Sandal-shaped teeth are found in species that feed on algae belonging to the orders of Siphonocladales (see volume 1, page 285) and Cladophorales (see volume 1, page 302). These algae contain cellulose in their cell walls. Blade-shaped teeth are found in many species that have a variable algal diet. A radula with blade-shaped teeth occurs, for instance, in *Elysia ornata*, one of the most commonly found snails of family Elysiidae and also the most frequently imported sacoglossa, which feeds exclusively on *Bryopsis*, (see Fact Sheet on page 116).

A brief overview of the biology of order Sacoglossa is certainly an excellent demonstration as to why certain marine animals are hard to keep in an aquarium. Without the knowledge of their feeding behaviour and the (respective) morphological adaptation to specific types of food, we have no way of understanding why some invertebrates will not accept certain foods under aquarium conditions. In order to enable Sacoglossa to survive in an aquarium, the correct type of food must be available to these animals. The summarized table on page 112 provides an overview of the algal diet of some sacoglossan species.

According to the latest taxonomic stand, order Sacoglossa contains 12 families, which are placed in three superfamilies. The following families are discussed below:

- Oxynoidae
- Volvatellidae
- Juliidae
- Elysiidae
- Calliphyllidae

Thuridilla sp. "KA5-GAS-19" in Coral Bay, Western Australia.
Photo: L. Newman and A. Flowers

Volvatella sp. "KA5-GAS-17".

Volvatella sp. "KA5-GAS-18".

Berthelinia cf. *rottnesti*.
Three photos: C. Bryce

Family Oxynoidae

Most species in this family are small and well camouflaged by the parapodia, and are therefore hard to spot. Oxynoids still have a small, thin and transparent shell. *Oxynoe viridis* is a delicate species from the Indo-Pacific, while *O. antillarum* inhabits calm bays and harbours in the Caribbean region, but there – like so many Sacoglossa – it is rarely seen. The feeding behaviour of *Oxynoe* species is rather complicated. From an evolutionary perspective it is essentially a dead end. With that, the theory of family Elysiidae having evolved from family Oxynoidea, is not sustainable (JENSEN, 1981). Another genus of this family is *Lobiger*. *L. souverbiei* has a cosmopolitan distribution in tropical seas, and it feeds on *Caulerpa racemosa* (HUMANN, 1994). *L. sagamiensis* also feeds on *Caulerpa* species and lives in the Indo-Pacific (JENSEN, 1993a).

Elysia bennettae at Heron Island, Great Barrier Reef.
Photo: L. Newman and A. Flowers

Algal diet of some sacoglossan species			
Family / Species	**Algae**	**Observed at**	**References**
Volvatellidae			
Ascobulla fischeri	*Caulerpa*	Western Australia	JENSEN (1993a)
Ascobulla ulla	*Caulerpa*	Florida	"
Volvatella bermudae	*Caulerpa*	Bermudas	"
Volvatella ventricosa	*Caulerpa*	Western Australia	"
Juliidae			
Berthelinia caribbea	*Caulerpa*	Florida	"
Berthelinia n. sp.	*Caulerpa*	Western Australia	"
Oxynoidae			
Oxynoe antillarum	*Caulerpa*	U.S. Virgin Islands	"
Oxynoe azuropunctata	*Caulerpa*	Florida	"
Oxynoe viridis	*Caulerpa*	Western Australia	"
Lobiger souverbiei	*Caulerpa racemosa*	Florida	JENSEN (1993a) HUMANN (1994)
Calliphyllidae			
Cyerce antillensis	*Penicillus, Udotea, Halimeda*	U.S. Virgin Islands Florida	"
Cyerce nigricans	*Chlorodesmis fastigiata*	Indo-Pacific	several references
Mourgona germaineae	*Cymopodia*	Florida	JENSEN (1993a)
Caiphylla mediterranea	Bryopsis	Florida	"
Elysiidae			
Elysia australis	*Enteromorpha, Cladophora, Rhizoclonium*	Western Australia	"
Elysia bennettae	*Chlorodesmis*	-	"
Elysia canguzua	*Codium*	Florida	"
Elysia chlorotica	*Cladophora, Vaucheria*	Florida	"
Elysia catulus	*Zostera*	Cheakspeake Bay	"
Elysia evelinae	*Biddulphila*	Florida	"
Elysia filicauda	*Acetabularia*	Western Australia	"
Elysia ornata	*Bryopsis*	Florida	"
Elysis papillosa	*Penicillus, Udotea, Halimeda*	Florida	"
Elysia patina	*Penicillus, Halimeda*	U.S. Virgin Islands	"
Elysia serca	*Halophila, Thalassia, Halodule*	Florida	"
Elysia subornata	*Caulerpa*	Florida	"
Elysia tuca	*Halimeda, Caulerpa*	Florida	"
Elysia crispata	*Caulerpa, Halimeda, Sargassum**	Western Australia	JENSEN (1993a), own observation see also page 109
Pattyclaya brycei	*Caulerpa*	Western Australia	JENSEN (1993a)
Elysiella pusilla	*Halimeda*	Phuket, Guam, Western Australia	"

Elysia thompsoni (left) and *E. expansa* from Western Australia. Photos: C. Bryce

Family Volvatellidae

Members of this family have a large shell into which the animals can withdraw completely. These sacoglossans are less than 1 cm long and are normally found among green algae from the genus *Caulerpa*. *Volvatella* is a frequently encountered genus of family Volvatellidae. Its species have a tiny, pear-shaped shell.

Family Juliidae
Bivalved gastropods

Opisthobranchs from this family have been known from fossil records for very long, but live specimens were not discovered until the late 1950s (Wells & Bryce, 1993). Species belonging to family Juliidae also carry a small shell with two wing-like epipoda, which was the reason why they were formerly referred to as bivalves. *Julia* and *Berthyelinia* are two genera in this family.

Family Elysiidae

This family is the largest among the sacoglossans and contains many beautiful species. According to Wells & Bryce (1993), at least 80 species from the Indo-West Pacific are known. Gosliner (1995) published a revision based on a phylogenetic analysis of the family and includes the former genera *Elysiella, Plattyclaya, Tridachia* and *Tridachiella* in *Elysia*,

Elysia* sp. "KA5-GAS-19" at Heron Island, Great Barrier Reef. This may be a colour variety of *E. bennettae.

Photo: L. Newman and A. Flowers

Two *Elysia diomedea* specimens in the Gulf of California.

Photo: S. W. Michael

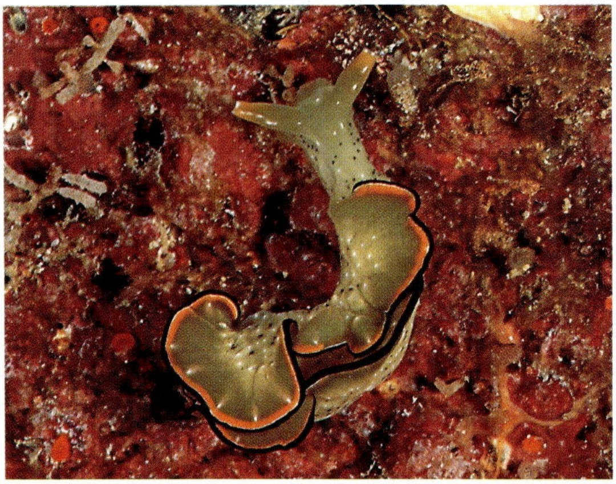

***Elysia crispata*, Cauliflower Sea Slug**
(Formerly known as *"Tridachia crispata"*)

Distribution: Caribbean Sea, including Florida and the Bahamas, southward to Venezuela.
Size: Up to 10 cm, however, usually smaller (3 to 5 cm).
Reproduction: Hermaphrodite with internal fertilization. The egg masses are deposited on a substrate. Successful reproduction in the reef aquarium has occurred. Mr. Jeff Zimmerman (Omaha, Nebraska, USA) was able to raise several broods of slugs in his reef aquarium. The aquarium had a volume of 90 gallons (340 litres) and was filtered over an oversized protein skimmer. The redox potential was 400-420mV, the temperature was 75-78 °F (app. 24-26 °C) and the internal water movement was 1350 gallon/hour (= 5100 l/h). *Caulerpa* and other algae had been used as an additional "algae filter". The aquarist started out with 4 adult slugs, which reproduced to a batch containing about 50 juvenile slugs. The juveniles grew at a rate equal to about 1/8 inch a month, and continue to reproduce, so the aquarium contained several generations of sea slugs. Many clear egg strands a couple of weeks before the juvenile slugs appeared always signalled the reproduction. The slugs seemed to reproduce sexually app. every 60 days and a typical life span of the animals was 90-120 days. See notes on feeding behaviour observed in the batch below.
Food: Feeds on green algae of genera *Caulerpa* and *Halimeda* (JENSEN, 1993a), as well as on *Sargassum* (pers. observation) and green filamentous algae (ZIMMERMAN, pers. comm.). Chloroplasts taken in are structured in to the body, and are responsible for variable colour patterns. Those cells where cleptoplastids have been included produce carbon compounds through photosynthesis. These compounds are used e.g. for the production of mucus, secreted by a foot gland (COLIN, 1978).

Jeff Zimmerman tells the following story about feeding observations on this tank-raised *Elysia crispata*-batch: "Feeding is quite easy, because they eat the green filamentous algae that tends to always grow on the sides and back of my tank. If I completely remove that algae, I tend to start losing the slugs. They are always cruising around looking for food, and will attack any algae that they happen to run across. After my lights have been on for 6-8 hours, they have a tendency to go up the side of the tank to get close

Photos:
Top: *Elysia crispata* on a sponge together with encrusting anemones. Photo: O. Gremblewski-Strate.
Centre: Specimen with unusually coloured body. Photo: S. W. Michael
Bottom: Specimen in an experimental aquarium.

to the lights, and then they "bloom" with the ruffles on their backs becoming much more pronounced. Typically a slug will stretch to be 3/4 of an inch long and 1/8 of an inch wide while they are eating, but will form an almost round shape (3/8 of an inch) while they are catching rays. It is quite amazing, I can tell when they're in "solar powered" mode when they crunch together to form a roundish ruffled shape".

Description: Lives principally in sea grass beds, but occurs also in reef habitats; normally not found deeper than 20 m. The edge of the mantel is frilled in a manner reminiscent of a cauliflower. The rhinophores are rather large, and a shell is absent. Rather variable body colouration, depending on particular algal diet.

Aquarium requirements: Basically rather delicate, but with a substantial longevity, provided correct conditions prevail. We have kept this species for many months in an aquarium by feeding it on a *Sargassum* species. During this period the body colouration changed from green to blue. Easy to keep if provided with enough algae for food. Not suitable for a coral-community tank, but should rather be given a small algae-tank especially designed for this beautiful slug.

GA: +/-; **SE:** +/-; **CO:** +

Photos:
Top: Close up of an adult *Elysia crispata* in the early light cycle.
Centre: Close up of an adult slug belly, this time later in the lighting cycle. The slug is no longer moving around, looking for algae. Instead it is in an almost round configuration in "solar mode." We are guessing that this position gives the inner part of the body (between the ruffles) the best shot at getting light. They stay in this position until the next lighting cycle, when they start cruising for algae again.
Bottom: Adult (right) and juvenile (left) *Elysia crispata* in the aquarium of Jeff Zimmerman.

Three photos: J. Zimmerman

***Elysia ornata*, Green Sea Slug**

Distribution: Probably with a circum-tropical distribution. Has been shown to occur in the Caribbean Sea, around Florida, Bermuda, Canary Islands, Red Sea, off Zanzibar, South Africa, Viet Nam, Indonesia, Japan, Guam, Marshall Islands, Hawaii, Huashine, New Caledonia, Fiji Islands and Australia (JENSEN, pers.comm.).
Size: Up to 5 cm.
Reproduction: Hermaphrodite, with internal fertilization. The egg masses are deposited on a substrate. In Atlantic specimens the egg masses are in spiral bundles. According to JENSEN (1992), spawning specimens have not yet been observed in the Indo-Pacific.
Food: Feeds on green algae of genus *Bryopsis*, especially on *B. pulmosa* according to JENSEN (1981). Radula teeth are blade-shaped (see SEM photo on page 110). The animal sucks for a period on the algal thallus and then – rather strangely – returns the cell fluid to the algae. Following that, suction is resumed but then the cell fluid is then passed on in to the digestive tract. Similar feeding behaviour (where the cell fluid is initially returned to the algae), occurs also in other sea slugs of this family (JENSEN, 1993b).
Description: Shell is absent. There is a large tissue lobe on each side of the dorsal region. This lobe has a black margin, and below it an orange-coloured band that is sometimes delineated from the black margin by a narrow, white line. Both rhinophores usually show the same colouration as the lobe. Many white and black spots are scattered over the body. Many *Elysia* with identically coloured dorsal lobes have been described, e.g. *E. marginata* which has caused confusion in regard to the valid name and synonyms. *E. marginata* must be considered as a synonym of *E. ornata* (JENSEN, 1992).
Aquarium requirements: It is essential that *Bryopsis* algae are available, and then the aquarium maintenance of this species is presumable easy (JENSEN, pers. comm.).

GA: +/-; **SE:** +/-; **CO:** +

Photos:
Top: *Elysia ornata* at Coral Bay, Western Australia.
Photo: L. Newman and A. Flowers
Centre: *Elysia ornata* in a display tank at Zoo-Welke, Lünen.
Bottom: *Elysia ornata* in our experimental aquarium.

which together with *Bosellia* and *Thuridilla* constitute three common genera in the family.

Elysiids lack shells, they have a thin slender foot and large parapodia that are leaf-like. *Elysia* spp. are very often green as the animals contains chloroplasts in their tissue, and many live in close associations with green algae from the genus *Caulerpa*. Well known to aquarists is *Elysia ornata*, which is frequently imported and distributed in the Indo-West Pacific and the Atlantic, although it is never common in the nature. The species is described in detail in the Fact Sheet page 116. Particularly impressive is *Elysia picta*, which inhabits the Caribbean Sea. It has coloured bands in shiny blue, green, red and yellow, and reaches a maximum length of 2.5 cm. Examples

Juvenile (top) and adult *Cyerce nigricans* specimen feeding on algae of genus *Chlorodesmis* (below), at Heron Island, Great Barrier Reef

Photos: L. Newman and A. Flowers

of other species from the Indo-Pacific are *E. bennetae*, *E. tomentosa*, *E. flava* and *E. pusilla*.

Elysia crispata (formerly known as *Tridachia crispata*) is distributed in the Caribbean region, where it lives in algae belts and on algae-covered rocks. Algae, and in particular *Caulerpa sertularoides* are its main food. Chlorophyll, removed from algae that have been consumed, is then deposited within the dorsal cells, where it produces nutrients for the snail via photosynthesis (TAYLOR, 1971), and is therefore most interesting from a biological perspective. The species is sporadically imported for the aquarium hobby (see Fact Sheet page 114) and has been raised successfully in reef aquariums.

The Mexican Dancer, *Elysia diomedea* (formerly known as *Tridachiella diomedea*) has a strong resemblance to *Elysia crispata*, but it occurs in the Gulf of California (Sea of Cortez) and along the Pacific coast of Central America up to Panama. Over its geographical range it is very often found in the inter-tidal zone, on rocky shores and on reefs, down to a depth of 30 m. Its characteristic features are black-yellow striped rhinophores. It feeds on algae and – like many *Elysia* spp. – deposits chlorophyll in its tissue.

The genus *Thuridilla* contains a number of species that in contrast to the species in *Elysia*, which are mostly greenish, are very colourful often with bright longitudinal stripes. Up to 1995, 13 species of *Thuridilla* were known. GOSLINER (1995) gives an overview of the genus and describes nine new species, which means that the genus now includes 22 known species. Most of the species are distributed in the Indo-Pacific. *Thuridiall carlsoni*, *T. hoffae* and *T. undula* are three common species. One species, *Thuridilla hopei*, is known only from the Mediterranean, while *T. picta* is known from the Caribbean, East Atlantic and the Mediterranean. One must assume that several new species will be described in the future. In contrary to most other elysiids these sacoglossans are commonly found crawling in the open and not always seen associated with the algae upon which they feed. We have no experiences with the genus in captivity, but assume that they are just as difficult to keep as most other sacoglossans.

Family Calliphyllidae

Beyond doubt, the most impressive sacoglossan is *Cyerce nigricans* of family Calliphyllidae. This species inhabits the Indo-Pacific, where it lives in shallow water areas, often also in tide pools. Its maximum size is 4 cm. A shell is absent and the dorsal region is covered by many attractively marked outgrowths that serve as respiratory- and swimming organs. In addition, these outgrowths also accommodate branches of the intestine and the digestive gland that can excrete poisonous substances when the animal is being disturbed. Its food consists of algae, especially the green algae, *Chlorodesmis fastigiata* (see Volume 1, page 292). The algal food consumed contains a poison that accumulates in the snail and can lead to its death and that of other organisms in the aquarium (HAY et al., 1989).

Apart from *Cyerce nigricans*, the genus also contains a number of other species. *C. cristallina* is occasionally found in algae zones of the Caribbean region, but this is rare (HUMANN, 1994). Another resident of the Caribbean is *C. antillensis*. It is commonly found on algae of genus *Penicillus* and sometimes also on *Utodea flabellum* and *Halimeda tuna* (JENSEN, 1981).

Order Anaspidea
Sea Hares

This order encompasses the family Akeridae, which contains only the genus *Akera*, and sea hares of family Aplysidae, with the dominating genus *Aplysia*, as well as the genera *Stylocheilus, Notarchus, Dolabella, Dolaberifera* and *Petalifera*. Sea hares are among the larges of all opisthobranchs. Some species can reach more than 60 cm in length and weigh more than five kilos. They have two large papapodia on a high back and either a substantially reduced shell that is totally covered by the mantel and parapodia, or a shell is completely absent. Characteristic features are two pairs of tentacles; large, fleshy anterior tentacles extending forward and a pair of posterior tentacles (rhinophores) extending backwards.

Stylocheilus longicauda **off Papua-New Guinea.** **Photo: S. W. Michael**

Aplysia parvula (left; photo: L. Newman and A. Flowers) at Heron Island and A. dactylomela (right) among sea grass at Heron Island, Great Barrier Reef.

Bushy, external gills on the hind body, which are typical for Opisthobranchia (order Nudibranchia), are absent. Instead, sea hares have internal gills, which are supplied with water through at dorsal aperture. The foot is very well developed, so that sea hares can slide rapidly along the substrate. The radula is equally well developed and can be used efficiently.

All sea hares are dedicated algae feeders, which makes them ideal for a coral reef aquarium. They feed principally on macroalgae that are cut off in sections with the powerful radula. From feeding on blue-green algae, the sea hares absorb toxins, such as "aplysiatoxin", which protect them against predators. It has been observed that fishes generally avoid eating sea hares. The mantel of sea hares contains several glands, one of which is called "the purple gland" as it secrets aplysiaviolin – a purple dyestuff that camouflage the animals when disturbed. The chemical is not produced by the sea hares themselves, but extracted from their nourishment (DEBELIUS, 1998).

Sea hares are, however, rarely available from the aquarium trade. *Aplysia dactylomela*, which can reach 40 cm in length, is often found among sea grasses in very shallow water. It occurs along all tropical coastlines, and to our knowledge it is only occasionally imported. The Californian Sea Hare, *Aplysia californica* is a giant that can reach a maximum size of 41 cm and weigh several kilograms. It can be found in the Gulf of California (Sea of Cortez), as well as along the Pacific coast of California, on rocks or sand bottom, in the inter-tidal zone. Just like other sea hares. The Californian Sea Hare can also eject a purple-coloured secretion when it is disturbed. It feeds on various green algae.

The reproductive capability of *A. californica* is enormous: it can release about 750,000 eggs per breeding season (KERSTITCH, 1989). This is not particularly unusual for sea hares that can produce more than 180 million eggs during their brief, one-year life span (WELLS & BRYCE, 1993).

Genus *Aplysia* also contains several smaller species, such as *Aplysia parvula* that has a circumtropical distribution, while *Aplysia juliana* is a small species from the Red Sea.

One of the larger sea hare in the Indo-Pacific is *Dolabella auricula* that can grow to a size of 30 cm. It is a common species, distributed from South-Africa to Fiji, and is most often encountered at night in shallow water areas on reefs. It too is able to discharge a purple-coloured liquid.

Stylocheilinus is yet another genus of sea hares containing several tropical species among which *S. longicauda* is one of the most common ones. It has an ex-

A big sea hares on the shallow reef flat of Heron Island, Great Barrier Reef secrets "aplysiaviolin" – a purple dyestuff that camouflage the animals when disturbed.

Pleurobranchus forskali (left; photo: L. Newman and A. Flowers) and *P. grandis* (right; photo: S. W. Michael).

tended distribution across the Indo-Pacific from South Africa to Japan and across the Pacific Ocean to the Sea of Cortex and the Pacific coast of Mexico. The species has a distinct colouration with thin, longitudinal lines and many small, blue spots all over its body.

Order Notaspidea Sidegill Sea Slugs

Side-gilled sea slugs have a gill plume on the right body-side between mantle margin and foot. This character makes them knows as "pleurobranchs". Very little is known about their requirements in a coral reef aquarium, but due to their food preferences one must expect that they are difficult to keep in captivity for a longer period of time. These slugs also possess two pairs of tentacles and an oral veil at the front of the head. The order is subdivided in three families:

- Tylodinidae
- Umbraculidae
- Pleurobranchidae

Family Tylodinoidea

This is the most primitive family in the order and contains species in the two genera *Tylodina* and *Anidolyta*, distributed in tropical and temperate waters. The slugs have an external, umbrella-shaped shell that does not cover the entire animal when the slug is expanded. The molluscs are, however, able to withdraw themselves completely into the shell. The foot is wide and the dorsal side of the body is often covered with warts. Like for most notaspideans, the food consists mostly of sponges.

Family Umbraculidae

This family contains only one species, *Umbraculum cincum*, which feeds on various sponges. It has a small, outer, flattened calcified shell, but is unable to withdraw itself into the shell.

Family Pleurobranchidea

This is the biggest family in Notaspidea, containing eight genera

Left: Umbrella Snail, *Tylodina corticales* (photo: K. Kates). Right: *Berthellina citrina* (photo: M. Cufer), feeding on sponges and stony corals.

that are largely tropical (WELLS & BRYCE, 1993). A shell is either absent or remains only in the form of a rudimentary, internal shell. These slugs are often totally enclosed by their mantle, and their skin usually contains calcareous spicules. They feed mostly on sponges, but some eats tunicates, corals, anemones and even prey on specimens of its own species. *Pleurobranchus, Berthella* and *Berthellina* are the most common genera, but little research has been done on this family. *Pleurobranchus forskali*, which reaches a length of only 4 cm, is probably the best known species in the Indo-Pacific. *P. grandis* is a beautiful red and white species from the Red Sea. *P. areoletus* is distributed in the eastern Pacific Ocean as well as rarely in the Caribbean, and occurs intertidally down to about 30 metres. It is a large species, reaching 15 cm in length. With its warty texture it is easily separated from other members of the family.

Berthellina engeli, which is distributed from California to the Sea of Cortez, Galápagos and Clipperton Island, is uniformly yellow or orange-coloured and has a nearly transparent body. Like many members of the family it can release sulphuric acid when it is being disturbed. *B. engeli* is usually found under flat coral blocks or rocks from shallow water down to 45 metres. The nocturnal *B. citrina* has probably also a circum-tropical distribution, and according to BERTSCH & JOHNSON (1981) it can be frequently found around the Hawaiian Islands. Its colouration is yellow to red. The maximum size range of this species is 2 to 5 cm. It has been observed feeding on sponges and stony corals, such as *Tubastraea* spp., *Porites lobata* and also on *Leptastrea purpurea*.

Pteropods - Sea butterflies

By Dr. Leslie Newman
School of Environmental Science & Management (EnvSM)
Southern Cross University, Lismore, NSW, Australia

The most abundant group of holoplanktonic gastropods (spending their entire life cycle in the plankton) are the pteropods, commonly known as sea butterflies or sea angels. As their name suggests (ptero - winged, pod-footed) these molluscs have successfully adapted to life within the water column as the molluscan foot has evolved into two wing-like projections. They are circumglobal in distribution being found in all oceans and are mainly epipelagic (occurring in the upper 200 m of the water column) although some species are meso- and bathypelagic (VAN DER SPOEL, 1967; 1976; VAN DER SPOEL & HEYMAN, 1983; VAN DER SPOEL et al., 1996).

The group is divided into two orders, the shelled Thecosomata and shell-less Gymnosomata. These two orders are vastly different in every respect with the exception that they possess "wings" that they use for swimming or flotation.

Pteropods have been notoriously difficult to study since they were traditionally collected in plankton nets which damaged their fragile shells or stretched and tangled their body parts beyond recognition. Samples were usually fixed with unbuffered formalin which can cause animals to violently contract and their delicate shells to dissolve completely. Most original descriptions were based on these unrecognisable animals. To further complicate matters, pteropods are also impossible to keep alive in captivity for more than a few days since they cannot behave as they would in situ. Shelled pteropods are known to sink immediately when they encounter a solid object and become entangled within their own mucus (HARBISON & GILMER, 1992). It follows that since gymnosomes feed mainly on thecosomes that they also

Sea Butterflies are holoplanktonic organisms that are sometimes captured by the million in plankton nets. This photo shows a plankton catch at Heron Island, Great Barrier Reef, in October 1989.

Photo: L. Newman and A. Flowers

Shells of
Sea Butterflies of suborder
Euthecosomata

Limacina inflata

Diacavolinia longirostris

Diacria rampali

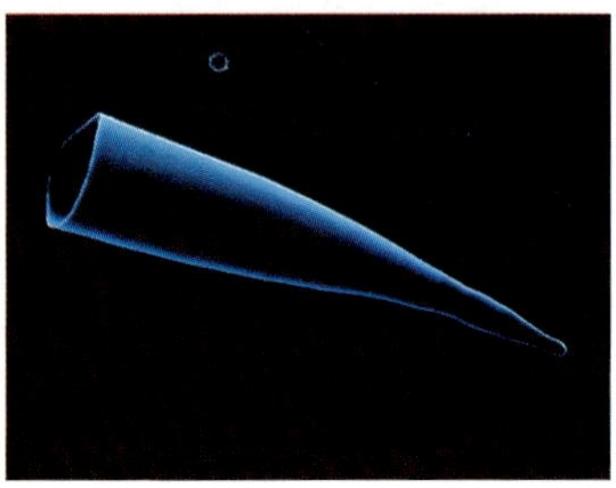

Creseis chierchiae SEM-Photo: L. Newman and A. Flowers

Systematic Overview of the Pteropods

Order Thecosomata
- Suborder Euthecosomata
  - Family Limacinidae
    - *Limacina**
  - Family Cavoliniidae
    - *Diacavolinia*, Cavolinia, Clio, Creseis*, Cuvierina, Diacria*, Hyalocylis*, Styliola*
- Suborder Pseudothecosomata
  - Family Peraclididae
    - *Peraclis**
  - Family Cymbuliidae
    - *Corolla*, Cymbulia*, Gleba*
  - Family Desmopteride
    - *Desmopterus*

Order Gymnosomata
- Suborder Gymnosomata
  - Family Pneumodermatidae
    - *Crucibranchaea, Pneumoderma*, Pneumodermopsis*, Schizobrachium, Spongiobranchaea*
  - Family Notobranchaeidae
    - *Microdonta, Notobranchaea**
  - Family Cliopsidae
    - *Cliopsis, Pruvotella*
  - Family Clionidae
    - *Cephalobrachia, Clione*, Fowlerina, Massya, Paedoclione, Thalassopterus, Thliptodon*
  - Family Hydromylidae
    - *Hydromyles**
  - Family Laginiopsidae
    - *Laginiopsis*

* treated in this manuscript.

would be difficult to keep in aquaria without their food source.

Only in the last 20 years have holoplanktonic molluscs been studied in situ with either SCUBA or by submersibles. As a consequence, invaluable information has been gathered regarding their unique biology.

Order Thecosomata

Thecosomatous pteropods have a thin, fragile, aragonite shell with two "wings" that project from the

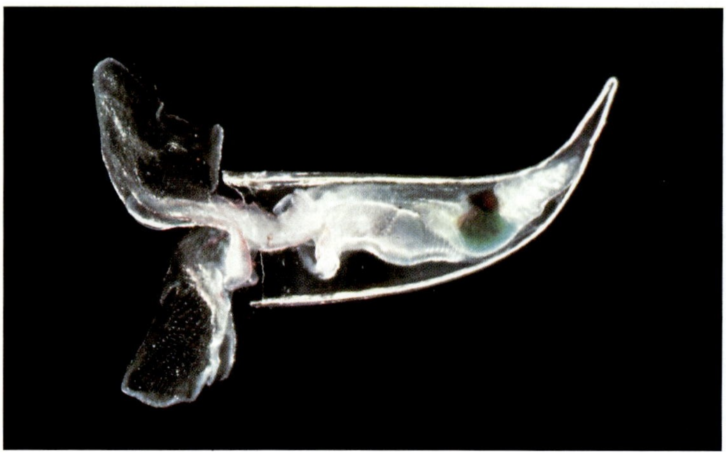

Creseis virgula (suborder Euthecosomata) with expanded wings.
Photo: L. Newman and A. Flowers

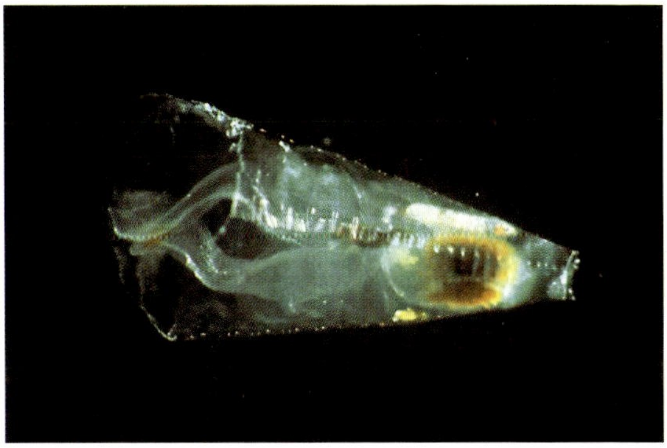

Shelled Sea Butterflies (suborder Euthecosomata): *Diacavolinia longirostris* with expanded wings (left) and *Hyalocylis striata* with retracted wings (right). **Photo: L. Newman and A. Flowers**

aperture. Most species can retract into their shell (suborder Euthecosomata) or they possess a large wing-plate and a transparent slipper or false shell in which the animal sits (suborder Pseudothecosomata). In both suborders, there are animals that are sinistrally coiled (to the left) and those that are bilaterally symmetrical. The shape of the shell, if present, is often bizarre but it serves as an important buoyancy aid.

The thecosomes are divided into two families, the Limacinidae and Cavoliniidae. The limacinids are considered to be the least specialised since the shell and body are sinistrally coiled. The shells are operculate and vary from high spired to having no spire at all in *Limacina inflata*. The shell is extremely thin and fragile and they rarely exceeds 8 mm in shell length.

In contrast to the limacinids, the cavoliniids are the most diverse family of thecosomes. The shell and body are bilaterally symmetrical and they lack an operculum. The shell shape is highly variable being needle-like, conical, vase shaped, triangular or globulose. Many species have spines or projections which serve to increase the surface area as to slow sinking. The shell is also surrounded by intricate transparent mantle appendages (only seen in undisturbed animals) which also aids in flotation by increasing the surface area, arresting sinking and stabilising the animal (GILMER & HARBISON, 1986). Often the protoconch or larval shell has a different shape or it is cast off at metamorphosis from juveniles to adults. Shell sizes range from less than 1 mm to a maximum of 30 mm in length.

The muscular "wings" protrude from the aperture and play an important role in locomotion, buoyancy and food collection. It is a general misconception that since thecosomatous pteropods possess "wings" they are constantly swimming. In situ observations have revealed that they are neutrally buoyant and spend most of their time motionless, suspended from their individual

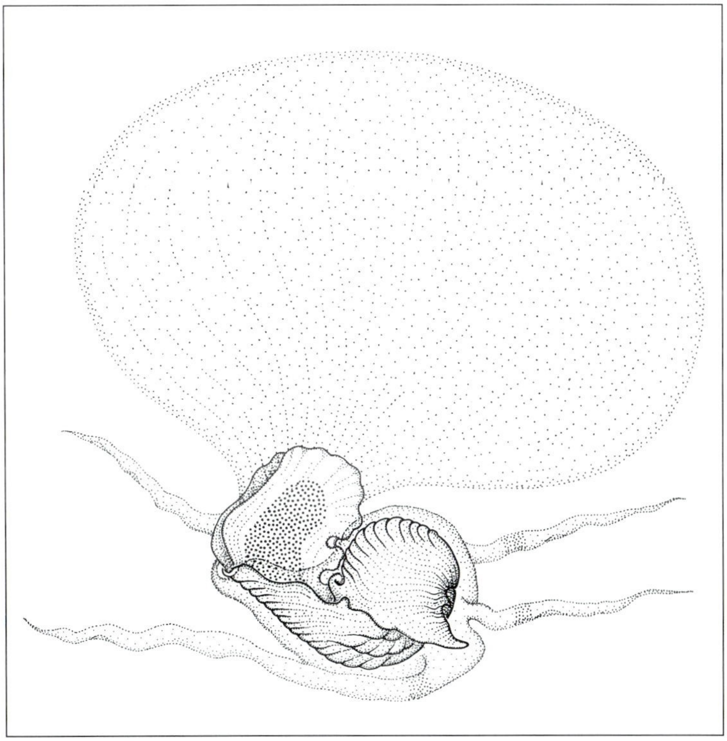

***Diacavolinia longirostris* suspended from its mucus collecting net and with extended, tentacular appendages of the mantle.**

Drawing: NEWMAN (1998b)

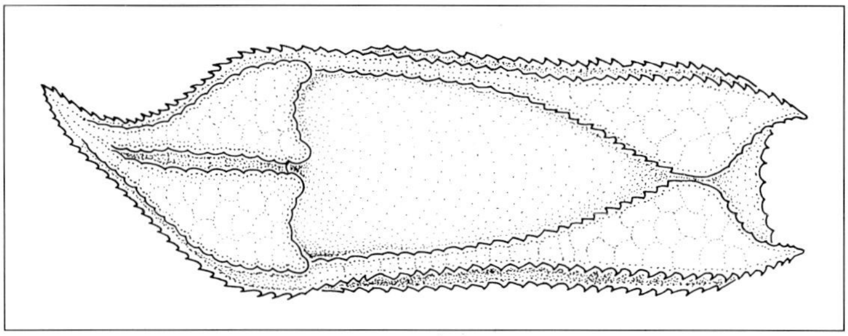

Transparent "false shell" (pseudoconch) of *Cymbulia sibogae.*
Drawing: NEWMAN (1998b)

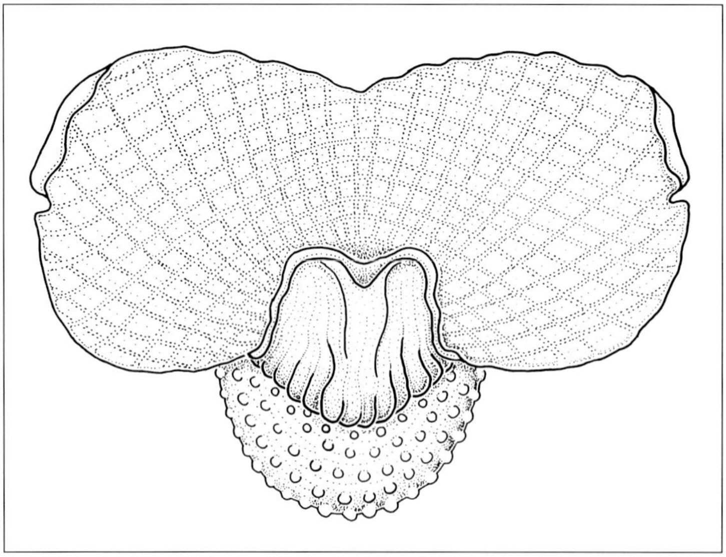

***Corolla ovata* with its large, muscular wing plate.**
Drawing: NEWMAN (1998b)

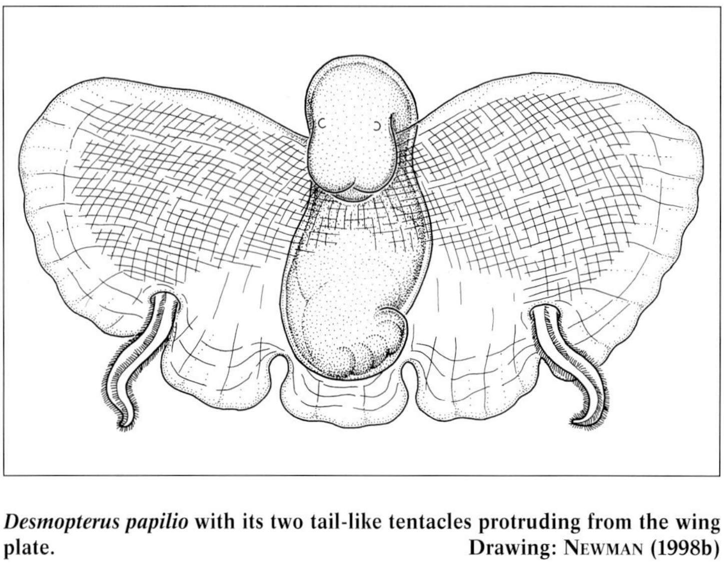

***Desmopterus papilio* with its two tail-like tentacles protruding from the wing plate.** **Drawing: NEWMAN (1998b)**

mucous feeding webs (GILMER & HARBISON, 1986; LALLI & GILMER, 1989). However, if the animals are disturbed, they detach from their web and use their wings to rapidly swim away.

The pseudothecosomes are vastly different from the euthecosomes and have evolved a much more efficient means to float within the water column. They are generally considered meso- and bathypelagic and many species are poorly know. The body is elongated and the "wings" are fused to form a large muscular wing-plate which increases the animal's surface area. All these molluscs possess an elongate proboscis which is formed from the footlobes and is used for food collection.

There are three families within the suborder Pseudothecosomata, the shelled Peraclididae, the "false-shelled" Cymbuliidae and the shell-less Desmopteridae. Peraclidids are thought to be the least specialised since they still possess a sinistral shell and are generally small, being less than 8 mm in shell length. Although their shell is similar to limacinids, the wings are typically fused into a wingplate and they have a proboscis. Some species dwell in deeper waters and are rarely encountered.

Further specialisation to a planktonic lifestyle is seen in the Cymbuliidae which are the largest of the pseudothecosomes. The wingplate is extremely large and the animals are highly transparent and elongate. They lack a calcareous shell and have instead a lightweight gelatinous "false shell" or pseudoconch composed of conchiolin, a cartilage-like material. The pseudoconch is not homologous to the shell in the euthecosomes but has evolved independently in this suborder. The pseudoconch is internal, encasing the viscera and varies from oval to triangular in shape. It is only loosely attached and many animals are collected without their pseudoconch attached. Cymbuliids are relatively large with a pseudoconch length of 80 mm.

One aberrant family, the Desmopteridae lacks a shell, pseudoconch or mantle cavity altogether. They have a large wingplate with two long trailing tentacles and a small body. LALLI & GILMER (1989) believe that they are in fact more closely related to the Gymnosomata.

Desmopterids are the smallest pseudothecosomes, only reaching 5 mm in diameter and remain the least known of the shelled pteropods since many species are believed to dwell in deeper waters.

It was not until observations were made in situ that feeding in pteropods was better understood. Most, if not all, thecosomes employ an external mucous feeding web many times larger in size that the animal itself. It can be deployed in a matter of seconds and is produced by the mantle lining and pallial gland on the wings. These molluscs catch small phytoplankton, zooplankton and detritus which are brought into the mouth by the lips that surrounding the mouth or, as in the pseudothecosomes, by the mobile proboscis. Prey are sorted before reaching the buccal cavity where there is a small radula and jaws. The gizzard is also used to grind up shelled prey such as diatoms. However, it is unclear if these molluscs are essentially omnivorous trappers or true suspension feeders (GILMER & HARBISON, 1986; HARBISON & GILMER, 1992).

Shelled pteropods are not well adapted to escape predation since all they can do is detach from their web and swim away. However, the pseudothecosomes use transparency to disguise themselves in the water column and are generally meso- or bathypelagic in nature. Thecosome predators include many species of commercially important fish such as herring and tuna, naked pteropods, seals, whales, and sea birds.

All pteropods are protandric hermaphrodites, developing first into males, then functional hermaphrodites and finally females. Copulation is not well understood, but many species employ a reciprocal transfer of spermatophores which the partner attaches to the aperture of its mate (LALLI & GILMER, 1989). The sperm are stored until the animals develops into its female stage. Free floating eggmasses are laid in the water column with hundreds or thousands of embryos and bilobed veliger larvae hatch within a few days. Several bathypelagic species are known to brood a small number of young in a modified mantle cavity.

The shelled pteropods are extremely important components of the oceanic zooplankton. Some species can become so abundant as to form dense swarms and it is not known whether these are for feeding and/or reproduction. *Diacavolinia longirostris* occasionally swarms off the north Great Barrier Reef of Australia and millions of shells are often washed up onto beaches (NEWMAN, 1987). Shelled pteropods are so abundant at times that they can directly influence the ecology of epipelagic communities since they are a seasonal food source for many fish species, as well as, whales in polar seas. Abandoned feeding webs are a major source of marine snow which is a primary protein source for small grazers such as copepods (ALLDREDGE, 1972).

The empty shells of thecosomes are also important to the regulation of the carbonate system in the world's oceans (BERGER, 1978; BETZER et al., 1984). The shells are composed of aragonite, a form of calcium carbonate which is more soluble than calcite. Pteropod shells dissolve while sinking. Dissolution of these shells contributes to the recycling of the carbonate system of the world's oceans. Sometimes pteropod shells reach the ocean floor and form deep-sea aragonite sediments or "pteropod oozes" formed from limacinids, creseids and the shelled heteropods (CHEN, 1964).

There are over 60 species of thecosomes in the world's oceans (VAN DER SPOEL, 1967; 1976;

Creseis chierchiae **(family Cavoliniidae with a spermatophore on the shell opening.**
Photo: L. Newman and A. Flowers

Shells of *Diacavolinia longirostris* on a beach at Lizard Island, Great Barrier Reef. **Photo: L. Newman and A. Flowers**

VAN DER SPOEL et al., 1996). They are most diverse in tropical and subtropical waters and can be extremely abundant in polar seas. Many species also exhibit diel vertical migration of up to 100 to 200 metres in a day. According to LALLI & GILMER (1989) limacinids are the most abundant mesogastropods in the world due to their capacity to swarm in vast numbers. Species identifications are based on shell surface sculpture (usually revealed by scanning electron microscopy) and shell morphology.

Order Gymnosomata

The shell-less or naked Gymnosomata are streamlined, active and swim by flapping the two "wings" attached to the lateral sides of their neck. Gymnosomes are commonly known as "sea-angels" due to their slender body shape and synchronised swimming motion. These molluscs are bilaterally symmetrical and the shell, mantle, and mantle cavity are lost in the adult stage. The body in divided into two regions, a head and tapering trunk.

The order is separated into seven families that have evolved different feeding strategies. The most diverse family is the Pneumodermatidae which have adapted to feed specifically on euthecosomes. All pneumodermatids possess acetabuliferous or suckered arms and are generally found in warmer seas. They are the only other group of molluscs besides the cephalopods to possess suckers on the buccal appendages. In contrast, the family Clionidae possesses one of the most common species found in all cold oceans, *Clione limacina*. This species is so common as to be known as "whale food". Clionids have adhesive buccal tentacles rather than suckered arms.

One aberrant family, Hydromyidae, do not resemble other gymnosomes. Instead, the body is rounded-oval and is surrounded by a tough, external integument similar to an external "shell". The head and wings can be withdrawn into an anterior opening and the animal resembles a fish egg when contracted. The remaining families have not been adequately studied and many animals are only known from preserved specimens. According to LALLI & GILMER (1989) gymnosomes remain the most poorly understood group of molluscs.

Uniquely, in gymnosomes, the molluscan foot is modified into two or three footlobes situated between the wings. The wings are highly muscular and swimming involves a synchronised sculling action. Rapid speeds can be

Live, shell-less Sea Butterflies

Pneumoderma heronensis. **Photo: L. Newman and A. Flowers**

Notobranchaea macdonaldi. **Photo: L. Newman and A. Flowers**

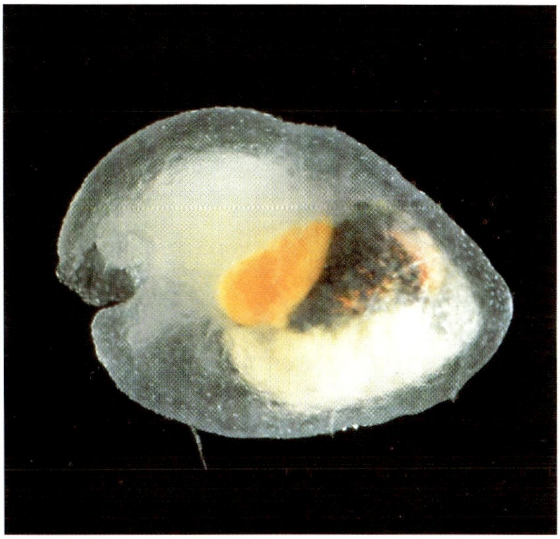

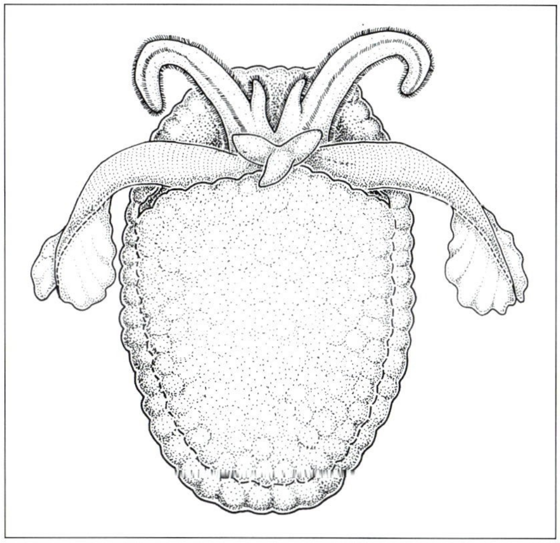

***Hydromyles globulosus*; a live specimen, where wings and mouth appendages are retracted, looks like a fish egg (left; Photo: L. Newman and A. Flowers). Yet, with its wings extended and mouth appendages protruding, it looks much more like a sea butterfly (Drawing: NEWMAN, 1998c).**

achieved to aid them in pursuit of their prey which consists primarily of the slower swimming shelled pteropods.

"Gills", if present, are situated laterally on the right side or at the posterior end of the trunk. The gills are most pronounced in the family Pneumodermatidae where the posterior gill may be elaborately fringed, thereby increasing its surface area. The function of these gills is not fully understood as there is no apparent vascularisation (VAN DER SPOEL, 1976).

Gymnosomes are voracious predators and have extremely complex buccal appendages. The buccal apparatus is usually invaginated and consists of a radula, proboscis, hook sacs and/or buccal cones or acetabuliferous arms which are evaginated during feeding. The buccal arms are employed in grabbing and holding the prey while the hooks are everted and used to pull the prey into the mouth. The proboscis is eversible and the radula, situated distally, is used to macerate prey tissue. The hook sacs are situated on either side of the radula and may have as many as 100 hooks per sac. The feeding efficiency of the gymnosomes is considered to be as effective as that of cephalopods and chaetognaths (MORTON, 1958).

Few pteropods can actively defend themselves against predation. Most gymnosomes are fast swimmers and can probably outswim many predators. However, VAN DER SPOEL (1992) noted that some species use a cloud of ink when disturbed to elude predators, much like a cephalopod. Also, *Clione limacina* has been found to produce a chemical that deters predation by fish. In a bizarre adaptation, an antarctic amphipod has been observed to hold this gymnosome on its back as an antipredatory device (MCCLINTOCK & JANSSEN, 1990; BRYAN et al., 1995).

LALLI & GILMER (1989) believed that these pteropods are simultaneous hermaphrodites and they are not protandric as are the thecosomes. The genital opening is situated on the right side below the wing and during copulation, the mating pairs attach ventrally. Fertilisation is reciprocal as the penis from each animal is placed in the genital aperture of the partner. Eggs are deposited in floating round or oblong egg masses, which are formed with the aid of the footlobes. Veliger larvae possess a bilobed velum and thimble shaped shell and hatch from the egg mass within a few days. The shell is cast off upon metamorphosis into polytrochous larvae. These larvae have three ciliary bands which they use for locomotion until the wings are fully developed.

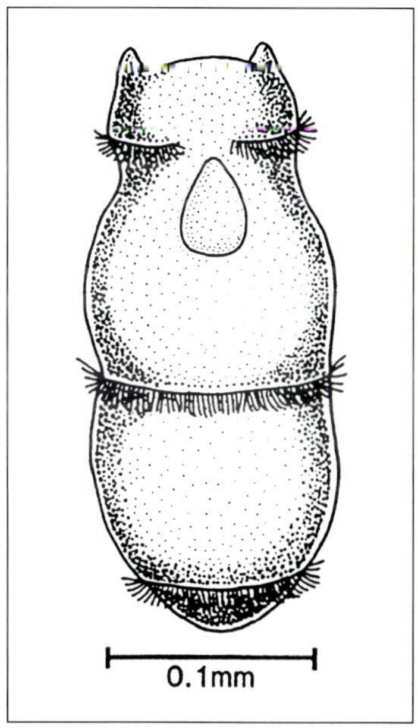

Polytrochus larvae of *Pneumodermopsis spoeli*.
Drawing: NEWMAN (1998c)

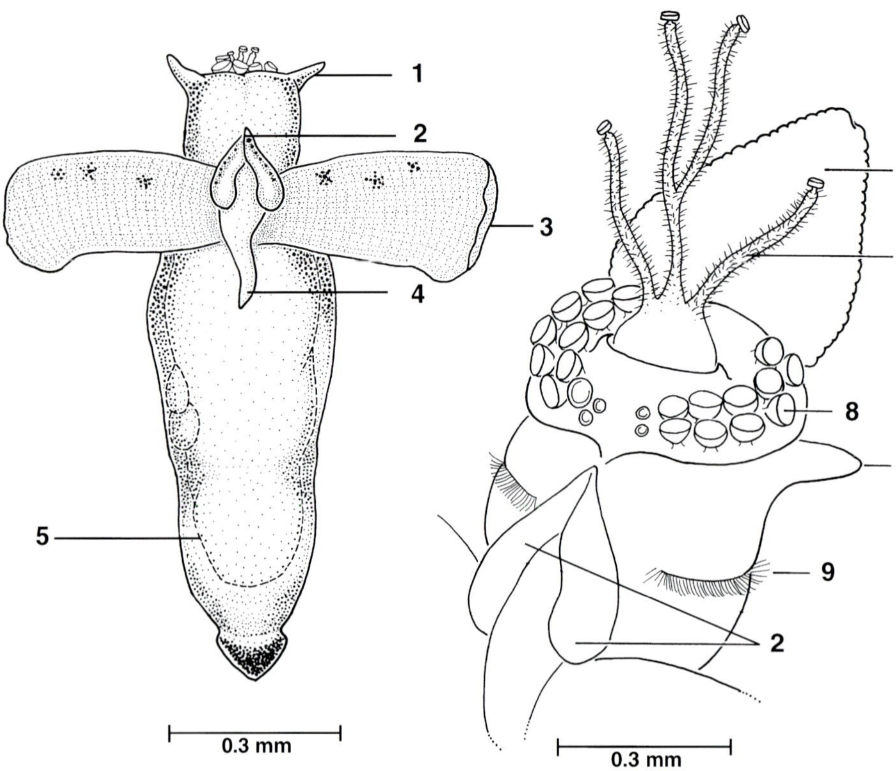

***Pneumodermopsis spoeli*: A - complete animal with its wings extended; B - head with everted mouth appendages. 1 - head tentacles; 2 - lateral foot lobe; 3 - wings; 4 - central foot lobe; 5 - viscera; 6 - proboscis; 7 - central oral suction arm; 8 - lateral oral suction cups; 9 - cilia ring. Drawing: NEWMAN (1998c)**

One species, *Hydromyles globulosa*, is known to brood its young within the mantle cavity and the "mother" eventually bursts open to release its veligers.

There are approximately 50 known species of gymnosomes in the world's oceans. Gymnosomes are solitary by nature compared to the more abundant and sometimes swarming shelled pteropods. The majority of the species are epipelagic and cosmopolitan, being commonly found in tropical and subtropical seas. TESCH (1950) believed that gymnosomes are generally rare, with the exception of *Clione limacina* and *Hydromyles globulosa* which are commonly found in high numbers. *Hydromyles globulosa* is round with an tough outer integument or false "shell" into which the animal can retract. When retracted this gymnosome looks just like a fish egg and may be overlooked in plankton collections.

Although, in general, pteropods are cosmopolitan in distribution and at times extremely abundant, little is known regarding their biology. Species identification is based primarily on the arrangement of the buccal appendages which are usually inverted and need to be carefully dissected out.

Order Nudibranchia True Naked Sea Slugs, Nudibranchs

Naked Sea Slugs of order Nudibranchia – mostly known as "nudibranchs" – are animals that quickly draw the attention of swimmers and divers on coral reefs, because many species have flushing colours, spectacular colour patterns and highly ornamental, artistic shapes. However,

Nembrotha sp. Photos: O. Gremblewski-Strate *Nembrotha* sp.

Nembrotha lineolata. Photo: P. Schupp

Robastra arika. Photo: L. Newman and A. Flowers

Tambja morosa. Photo: O. Gremblewski-Strate

Tambja verconis. Photo: M. Cufer

This *Glossodoris* species is an example of the magnificent colouration and elegant shape of nudibranchs. Photo: S. W. Michael

the bright colours are often a signal to the immediate vicinity: "Caution: I am poisonous!". Some nudibranchs are indeed poisonous, but in other species such conspicuous colouration is absent; instead, they have specific colours that provide camouflage. Some species, such as the Spanish Dancer, *Hexabranchus sanguineus* reach a substantial size, together with a weight in excess of 1 kg. On the other hand, there are those that are microscopically small or live as parasites. In a coral reef aquarium these can sometimes become a nuisance.

As indicated by their colloquial name, a shell is absent in naked Sea Slugs. Instead, they can have calcareous spicules embedded in their skin. Internal gills are absent, but they have developed secondary respiratory organs that are located at the posterior end of the body, either in the form of bushy- or finger-like appendages around the anus or as folds along the sides of the body. The foot is powerful, oblong and without outgrowths.

Naked Sea Slugs or nudibranchs inhabit all seas, from shallow inshore waters down to great depths. A few species are pelagic; as for instance those of family Glaucidae. However, most are tied to a substrate. Occasionally one can see egg clusters appearing underneath a slug, which are almost as decorative as the animals themselves

With their magnificent colouration and exotic appearance, nudibranchs are also attractive to marine aquarists. Uncountable numbers of attempts have been made to keep these animals in coral reef aquariums. With the exception of two species, we do not know of any attempts that have been successful. The reason for such numerous husbandry failures is the specific food requirements of nudibranchs.

Nudibranchs are among the most highly developed food specialists in the Animal Kingdom, and generally little is known about their food requirements. In general nudibranchs are therefore very difficult to keep in captivity. Corals, sponges, moss animals (bryozoans), hydroids and other invertebrates are on the menu of these animals. Often particular species require a species-specific type of food. For instance, one nudibranch may feed on a specific sponge species, and makes no attempt whatsoever to feed on another one, while another species may well feed exclusively on moss animals. This then presents aquarists with so far unsolvable problems.

Nudibranches kept in an aquarium do, just like in the wild, completely depend on their special food sources for survival. If they are not fed correctly in captivity, they will die very soon. We have seen nudibranchs kept in home reef tanks several times, but in nearly every case the animals disappeared within a few days. The conclusion is that nudibranchs – with few exceptions – should not be imported, sold or purchased.

A protective shell is absent in nudibranchs. Instead, many species have calcium carbonate spicules (sclerites) in their skin. Seen here is *Platydoris cruenta* at a depth of 8 m, in the Philippines. Photo: P. Schupp

Some nudibranchs are real camouflage artists. *Notodoris minor* (left; photo: O. Gremblewski-Strate) is barely distinguishable by its colour from the sponge it is feeding on. *Scleodoris rubricunda* (right; photo: L. Newman and A. Flowers) is imitating a sponge.

Suborder Doridoidea

Numerically this suborder is the largest group within order Nudibranchia. There is considerable variability among the numerous species, with a size range from 1 cm to more than 30 cm. The mantle extends beyond the sides of the body and encloses the entire animal. Commonly, there are two chemosensory rhinophores located on the head. The dietary requirements extend from moss animals via sponges, tunicates, and worms to other opisthobranchs. The systematic in this suborder is difficult and much debated. Revisions have lead to the establishment of new families and regrouping of species among families.

The **family Polyceridae** is diverse and contains a large number of tropical species. The body is long and slender with sizes that vary from 1 to 7 cm. Polycerid nudibranches feed on bryozoans, tunicates or other nudibranches. Some species can swim freely over short distances and there are also species that can produce bioluminescent light when being disturbed (WELLS & BRYCE, 1993). *Nembrotha, Roboastra* and *Tambja* are widely distributed genera with several beautifully coloured species.

Family Gymnodorididae contains only about 20 species, and most of these belong to genus *Gymnodoris*. These are carnivorous nudibranchs, feeding on other naked Sea Slugs and sacoglossan molluscs.

The species in **family Notodorididae** (formerly Aegridae) have a tough skin that resembles plastic. They feed on sponges. Some species can reach substantial sizes of 12 cm or more. *Notodoris* is a frequently occurring genus of this family.

Family Hexabranchidae is monotypic with the Spanish Dancer, *Hexabranchus sanguineus* as the only valid species. Others have

Nudibranchs are hermaphrodites, which exchange eggs and sperm with each other. The photo on the left (photo: L. Newman and A. Flowers) shows a "pair" of *Tambja* sp. during copulation, at Heron Island, Great Barrier Reef. The photo on the right (photo: S. W. Michael) shows the intertwined bands of egg masses deposited by *Notodoris gardineri* in the waters off Papua-New Guinea.

Halgerda carlsoni. Photo: L. Newman and A. Flowers

Halgerda terramtuensis. Photo: P. Schupp

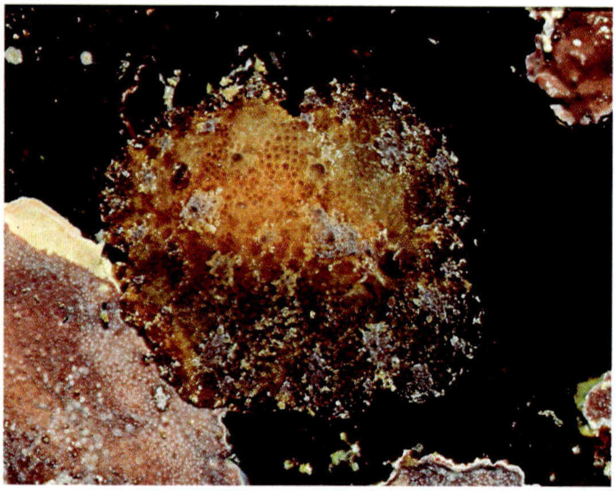

Discodoris sp. Photo: L. Newman and A. Flowers

Jorunna funebris.

Apheolodoris varia.

Neodoris chrysoderma.

Photos: K. Kates

also been described but they are only colour varieties of the Spanish Dancer. *H. sanguineus* is distributed widely in the Indo-Pacific, and according to ERHARDT & MOOSLEITNER (1995) it is also supposed to occur in the Atlantic Ocean. It feeds principally on Sponges, but also on other molluscs, echinoderms and tunicates. Although this species is often available in the aquarium trade, it should really not be purchased. So far feeding the Spanish Dancer under aquarium conditions has not been successful, and inevitably it dies quickly.

Family Chromodoridae is a large family of nudibranches containing more than 300 described species distributed in temperate and tropical waters all over the Indo-Pacific. Chromodorid nudibranches feed on sponges and are often found only on the species on which they feed. Sponges produce many toxins, which often are absorbed by the nudibranches and stored in mantel glands and used to repel predators. *Chromodoris* is the dominating genus in the family. It is common in all tropical seas. Most species reach a maximum length from 1 to 5 cm, but a few may get as large as 12 cm. Their colour patterns are brilliant with stripes or dots in contrasting bright colours as can be seen from the colour plates on pages 135-139. Because of their attractive colouration these nudibranchs are often collected and available in the aquarium trade. Even though they may appear healthy and happy in aquarium shop's display- and sales tanks, they should not be purchased. In an aquarium they need their specific sponge diet, and when this is not available they usually die within a few days. Other genera from this family are *Adeadoris, Cadlina, Cadlinella, Digidentis, Glossodoris, Hypselodoris, Mexichromis, Noumea, Risbecia and Verconia*. Species of these genera are also very attractive and do occasionally occur in the trade. They are, however, just as difficult to keep in captivity as those in *Chromodoris* and should neither be collected nor imported or bought.

Notodoris minor. **Photo: L. Newman and A. Flowers**

Glossodoris sp. **Photo: L. Newman and A. Flowers**

Miramira magnifica. **Photo: M. Cufer**

Family Kentodorididae contains the two genera *Kentodoris* and *Jorunna*. One species, *Jorunna funebris* is relatively common in the trade, but difficult to keep. More information on this species is found in the Fact Sheet page 140. The genus *Halgerda* was formerly included in the family Diodorididae (an old family that has been revised and split into new families), but is now transferred to its own family **Halgerdidae**. The genus contains a number of beautiful, stout species, which feeds on sponges.

Family Dendrodoridae contains the genera *Dendrodoris*,

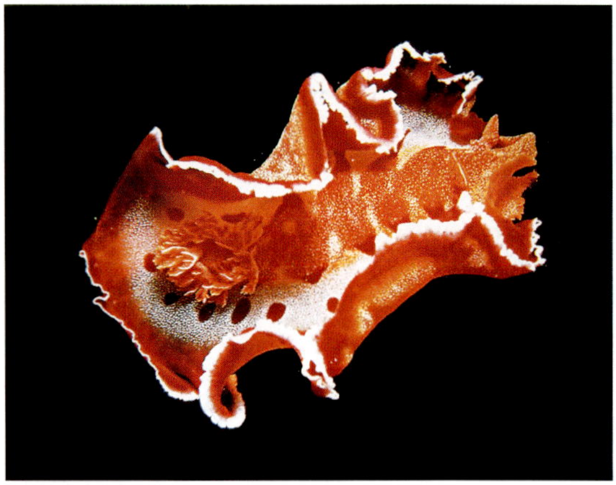

Two colour varieties of the Spanish Dancer, *Hexabranchus sanguineus*. Photo (right): L. Newman and A. Flowers

Glossodoris cruentus (left) and *Reticulida fungia* (right). Photos: L. Newman and A. Flowers

Phyllidiella pustulosus (left) and *Ceratophyllia* sp. (right). Photos: L. Newman and A. Flowers

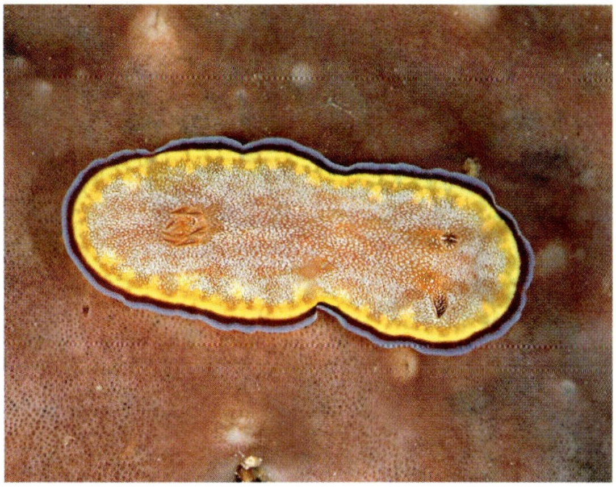

Chromodoris albopuretatus. **Photos: L Newman and A. Flowers** *Chromodoris bullochi.*

Chromodoris cf. *africana.* **Photos: L Newman and A. Flowers** *Chromodoris* cf. *strigata.*

Chromodoris coi. **Photo: L Newman and A. Flowers** *Chromodoris elisabethina.*

Chromodoris kuiteri. Photo: M. Cufer

Chromodoris kuniei. Photo: S. W. Michael

Chromodoris lochi. Photo: S. W. Michael

Two colour morphs of *Chromodoris lubbocki.* Photo: L. Newman and A. Flowers

Chromodoris magnifica. Photo: E. Svensen

Chromodoris quadricolor. Photo: S. W. Michael

Chromodoris tritos.

Chromodoris sp. Photo: L. Newman and A. Flowers

Chromodoris sp. Photo: P. Schupp

Chromodoris splendida. Photo: K. Kates

Chromodoris tinctoria.

Chromodoris westaustraliensis.

Photos: L Newman and A. Flowers

Phyllidia varicosa at the Maldives and at Madang, Papua-New Guinea (photo right: L. Newman and A. Flowers).

Phyllidia babai. Photo: P. Schupp

Phyllidia coelestis. Photo: S. W. Michael

Phyllidia elegans. Photo: O. Gremblewski-Strate

Phyllidia sp. Photo: S. W. Michael

Phyllidia ocellata. Photos: S. W. Michael *Phyllidia* sp.

Phyllidia sp. Photos: P. Schupp *Phyllidia* sp.

Phyllidia sp. Photo: P. Schupp

Phyllidia willani. Photo: L. Newman and A. Flowers

Photo:
Two specimens feeding on *Xestospongia* sp. in an aquarium.

***Jorunna funebris*, White Sea Slug**

Distribution: Widespread in the shallow waters of the Indo-Pacific; from the Red Sea and the Maldives to Australia and Japan. Also found in temperate waters.
Size and growth form: To 10 cm in length, temperate water specimens are smaller.
Reproduction: Unknown.
Feeding: Feeds on sponges. We have observed the slug to attack and feed on the "Hard Blue Sponge", *Xestospongia* sp. ("KA5-POR-16" described in volume 3, page 61, as well as seen in the photo to the left). Other authors such as GOSLINER et al. (1996) reports the slug to feed on the "Soft Blue Sponge", *Haliclona* sp. ("KA5-POR-15", described in volume 3, page 61).
Description: Body white with large black or dark brown-spotted patches. Two black rhinopores anteriorly. Easily recognised and cannot be confused with other species.
Aquarium requirements: Occasionally imported to the aquarium trade. Must be fed with "hard blue sponges" in order to survive. If not provided with such food, the slug will vanish during a few days.

GA: -; **SE:** -; **CO:** +/-

Photo: *Phyllidia varicosa* from Thailand.

Phyllidia varicosa
The species *"Phyllidia arabica"* has now been synonymized with *P. varicosa*; DEBELIUS, 1998).

Distribution: Wide spread in the Indo-Pacific; from the Red Sea and South-Africa to Japan, Australia and Hawaii. Probably the most common and widespread member of the genus.
Size: Up to 12 cm long.
Reproduction: Unknown.
Feeding: Like all phyllidiid nudibranchs, *Phyllidia varicosa* feeds on sponges and extract and accumulate chemicals from the food to use as its own defence. If disturbed the chemicals are released as a milky fluid (see below).
Description: The body is pale blue with longitudinal ridges that possess yellow tubercles. The body is black in between the ridges. The rhinopores are yellow. The underside of the foot has a black line. Can easily be confused with *P. coelestis*, which does, however, has an Y-shaped ridge of yellow-capped blue-grey pustules (DEBELIUS, 1998).
Aquarium requirements: Relatively common in the trade, but very hard to keep due to its food preferences. Must be fed with sponges. Do, however, take great care: The poisonous fluid released when disturbed is a pungent toxin that can ruin an aquarium in a short time!

GA: -; **SE:** -; **CO:** -

Doriopsis and *Doriopsilla*. All genera include only a few species that all feed on sponges. Radula and jaws are absent in these Naked Sea Slugs.

There are about 70 species in **Family Phyllidiidae**, most of which are found in the Indo-Pacific. Some species occur in the Caribbean, in the Mediterranean Sea and at abyssal depths. Genus *Phyllidia* is the dominant genus in this family, and it is distributed widely throughout the Indo-Pacific. *Phyllida varicosa* is a very common species that is often seen in the trade, (see Fact Sheet page 140). Most species reach an average length of 4 to 5 cm, but some grow to 8 cm. They are like many other nudibranchs, highly specialized sponge feeders that are frequently collected and encountered in the aquarium trade. Consequently, they too should not be purchased, because their survival chances in an aquarium are minimal. *P. varicosa*, and probably other species of the genus, produces strong poisons from their nutrition. These toxins are released under stress and can ruin an aquarium completely.

Suborder Aeolidinia

Aeolidinia is the second largest suborder in the order Nudibranchia and contains about 20 families (DEBELIUS, 1998). The species all have a long, slender body with numerous outgrowths protruding from the mantle, referred to as cerata, which are extensions of the digestive gland. Their skin is rather thin and so facilitates gas exchange. Two chemically sensitive rhinophores and two tactile tentacles are located on the head. The length of the various species ranges from microscopically small up to 15 cm. These are very rapidly moving Naked Sea Slugs that feed on cnidarians.

Family Flabellinidae contains many colourful species in the genus *Flabellina*. The species feed on hydroids and are usually seen crawling on the colony upon which they feed. Many species feed only on a distinct species of hydroids, which makes it impossible to offer them food in the reef aquarium. *Flabellina rubrolineata* is just one example of such. This species, which grow to 4 cm in length and is widely distributed in the tropical Indo-Pacific, is usually found on the Hydrozoa *Halocordyle disticha*, which is its preferred food.

Family Glaucidae is mainly known for its pelagic (in open water) living species of genus *Glaucus*. *G. atlanticus* is a circumtropical species that lives up side down on the surface film of the ocean. It maintains abundance by swallowing air bubbles. The species feeds on the floating colonial hydrozoan "Portuguese man-o-war" (*Physalia physalis*) and on other pelagic hydrozoans. The slug can recycle the nematocysts of the hydrozoans and use them for its own defence (DEBELIUS, 1998).

Family Fecelinidae contains slender, beautifully coloured and highly advanced species in the genera *Phidiana, Phyllodesmium, Pteraeolidia*, as well as many others.

A very interresting story is the connection Naked Sea Slugs have to the aquarium hobby,

This *Glaucus* species is an example of a pelagic nudibranch. Photo: M. Cufer

A beautiful *Phidinia* species. Photo: M. Cufer

through the discovery of the glass anemone-feeding Sea Slug, *Berghia verrucicornis* in **family Aeolidiidae**. This is a tiny Naked Sea Slug from the Caribbean region, which – like many other Naked Sea Slugs – accommodates zooxanthella of genus *Symbiodinium*. It takes up the zooxanthellae with its food, which consists exclusively of glass anemone of genus *Aiptasia*. This Sea Slug has the potential to control aquarium populations of this dangerous sea anemone. With some knowledge of the biology of this slug it can be kept and bred in an aquarium and the progeny distributed to other aquarists. For that reason, a rather comprehensive Fact Sheet has been dedicated to glass anemones-feeding, Naked Sea Slug (see pages 143-145), written by Prof. Dr. Stephen Kempf. More comprehensive information is available from CARROL & KEMPF (1990, 1994), KEMPF (1991) and KEMPF & BRITTSAN (1996).

Another species in the same genus, *Berghia australis*, is known from southern Australia and New Zealand and just like *B. verrucicornis* does this species feed on small anemones.

Nudibranchs of suborder Aeolidinia have a slender, elongated body, covered by many cerata (mantle outgrowths), as seen here in *Pteraeolidia ianthina*. Photo: K. Kates

Suborder Dendronotina

This group contains only a few species that are characterized by raised sheaths around the base of the rhinophores, an oral sail and with only small cerata, or none at all. According to WELLS & BRYCE (1993), the oral sail is used to capture small crustaceans, but most species feed on cnidarians. This suborder also contains *Melibe bucephala* which is of interest to the aquarium hobby. It feeds on algae and small crustaceans (see Fact Sheet on page 145). Species of family Tritonidae have two secondary, branched gills located at the edge of the mantle. *Tritonia* species are normally found on gorgonian corals.

Berghia verrucicornis
Aiptasia-eating Nudibranch
(Text and photos by Prof. Dr. Stephen Kempf)

Distribution: The Caribbean, found in sub-tidal, shallow reef flats. *B. verrucicornis* is also reported from the Mediterranean. It is uncertain if these reports really describe the same species as found in the Western Atlantic or if they point to a closely related species.
Size and growth form: Very small and in general hard to see in the natural environment. In the field, large adults usually have a total length (including the oral tentacles) of 15 to 20 mm. In closed aquarium systems, well fed adults may reach 30 to 35 mm in length. Most of the dorsal surface of the body is covered with tufts of cerata 3 to 8 mm in length. The oral tentacles are about 5 to 8 mm long. Newly metamorphosed juveniles are about 0.5 mm in length.
Reproduction: Adults are reciprocal hermaphrodites that are not capable of self fertilization. Egg masses are laid as a white spiral string that contains a twisted thread of egg capsules. Each egg capsule contains one fertilized egg (CARROLL & KEMPF, 1990; KEMPF & BRITTSAN, 1996). See notes on culture below.
Feeding: The larval stage hatches after 9 to 10 days at 25 °C and is non-feeding. Larvae will metamorphose 1 to 3 days after hatching (CARROLL & KEMPF, 1991; KEMPF & BRITTSAN, 1996). Our observations suggest that the juveniles and adults feed only on *Aiptasia* spp. Other species of anemones are ignored. If the *Aiptasia* harbor algal symbionts (i.e. zooxanthellae), the feeding results in these algae being retained in cells lining the ceratal digestive diverticula. Uptake of the algal cells is by phagocytosis, each intracellular symbiont being contained in its own vacuole. Natural populations of the nudibranch always contain a high concentration of algal cells in their cerata. This population of algal symbionts appears to be in a state of constant flux, "older" symbionts being expelled by exocytosis while new ones are phagocytosed as the *Berghia* feeds. As a result, starved *Berghia*, or *Berghia* fed non-symbiotic *Aiptasia*, will lose their symbionts over a 5 to 6 day period as the algal cells are exocytosed and passed out of the animal in its feces. Field collected *Berghia* always have populations of symbionts in their cerata due to their constant feeding on symbiotic *Aiptasia* (KEMPF, 1991).

Photos:
Top: Adult pair of *Berghia verrucicornis*; each individual is 1,5 to 2,0 cm in length and both nudibranchs has fed heavily on *Aiptasia* and showing the brown cerata, full of zooxanthellae.
Bottom: Typical egg-mass shortly after deposit.

Description: Elongated body with many cerata on the dorsal surface. In well fed animals the cerata are dark brown due to the presence of symbiotic zooxanthellae in cells lining the ceratal digestive diverticula. A pair of rhinophores is present just anterior to the foremost cerata and two long oral tentacles extend from the oral veil at the anterior end.

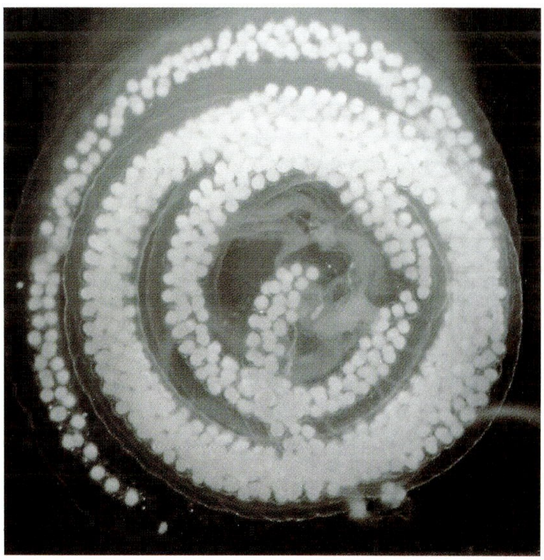

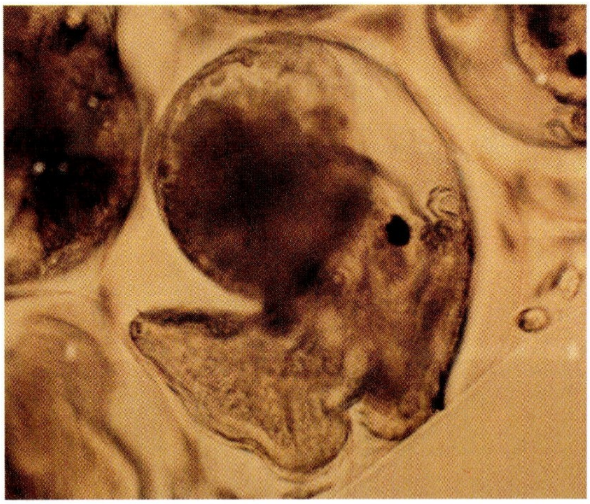

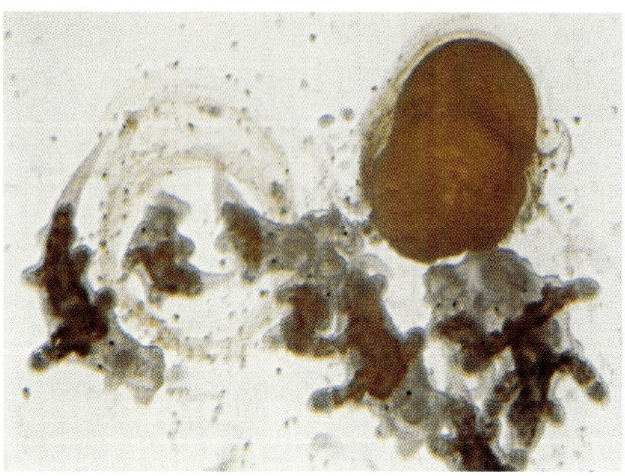

***Berghia verrucicornis* (continued from page 143)**

Aquarium requirements and notes on cultivation: As is well known, species of the anemone *Aiptasia* can be an extremely difficult and tedious problem in the marine aquarium. While injections with acid (5 % HCl) and/or the use of butterfly fish may be used to control populations of *Aiptasia* to some extent, the introduction of *Berghia verrucicornis*, a voracious *Aiptasia* predator, should be seen as a possible new method for biologically controlling outbreaks of this anemone. According to KEMPF & BRITTSAN (1996) the introduction of adult specimens to an aquarium infected with *Aiptasia* will sometimes, but not always, result in the successful establishment of *B. verrucicornis* after a few months. A better approach is to culture the nudibranch. This is a typical task for the local aquarium union where nudibranchs can be spread among the members. KEMPF & BRITTSAN (1996) have described the method of cultivation in detail. We strongly recommend that readers study this paper, as well as the other related papers listed on page 142. Below a brief summary of the method for cultivating *B. verrucicornis* is given.

Culturing *Berghia* is not difficult; however, it is extremely important that new glass bowls and other culture ware be used since small amounts of seemingly innocuous contaminants (e.g. soap, trace element solutions, etc.) may kill the embryos or larvae. All culture ware should be washed only with fresh water (i.e. no soap). Water from an established aquarium that has been filtered through a very fine filter (0.45 micron filtration is preferable) should be used for all cultures. Freshly prepared artificial sea water should not be used since it will not support complete development of the embryos and larvae. Adult *Berghia* are maintained in 300 to 500 ml glass bowls (two animals per bowl). These animals are fed *Aiptasia*, with bowl water and food being changed daily. It is sometimes advantageous to cut up the

sPhotos:

Top: A larvae just before hatching from the egg-mass, about 11 days after oviposition. The larvae are about 250 μm in length.

Centre: A single, juvenile about 7 days after metamorphosis. The brown colour on the back of the animals proofs that it has been feeding on Aiptasia.

Bottom: A group of feeding juveniles. The large yellow-brown structures are the remains of an Aiptasia pallida. 7 to 8 snails are feeding and the dark colour of the cerata again proves that the symbiotic algae are accumulating.

anemones before adding them to the adult cultures. The bowls are kept at 20 to 25 °C. Young adults, if well fed, can be expected to lay one egg mass every 1 to 2 days. Older animals produce eggs more slowly. Egg-masses are laid either attached to the bowl or floating at the water surface. Attached egg masses may be carefully scraped off the bowl with an old plastic credit card. These egg masses are transferred to a culture beaker containing 300 to 500 ml of filtered aquarium water. The egg mass culture should be aerated such that the egg masses are gently lifted off the bottom and circulated. Too much aeration will result in abnormal development of the embryos. The egg mass culture container and water are changed daily. The development of the embryos takes about 10 days at 25 °C. Aeration of the egg mass cultures should be decreased after 7 days such that the egg-masses are no longer lifted off the bottom of the container. At 9 days the egg masses are transferred to a small bowl with about 1 cm of filtered aquarium water and are artificially hatched by pumping them in and out of a short Pasteur pipette. This pumping is continued until the egg capsules are separated from the egg string. Larvae will generally be released from the egg capsules during this process, and those that are not will usually break out of the capsules on their own soon thereafter. The larvae are transferred to a metamorphosis culture containing 300 ml of water that has been previously seeded with very tiny *Aiptasia*. A few flakes of ethyl alcohol sprinkled on the surface of the culture water will prevent the larvae from getting stuck at the water surface. The metamorphosis culture is covered with parafilm and left undisturbed for 5 days. After 5 days the newly metamorphosed juveniles will appear as tiny (~ 0.5 mm) white slugs, many of which will be clustered at the bases of the anemones. These juveniles will soon start to feed on the anemones. Once feeding begins the culture water should be changed every 2 to 3 days. This is generally accomplished by carefully pouring the water out of the bowl and then gently adding new water. New anemones should be added to the culture as old ones are consumed. As the juveniles grow it is advantageous to split the culture into additional bowls with fewer juveniles per bowl. At a size of 4 to 5 mm the juveniles may be introduced into the aquarium.

Melibe bucephala
Two-headed Tethys Slug
(Text and photo: Robert Brons)

Distribution: Indo-Pacific. Rather common in the Red Sea in shallow water (1 to 3 m depth) at protected locations, where there are sea grass beds.
Size: Specimens in the sea usually grow to a size of 10 to 12 cm; however, aquarium specimens may reach a size of up to 18 cm.
Reproduction: Hermaphrodite.
Food: According to VINE (1986) this slug feeds on algae and detritus; however, under aquarium conditions it has been observed that small crustaceans (mainly copepods and amphipods) are eaten.
Description: Greenish or brownish body colour, but transparent. The large, branched cerata give this slug the appearance of sea grass.
Aquarium requirements: Easy to keep in an aquarium with live rocks. Specimens, collected during December and January in the Red Sea, off Eilat (Israel) and transferred to an aquarium, produced many slugs young that grew in to adults within three weeks. The adults died at the end of the breeding season.

GA: +; **SE:** +/-; **CO:** +

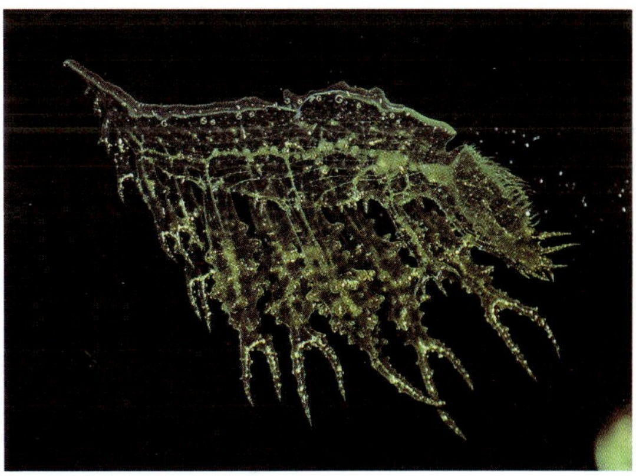

Photo:
Melipe bucephala.

Subclass Pulmonata – Air breathing sea slugs

Most members of this subclass live either on land or in freshwater. Consequently, this group is not of particular interest to the aquarium hobby. Air-breathing sea slugs do not have gills; instead, their mantle cavity has become modified in to a primitive lung. This adaptation enables them to live on land. Today there are more than 20,000 species, distributed throughout subtropical and tropical regions. Individual species, like our local garden snail of genus *Arion* also occur in cooler climates. Those few species that are found in the sea usually occur among rocks, in shallow water of inter-tidal zones. They belong to order Systellommatophora, and some are strongly reminiscent of limpets (order Docoglossa), whilst others have spirally coiled shells and live mainly in rock crevices in the intertidal zone.

Class Scaphopoda Tusk- or Tooth Shells

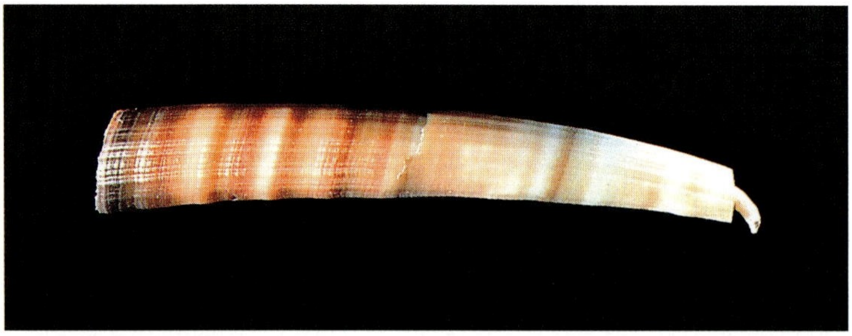

Shell of *Dentalium formosum*. **Photo: C. Bryce**

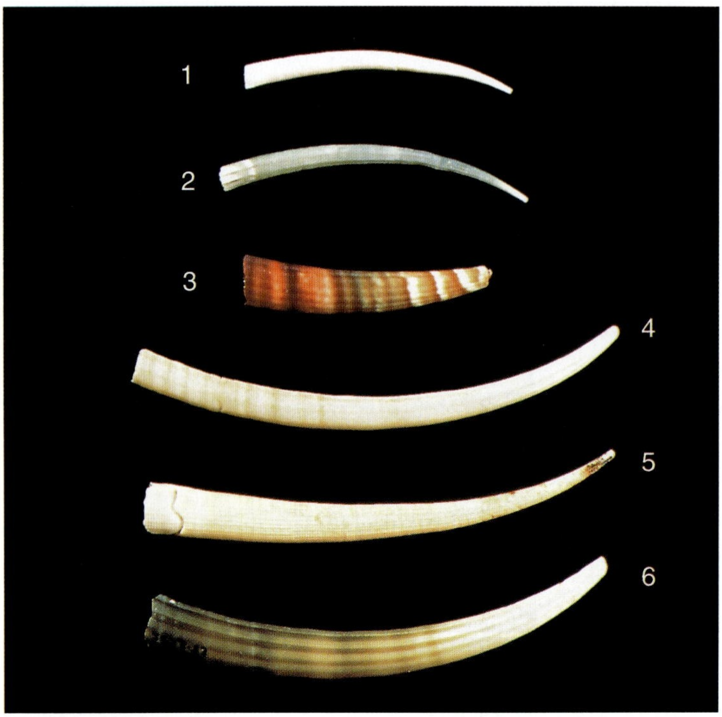

Tusk shells of genus *Dentalium* from the Indo-Pacific: 1 - *D. elephantinum*. 2 - *D. vernedei*. 3 - *D. javanum*. 4 - *D.* cf. *formosum*. 5 - *D. aprinum*. 6 - *Dentalium novaehollandiae*. **Photo: C. Bryce**

Scaphopods – commonly knowns as tusk shells, tooth shells, elephant teeth or sea teeth – are an ancient group of animals, that has remained largely unchanged since it first appeared during the Ordovician Period about 500 million years ago. All of the approximately 350 species are marine and live buried in sand- or mud bottom. The snails are enclosed in a tusk-like shell. Scaphopods have a very simple and primitive body structure. Gills are absent, but the mantle that lines the inside of the shell can takes up oxygen directly from the water circulating inside the shell. The foot, at the buried end of the shell, can create a small hollow space. The head and foot project from the anterior end of the shell into this space where tiny club-shaped tentacles, which are used as sensory organs as well as for food capture, collect tiny food organisms, particularly foraminiferans. The cilindrical foot is used for burrowing and in some species it can form an anchoring disk.

Dentalium dendale is very common along the Norwegian coast. Digging in to the muddy bottom can produce thousands of these animals. Genus *Dentalium* has a worldwide distribution, and several species live on coral reefs. Most of these are smaller than 5 cm long. So far we are not aware that tooth shells have been imported specifically for coral reef aquariums. The main reason for that is probably the fact that they live buried in the substrate, where they are difficult to collect without digging in shallow water or without collecting with special equipment.

Most clams live in shaded areas, however, a few have adapted to high light intensities, as for instance the giant clams of family Tridacnidae. Shown here is *Tridacna maxima* off the Chuuk Islands (Micronesia), at a depth of 4 m.
Photo: P. Schupp

Class Bivalvia - Clams, Mussels, Scallops, and Oysters

Bivalves are well known to most people. Just like the shells of marine gastropods, bivalve shells (or valves) are equally in demand as collectors' items. This class encompasses more than 15,000 species, and so here too we will have to confine our discussion to a very limited number of species. We have decided to focus our attention on those bivalves that are most conspicuous on coral reefs and that have the potential of being introduced in to our aquariums. The systematic review (page 148), compiled after LINDNER (1976) and VINE (1986), lists an overview down to order level.

Bivalves (clams, mussels, scallops, oysters) can be well camouflaged on reefs, such as this Cock's Comb Oyster, *Lopha cristagalli* (left; Photo: S. W. Michael). Many different organisms tend to establish themselves on their valves (shells). The mantle of bivalves are often beautifully coloured with outgrowths and light-sensitive compound eyes (ocelli), as seen here in this *Spondylus* species (right), photographed during a night dive on the Great Barrier Reef.

General Biology

Class Bivalvia

Subclass Palaeotaxodonta*
Subclass Cryptodonta*
Subclass Pteriomorphia
  Order Arcoida
  Order Mytiloida
  Order Pteroidea
Subclass Heterodonta
  Order Veneroida
  Order Myoida
Subclass Anomalodesmata*

*not dealt with here

Mussels, clams, scallops and oysters are laterally flattened and have a left and right half-valve connected to each other by means of a hinge-like ligament. The shape of this ligament is one of several important systematic characteristics in bivalves. The valves can have the most extraordinary shapes, from round or heart-shaped completely symmetrical to oblique, drawn-out to a point or distorted shapes with totally different shell halves. The valves grow in conjunction with an increasing size of the animal inside. The oldest part is usually the pointed end, commonly referred to as umbo. In some instances the valves can be closed with the aid of cardinal teeth, which are small serrations along the edge of the shell near the umbo. In some groups the valves are covered by a tough, corneous skin (periostracum), which consists of a cartilage-like substance (conchiolin). On the inside of the valves there are adductor muscles that can pull the valves together and keep them shut tightly if need be.

Bivalvia is often divided in three major groups separated by the nature of their gills: "Protobranchs", "Lamellibranchs" and "Septibranchs". Earlier these groups were looked upon as three subclasses, but to day the groups do not have any taxonomic status. In the primitive "Protobranchs" the gills consist of two roes of filaments hanging down from the mantel. In "Lamellibranchs" there are also two pairs of gills, which in addition to the exchange of gasses, also collect and sort food particles. In "Septibranchs" the gills are reduced to muscular partitions in a gill chamber. The gill do of course serve as respiratory organs, but do - strangely enough - also tell us a story about the evolution of the bivalve feeding.

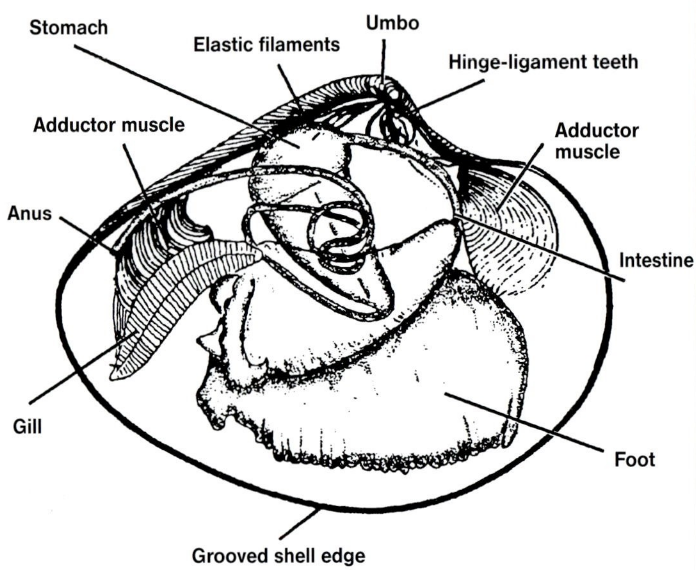

Basic body structure of clams.

Early bivalves from Ordovicium and Cambrium were "Protobranchs" buriers living in soft substrata in shallow waters. Most "Protobranchs" living to day are selective deposit feeders that live buried in the mud and feed by the use of two tentacles, which are elongations of the margins of the mouth. Each tentacle is associated with a large fold composed of two labial palps, one on each side of the mouth. During feeding the tentacles collect deposit material (detritus), which adheres to the mucus-covered surface of the tentacles. The food is transported to the labial palps, which sort the food. "Protobranchs" are most common in deeper waters, but some genera also exist in shallow habitats. At some point some "Protobranchs" developed another, most successful way of feeding, and evolved into the "Lamellibranchs".

"Lamellibranch" bivalves are filter feeders. Their respiratory gills also function as filters and the gill cilia, which originally served to keep the gills clean, in this group also transport the food captured in mucous to the labial palps and mouth. The "Lamellibranchs" and their way of using the gills both for gas-exchange as well as feeding were immensely successful and have undertaken an explosive evolution. To day they are the dominant group of mussels.

The diet consists of small plankton, mostly of phytoplankton. Particles that are too large as food are coated with mucus and released back in to the open water. In the stomach of bivalves there is a peculiar rod-like crystalline style, made up of a protein matrix, which secretes digestive enzymes. A simple heart and vascular system is present, while the head is strongly reduced. Many bivalves are equipped with tentacles located along the edge of the mantle, and they also have eyespots with which they can perceive light.

The foot is usually well developed, and can be extended through the valves toward the out side. It is used primarily for burrowing and locomotion, but many bivalves also use a mucus from a byssus gland at the base of the foot. This mucus reacts with minerals in the water and secretes fila-mentous threads (byssal threads), with which bivalves adhere to the substrate. Those readers in temperate latitudes are no doubt familiar with mussels that use this method of attaching themselves to pilings and floating buoys and there gradually establish massive colonies containing thousands of animals. On the inside of valves there is a layer of mother of pearl, which renders the insides smooth and shiny. When a foreign particle becomes lodged between shell and mantle, it can lead – in some cases – to the formation of a pearl. Such pearls are angular and only of minor value. However, occasionally one can find very beautiful, round pearls that are highly prized as jewellery.

An oyster (possibly a *Lopha* species), seen here on an Indonesian reef, releasing its sex products in to the open water where fertilisation takes place.
Photo: S. W. Michael

Most species have separate sexes, while some are hermaphrodites – including some oysters and scallops. The sex products are as a rule released in to the open water, where fertilization takes place. The zygote develops into a free-swimming trocophore larvae followed by a veliger larva, the common scheme among molluscs.

Many bivalves live buried in sand or some other substrate. They can extend a long tube (siphon) which facilitates water movement in and out of the animal, and simultaneously also supplying providing food. The beloved blue-mussels from the genus *Mytilus* live in large aggregations attached to rocks or floating objects in shallow waters. Some are associated with stony corals, others with gorgonians or black corals. A few species drill in to corals and rocks, where they become totally encapsulated and protected against the environment around them. Invariably, these types of bivalves are intro-

Ark clams of genus *Arca* are often introduced in to a coral reef aquarium attached to live rocks or to the base of coral colonies (left). If such a clam is removed from its hiding place one can readily recognise the typical ark-shape of the valves.

Clam and mussel shells of families Arcidae and Mytilidae from the Indo-Pacific. Family Arcidae: 1 - *Barbatia helblingi*, 5 cm. 2 - *Barbatia amygydalumtostum*, 4 cm. 3 - *Arca avellana*, 4 cm. 4 - *Anadara vellicata*, 2 cm. 5 - *Barbatia tenella*, 3 cm. 6 - *Barbatia plicata*, 3 cm. Family Mytilidae: 7 - *Septifer biocularis*, 3 cm. 8 - *Modiolus* sp., 3 cm. 9 - *Lithophaga* sp., 3 cm.
Photo: C. Bryce

duced in to our aquariums in conjunction with live rocks.

There are even a very few bivalves that live as commensals on burrowing echinoderms and crustaceans. One species of *Entovalva* live in the gut of sea cucumber and is the only bivalve known to be parasitic (Ruppert & Barnes, 1994).

Perhaps the strangest of all bivalves are those species found in association with deep-sea hydrothermal vent sites. The deep-sea mussel species-complex, *Bathymodiolus*, is a dominant component of hydrothermal vent communities on the Mid-Atlantic Ridge. In common with other vent bivalves (as well as with other hydrothermal invertebrate groups such as the Pognophoran worm *Riftia pachyptila*, see volume 3 page 158), *Bathymodiolus* is nutritionally dependent upon a bacterial symbiosis utilising the reduced sulphur compounds released by these submarine hot springs, although there is evidence that these mussels are also capable of ingesting microscopic food particles which they trap by conventional filter feeding. These bivalves take part in the only animal community known on Earth that live completely independent of energy from the Sun, communities that might shed new light on the origin of life on Earth.

Order Arcoida

This order contains three families, where Ark mussels family Arcidae are important to aquarists and will be discussed in detail.

Family Arcidae - Arks

Most of the approximately 200 species in this family are tropical. They are characterized in that the valves are straight along the hinge-ligament. Most have patterns of distinct, radiating ribs, and grow to a maximum size of about 5 to 10 cm. In shape they are reminiscent of an ark, which is also the basis of their scientific- as well as popular name.

It is not uncommon for *Arca* species to find their way in to coral reef aquariums by being attached to live rocks (see Volume 1, pages 186-187). *Arca* spp. grow to 10 cm and is anchored on to some hard substrate – often under rocks or attached to corals – by means of their byssal threads. Along the edges of the shells there are a great many cardinal teeth. *Arca ventricosa* is distributed widely throughout the Indo-Pacific and in the Red Sea and is especially abundant embedded in huge stony coral colonies from the genus *Porites*,

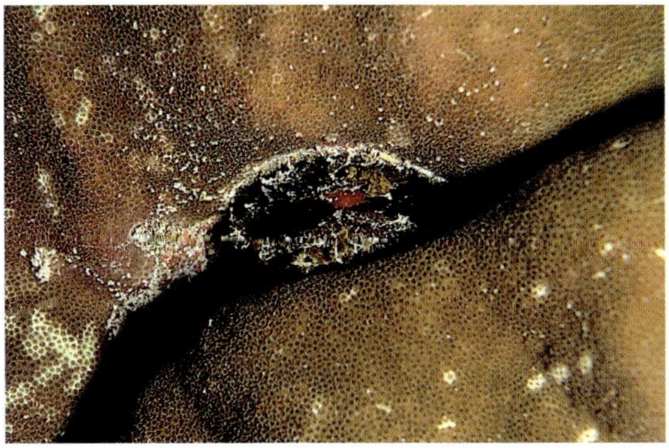

Typical appearance of a live *Arca ventricosa* encased in a *Porites* colony (left). Right: An *Arca* species at Heron Island, Great Barrier Reef (photo: L. Newman and A. Flowers).

where it co-exists with other species in the genus. Two species common in the Caribbean region are *A. imbricata* and *A. zebra*. The latter shows an attractive zigzag pattern on its valves.

Barbatia species also live on hard substrate, however, they are smaller than *Arca* species and have a rather wide byssal opening. *B. lacerta* occurs throughout the Indo-Pacific. *B. domingensis* is frequently found in the Caribean Sea, where it lives below the low tide mark, usually attached in small holes or in rocky caves.

Anadara species have particularly strong valves. Byssal threads are absent since these bivalves live on soft substrate where they do not need to attach themselves. These arks are somewhat reminiscent of cockles or heart clams (family Cardiidae; see page YXY), but are distinguished from the latter by a long hinge-ligament line that keeps the valves together. *A. granosa* is a rather common species in the Indo-Pacific. There are several species in the Caribbean region, where *A. brasiliena* and *A. ovalis* are the two most common ones.

Order Mytiloida

This order includes the families

- Mytilidae and
- Pinnidae

Family Mytilidae - Mussels

This family encompasses more than 30 genera and has a wide geographical distribution. Generally, mussels grow best in shallow water areas, often on a wide range of submarine structure, such as wharf pilings and similar man-made marine facilities, where the salinity is lower than out in the open sea. Their growth can be very rapid, so that they can become established over large areas within a short period of time. This characteristic is being utilized commercially in aquaculture facilities. Mussel farms have been established all over the world to breed and raise mussels for human consumption.

The best-known species is the Blue Mussel, *Mytilus edulis*, which has a worldwide distribution. Subspecies of *M. edulis*, include e.g. *M. edulis planatus* occur in the southern hemisphere.

The Green Mussel, *Mytilus smaragdina* occurs widely in coastal waters around Sri Lanka, off India and throughout Southeast Asia. This is a very prolific species

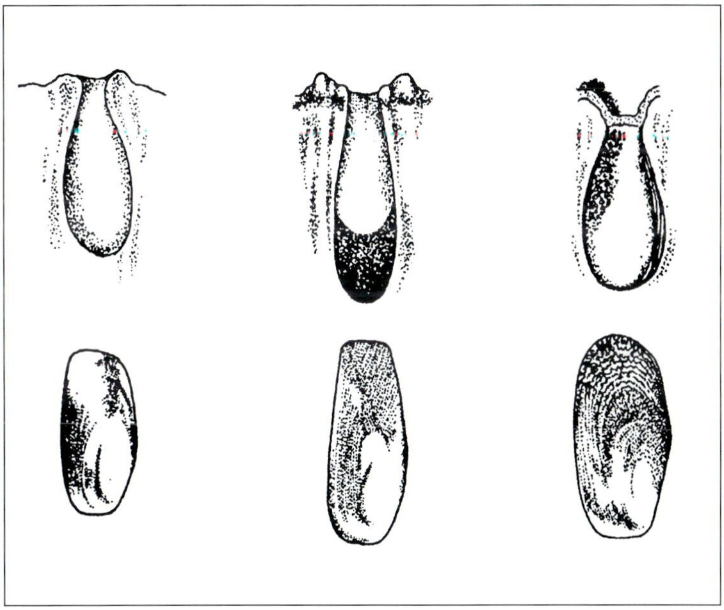

Drill holes (top) and valves (below) of *Lithophaga cumingiana* (left), *L. hanleyana* (centre) and *L. lima* (right). **Drawings after VINE (1986)**

Mytilus smaragdina
Green Mussel

Synonyms: *Mytilus viridis* and *Perna viridis*.
Distribution: Indo-Pacific; Sri Lanka, India and particularly common throughout Southeast Asia.
Size: Up to a length of 6.5 cm.
Reproduction: Separate sexes. Sexual maturity is reached in about one year. Spawns during the tropical winter; extremely prolific breeder. The planktonic larval stage lasts for three weeks.
Food: Filter feeder.
Description: Occurs in large aggregations in estuaries and coastal waters, which are rich in organic material and plankton, have a salinity of 20 to 30 ‰ and a temperature range of 20–28 °C. This mussel is being bred in aquaculture facilities in Cambodia, Indonesia, Malaysia, the Philippines, Thailand and in Viet Nam. It is an important food source for local populations in southeastern Asia (Ling, 1977). The valves are elongated, laterally compressed, with a relatively blunt apex, a long straight lateral margin, large concentric growth lines, coloured emerald green with the most intense colouration along the outer margins.
Aquarium requirements: Frequently imported, however, difficult to maintain in a "normal" coral reef aquarium. Best kept in a brackish water- or mangrove aquarium, with a salinity of 20 to 30 ‰. Needs plenty of suspended food.

Photo:
Group of Green Mussels *Mytilus smaragdinus* in an aquarium shop sales tank.

GA: +/-; **SE:** +/-; **CO:** +

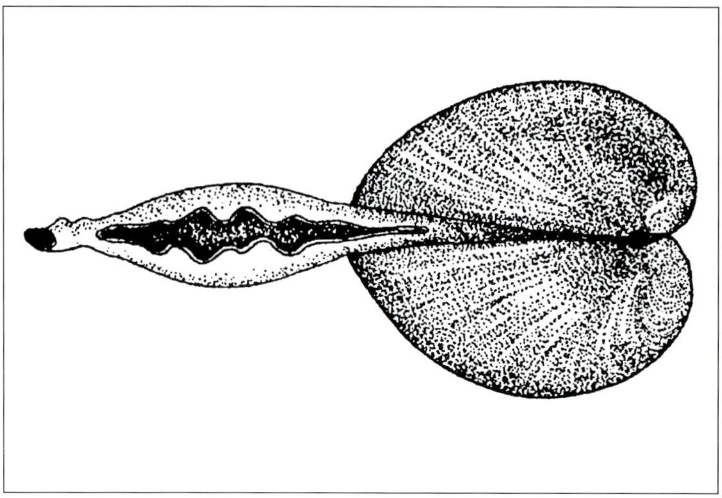

Fungiacava eilatensis. **Drawing after Vine (1986)**

and is being bred on a large scale. It is an important provider of protein in the human diet in that part of the world (Ling, 1977) and is served as a part of many delicious dishes among others in the Thai-cuisine. This species also appears frequently in the aquarium trade. With its emerald-green colouration it appears to be a very decorative species for coral reef aquariums (see Fact Sheet above), but cannot be kept for long unless provided with sufficient food and salinity lower than most other reef animals can tolerate.

Tropical relatives of the common mussel are also found in genera *Brachidontes* and *Septifer*. *Septifer* species are numerous on

Pinna bicolor is one of the most commonly occurring razor clams, (left) in its natural habitat and (right) prepared valves.
Photos: C. Bryce

coral reefs and can well be imported together with live rocks. *Septifer biocularis* grows to a length of 4 cm and is common in the Indo-Pacific. It has an attractive shell, varying in shades of green to orange, but it can also appear brown because the valves are covered by periostracum.

Modiolus species live firmly attached to corals or live rocks. They are reminiscent of a smaller version of the common mussel, with a long hinge-ligament and hairs often cover the valves. *Modiolus philippinarum* is frequently seen in the Indo-Pacific. *Musculus* species can be commonly found in all seas. They are, in contrast to *Modiolus* species, small and have a thin shell. *Musculus lateralis* lives in the Caribbean region, often in association with tunicates.

In addition, family Mytilidae is also represented in the Caribbean region by several other genera; *Brachidontes* is common with its species *B. exustus* and *B. citrinus* on marine pilings. *Ischadium recurvum* is characterized by an umbo that is bent in to a kind of hook. It lives together with colonies of oysters in areas where the salinity is low, and in general, many mytilids prefer a low salinity.

Members of genus *Lithophaga* that are very common on coral reefs, have a special significance for coral reef aquarists. With the aid of secreted acid, this mussel can etch out a hole in corals or calcareous rocks that fits precisely around its shell (lithos, Greek = rock and phagein, Greek = eating). This mussel tends to bore in to the stony coral, preferably in to *Porites lobata*. Such an association is disadvantageous for the coral since it distinctly weakens it (Scott & Risk, 1988) and *Lithophaga* spp. are in this way natural destructors of the coral reefs. The mantle in *Lithophaga* species is blood red and can extend for a few centimetres from the hiding place of the mussel. If a live rock is split in to two sections there are often large holes that were once occupied by *Lithophaga* (also refer to Volume 1, page 169). Individual species can be identified on the basis of shell shape. We found dense populations of *Lithophaga* together with several specimens of *Tridacna crocea* on the inner reef flat of a reef in Thailand. The blood-red inhalant and exhalant siphons were visible as small red tubes protruding from the rocks and in between them the colourful mantles of the giant clams, a truly beautiful and spectacular sight.

Family Mytilidae also contains the atypical *Fungicava eilatensis*, which is found exclusively on stony corals of family Fungiidae, where it bores in to the calcium carbonate skeleton of the coral. In this species the mantle excretes substances used for boring and it also produces aragonite. The latter fills up the hollow space around mussel so that it is totally encapsulated by the coral. So far we are not aware that this mussel has ever been discovered in an aquarium, but it is quite likely to be present in some of the frequently imported Fungiidae.

Family Pinnidae - Pen Shells, Razor Clams

Pen shell or pen-shelled mussels (sometimes also referred to as razor clams) have wedge-shaped valves and can be reminiscent of a ham. The valves are large, triangular and thin. Most of the approximately 200 species live with the pointed part of the shell buried or are anchored to rocks. With a length of 30 cm, pen shells can be rather large. They are usually found in shallow water with a sandy bottom and among marine grasses. In that case they are overgrown by algae, or they may be covered by mud, so that the attractive shell remain totally hidden.

The most common genus is *Pinna*. Well-known and in demand is the large *Pinna nobilis* that is native to the Mediterranean Sea. It lives on a sandy bottom, and with its length of 40 to 60 cm it is indeed an impressive sight. However, because it is also a desirable collector's item its population numbers have seriously declined.

Pinna carnea at Florida.

Pinna bicolor has a greyish shell with shades of red. It is common throughout the Indo-Pacific, while *P. carnea* lives in the Caribbean region.

Genus *Atrina* is generally found on a sandy bottom among corals, where these mussels protrude partially from the bottom. *Atrina* species have shorter and wider valves than *Pinna* species. *Atrina vexillum* occurs sporadically in the Indo-Pacific, while *A. seminuda* and *A. serrata* are relatively frequent species in the Caribbean Sea. So far we have never seen pen shells in the aquarium trade, nor are we aware that they have ever been imported together with live rocks. However, since they occur in the shallow water of tropical seas, there is a certain possibility that they may occasionally appear in the trade. Pen shells are for sure difficult organisms to keep alive in captivity. They would need a special tank with a deep sand-bed and an almost continuous flow of minute organic food and are organisms for the advanced and experienced aquarist only.

Order Pteroidea - Winged Bivales: Oysters, Scallops, File Shells

This order is made up of ten families, which are contained in five superfamilies. We will discuss the following families in this volume:

- Pteriidae
- Malleidae
- Isognomonidae
- Pectinidae
- Plicatulidae
- Spondylidae
- Anomiidae
- Limidae
- Ostreidae

Atrina vexillium is rather frequent on sandy bottom along reef flats among corals - seen here at the Maldives. Its shell is rounder than is common among razor clams. The mantle has an attractively striped pattern.

Family Pteriidae - Pearl Oysters

Pearl oysters of family Pteriidae have a long and sharply pointed, hinge-ligament, which is a most useful character for identification. As is indicated by their popular name, pearl oysters produce pearls. Some species, such as *Pinctada maxima* are collected and then reared in oyster farms, where a foreign particle is inserted between shell and mantle. After several years these particles will have become highly valued culture pearls (Wells & Bryce, 1988). But it is rare to find natural pearls in pearl oysters.

The most widespread genera of this family are *Pteria* and *Pinctada*. The Penguin Pearl Oyster, *Pteria penguin* is very common in the Indo-Pacific. Usually it lives in association with black corals, along reef slopes with strong currents, often in caves or under overhangs. It is not suited for a community tank, but it is an attractive animal for experienced aquarist, to be kept in dedicated cave aquariums (see Fact Sheet on page 156).

A common inhabitant in the Caribbean Sea is *Pteria colymbus*. It has a distinctly projecting wing located at the hinge-ligament, and it grows to a size of up to 7.5 cm. It too lives in areas with a strong current and is associated with cnidarian animals. *Pinctada margaritifera* is distributed in the Indo-Pacific. It can occur as solitary specimens or live in colonies (Erhardt & Moosleitner, 1995).

Family Malleidae - Hammer Oysters, Finger Oysters

The species in this family look similar to those in family Isognomonidae, but instead of living on mangroves, they live in shallow water, on rocks and among corals. Because of the extremely long hinge ligament, the shell of

Pteria sp. "KA5-BIV-01" in the Red Sea. Photo: S. W. Michael

Pteria sp. "KA5-BIV-02" in the Red Sea. Photo: R. Brons

Pinctada margaritifera is usually encountered among rocks and coral rubble. Photo: L. Newman and A. Flowers

this oyster looks like a hammer. There are several genera, but the one most commonly encountered in the Indo-Pacific is genus *Malleus*. Its species are rather common on live rocks or on the underside of coral colonies. They are quite hardy and durable in a coral reef aquarium. Normally, the shell is thin, longish with a dark colouration, which is the reason why they are often overlooked by aquarists.

The genus *Vusella* encompasses species, which live totally hidden inside sponges. A good example of this is *Vusella vusella*, which occurs in the Red Sea.

Hammer oysters of genus *Malleus*, seen here is *Malleus* sp. "KA5-BIV-03" attached to the shell of a giant clam, often appear in coral reef aquariums.

Family Isognomonidae - Wood Oysters or Wood Scallops

These bivalves grow on the roots on mangroves, and so they are often referred to as wood scallops or wood oysters. But some species occur on and under rocks in shallow water. These species are then also often found on live rocks and so they appear sometimes in a coral reef aquarium. In the Caribbean Sea *Isognomon radiatus* is particularly common in areas where the water is rich in

Photo:
Penguin Pearl Oyster at the Maldives, hanging from Black Corals.

Pteria penguin
Penguin Pearl Oyster

Distribution: Widely distributed throughout the Indo-Pacific.
Size: Up to 25 cm.
Reproduction: Separate sexes. Sexual products released in to open water.
Food: Filter feeder.
Description: Often found attached to black corals along reef slopes or in caves. Rather common in some areas. Hinge-ligament straight. A groove on the right valve (half-shell) with protruding byssal threads. A long, winged lobe located on the hind body. Valve interior dark blue, almost black and externally often covered with epiphytic organisms.
Aquarium requirements: Only occasionally imported. Requires a dedicated aquarium with adequate current and water containing minute plankton and suspended food particles. A cave aquarium comes very close to meet such requirements. Should be kept by experienced aquarists only.

GA: +/-; **SE:** +/-; **CO:** +/-

sediments and nutrient particles. Thousands of specimens can become established in large areas like that.

Family Pectinidae - Scallops

Scallops of family Pectinidae have an almost totally flat left valve, while the right one is more or less arched. This is a morphological characteristic of this group. Many scallops have well-developed eyespots, which are light-sensitive cells located along the margin of the mantle. In addition, the mantle carries tentacles with statocysts that are cells used for detecting movements and vibrations. Together, these two characteristic features provide an effective mechanism with which dangers can be detected quickly. Moreover, scallops are mobile and can quickly swim away from predators, by flapping movements of their valves.

Scallops have a worldwide distribution and contains 12 genera with about 350 species (Wells & Bryce, 1988). The majority occurs in tropical seas, often in moderate deep waters on a sandy bottom, where they lie with their flat valve pointing up. Some species are periodically attached with byssal

Group of *Isognomon radiatus* in Miami, Florida.

Razor clam- or Pen Shells of family Pinnidae, order Mytiloida (1 and 2), as well as pearl oyster shells of families Pteridae (3 and 4) and Isognomonidae, order Pteroidea, from the Indo-Pacific (5 – 7): 1 - *Streptopinna saccata*, 12 cm. 2 - *Pinna muricata*, 12 cm. 3 - *Pteria penguin*, 10 cm. 4 - *Pinctada margaritifera*, 10 cm. 5 - *Isognomon perna*, 5 cm. 6 - *Isognomon legumen*, 9 cm. 7 - *Isognomon isognomum*, 4 cm. Photo: C. Bryce

Scallops have rather flat, but well-structured valves. The mantle if frequently very colourful but visible only when the scallop is undisturbed and the valves are open. Photo: S. W. Michael

Lyropecten nodosus
Knotty Scallop

Distribution: Western Atlantic: Florida Keys, Texas coast and Mexico, West Indies, Surinam coast and Brazil.
Size: Maximum diameter 15 cm.
Reproduction: Separate sexes. Eggs and sperms are released in to open water.
Food: Filter feeder.
Description: Valves fan-like; nearly identical in shape, with seven to nine large radial ribs covered by many large knots; numerous cord-shaped, secondary radial ribs present. Hinge-ligament straight and without teeth. Colouration of valves varies from light red and dark orange to chestnut brown. Mantle very colourful with tentacles and blue ocelli.
Aquarium requirements: Occasionally imported, but basically rather difficult to keep. Requires a special aquarium with ample currents and suspended minute food particles in the water. Suitable only for experienced aquarist.

GA: +/-; **SE:** +/-; **CO:** +

Photos:
Top left: Knotty Scallop in an aquarium.
Top right: Close-up photo of valve opening.
Centre: Close-up photo of mantle with tentacles and ocelli.
Bottom: Preserved valves.

Pedum spondyloideum
Iridescent Scallop, Coral Clam

Distribution: Widely distributed throughout the Indo-Pacific.
Size: Up to 5 cm.
Reproduction: Separate sexes. Eggs and sperm are released in to open water.
Food: Filter Feeder.
Description: Normally occurs in groups deeply embedded in stony corals of genus *Porites*, so that only valve margins and mantle are visible. Right valve concave and with a large byssal groove, left valve mostly flat. Outside of right valve covered by delicate concentric lines; left valve with protruding radial ribs and distal (away from the body) spines (VINE, 1986). Mantle red and purple-coloured, due to embedded pigments to block out ultra-violet radiation. Small red ocelli normally present.
Aquarium requirements: Normally not available from the aquarium trade, and therefore hardly any published experiences available. Since this species is adapted to high illumination levels in shallow reef areas it should be easy to keep in an aquarium. It would, however, be necessary to feed the animals with minute suspended food particles and/or minute live plankton.

GA: 0; **SE:** 0; **CO:** +

Photos:
Top: Group of Iridescent Scallops embedded in a *Porites* colony in Thailand. Photo: E. Svensen
Centre: Close-up photo of one of the specimens in photograph seen at the top. Photo: E. Svensen
Bottom: Most commonly occurring colour variety in the Maldives.

Chlamys **sp. at Heron Island, Great Barrier Reef.**
Photo: L. Newman and A. Flowers

threads while others are permanently anchored to a rocky substrate. On a reef, scallops are often difficult to see as they often stay under rocks in shallow water or on sandy substrate at deeper sections of the reef.

Genus *Chlamys* occurs in all seas. The left half of the shell is more arched than the right one. *Chlamys asperrinus*, *C. australis*, *C. leopardus* and *C. funebris* are common species in the Indo-Pacific. *C. squamosa* lives in association with corals in the Red Sea, while *C. varia*, *C. islandica* and *C. distorta* are species from Northern Europe.

Gloripallium pallium is another species that lives commonly under rocks in shallow lagoons in the Indo-Pacific. It is characterized by a strongly ribbed shell structure and a reddish marbled colour pattern. Scallops of genus *Aequipecten* are represented in the Caribbean Sea. There, *A. irradians* is the most frequent species. It occurs along the coastline of northern Florida; however, *A. gibbus* is found throughout the entire Caribbean region, but it is rare in some localities. On the other hand, *Lyropecten nodosus* (formerly known as *Chlamys nodosus*) is common throughout the Caribbean Sea, and it is sporadically imported for the aquarium trade (see Fact Sheet on page 158)

Species of genus *Pedum* are sometimes seen in coral reef aquariums. Unlike the majority of scallops, they do not live free but in association with stony corals of genus *Porites*, where they bury so deep that only the shell opening and the mantel are visible. The most frequently seen species is *Pedum spondyloideum*, which has an unusually beautiful mantle colouration. In some Indo-Pacific localities, especially in Thailand, it can be rather common with hundreds of specimens inhabiting each *Porites* colony (see Fact Sheet on page 159).

Family Plicatulidae

This family is closely related to the scallops of family Pectinidae. Its members have the ability to camouflage themselves perfectly, whereby the right half shell is firmly attached to a substrate. Colouration is often light, but the shell becomes overgrown by calcareous algae and other organisms so that its natural colours are soon covered up. These scallops were introduced in to our aquariums together with live rocks, and several months passed before we discovered them. The only genus in this family is *Plicatula*.

Two specimens of the Variable Spiny Oyster, *Spondylus varius*, at the Maldives.

Scallop shells of family Pectinidae (1 – 4) and Spiny Oyster shells of family Spondylidae (5 – 8) from the Indo-Pacific: 1 - *Chlamys pallium*, 4 cm. 2 - *Semipallium tigris*, 4 cm. 3 - *Chlamys rastellum*, 2 cm. 4 - *Chlamys corsucans*, 2 cm. 5 - *Spondylus candidus*, 4 cm. 6 - *Spondylus nicobaricus*, 7 cm. 7 - *Spondylus linguafelis*, 15 cm. 8 - *Spondylus multimuricatus*, 5 cm.

Photo: C. Bryce

Family Spondylidae - Thorny Oysters

Thorny oysters are among the most attractive bivalves on coral reefs. This is not so much due to the shell as such, but rather because the fantastic epiphytic growth that always becomes established on it, as well as because of magnificent mantle colouration. The shell halves are differently formed. While upper one is flat, the lower one is strongly convex. Both valves are covered with many spines and other outgrowths.

Thorny oysters are usually hanging in caves along reef walls or below overhangs, where lighting is low but there is a strong current, i.e. ideal locations for epiphytic growth, which make for perfect camouflage of the shells of thorny oysters. Unfortunately the attractive, epiphytic growth does not survive very well under aquarium conditions and new growth does not establish itself, because invariably it involves delicate sponges, gorgonians and tunicates that cover the valves. The majority of these organisms do not survive the transport to the aquarium trade, and there is a vast difference in appearance between thorny oysters in an aquarium and those in the sea. Occasionally thorny oysters appear in the aquarium trade, and they have shown to be rather difficult to keep in a coral reef aquarium.

The best-known genus is *Spondylus*, which occurs exclusively in tropical seas. There are several *Spondylus* species in the Indo-Pacific, but species identification is often rather difficult, because spines and colouration of the valves can be highly variable. No doubt, that may have led to duplicate (erroneous) descriptions.

Spondylus varius is distributed widely in the Indo-Pacific, where it lives mostly in caves or below overhangs. Its mantle is magnificent and displays a substantial variability in colouration. The maximum diameter of the shell is 20 to 30 cm. We are not aware whether *S. varius* has ever been kept in a coral reef aquarium.

Spondylus princeps lives in the shallow water of reef flats and in coastal waters of the Indo-Pacific and the western Atlantic. It is imported for the aquarium trade more frequently. According to ERHARDT & MOOSLEITNER (1995) this species reaches a maximum length of 8 cm. The valves are covered with strong spines. It is rather hardy when kept in an aquarium, but – like all bivalves without zooxanthellae – is requires minute food particles suspended in the water.

Spondylus americanus occurs in the Caribbean region, where it grows to a size of 10 cm. Its shell

***Spondylus* spp.**
Thorny Oysters or Spiny Oysters

Distribution: Widely distributed throughout the Indo-Pacific and in the Caribbean as well as in temperate seas.
Size: Variable with species, from a few cm up to more than 30 cm across.
Reproduction: Separate sexes. Eggs and sperms released in to open water.
Food: Filter Feeders.
Description: Common on most reefs. Usually found in caves or under overhanging rocks. The valves are unequal; the upper usually with many projecting spines, the lower attached to the substratum. Hinge with prominent teeth. The mantle colour varies a lot and is often very colourful. The valves are commonly overgrown with other invertebrates, which gives the oyster a perfect camouflage. Colourful compound eyes and many brightly coloured tentacles are found along the edge of the mantel.
Aquarium requirements: Small specimens are occasionally available in the aquarium trade, but basically rather difficult to keep. Should be given a special aquarium with overhanging rocks and a lot of current. Requires a lot of food at regular intervals, such as suspended particles of other clams, shrimps or crabs or minute live plankton.

GA: +/-; **SE:** +/-; **CO:** +/-

Photos:
Top: Group of typical *Spondylus* sp. in a reef wall in the Maldives.
Middle: *Spondylus* on display in Seashell Museum, Phuket, Thailand.
Bottom: *Spondylus varius* at the Solomon Islands (left; photo: J. Macaré) and at the Maldives (right; photo: S. W. Michael).

is highly variable in colouration: white to pink or creamy yellow. Adult specimens have powerful, erect spines (up to 5 cm long) on their shell. *Spondylus americanus* is not common and is usually found only in exposed reef areas of southern North America, southward to Brazil.

Just how magnificently coloured and structured the mantle of *Thorny Oysters* can be, is well illustrated by these two close-up photographs (top photo: L. Newman and A. Flowers; middle photo: P. Schupp).

Family Anomiidae - Jingle Shells or Venus' Toenails

These bivalves are often found on the underside of corals and rocks. They are relatively inconspicuous, and are not immediately noticed. However, this family has one particular characteristic as one of the valves has a hole from which byssal threads protrude. The genera *Anomia* and *Placuna* are forms that live in rather shallow water.

Family Limidae - File Shells

File shells of family Limidae are well known within the aquarium hobby. They are characterized by the presence of many intensely red (or shaded in red) tentacles along the margin of the mantle, and chalky white, rather brittle valves. Together with scallops they are the only bivalves that are able to actively swim by a series of rapid movements of their half-shells. *Lima* and *Limaria* species are regularly imported. Both genera have a cosmopolitan distribution.

In genus *Lima* both valves are equal, carry distinct radial ribs, and have a flat shape. It has become customary in the aquarium hobby to refer to all file shells as *Lima scabra*, which is incorrect. *Lima scabra* is a Caribbean species that is often imported. It is a very decorative mussel, but rather difficult to keep in a coral reef aquarium, and so it belongs in to a special aquarium. In Puerto Rico we have collected specimens in shallow water, where they lived in depressions or hollow

To the right: *Anomia* sp. - seen here at the Maldives - usually lives under rocks or among corals.

Lima scabra
Flame Scallop, Red Caribbean File Clam

Distribution: Widely distributed throughout the Caribbean Region.
Size: Up to 8 cm.
Reproduction: Separate sexes. Eggs and sperms released in to open water.
Food: Filter Feeder.
Description: Lives in rocky crevasses in shallow water as well as down to a depth of at least 130 m; often in groups. The long, red tentacles normally protrude from crevasses. Can be attached by means of byssal threads. Unattached specimens are able to swim over short distances by alternately opening and closing the valves. Valves are identically developed. Mantle intensely red with long, deep red tentacles. Delicate species that is easily damaged. Dependent upon availability of hiding places in crevasses.
Aquarium requirements: Often available in the aquarium trade, but basically rather difficult to keep. Should be given a small special aquarium with sedate conditions and many crevasses available among the aquarium decoration. Requires a lot of food at regular intervals, such as suspended particles of other clams, shrimps or crabs or minute live plankton. Must not be kept together with fishes that nibble on epiphytic growth in the tank.

GA: +/-; **SE:** -; **CO:** +/-

Photos:
Top: *Lima scabra* in natural habitat off Puerto Rico.
Bottom: Specimen on sand bottom in an aquarium.
Cover photo of this Volume: two specimens in an aquarium.

Limaria orientalis
Eastern File Clam

Synonyms: *Limaria vigens* and *Promantellum vigens*
Distribution: Indo-Pacific; details unknown.
Size: Up to 5 cm.
Reproduction: Separate sexes: Eggs and sperms released in to open water.
Food: Filter feeder.
Description: Lives under rocks and corals, often on soft bottom. The systematic classification of genus *Limaria* is not yet completely resolved. *L. orientalis* has uniformly pale or light-red tentacles, while those of *L. fragilis* are red with white bands. Yet, there is some debate whether they are two distinct species or only colour varieties of one species. Valves different from those of *Lima* species: they are thin, small, oval, higher than they are long and they break very easily.
Aquarium requirements: Often available in the aquarium trade. According to our experiences, it is a hardy species, but will hide under or behind decoration so that it is rarely seen in an aquarium. Should be kept in a smaller tank together with other delicate animals, which also require somewhat more nutrient-rich water than is normally found in a coral reef aquarium. Valves and tentacles are distinctly fragile. Must be handled very delicately!

GA: +; **SE:** +/-; **CO:** +/-

Photos:
Top: *Limaria orientalis* in a shop display aquarium.
Middle: Specimen in an aquarium. Photo: S. W. Michael
Bottom: When a boulder was removed from the inshore reef around Heron Island, GBR, these two specimens of *Limaria orientalis* appeared showing their typical habitat on soft bottom under boulders.

Limaria fragilis. Photo: L. Newman and A. Flowers

Lima sp. "KA5-BIV-04". Photo: C. Bryce

Limaria sp. "KA5-BIV-05". Photo: C. Bryce

spaces among algae and corals (see Fact Sheet on page 164).

The genus *Lima* contains an unknown number of species. The most commonly referred to species is *Lima lima* that lives in the Indo-Pacific as well as in the tropical Atlantic and in (adjacent) subtropical seas. This taxon probably involves several species, however, the taxonomic details have not yet been resolved. BOSCH et al. (1995) are of the opinion that reports of *Lima lima* and/or *L. vulgaris* from the Arabian Gulf refer to *Lima sowerbyi*. PRASHAD (1932) believes that *Lima lima* occurs only in the Indo-Pacific.

In species of genus *Limaria* the shell is very thin, egg-shaped and higher than it is long. *Limaria* species are reminiscent of *Lima* species, but they have very sticky tentacles that readily remain attached to an aquarist's fingers and then tear off. These are delicate bivalves that must be handled with the greatest of care. Gradual acclimatisation to aquarium conditions is absolutely essential. Most frequently available in the aquarium trade is *Limaria orientalis* (synonyms: *L. vigens and Promantellum vigens*) from the Indo-Pacific. It is often traded erroneously as *Lima scabra* (for further details refer to Fact Sheet on page 165). In the sea *Limaria* species live very much like *Lima* species, i.e. on sandy bottom under coral blocks. Another species from the Indo-Pacific is *Limaria fragilis* (synonyms: *Limaria linguatula, L. parafragile* and *L. basilanica). Limaria pellicida* occurs in the Caribbean Sea (for further details about this group refer to ABBOT, 1991, and LAMPRELL & WHITEHEAD, 1992).

Family Ostreidae - Oysters

Oysters are invariably associated with gastronomy and the perception of delicate, exotic meals. In fact, many different oyster species are commercially cultivated on a grand scale just for that very purpose, in tropical latitudes but also in the temperate waters of European coasts, along eastern

Cock's Comb Oyster, *Lopha cristagalli*, - seen here at the Maldives – has a zigzag-shaped shell edge along the opening. The valves (as seen here) are overgrown by a red sponge of genus *Microciona* (right photo: S. W. Michael).

Australia, around New Zealand, off North America and in many other localities. In addition, there are vast natural oyster banks in several regions of the World's oceans. In tropical regions the small *Saccostrea cuccullata* is common in coastal waters, where dense colonies are formed in the belt between high and low water marks. However, we have not yet encountered this species in the aquarium trade.

In oysters of family Ostreidae the right half-shell is often firmly attached to the substrate. This valve is usually flat, while the left one has a rounded off shape, and so in appearance they are similar to thorny oysters of family Spondylidae. True oysters belong to genus *Ostrea*, and several species are utilised commercially. *Ostrea edulis* occurs from the northern Atlantics southward to the Mediterranean Sea. It is well known from the commercial oyster trade for human consumption. The Caribbean Sea is inhabited by *O. equestris*, which – in

File shells of family Limidae (1 and 2), oyster shells of family Ostreidae (3 and 4) and chamid shells of family Chamidae (5 and 6): 1 - *Ctenoides ales*, 6 cm. 2 - *Lima fragilis*, 3 cm. 3 - *Saccostrea cuecullata*, 6 cm. 4 - *Dendrostrea folium*, 8 cm. 5 - *Chama pacifica*, 8 cm. 6 - *Chama lazarus*, 8 cm. Photo: C. Bryce

contrast to other oysters – thrives in water with normal salinity.

On coral reefs it is genus *Lopha* with the beautiful Cock's Comb Oyster, *L. cristagalli*, that is highly conspicuous. It is readily recognisable by a typical zigzag pattern of valve margins. The maximum size of this species is 10 cm. It is distributed widely throughout the Indo-Pacific, where it is often seen in groups, attached to black corals or below overhangs in subdued light, down to a depth of 40 m. *Dendostrea frons* occurs in the Caribbean Sea, and it too has a zigzag-shaped shell opening. It grows on Gorgonian branches and dead corals, but it can also be attached to other substrates. Like *Lopha cristagalli*, it too is usually overgrown with sponges or other organisms.

Order Veneroida - Venus Shells and related forms

From among this order, we will discuss only the following families:

- Chamidae
- Cardiidae
- Tridacnidae
- Tellinidae
- Veneridae

Family Chamidae - Jewel Box Shells

In species of this family the valves are unequal; one half-shell has firmly grown on to some substrate, usually a coral rock. The other half is equipped with spines and can feature magnificent colours. We found species belonging to this family often in shallow lagoons at the Maldives, but they were always attached to the underside of coral boulders.

Genus *Chama* contains many species. *C. lazarus* occurs around the Philippines and is probably also widely distributed throughout the entire Indo-Pacific. It possesses strong spines that can grow to a length of 9 cm. *Chama pacifica* is a common species of the Indo-Pacific and the Red Sea. The species has only short spines that are distributed unevenly over the shell. *C. macerophylla*, *C. sinuosa* and the smaller *C. sarda* live in the Caribbean region (Colin, 1978).

Cockle shells of family Cardiidae from the Indo-Pacific: 1 - *Acrosterigma elongatum*, 8 cm. 2 - *Acrosterigma alternatum*, 6 cm. 3 - *Acrosterigma reeveanum*, 6 cm. 4 - *Acrosterigma dupuchense*, 5 cm. 5 - *Acrosterigma fultoni*, 3 cm. 6 - *Fragum erugatum*, 2 cm. 7 - *Fragum retusum*, 5 cm. 8 - *Fulvia tenuicostata*, 6 cm. 9 - *Fulvia aperta*, 5 cm. 10 - *Fragum hemicardium*, 7 cm. 11 - *Fragum unedo*, 6 cm. 12 - *Fragum fragum*, 4 cm. Photo: C. Bryce

One oddity here is the distinguishing difference between genera *Chama* and *Pseudoichama*. While in *Chama* species the **left** half-shell grows firmly to the substrate, it is the **right** one in *Pseudochama* species.

Chama spp. are imported to the reef aquarium in association with live rocks. Often they are so well camouflaged that they are really hard to discover. See more information on *Chama* in the Fact-Sheet on page 169.

Chama **spp.**
Jewel Boxes

Distribution: Widely distributed throughout the Indo-Pacific and the Caribbean Region.
Size: Variable with species; many species only 3-4 cm across, others larger reaching a size of 10 cm.
Reproduction: Separate sexes. Eggs and sperms released in to open water.
Food: Filter feeder.
Description: Valves are unequal; the left valve attached to the substratum the right one exposed and often equipped with short or longer spines. The right valve is often overgrown with other invertebrates in a way that makes the animal almost invisible. Attached to coral boulders on the reef flat or to coral rock in caves and crevices. The mantel is colourful; often bright yellow or red. Several species exist; *C. pacifica* is widespread in the Red Sea and Indo-Pacific while *C. macerophylla* is a common species of the Caribbean.
Aquarium requirements: Occurs on live rocks and is imported to the reef aquarium in this way. The molluscs are so well camouflaged that they can be very hard to detect. Requires a lot of food at regular intervals, such as suspended particles of other clams, shrimps or crabs or minute live plankton. Perfect for a small, special "bivalve aquarium" where the animals can be kept in peaceful conditions together with other delicate filter feeders that requires an equal amount of suspended food.

GA: +/-; **SE:** +/-; **CO:** +

Photos:
Top: Two *Chama* sp. are perfectly camouflaged on a live rock.
Bottom: Close up of a jewel box with a yellow mantel.

Family Cardiidae
Cockles

Cockles are among the best known and most numerous bivalves living on coral reefs as well as in more temperate seas around the world. However, they appear rather rarely in the aquarium trade, which may be due to their mode of life in sandy substrate. With help of a strong foot, cockles are readily able to bury in to sand. This affords them excellent protection against predators. In order to find live cockles one has to dig for them. Those lying on top of the sand are mostly empty shells. The foot is also used for locomotion and it is so powerful that these bivalves can actually hop over the sand in order to escape from a predator. Cockles are decorative. Both valves are equal and are marked with distinct ribs.

Trachycardium species live in the Caribbean region, while those of genera *Acrosterigman*, *Fulvia* and *Fragum* are common in the Indo-Pacific. A special position within this family is occupied by genus *Corculum*, which contains about 10 species (KNOP, 1996). These are the only bivalves (except for the giant clams of family Tridacnidae) that have symbiotic algae in their tissue. This then enables them to live as autotrophic or partially autotrophic organisms with light as energy source. The best-known species in this family is *Corculum cardissa*. It occurs in the central Indo-Pacific, from the Amami Islands near Japan to Indonesia and Australia, where it lives on a sand bottom in the shallow water of reef flats and in inter-tidal zones, exposed to very high light intensity. The shell can be shaped differently from one specimen to the next. *Corculum cardissa* is often rather difficult to detect in its natural habitat (KAWAGUTI, 1950).

Corculum cardissa **(family Cardiidae) and the about 10 other species in the genus are the only species of bivalves, except for the giant clams (family Tridacnidae) that contain symbiotic algae. Their shells are very thin in order to penetrate light.** **Photos: D. Knop**

Although there are zooxanthellae present in the mantle of *Corculum cardissa*, the mantle itself does not extend beyond the two valves like in *Tridacna*. Light must first penetrate the shell before it reaches the zooxanthellae, and the shell is rather thin and almost transparent. Without such adaptation life with symbiotic algae would not be possible. Although this is an interesting bivalve for the aquarium, we are not aware that *Corculum cardissa* has ever been imported.

Family Tridacnidae - Giant Clams

The bivalves best known in the aquarium hobby are – beyond doubt – the giant clams of family Tridacnidae. For many aquarists these magnificently coloured clams are the actual landmarks of a coral reef aquarium. Now there is even special literature on this topic (e.g. KNOP, 1996). For some years now giant clams are being bred commercially in aquaculture facilities in some regions of the world, as food for human consumption as well as for the aquarium trade.

Family Tridacnidae encompasses the genera *Tridacna* and *Hippopus*, with a total of nine species that occur only in the Indo-Pacific. When viewed phylogenetically, it is a young family, which can be traced back only as far as the Eocene Period, which commenced about 60 million years ago. Giant clams of Family Tridacnidae are highly specialized. In comparison to other bivalves, they have undergone substantial morphological modifications, in order to live in shallow water on coral reef and in symbiosis with zooxanthellae.

General Biology

Giant clams can be attached to the substrate with byssal threads, but they can also simply lie unattached on the bottom. They can bore in to rocks or grow among branching corals. Specimens that live in very shallow water can even become exposed to air for brief periods during low tide, without sustaining any apparent harm. However, on the Great Barrier Reef we saw a number of dead giant clams of genera *Tridacna* and *Hippopus* on reef flats

The beautifully coloured mantle is one of the reasons why giant clams are so popular with marine aquarists. This photo shows what is possibly a *Tridacna squamosa* on the Great Barrier Reef. Species identification based on mantle colouration is limited by the fact that this can vary within a species and may also be genetically as well as environmentally dependent. **Photo: M. Cufer**

that became frequently exposed. This could suggest that not all specimens can handle exposure to air equally well, and that extreme low tide is actually very stressful for giant clams.

Giant clams are simultaneous hermaphrodites, which means that each specimen is male as well as female, and so releases sperm and also eggs. If the sex products were being released at the same time there would be the risk of self-fertilisation. Yet, considering the development of the species, self-fertilization would be a disadvantage. Larvae of giant clams that originated from self-fertilisation do not develop. Therefore, sperm cells are normally released first, and the egg cells are not released until some time later. With such mechanism in place, self-fertilization is largely avoided.

Reproduction takes place normally during the warmest season, but it is not an optimal temperature alone that stimulates the development of sex cells. Phase of the moon, formation of pheromones in the clam as such as well as a series of other seasonally-rela-

Tridacna squamosa **off Papua-New Guinea at a depth of 4 m (left) and at the northern Mariana Islands at a depth of 6 m (right).** **Photo: P. Schupp**

A specimen of *Tridacna crocea* secrets egg cells during spawning in an aquarium.

ted phenomena participate in this complex interaction (Braley, 1984; Heslinga et al., 1984). The reproductive productivity of giant clams is enormous. One mature *Tridacna gigas* specimen can release several hundred million sex cells per day during the reproductive phase.

The fertilized egg divides very rapidly. Within a few hours the blastula stage has already several hundred cells. In the course of the first 12 hours after fertilization trochophore larvae develop that have a ring of cilia (ciliated band, prototroch) which functions like a propeller. Movement of the larvae follows a particular scheme. They swim along the surface in slowly moving circles, remain (stationary) for a while, and then drift downwards again in the water column.

The veliger stage normally commences barely 48 hours after fertilization. The veliger larva develops gradually and three days after fertilization it has a distinct foot, an intestinal tract, an anal opening, digestive glands, interconnected shell-halves, cilia (velum) for swimming and fat droplets as energy reserve. In principle, the barely 0.2 mm long veliger larva is already a miniature giant clam. After another three days the larva develops in to a prediveliger larva whereby the foot continues to grow and becomes about twice as long as the larva as such. This is the first step of the metamorphosis to a juvenile, firmly attached giant clam. However, during this entire period the mortality of larvae is very high. Only a small number of them survive; most become victims of the many predators present on a reef.

Experimental laboratory tests have revealed that the symbiotic relationship with zooxanthellae can commence already from the third day after fertilization, i.e. during the veliger stage of the larvae. However, it is not exactly known when this relationship starts in the sea. Symbiotic algae, which are of great benefit to the clam throughout its entire life, are not present from the beginning, that is, they are not present in the larvae, as is the case in many corals. Giant clams must acquire the zooxanthellae; more specifically they themselves must "capture" them. As was mentioned in Volume 1 (page 322), the symbiotic algae have a free-swimming stage. That is when giant clams "capture" the algae by feeding on them (phagocytosis), since phytoplankton is part of the normal diet of giant clams. The free-swimming flagellates are easily filtered out of the water by giant clams. This process takes place already when giant clams are still quite small. At that stage something extraordinary happens: for reasons still unknown, the free-swimming zooxanthellae have the ability to survive digestion inside the stomach of giant clams (Fitt et al., 1984, 1986). The microscopic algae can stay inside the stomach of a giant clam without getting digested.

Metamorphosis from larva to juvenile, sessile clam can – under optimal conditions – commence a few hours after fertilization or be delayed for a week if conditions are unfavourable, but tests in aqua-cultures have shown that there can be substantial variations in metamorphosis time-span. A short time after metamorphosis one can see how algal cells spread out over the mantle of a giant clam (Fitt & Trench, 1981). The juvenile clams collect free living zooxanthellae and accumulate these in the tissue surrounding both mantel openings (Knop, 1996) Soon thereafter a large population of algae has become established. A five-year old *Tridacna derasa* contains more than 500 million zooxanthellae (Heslinga et al., 1990). In contrast to cnidarian animals where the algal cells live inside endodermal cells, the zooxanthellae in giant clams establish themselves in the ectoderm, that is, they live in the outer-most cell layers (Trench et al., 1981). Just how these symbiosis relationships originated millions of years ago is uncertain. The most likely scenario is that the flagellates wandered in to the cells around the inhalant opening of the clams by accident (Yonge, 1936).

Little is known about the longevity of giant clams, but presumably some specimens can get very old (may be several hundred years). Rose (1965) assumes that there is an average growth of about 5 cm per year. We have kept *Tridacna squamosa* in our tank for 10 years. During this period the clam grew from a length of 18 cm up to more than 50 cm length. Similar growth rates have been observed in aquariums worldwide. We have also kept a juvenile specimen of *Tridacna gigas* in a small experimental aquarium (NILSEN, 1999, 2000) and measured the growth regularly. During a period of 420 days the clam increased from a start weight of 330 grams to a weight of 1075 grams, which means that on an average basis the clam added 1.77 grams to its weight daily!

Apart from *Corculum* species of family Cardiidae, giant clams are the only bivalves that contain symbiotic algae. The zooxanthellae and pigments give the mantle an intense colouration, whereby each clam has its own, individual pattern. Therefore, it is not possible to identify *Tridacna* species on the basis of mantle colouration alone. Coloration is genetically as well as environmentally linked. Even among the progeny of the same parents there is great variability in mantle colouration, and only in particular locations can mantle colouration be used for species identification.

An interesting feature of the symbiotic relationship between giant clams and algae are the beautiful hyaline organs located in the mantle. These are specific sites with a particularly high concentration of zooxanthellae, which can be seen as small, fluorescent crystalline dots and spots.

There is little doubt that for their survival, giant clams are totally dependent upon the zooxanthellae and their photosynthesis. Although giant clams can, to a certain extend, filter phytoplankton from the water and then utilize that as food, but the photosynthetic products of zooxanthellae make up the essential component in energy uptake. In most cases symbiotic algae provide all of the required energy (FISHER et al., 1985). Therefore, in most species the organs for food intake show signs of degeneration, which is also an indicator of the dependence on symbiotic algae. It is interesting to note that giant clams – just like many corals – can absorb dissolved organic compounds directly from the water via their mantle (FRANKBORNER, 1971; GOREAU et al., 1973). The advantage for giant clams of the symbiosis with algae has many similarities to the symbiotic relationships in corals. The zooxanthellae can recycle vital nutrients and so they are particularly well adapted to the nutrient-deficient environmental conditions on coral reefs.

Relative to their shell, the internal organs of giant clams are – when compared to those of other bivalves – reversed, which is due to the light requirements of the zooxanthellae. The mantle is well developed and protrudes far beyond the edges of the shell. In addition, the mantle has two distinct openings: one wide opening through which water flows in (inhalant opening), and a small, funnel-shaped exhalant opening. The mantle extends beyond the valves along the dorsal side, while valves along the abdominal side are inter-connected via the hinge-ligament. While other bivalves usually have several muscles, only two strong muscles have remained in giant clams. Another characteristic are the strangely massive kidneys, which can make up about 15 % of the weight of all internal organs. The kidneys remove toxic heavy metals, such as zinc, copper, mercury, cadmium and lead from the water. That means that giant clams can be used as indicator organisms for heavy metal pollution in seawater (DENTON & WINDSOR, 1986).

The shell structure is decisive

Hippopus hippopus **among sea grass in very shallow water at Green Island, Great Barrier Reef.**

Tridacna crocea **(left) at a depth of 5 m, and** ***Tridacna maxima*** **(right) at 6 m depth off Chuuk Islands (Micronesia).**
Photo: P. Schupp

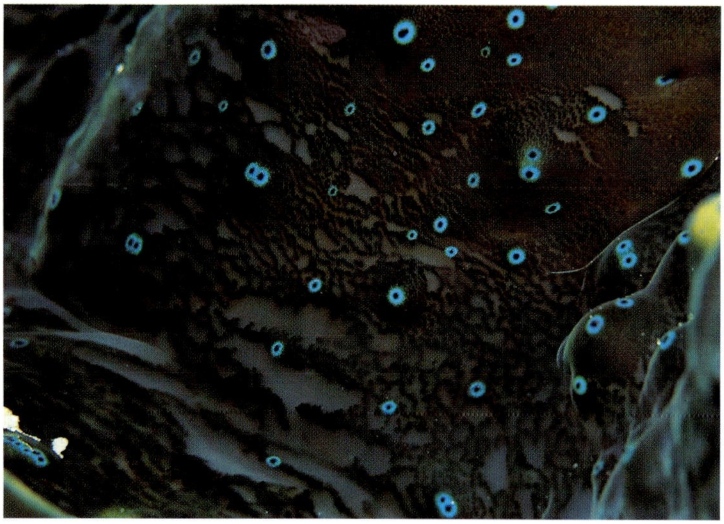

Hyaline organs, seen here in a *Tridacna gigas*, are spots on the mantle containing an extremely high concentration of zooxanthellae.
Photo: S. W. Michael

for species identification. In some species the shell is significantly folded and is covered by outgrowths, while the shell structure of other species is far smoother. The outside of the shell can be heavily overgrown, the principal reason why so many other interesting invertebrates are often inadvertently introduced in to coral reef aquariums. Sometimes giant clams also produce pearls, which can reach enormous sizes (COBB, 1939). In the tropical regions of the world giant clams are treasured as food, and they are eaten raw, cooked, fried or grilled. This has lead to excessive exploitation of these clams in certain areas.

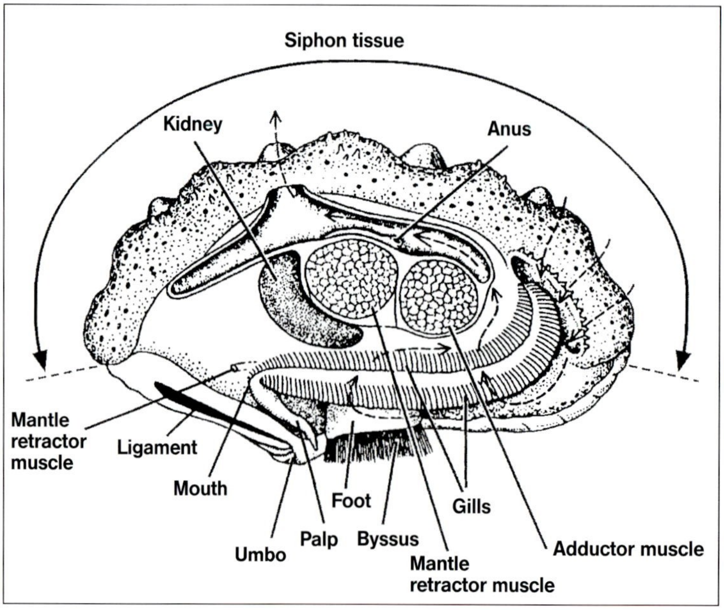

Schematic drawing of exposed internal organs of giant clams of genus *Tridacna*. The dashed arrow lines indicate the direction of water flow inside the mantle cavity.
Drawing after YONGE (1975)

Giant Clams in the Coral Reef Aquarium

In the early days of marine aquaristic, giant clams could not be kept alive in captivity at all. Like stony corals and most other calcifying organisms they were doomed. As techniques improved, the clams proved to be rather easily kept and many wild-caught specimens were frequently available in the trade. Big specimens of *Tridacna squamosa* were exported, one animal in one box, and were kept for years in captivity by enthusiasts in many countries. Now the situation is very different. Due to a worldwide protection of giant clams and CITES regulations, wild-caught specimens are relatively rare in the trade. There are differences from country to country, but in 2001 less than 25% of all clams sold are wild-caught. Instead farm-raised clams are being sold for commercial purposes and compose the vast majority of the giant clams sold for aquarium purposes.

Giant clams can be kept alive in a coral reef aquarium only if they are given a high illumination intensity and optimal water quality. If lighting is too low, they will gradually die. On the other hand, we have also observed that giant clams can react negatively when they are transferred from a location with moderate light intensity to one with very intensive light, where they had obvious problems tolerating the new light intensity. Consequently, one must always proceed rather cautiously when acclimating giant clams to aquarium conditions.

Although it is unknown (and probably highly variable) to what an extent the giant clams utilise phytoplankton in the wild, feeding with live phytoplankton does seem to have a positive influence on the durability of giant clams in captivity. Although corals reefs in general are regarded to be poor in nutrient, minute plankton is rather abundant and might very well be a vital food supply to clams, corals and other invertebrates. Live phyto- and zooplankton can be cultivated at home and added to the system drop wise under control, which will probably almost always have a positive effect on the well being of the system.

Juvenile clams are attached to substrate with the aid of byssal threads that are excreted from the foot. In *Tridacna squamosa*, *T. gigas* and *T. derasa* this attachment loosens when the clam grows. **If one wants to transfer an attached juvenile clam to another site in the aquarium, one has to detach the clam very gently.** If the clams is roughly pushed or ripped out so that the byssal threads are torn, that – in most cases – will damage the foot, and in our experience that means certain death for the animal. Therefore, it is highly advisable to enable a clam to become attached to a small rock. If such a clam needs to be moved, it can be done easily simply by moving it together with the substrate (rock) it is attached to.

A giant clam that has been injured spreads its mantle in a particular way. The valves are opened unnaturally wide, the mantle is tightly wedged between the shell halves and the exhalant opening is wide open, called "gaping". In our opinion, this is a sure sign that the clam has been injured and will die.

Giant clams do also react rather negatively, when they are being overgrown by filamentous algae or other organisms, such as glass- and other small anemones; however, when they given optimal aquarium conditions they can live for a long period of time. In the presence of an adequate calcium concentration they often grow several centimetres per year (YONGE, 1980).

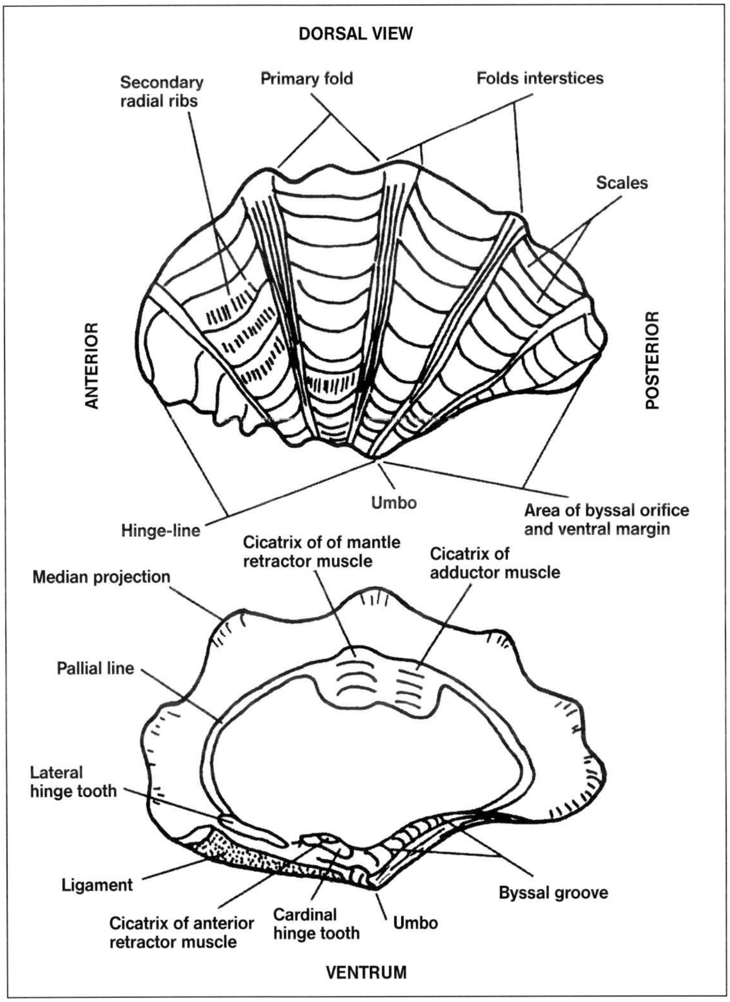

Schematic drawings of shell characteristics of giant clams of genus *Tridacna*. Drawing after ROSEWATER (1965)

Parasites and enemies

It goes without saying that giant clams must not be kept together with fishes that normally include them in their diet, e.g. large wrasses and similar fishes. Yet, we had not expected that the small Molly Miller, *Scartella cristata*, from the Caribbean region could also inflict damage to giant clams. As soon as this fish had been introduced in to the aquarium with a *Tridacna squamosa*, the clam closed up and remained in that condition for several weeks. That, of course, caused the zooxanthellae to disappear and the colouration of the mantle was substantially lost (bleached out). Purely by chance we discovered that the small blenny picked on the edge of the mantle. The fish was immediately removed and a few minutes later the clam opened up again. Within a period of several months the zooxanthellae population was re-established, and the clam obtained its normal mantle colouration again. How was the clam able to detect so rapidly that the danger had passed? May be an excellent example of the effect of chemical substances between predator and its prey that can indeed occur in nature.

There is also a number of small parasites that attack giant clams. The small snails of the family Pyramidellidae (see Fact Sheet on page 101) frequently live on giant clams where they breed and multiply to large populations. Jan Olsen of Aqua Design, Oldenburg,

The deep orange-coloured bristle worm *Oenone fulgida* is – with a diameter of about 2.5 mm – very thin, but it can reach a maximum length of 10 to 30 cm. It is a specialised to feed on molluscs, e.g. giant clams and snails of genus *Astraea.* This worms suffocates the snails by covering them with drops of mucus. Once a snail is dead, the worm removes the operculum and eats the snail (left). The worm can bore perfectly round holes in to the valves of giant clams close to the umbo (right), and then feed on the clam's tissue. Giant clams can plug up such holes by using a protein matrix and a calcium carbonate plug, however, the worm weakens them to such an extent that they ultimately die of an infection.
Photo: J. Sprung

Germany removed 38 snails from a single specimen of *Tridacna maxima*. Other predatory snails can also attach giant clams. Among these are *Vexillum plicarium* (family Costellaridae), *Cymatium* spp. (family Ranellidae), *Chicoreus* spp. and other snails of family Muricidae as well as members of families Fasciolaridae and Buccinidae. One subspecies of flatworm from the family Stylochidae (see Volume 3, page 76) has been reported to attack *Tridacna gigas* (KNOP, 1996). The bristle worm *Oenone fulgida* attacks not only giant clams, but also other bivalves and gastropods (see photos above). Boring sponges from the genus *Cliona* attack the shell of giant clams and weaken it a state that makes the clam much more vulnerable to other parasites. As for stony corals boring green algae are also known to attack giant clams. The microscopic algae will in a severely attacked clam give the inside of the valves a greenish colour and will, just like the boring sponges, over time weaken and harm the clam. Several crustaceans, especially some crabs from the family Dromiidae (see Volume 3, page 316) and family Xanthidae (see Volume 3, page 328), can harm giant clams. There are, however, also examples of commensal crustaceans, such as pontoniin shrimps, that live permanently inside giant clams without harming them at all.

KNOP (1996) reports on a protozoa infection that killed giant clams in his aquarium. The clams got severely infected by protozoans resulting in several white spots appearing on the clam's mantle, a state now known as "Whitespot Disease" or "WSD-infection". The infection is, accor-

As we have outlined in the Fact Sheet on page 101, snails of family Pyramidellidae can tap the mantle of giant clams for food, causing them to die. When we visited "H_2O – Tropicals" - the former aquarium shop of Jeff Macaré in the vicinity of Los Angeles – we discovered a few giant clams that had been attacked by tiny snails of family Trochidae. These snails were located close to the byssal opening. We were unable to determine whether the snails were causing any damage to the clams. In any event, such observations should always be sufficient reasons for concern!

Above to the left: Shell of *Tridacna crocea* attacked by boring sponge.
Above: The mantel of this *Tridacna squamosa* shows clear signs of white spot disease as well as bleaching.
Left: Three specimens of *Tridacna crocea*; the back one is healthy, the specimen in the middle is gaping and is dead, while the right specimens shows severe signs of mantle bleaching. Photos: D. Knop

ding to KNOP (1996), always fatal, death occurring from a few weeks to three months after the outbreak. The Protozoa that causes the infection is probably a subspecies of a known species in the genus *Perkinsus*, a group of protozoan that is known to cause diseases in commercial mussel farms.

Bacterial infections can also attack giant clams. *Vibrio* spp. as well as *Aeromonas* spp and *Plesoimonas* spp. are known to attack larvae and juvenile clams less than two months old, according to KNOP (1996) especially in aquariums which contain a lot of detritus. The results are always fatal. Other bacteria, probably again belonging to the genus *Vibrio*, attack juvenile as well as adult clams. Several causes seem to lead to bacterial infections; stress during catch and transportation, poor acclimatisation and changing of the aquarium environment are just three. It is also very likely that some clams can carry bacteria into the aquarium and spread infections. There are several incidents that new *Tridacna maxima* have been added to reef tanks resulting in the death of the new as well as of all other giant clams in the system. According to personal communications with several aquarists, *T. maxima* seems to be a lot less durable than the other species of giant clams commonly found in the trade, a fact which might point to the species being more vulnerable to bacterial infection than other species.

Perhaps the most common problems with giant clams in reef aquariums are bleaching of the mantle, either partially or completely loses of the symbiotic algae. Lack of nutrient for the algae, such as a too low concentration of nitrogen, insufficient light or a changing of light are the main reasons for this, all causes that in principle can be controlled by the aquarist himself.

Brood stock of *Tridacna maxima* at the Wau Island Mariculture Facility (Marshall Islands). This photo demonstrates clearly how variably coloured the mantle can be in different specimens of the same species.

Photo: J. Macaré

Commercial Breeding of Giant Clams

One of the most interesting developments in the aquarium hobby may well be the utilization of animals produced in aquaculture facilities. In many parts of the world there are now aqua-farms that specialize in the commercial rearing of giant clams. The reason for this commercial production is the fact that natural clam populations in many localities were being decimated due to excessive harvesting for human consumption as well as the shell craft industry. In Asia – towards the end of the 1970's - 500 tons of giant clams were being eaten every year (Erhardt & Moosleitner, 1995). The shells of giant clams have been (and are to this day) ornamental objects high in demand for jewellery and other decorative purposes, especially the curio trade. The trade is now controlled by the Washington Convention on the Importation and Trade of Endangered Species (CITES), where Giant clams, together with stony corals are listed in Appendix 2. But this control only concerns the trade, and in many tropical countries the production of those decorative items is still going on (Knop, pers. comm.).

This is a size comparison between 4 ½ months old juvenile giant clams raised in an aquaculture facility, and a fully-grown *Tridacna gigas*.

Photo: R. D. Bradley in "Revisto da Phillippine Airlines", 1/92

The first successful artificial rearing of giant clams involved *Tridacna derasa* at the Micronesia Mariculture Demonstration Center (MMDC) at Palau (Micronesia, now known as PMDC – "Palau Mariculture Demonstration Center") under the supervision of Dr. Gerald Heslinga. From modest beginnings during the years 1979 to 1982, when only 30,000 juvenile clams were produced, production increased to 100,000 clams in 1984 and to more than 250,000 clams in 1985. Similarly, the University of Papua-New Guinea (Motupore Island Research Center) as well as several institutions in Australia, including the University of New South Wales (Australian Center for Agricultural Research; ACIAR) and the James Cook University in Townsville (Northern Queensland) participated in giant clam breeding pilot projects. Gradually aqua-farms were established for breeding and research on the aquaculture of giant clams, especially in Micronesia and throughout the central Pacific, including the International Center for Living Aquatic Resources Management (ICLARM) on the Solomon Islands and in the Philippines (Silliman University Marine Laboratory and Marine Science Institute at University of the Philippines, MSI).

As is common for such ventures, problems did arise; most notably financing these projects presented the greatest challenges; however, it had not been anticipated that the aquarium hobby would eventually emerge as the most important market. Although the commercial production of giant clams for human consumption and conservation of species had been the initial goal, in reality this proved to be more problematic than had been expected. Initially it was thought that commercial breeding of giant clams would cause the price of clams caught illegally in the wild to drop and that this trade would gradually cease. Unfortunately, this turned out not to be true! Restaurants want to buy large, fleshy clams at favourable prices, which cannot be achieved through commercial aquaculture, since a large clam must grow over many years.

After six months in a rearing tank, juvenile clams - here *Tridacna squamosa* - have reached a size of about 2 cm. At that point they are transferred to the protective environment of underwater farms for further growth.
Photo: "Marine Science Institute" in "Revisto da Philippine Airlines", 1/92

Hundreds of juvenile *Tridacna maxima* reared at the Wau Island Mariculture Facility (Marshall Islands).
Photo: J. Macaré

Left: Preparation to trigger artificial release of gametes in *Hippopus porcellanus.* Here a syringe is being filled with the hormone Serotonin. Right: Injecting Serotonin in to gonadal tissue of the clams, the progeny of the first artificially reared (F1) generation.

Left: The spermatozoa, released soon after the hormone injection, are being collected and are stored in a separate container. Right: The subsequent release of eggs must be considered a milestone in the breeding of giant clams, since it was the first time that had been achieved with a captive bred specimen of the rare *Hippopus porcellanus.*

Left: Immediately upon release of the first egg cells the giant clam was isolated and placed in sperm-free seawater. Right: As soon as the egg density had reached a certain level, a specific volume of sperm-containing water was added in order to facilitate fertilisation.

Six Photos: D. Knop

Beyond that, it was also thought that with the help of commercially produced clams, coral reefs in those areas where the clam populations had been severely decimated could be re-introduced. That too turned out to be incorrect, certainly in areas around the Philippines (KNOP, pers. comm.), since the released giant clams were also poached and sold. The result is that clam farms have discovered the aquarium hobby as the most important and so far the best market. Importers buy the small clams at relatively high prices, which enables clam farms to survive economically. According to KNOP (pers. comm.) in 1996 probably all commercial aquaculture facilities for giant clams were vel (e. g. Marine Science Institute, University of the Philippines, MSI; KNOP & ROA, pers. comm.). The large clam farming facility in the International Centre for Living Aquatic Resources Management (ICLARM) on the Solomon Islands has shut down from local political reasons.

Rearing giant clams in aquaculture facilities is briefly described below. The process starts out by collecting brood specimens of different species from their natural habitat. Initially, these were only the large species *Tridacna derasa* and *T. gigas* that were the easiest to rear. But then eventually there was also success with the remaining species, except for *T. mbalavuana* (= *Tridacna te-*

shown that it is relatively easy to induce sperm secretion, but it can be difficult to stimulate captive specimens to release egg cells. Normally, sperm cells are – as usual - released, but egg cells are being retained (our observations have revealed that there appears to be a similar tendency in the reproduction of stony corals when kept in an aquarium). This may well be the most difficult biological problem in the captive reproduction of giant clams (KNOP, pers. comm.). In order to stimulate the production and release of eggs, the clams are exposed to an artificial stress situation, by raising the temperature to more than 30 °C through direct exposure to the sun for a brief period

Brood stock of *Tridacna squamosa* (left) and *Tridacna gigas* (right) in brood containers at the Wau Island Mariculture Facility (Marshall Islands).
Photos: J. Macaré

totally dependent on the aquarium hobby and so on us, the aquarists, for their survival. The aquarium trade is the most important market for these aqua-farms. This then means that aquarists who purchase giant clams reared in commercial breeding facilities, contribute indirectly to the protection of natural populations.

But due to missing exportation licenses, for many farms the clam farming has not been profitable. The strict trading prohibitions that protect wild stocks of Giant Clams are in some countries also applied to farm raised clams and drastically limit their market. Consequently some farms have stopped production (e. g. Silliman University Marine Laboratory) or reduced it to an extremely low le-

voroa) and *Tridacna rosewateri*, which are very rare in the wild. Nowadays, the species most commonly bred are *T. derasa*, *T. gigas*, *T. squamosa* and especially *T. crocea*. The rare species *Hippopus porcellanus* has only been bred in three locations: Silliman University (Philippines), ICLARM-CAC (Solomon Islands) and recently in MSI, University of the Philippines by Hilly Ann Marie Roa.

The clams are kept on land in large vats with flowing water. Selection of the right brood stock is critical since sexually mature specimens must be able to release sperms as well as eggs. As already mentioned above, giant clams are simultaneous hermaphrodites, however, it has been

(KNOP, 1996). In this procedure, 33 °C is a critical point, which should never bee exeeded (ROA, pers. comm.). To induce spawning, an injection of the hormone Serotonin into the gonadal tissue has shown to be a more practical procedure than the use of macerated gonadal tissue. Serotonin has therefore become part of the routine procedures in Giant Clam farming.

For spawning, the broodstock clams are kept in a shallow tank. Many clams may expel sperm, but normally only a few spawn egg cells. As soon as a clam starts to expel eggs, it immediately has to be transferred to another container filled with filtered seawater, where a small portion of the sperm containing water will

Of the nine known species of giant clams, only *Tridacna crocea* is commonly encountered in the central Indo-Pacific.
Photo: S. W. Michael

be added (KNOP, 1996). A higher concentration of sperms inhibits the development of the eggs. KNOP (1996) described the Giant Clam breeding procedure in detail. The developing larvae are usually kept in aquariums with a volume of about 4,000 l (ERHARDT & MOOSLEITNER, 1995). Depending on the species, the clams stay in the shallow tanks of the land based hatchery for 1 to 3 years. Afterwards they will be brought to the ocean nursery, where they grow out to a marketable size of 5 to 6 cm, which takes several years, depending on the species (ROA, pers. comm.). Rearing giant clams for human consumption takes some additional years. After seven years, *T. maxima* will have reached a size of approximately 16 cm, a *T. gigas* will have grown to about 30 cm. During this period, clam breeders have to cope with many problems, e. g. bureaucratic controls that are not adapted to the sale and export of giant clams, thefts in rearing facilities and parasitic infestations (e. g. by snails of family Pyramidellidae).

As in most aquaculture ventures, rearing Giant Clams is a time-consuming, complicated and expensive process. For detailed information about the aquaculture of Giant Clams refer to BRALEY, 1984; CALUPONG, 1992; COPLAND & LUCAS, 1988; HESLINGA & FITT, 1987; HESLINGA & WATSON, 1985; HESLINGA et al., 1984, 1990; KNOP, 1996; SHOKITA, 1991.

Giant Clam Species

Family Tridacnidae contains only the genera *Hippopus* and *Tridacna*, with a total of nine known and described species. Seven of these are currently of interest to the aquarium hobby. In some instances, mantle colouration can aid species identification, but this contains many uncertainties. Usually other characteristics have to be used in order to identify a species. The identification key on page 184 can be of help.

Hippopus hippopus, or the Strawberry Clam, (see Fact Sheet on page 185) is probably the most primitive giant clam. This species lacks hyaline organs, which could be indicative of this. The species is usually found on a sandy bottom, from the shallowest water down to a depth of 6 to 7 m. We found *H. hippopus* densely covering sand bottom in the lagoon of Flinders Quay in the Coral Sea where they lived together with large stands of soft corals of genus *Xenia*. Juvenile specimens are attached with byssal threads, while older ones live unattached. The mantle does not exceed the edge of the shell, which is thick and has a peculiar shape. Perpendicular ribs with red markings run along the outside of the shell, which is the reason for its colloquial name Strawberry Clam.

Hippopus porcellanus is generally called the Porcelain Clam, and has a strong resemblance to *H. hippopus*; however, it has a thin and more spherical shell. A series of small, branched papillae are located around the inhalant opening. This feature is the best species characteristic and distinguishes it from *H. hippopus*, which lacks papillae. In addition, *H. porcellanus* has a smoother shell and a distinctly restricted distribution. It occurs in large in umbers only in waters around the southern part of the Philippines. The difference in geographical distribution is the principal reason for establishing two separate species instead of considering it as a subspecies of *H. hippopus* (ROSEWATER, 1982). It rarely ever reaches the aquarium trade.

Tridacna gigas (see Fact Sheet on pages 190) is the largest of all clams. It is reported to have killed humans, by wedging their body parts between the half-shells and then closing down on them. This is the reason for the unattractive designation as "Killer Clam", but these are doubtful, unsubstantiated reports (JOHNSON, 1930). It is also listed in the most interesting and informative book by HALSTEAD et al. (1990) "A Color Atlas of Dangerous Marine Animals" next to a page covering "Polar Bears", which is – mildly expressed – a bit of a dramatic exaggeration! *Tridacna gigas* is a beautiful and impressive mollusc that would not hurt anyone (but it is of course not necessary to put your hand or food in between its large valves). The species usually lives on sandy substrate among coral formation or among scattered coral growth in shallow waters. Adult specimens – because of their obvious size – cannot be mistaken for other species. This clam does not drill into a substrate, as seen in *T. crocea* and *T. maxima*, but lie loosely on the substratum. The shell is usually open during the day, unless the animal is being disturbed, when it would close rapidly, however, unlike other giant clams it cannot close its valves completely.

Porcelain Giant Clam, *Hippopus porcellanus*, at the Waikiki Aquarium in Honolulu, Hawaii.

The distribution area of *T. gigas* in the central Indo-Pacific is restricted. Although reports suggest that this species occurs in the same localities as *Hippopus hippopus*, its distribution is more in the centre of this region. *T. gigas* has been heavily exploited, since it is a highly prized delicacy, so that population numbers have substantially declined in certain localities. In fact, in some areas it appears to have become already extinct (like *T. derasa*); this includes the area around Guam, Yap, Truk, Pohnpei, Kosrae as well as parts of southern Japan and the Philippines (HESLINGA et al., 1990). Efforts are now being taken to breed and raise this clam in farms and re-establish it back in the sea. The only location where *T. gigas* is still numerous is the Great Barrier Reef. In Australia law pro-

***Hippopus porcellanus*: External view of shell with barely indicated contours only (left), and inside of the shell (right). It is interesting to note the length/height ratio. The height of the shell is nearly identical to its length, a characteristic that is commonly to both *Hippopus* species.** **Photo: D. Knop**

Identification Key for Giant Clams

(Modified from KNOP, 1996)

1	a	The mantel is overhanging the shelf rim	2
1	b	The mantel is not overhanging the shelf rim	5
2	a	The byssal orifice is narrow	3
2	b	The byssal orifice is wide	4
3	a	The valve lacks scales and currugations, the rim of the valve in large specimens has large, pointed inward bends. There are not tentacles along the rim of the inhalant mantel siphon, the mantel is often ochre with several metallic blue hyaline organs .... *Tridacna gigas*	
3	b	The valve is smooth without scales and corrugations, the inhalant siphon with a ring of tentacles .... *Tridacna derasa*	
4	a	The valve has long, widely interspaced scales stretching from the upper rim to the bottom of the shell .... *Tridacna squamosa*	
4	b	The valve is asymmetrical and has plate-shaped scales that are growing densely in rows on the upper half of the valve, but only found sparsely (or even missing) on the lower half. The mantel is often metallic blue or green .... *Tridacna maxima*	
4	c	The valve has only a few long scales on it's upper part, the rest of the valve is smooth, but has file that resemble horizontal corrugations. The byssal orifice is very wide. The mantel is often metallic blue .... *Tridacna crocea*	
4	d	The valve has very large scales. The shell is in contrast to the other species of giant clams, very thin .... *Tridacna rosewateri*	
5	a	The byssal orifice without a toothed rim .... *Tridacna mbalavuana*	
5	b	The byssal orifice with a toothed rim, narrow and slit-shaped	6
6	a	Inhalant orifice without a ring of tentacles .... *Hippopus hippopus*	
6	b	Inhalant orifice with a ring of tentacles .... *Hippopus porcellanus*	

tects this species, so that on some reefs it can be found again in densities of up to 30 specimens per hectare (CRAWFORD & NASH, 1986).

Tridacna derasa (see Fact Sheet on page 187) is generally considered (together with *T. gigas*) one of the "true" Giant Clams, yet the largest specimen ever found was only slightly longer than 50 cm, i.e. substantially less than the maximum length of *T. gigas*. The species is distributed in the central Indo-Pacific like *T. gigas*, but is found as far southeast as to the Lord Howe Island.

Tridacna derasa is easy to identify, since the shell is usually very smooth with virtually no outer shell sculptures. This is probably related to the fact that the species lives in areas of strong water movement along the outer edge of barrier reefs. Specimens that live in calmer waters do, however, develop a distinct outer shell sculpture. The inaccessible habitat of this species is the principal reason why wild specimens are rarely ever found in the aquarium trade. *Tridacna derasa* was, however, the first giant clam species ever to be bred in captivity. Nowadays it is bred on a large scale and so it occurs quite regularly in the trade. Like *T. gigas* the growth rate in the reef aquarium is phenomenal. We have seen huge specimens with bright iridescent blue-striped mantel colours in the Great Barrier Reef. The bluish colouration seems to be a typical colour among many natural populations, while farm raised specimens tend to develop a brownish striped mantel.

Tridacna squamosa (see Fact Sheet on pages 194) was regularly available in the aquarium trade in Europe from the beginning of the eighties, when large specimens were sold and very successfully kept in reef tanks. These days such specimens are very rarely seen. Wild caught Giant Clams are now protected and are replaced in the aquarium trade with farm-raised.

Tridacna squamosa can be readily recognized by its large and regular, plate-like outgrowths on the shell. Mantle colouration is often brown, yellow or shades of ochre, but is normally not at all as fluorescent as that of *T. maxima* or *T. crocea*. The species is widely distributed in the Indo-Pacific including the Red Sea.

Tridacna squamosa is easy to keep in a coral reef aquarium, provided there is a high light intensity and optimum water quality. Growth is rapid and so this clam must be given sufficient space to grow. Newly imported specimens often carry symbiotic shrimps of family Palemonidae, and occasionally crabs of family Pinniheridae.

Because of its species name, aquarists tend to believe that *Tridacna maxima* (see Fact Sheet on pages 192) is the largest of the Giant Clams. However, it reaches only a length of 35 cm, which makes it one of the smaller species. Yet, *T. maxima* is the Giant Clam with the largest distribution area. This species occurs throughout the entire Indo-Pacific, from East Africa and the Red Sea to the Pitcairn Island. Northward its range extends to southern Japan, but it does not occur around the Hawaiian Island. The southernmost occurrence is around Lord Howe Island, off the east coast of Australia, and off Durban in South Africa.

The two valves of *Tridacna maxima* have an asymmetrical shape

Hippopus hippopus
Hippopus- or Strawberry Clam

Distribution: Central Indo-Pacific; from Thailand eastward to the Marshall- and Gilbert Islands as well as from Japan southward to the Great Barrier Reef and Western Australia. According to KNOP (1996) this species has recently become extinct around Tonga, the Fiji Islands and Samoa. Listed as an "endangered species".
Size: Up to 45 or 50 cm, but normally only to 40 cm.
Reproduction: Simultaneous hermaphrodite. Gametes released in to open water, where fertilization takes place.
Food: Feeds on the products of its zooxanthellae as well as on phytoplankton and possibly also from dissolved organic compounds..
Description: Lives normally on sunny reef flat in shallow water down to a depth of 6 m. Adult specimens lie unattached on the bottom, but are able to maintain the correct position due to their weight. Juveniles live usually attached to corals or coral rubble. The wide based of the shells is shaped like a horse's hoof (hippos, Greek = horse and pous, Greek = foot). The shell with strawberry-coloured markings. It is thick, with strongly developed ribs and has a small byssal opening. The mantle does not protrude beyond the edges of the shell. Colouration of the mantle along the shell edge has an irregular, sprinkled pattern; the centre is normally dark-greenish yellow.
Aquarium requirements: Relatively rare in the trade. Easy to keep, but requires a high illumination intensity. Very well suited for sand zone aquariums. Grows fast and needs space where it can expand.

GA: +; **SE:** +; **CO:** +

Photos:
Top: *Hippopus hippopus* specimen partially exposed out of water, on a reef flat at Green Island, Great Barrier Reef. The valves are often overgrown with many organisms.
Centre: Three specimens at Low Island, Great Barrier Reef, during low tide.
Bottom: Strawberry-coloured markings well visible on this prepared shell.

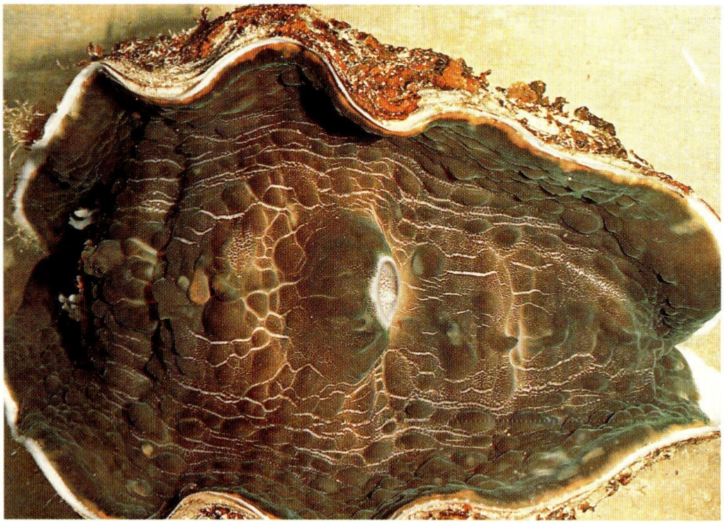

***Tridacna mbalavuana* (= *Tridacna tevoroa*), dorsal view.** **Photo: J. Norton**

and are extended forward, a feature that had earlier lead to the species name *elongata*; however, this is no longer valid. The shape of the valves sets this species apart from *T. gigas*, *T. derasa* and *T. squamosa*, but it is not always easy to distinguish from the closely related *T. crocea*. *T. maxima*, however, has stronger and more conspicuous structures on its valves, and the byssus openings are differently shaped in both species. The mantle of *T. maxima* is usually intensely green, blue, lilac or brown with many intermediate shades and numerous hyaline organs are also present. This clams bores into its substrate and attaches itself with byssal threads. It is very common in shallow water where it in some locations can colonise the shallow reef flats densely..

While the other species of giant clams commonly found in the trade are all very durable and can live for many years in the reef aquarium, *T. maxima* is less durable and especially vulnerable to bacterial infections.

Tridacna crocea (see Fact Sheet on pages 188) is closely related to *T. maxima*; however, it is smaller and reaches a maximum length of only 15 cm. This then makes it the smallest of all giant clams. Shell structures are only weakly developed, and compared to *T. maxima* its shells appear almost smooth. *T. crocea* is the giant clam that drills deepest in to its substrate and can be totally encapsulated in corals and live rocks. There is no consensus about its distribution in the Indo-Pacific. ROSEWATER (1965) limits its distribution westward to the Nicobar Islands. *T. crocea* has been shown to be present in early fossil material from East Africa and the Aldabras (CRAME, 1986; TAYLOR, 1971), but so far it has not been reported to be alive in that part of the Indian Ocean. Its distribution extends eastward to the Mariana- and Fiji Islands. The valves have six to ten wide radial ribs, of which three or four are very pronounced. The mantle is usually coloured blue or green. *T. crocea* is often imported, and in a well-maintained aquarium it can be live and thrive for many years.

Tridacna mbalavuana was discovered only in 1988 near the eastern Lau Islands of Tonga and Fiji (LUCAS et al., 1990), and originally described as *Tridacna tevoroa*, (which is now reckoned as a junior synonym of *T. mbalavuana*). The species is very rare, usually found in outer reefs in 14 to 33 m depth, which is far deeper than accepted by the other species of Giant Clams (LEWIS & LEDUA, 1988).

T. mbalavuana is closely related to *Tridacna derasa*, but distinct from the latter by a folded mantle that is covered by many wart-like protuberances. The mantel is dark grey-brown in colour and has relatively few irrodophore cells that protect the animal from UV-radiation, an adaptation to a life in deeper waters. The wide gaping mantle as well as its rugose texture are other adaptations to moderate light intensities in deeper waters. In a study made by KLUMPP & LUCAS (1994) it was found that this species showed photoadaptations at the greater depths of its range. The authors also suggest that *T. mbalavuana* may rely more on filter feeding than the other species of *Tridacna.*

As in *Hippopus* species, the mantle does not extend beyond the edge of the shell. The inhalant openings have strong, protective tentacles. The shell is thin, has

Holotype of *Tridacna rosewateri*. Photo provided by D. Knop, with permission from "La Conchiglia", Rome, Italy.

Tridacna derasa
Smooth Giant Clam

Distribution: Indo-Pacific; from Cocos-Keeling Atoll northward to the China Sea; from the Philippines southward to Lord Howe Island and eastward to the Fiji Islands. Has become extinct in many localities, and is listed as an "endangered species".
Size: Up to 50 cm in length.
Reproduction: Simultaneous hermaphrodite. Gametes are released in to the open water where fertilization takes place. This was the first Giant Clam reproduced successfully on breeding farms.
Food: Feeds on the products of its zooxanthellae as well as on phytoplankton and possibly also from dissolved organic compounds..
Description: Lives commonly in exposed reef sites, e.g. along the edge of barrier reefs. Loses the byssal threads rather early and then lives unattached. Heavy, smooth shell, only with low primary- and secondary ribs. In aquarium specimens the shell can be rougher, either because of lacking exposure to oceanic conditions or as a result of genetic characteristics (Delbeek & Sprung, 1996). Massive umbo and long hinge-ligament. Inhalant opening with tentacles. Colouration of mantle in juvenile specimens mostly green with orange stripes; in adults the mantle is covered by yellow, black, blue or white spots.
Aquarium requirements: Very easy to keep. Grows quickly in an aquarium and can double or even triple its size within a year. This is probably the most widely available, captive-bred giant clam.

GA: +; **SE:** +; **CO:** +

Photos:
Top: Tridacna derasa in the Ribbon Reef, GBR.
Centre: Colouration of the mantle changes with increasing growth.
Bottom: Typical appearance of an adult specimen.

Tridacna crocea
Boring Giant Clam

Distribution: Central Pacific: from the west coast of the Malayan Peninsula, the Rowly Shoals, western and northern Australia, northward to Japan, southward to the Great Barrier Reef and New Caledonia and eastward to the Mariana- and Fiji Islands.
Size: Maximum length 19 cm.
Reproduction: Simultaneous hermaphrodite: The gametes are released in to the open water, where fertilization takes place.
Food: Feeds on products from its zooxanthellae as well as on phytoplankton and possibly also from dissolved organic compounds..
Description: Lives totally encased in rocks or corals, so that only the mantle remains visible. Can occur in large aggregations with several specimens per square metre. Because of its totally enclosed mode of life the shell is rather smooth and delicately sculptured. Specimens that have grown up in an aquarium and where unable to bore in to solid substrate, display a coarser sculptured shell. The mantle is magnificently coloured and shines in shades of deep blue, green and/or yellow. The crab *Xanthasia murgiera* lives commensally with *T. crocea* (ROSEWATER, 1965).
Aquarium requirements: Generally easy to keep, but requires a very high illumination intensity. Wild-caught specimens often become stunted and regress because of injury to their byssal gland, which is invariably due to improper handling while the animal was collected. When purchasing this clam it is imperative to make sure that the specimen under consideration is firmly attached to a rock.

GA: +; **SE:** +/-; **CO:** +

Photos left:
Top: Typical, unattached living aquarium specimen showing standard mantle colouration.
Centre: Specimen at the Solomon Islands
Photo: J. Macaré
Bottom: Ventral view of shells with the large byssal opening.
Photos right: Different mantle colourations of *T. crocea*.
Photos: S. W. Michael

Tridacna gigas **- Large Giant Clam**

Distribution: Central Indo-Pacific; from Thailand via the Philippines eastward to Micronesia and the Fiji Islands, northward to Okinawa, Japan and Bonin Island, as well as southward to the Great Barrier Reef and New Caledonia. A few specimens have also been seen at Rowley Shoals and Scott Reefs in Western Australia (WELLS & BRYCE, 1988). Has already become extinct in a number of localities; is listed as an "endangered species".
Size: Maximum length 137 cm.
Reproduction: Simultaneous hermaphrodite. The gametes are released in to the water, where fertilisation takes place.
Food: Feeds on the products from its zooxanthellae as well as on phytoplankton and possibly also from dissolved organic compounds.
Description: Lives in shallow water, such as lagoons or on reef flats. The umbo of the shell is central, byssal opening small. The radial ribs consist of 4 – 6 convex scultures, of which 4 are very large. Fine, concentric growth lines interrupt secondary radial ribs. Hinge-ligament slightly longer than half of the shell length. Mantle normally yellowish, with deep blue hyaline organs that are surrounded by a blue-green ring. Juvenile specimens can be mistaken for *T. derasa*. The shrimps *Anchistus custodies*, *A. demani* and *Conchodytes tridacnae* live in symbiosis together *T. gigas*.
Aquarium requirements: Most specimens available from the aquarium trade come from clam breeding farms. Juveniles are usually only a few centimetres long, but growth is very rapid and so they must be provided with sufficient space. Basically, easy to keep, but a high illumination intensity is required. *T. gigas* grows extremely fast under favourable conditions – we have measured a growth of 1.77 grams/day in an experimental and well illuminated aquarium – the clam therefore requires a lot of space to expand.

GA: +; SE: +CO: +

Photos left:
Top: Typical mantle colouration of juvenile specimen.
Centre: Adult specimen photographed off Papua-New Guinea. Photo: S. W. Michael
Bottom: Dorsal view of the shell.
Photos right:
Top row: Adult specimens at Green Island, Great Barrier Reef (left: closed; right: open).
Centre: Three commercially bred juveniles in a dealer's aquarium.
Bottom: Juvenile specimen in an aquarium.
Photo: S. W. Michael
Right: Specimen photographed off Papua-New Guinea. Please note size comparison with a diver.
Photo: S. W. Michael

Tridacna maxima
Variable Giant Clam

Distribution: Widely distributed throughout the Indo-Pacific; from the Red Sea and East Africa, eastward to the Marshall- and Gilbert Islands, as well as Polynesia, northward to southern Japan and southward to Great Barrier Reef, New Caledonia and Western Australia.
Size: Maximum length 35 cm.
Reproduction: Simultaneous hermaphrodite. The gametes are released in to the open water, where the eggs are fertilized.
Food: Feeds on the products of its zooxanthellae as well as on phytoplankton and possibly also from dissolved organic compounds.
Description: Bores deeply in to the substrate, but without becoming totally encased. Anchors itself with byssal threads, and prefers to live in shallow water. Can occur in large numbers in certain areas, e.g. in Polynesia, where more than 60 specimens have been observed per square metre (DELBEEK & SPRUNG, 1996). Shape and sculpture of shell as well as the colour of the mantle is rather variable. Both valves are strongly asymmetric and elongated. Hinge-ligament distinctly shorter than ventral margin; large byssal opening. Scales of shell slightly reminiscent of *T. squamosa*, however, the scales are smaller and closer together. Colouration of mantle is usually blue, but can also be green or other colour shades; also sometimes with yellow or yellowish stripes. Inhalant opening with tentacles. The shrimps *Conochodytes tridacnae* and *Anchistrus demani* live symbiotically with *T. maxima* (ROSEWATER, 1965).
Aquarium requirements: Most commercially available specimens come from clam breeding facilities. Not as durable as the other common species of Giant Clams, probably due to a sensitivity to bacterial infections. Requires careful acclimatisation and very high illumination intensity. Most suitably placed among aquarium decoration and corals.

GA: +/-; **SE:** +/-; **CO:** +

Photos left:
Top: Typical aggregation of *T. maxima* on a reef flat at Rangiroa, Polynesia. Photo: S. W. Michael
Centre: Specimen photographed at the Solomon Islands. Photo: J. Macaré
Bottom: External view of shell.
Photos right:
Top row: Specimens photographed at the Maldives (left) and Papua-New Guinea (right). Photo: S. W. Michael
Centre row: Specimen embedded in a brain coral in Papua-New Guinea (left; photo: S. W. Michael) and a specimen surrounded by soft corals at the Solomon Islands (right; photo: J. Macaré).
Bottom row: Specimens at the Solomon Islands. Photo: J. Macaré

Tridacna squamosa
Scaly Giant Clam

Distribution: Widely distributed in the Indo-Pacific, from the Red Sea and East Africa eastward to the Marshal- and Gilbert Islands as well as Polynesia and Pitcairn Island (PAULAY, pers. comm.), northward to southern Japan as well as southward to the Great Barrier Reef, New Caledonia and Western Australia.
Size: Maximum length 40 cm.
Reproduction: Simultaneous hermaphrodite. The gametes are released in to the water, where the eggs are fertilized.

Food: Feeds on the products of its zooxanthellae as well as on phytoplankton and possibly also from dissolved organic compounds.
Description: Lives in protected sites, often among corals, but on a sand bottom in shallow water (down to a depth of 15 to 20 m). Only loosely attached with byssal threads. Our experience has shown that adults can live in an aquarium without being anchored down. This species can be distinguished from other giant clam species by the scaly sculpture of its shell. Both valves (shell halves) usually symmetrical and with large radial ribs. Byssal opening smaller than in *T. maxima* and *T. crocea*. Mantle protrudes distinctly beyond the edges of the shell; colouration of the mantle is brownish or yellowish-brown with blue or blue-green spots and stripes, which can be numerous along the mantle margin. Many hyaline organs present. Inhalant opening with tentacles.
Aquarium requirements: Very hardy aquarium specimen, which has been kept in aquariums for more than ten years by many aquarists. It grows rather rapidly and requires sufficient space as well as a high illumination intensity. Most commercially available specimens come from clam breeding facilities, but those less than 3 cm long are almost impossible to adapt to aquarium conditions. The large specimens available in the aquarium trade during the 1980's were invariably wild-caught, but fortunately this practice has now ceased.

GA: +; **SE:** +; **CO:** +

Photos left:
Top: Specimen in an aquarium owned by Flemming Jørgensen, Larvik, Norway, where it lived and thrived for more than 10 years.
Bottom: Close-up photo of inhalant opening; tentacles are well visible.
Photos right:
Upper row: Juvenile specimen in an aquarium (left; photo: S. W. Michael) and specimen among branching *Acropora* corals, in a shallow lagoon at the Maldives (right).
Centre row: Adult specimens on the Great Barrier Reef.
Bottom left: Sometimes the mantle is coloured blue, as in this specimen on a sand bottom in a lagoon at the Maldives.
Bottom right: External view of prepared shell.
Photo: D. Knop

sharp edges and is only slightly arched. Weak indentations along the shells edge fit almost perfectly in to each other, so that the shell-halves can be tightly closed. The indigenous population of the Fiji Islands call this clam "tevoro", which means "devil". *T. mbalavuana* grows to a length of up to 55 cm, that makes it one of the larger Giant Clams. To our knowledge *T. mbalavuana* has not yet been kept in captivity and nothing can be reported on its durability in the reef aquarium.

Tridacna rosewateri was not discovered until 1991 off Mauritius in the Indian Ocean (SIRENKO & SCALATO, 1991). It has a particularly thin shell that is rather vulnerable to predator attacks. Instead, the shell carries large plate-like outgrowths that could represent defensive structures (KNOP, 1996). It is a rather small species. Specimens on hand for examination measured from 6.7 cm to 19.1 cm, with a normal length of 15 cm. *T. rosewateri* specimens have not yet been kept in aquariums. May be one day this clam can be captive-bred so that specimens become available for the aquarium trade.

Family Tellinidae

With a strong foot, the species in this family are able to bury themselves in coral sand. Usually they stay just below the surface of the sand and from there they extend their long siphon above the substrate. The siphon gives these bivalves access to fresh water and with it they also gather food from the sediment. The shells often have attractive pastel colours, which make these bivalves desirable collectors' items. Since they live buried in sand, there is not much of a chance that they will appear in the aquarium trade. However, should they ever become available they must be kept in a special aquarium with a deep sand layer over the bottom, so that they can bury themselves. Little is known about the systematics of this family. Currently, genus *Tellina* is being used merely as a collective name for at least 40 genera and subgenera of the family.

Valves of Venus shells (family Veneridae) from the Indo-Pacific: 1 - *Tapes dorsatus*, 10 cm. 2 - *Tapes variegates*, 5 cm. 3 - *Circe scripta*, 5 cm. 4 - *Circe sulcata*, 3 cm. 5 - *Circe lenticularis*, 5 cm. 6 - *Pitar citrinus*, 6 cm. 7 - *Tawera lagopus*, 3 cm. 8 - *Sunetta contempta*, 4 cm. 9 - *Lioconcha ornate*, 5 cm. Photo: C. Bryce

Family Veneridae - Venus Shells

Venus shells are generally well known. The shell of dead animals can be found along sandy beaches or in the mud of tide flats by the thousands, even along European coasts. Numerically, this family is among the largest groups in Class Bivalvia. In fact, some species occur in such abundance that they are harvested for human consumption. Many species have strongly rounded shells with distinct sculptures, but there are also species with thin, brittle shells. The variability within this family is large. Venus shells live in shallow water, buried in sand or mud, often in harbours and intertidal zones. Usually they can be found immediately below the substrate surface, but they may also be lying directly on the bottom. Often part of the shell and the siphon protrude through the bottom surface.

Lioconcha species are common in coral sand; they can be readily recognized by a zigzag pattern on their shell. A similar, but less distinct marbled pattern can be found on *Cirpe scripta*, which – together with *Gafrarium* species – are among the most common venus shells on coral reefs. This is also the same habitat were *Callista* species are often encountered. The small *Irus macrophyllus* usually lives attached to corals and coral rocks, so it occasionally finds its way in to coral reef aquariums. Other genera of this family are: *Periglypta*, *Dosinia*, *Paphia*, *Venerupis*, *Bassina*, *Placamen*, *Anomalocaria* and *Antigona*, just to mention a few. Venus shells are rarely ever imported for the aquarium trade, but aquarists should be able to keep them in a coral reef aquarium, especially when they are given a special aquarium or when kept together with other burrowing and sand-dwelling organisms in a sand zone aquarium, as described in Volume 1.

Venus shells live predominantly in a sandy bottom. Seen here at Myrion Island (Western Australia) is a *Periglypta reticulata* burying in sand.
Photo: C. Bryce

Order Myoida

From among the seven families that make up order Myoida, only family Pholadidae is of some interest to aquarists.

Family Pholadidae - Angelwings, Piddocks

In family Pholadidae we encounter a group of bivalves that are often introduced in to coral reef aquariums together with live rocks and corals. Typical for the 16 genera in this family is genus *Pholas* (Volume 1, drawing on page 186). These bivalves bore into calcareous rocks. If one takes a close look at the underside of stony corals, one often discovers these boring bivalves, or at least the oval holes created by them. Mantel or siphon protrudes from a hiding place and are readily visible when the clam opens up. Another peculiarity of this family is the fact that the shell-halves consist of unequally formed shell plates. This increases flexibility of the valves, which is advantageous for the way these bivalves live.

Class Cephalopoda - Octopus, Squid, Cuttlefish, Nautilus

The cephalopods represent a most unusual group of animals. It encompasses (pearly) nautiluses of subclass Nautoiloidea and octopuses, squids and cuttlefishes of subclass Coleoida. Cephalopods are the most advanced invertebrates, but – at the same time – they also have the shortest life expectancy. In most of the octopuses, squids and cuttlefishes the longevity is little more than a few month to a year. Moreover, these animals are also shrouded in a certain mystique. The folklore of many countries includes stories about giant octopuses and squids that allegedly rose up to the surface from great depths and dragged sailing ships and their crews under water. Of course, these are fantastic fables, yet there are – nevertheless – giant squids living in the deep sea. The largest known specimen drifted on shore near Ranheim along the Trøndelag coast of Norway in 1954. The species, which belonged to the genus *Architeuthis* had a length of 18 m including the tentacles. Another example of a deep sea species is *Vampyroteuthis infernalis*, a squid that lives at depths of 300 to 1,500 m (NESIS, 1987). However, in this case it is not a giant species, but a small one with maximum length of 20 cm. Those who are interested in the giant squids should read the books by ELLIS (1996, 1998).

About 700 cephalopod species are currently known. They occur in all seas. Most lead a pelagic life. During the Mesozoic Period, 65 million years ago, cephalopods were the dominating animals, and then also included the Nautiloida (pearl boats or pearly nautiluses), Ammonites and Belemnites. Most of these had an outer shell. Today, there are only a few nautilus species left. Most of the recent octopuses, squids and cuttlefishes evolved from (now extinct) Belemnites, which had an internal shell, a feature that has reappeared again in the recent Coleoida.

General Biology

Octopuses, squids and cuttlefishes all have well-developed internal organs that include a brain, a well-developed nervous system, a highly developed pair of eyes, a powerful rasping tongue and beak-like jaws. In octopuses eight arms are located at the anterior section of the animal. Squids and cuttlefishes have five pairs of arms, four of which are short and heavy and normally called "arms". The fourth pair down from the dorsal side is called tentacles, which are much larger than the arms. Both arms and tentacles are normally equipped with suction cups and used for capturing prey, which is then restraint by the action of the suction cups. In large, deep-sea species the suction cups can have the size of small plates. Some species live a pelagic life and have developed lateral fin appendages used for swimming; however, all cephalopods can flee rapidly, using a jet propulsion system. For that, water is forced – under great pressure - out through a mantle opening, so that the animal can propel itself instantly in to the opposite direction. In conjunction with such escape, coleoid Cephalopoda can squirt out a coloured liquid to create a "smoke screen" that irritates a predator and so persuades it to save its own life. This "ink" is excreted from an "ink sac", which has an opening near the hindgut. Under the name of "sepia" painters used this ink in early years. Another camouflage effect deployed by many cephalopods is the ability to change colours. Such colour changes can be effected when pigment cells located in the skin are contracted or expanded, respectively,

Cuttlefishes, squids and octopuses, like this Common Octopus, *Octopus vulgaris* have highly developed eyes with crystalline lenses. They are even capable of opening the pupils variably wide in either eye; the width of the pupil on the bright side is reduced and the one on the darker side is wide open. Photo: Prof. Dr. P. Wirtz

Cuttlefish eggs attached to a branching sponge (left; photo: K. Kates) and octopus eggs underneath an overhang (right; photo: Dr. T. Paulus).

resulting in the pigments in affected cells being either concentrated or diluted.

Most cephalopods are excellent swimmers, including those species that live mostly on the bottom, in holes or in other depressions. Their capability to react quickly is based on a highly developed nervous system that consists of giant nerve fibres (axons) in contact with synapses, so that a nerve impulse is transmitted to the brain very rapidly. Cephalopods are the only invertebrates with eyes that incorporate a lens apparatus, as found in vertebrate animals. In humans, focusing is accomplished by adjustment of lens shape, but in cephalopods the entire lens in the eye is moved when focusing. Their excellent visual acuity and the quick response capability, enables cephalopods to feed on swimming fishes and crustaceans. That is a type of prey, which is normally hard to catch. Fishes are a nutritious (high protein) type of food that requires efficient metabolism. In order to be adapted to that cephalopods have developed an equally effective circulatory system, which includes a special branchial (gill) heart, arteries and veins.

Cephalopds have separate sexes. Males fertilise females by transferring spermatophores (sperm packages), either placing the package inside or outside the mantel cavity. For that purpose, the end of one or two arms has become modified in to a hectocotylus in some groups. The eggs are large and nutrient-rich. The young, who are essentially miniature versions of their parents, often undergo a pelagic larval stage. Some cephalopods, e.g. calamari of family Loliginidae, as part of their pelagic life often migrate over vast distances to specific bays or fjords in order to breed there. During such migrations vast numbers of them are often caught by commercial fisherman for human consumption. Moreover, in many parts of the world, octopuses, cuttlefishes and squids are in great demand as culinary specialties.

Octopus at the Waikiki Aquarium (Honolulu, Hawaii), attached with its suction cups to the front glass of a display tank. The tentacle-carrying head and the visceral sac that accommodates all organs are clearly visible.

Nautilus and *Allonautilus* are the only surviving genera of subclass Nautiloidea. It had been assumed for a long time that extinct pearly nautiluses lived in deep cold seas. However, recent studies on the oxygen isotopes of fossil nautiluses have revealed this to be incorrect; instead, prehistoric nautiluses lived in shallow, warm water.

Nautilus pompilius **at Papua-New Guinea.** **Photo: S. W. Michael**

General Biology

Nautiluses have a spirally coiled shell that is divided internally in to many small chambers, which provide neutral buoyancy for the animal. The body is located in the most anterior chamber. The shell is white, covered by reddish brown stripes. Nautiluses have many arms, but suction cups are absent. The eye is a depression without a lens. The mantle cavities in the lower part of the shell contain two feather-like (branchial) gills and lead in to bipartite funnel. When water is forced through the funnel by muscular action, a nautilus can jet through the water, but not very fast. Pearly nautiluses do not posses an ink gland and so are unable to camouflage themselves by releasing ink in to the water.

Pearly nautiluses capture fishes and crustaceans. For a long time it had been assumed that the recent *Nautilus* species live constantly in deep water. Yet, observations on *Nautilus belauensis* at Palau in the Pacific Ocean have shown that this animal migrates vertically at night from 200 m to 400 m (water temperature of only 10 °C) up to 100 m depth, where the water has a temperature of about 20 °C. During the day the animal returns to deeper water.

In contrast to other cephalopods that have a short life expectancy, pearly nautiluses can reach longevity of up to 20 years according to Wells & Bryce (1988). However, the longevity of Nautilus needs to be confirmed through further field or aquarium studies. Today we know five *Nautilus* species (Saunders, 1981b). All of them occur in the central Indo-Pacific.

Nautilus pompilius is the best-known pearly nautilus. It lives around the Andaman Islands, the Philippines, in Indonesia and off northeastern Australia, which gives this species the largest distribution area of all *Nautilus* species. It occurs only below a depth of 60 to 70 m and it can descend down to depths of around 600 m.

The distribution of *N. belauensis* is confined to the waters around Palau (Saunders, 1981a), that of *N. macromphalus* to New Caledonia and that of *N. stenomphalus* to the Great Barrier Reef. A possible fifth species, i.e. *N. repertus* lives endemic off Western Australia, but for the time being its validity is uncertain. The former *N. scrobiculatus*, which is restricted to Papua-New Guinea, is now transferred to a new genus, *Allonautilus* (Ward & Saunders, 1997), so there are now two genera of living nautilus species.

The eyes of pearly nautiluses are only depressions without lenses.
Photo: P. Schupp

Pearly Nautiluses in the Aquarium

Generally speaking, nautiluses are rather difficult to keep alive in an aquarium. Caring for them is a task for specialists or very experienced and knowledgeable aquarists. These animals need to be given low water temperatures of 16 to 24 °C. They do not survive for long in aquariums with higher temperatures. CARLSON (1979) reports that a nautilus kept at 27 °C, died within 48 hours. Moreover, lighting has to be subdued since *Nautilus* species in the sea live at depths were there is little light or it is almost totally dark. The Waikiki Aquarium is among the leading institutions in the care and husbandry of pearly nautiluses.

Nautilus pompilius and *N. belauenmsis* have been bred in this facility. At the Waikiki Aquarium each Nautilus is fed three-times a week with a shrimp weighing 18 g. According to CARLSON (1987) fresh tuna meat, the carapaces of crabs and lobster meat are usually accepted, but the Samoa Crab, *Scylla serrata* as well as the Marbled Shrimp, *Saron marmoratus* are always rejected. Consequently, *Nautilus* species appear to be quite selective in terms of the food they accept. Excellent overviews of the aquarium requirements for keeping pearly nautiluses, including listings of public aquariums and other facilities that keep these animals, are available in ARNOLD & CARLSON (1986), CARLSON (1979, 1984, 1987), CARLSON et al., (1992) and LANDMAN et al., (1994). Nautilus shells are attractive to humans and many specimens are killed each year for this reason. Indonesia had banned the collection and export since 1987 but many other countries still collect and kill them. In New Caledonia over a period of two years or less, 10,000 live Nautilus were captured and killed for their shells.

Breeding Pearly Nautiluses, *Nautilus belauensis* in an aquarium

Although many attempts to keep *Nautilus* spp. alive in public aquariums have been tried with more or less success, the first successful breeding of *Nautilus belauensis* happened in 1985 at The Waikiki Aquarium, Honolulu, Hawaii by the supervision of the aquariums's director Dr. Bruce Carlson, and was based on a study of the animal's biology in the nature (CARLSON et al., 1986).

Adults *Nautilus belauensis* were caught at Ngmelis Island (Palau, Federated States of Micronesia) in order to establish sufficient brood stock.

Photo: Dr. B. Carlson

These pearly nautiluses were individually marked and transferred to an aquarium, as described by CARLSON (1987).

Initially the nautiluses were being kept at a water temperature of 24 °C, which was about 4 degrees higher than at their capture site. The first copulations were observed after a few weeks (left; specimen on the right is the male; photo: Dr. B. Carlson); however, the female deposited only unfertilised eggs. The water temperature was lowered to 12 to 15 °C over a period of nine months, so that egg laying ceased. Then the temperature was raised to 18 - 21 °C, and copulations re-occurred and fertilised eggs were deposited. The nearly cylindrical egg capsules were attached to rocks (right; egg capsule of *N. pompilius*).

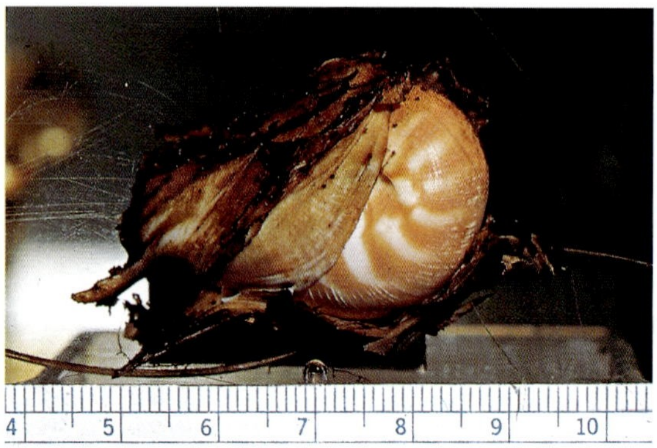

The upper corner of an egg capsule has become detached (left) and the first embryo ever seen by a human hatched on 4 March 1995 (right; photos: Dr. B. Carlson).

Young pearly nautilus one day after hatching (left; photo: T. Kelly) and a slightly older juvenile specimen (right).

This subclass contains four orders, of which we will discuss the following three orders only:

- Sepioidea
- Teuthoidea
- Octopoda

A fourth order, Vampyromorpha, contains only deep-sea species that are not of interest to the aquarium hobby.

Order Sepioidea - Cuttlefishes

This order contains the families Spirulidae and Sepiidae; but we will deal only with cuttlefishes of family Sepiidae. Cuttlefishes are "ten-armed" cephalopods with a flat, depressed body, an internal – usually oblong and porous – shell and fin appendages used for swimming. Two of the arms are called tentacles and are extra long and equipped with suction cups. They are used for capturing prey and can be retracted in to the body cavity of the animal. Although cuttlefishes spend a fair bit of time on the bottom, they are nevertheless excellent swimmers and can often be seen on coral reefs (especially at night).

There is about 100 species in family Sepiidae, where genus *Sepia* contains some commonly known species. *Sepia officinalis* from the Mediterranean Sea is probably the most widely known species, since it has already been bred in aquariums for some time now (OVERATH, 1981). *Sepia latimanus* occurs in the Indo-Pacific, from the Dead Sea to southern Japan and southward to the Great Barrier Reef. It lives on coral reefs down to a depth of 30 m, and grows to a length of about 15 cm. If one can refer to cuttlefish husbandry as being at all "easy", it could only apply to *S. latimanus*. However, even this species requires a special aquarium. It feeds mainly on crustaceans, but will also take small fishes. There is one problem though that can arise while trying to adapt cuttlefishes to "dead" food: usually these cephalopods have a strong hunting instinct, and therefore will accept only food that moves. According to CARLSON (1983), cuttlefishes that were initially fed with live shrimps would later also accept dead shrimp that were dropped directly in front of them.

The Giant Cuttlefish, *Sepia apama* occurs along the Great Barrier Reef, and grows to a size of up to 100 cm and attains a weight of more than 5 kg. In stark contrast to *S. apama* is the rather rare and very small (up to six cm) *Metasepia pfefferi*, which is found from Western Australia to Queensland. *Sepia dollfusi* and *S. pharaonis* inhabit the Red Sea where they are rather common. However, according to VINE (1986) several species are listed to occur in that region of the Indo-Pacific. *S. pharaonis* also occurs at the Maldives, but according to ERHARDT & MOOSLEITNER (1995) it is absent from East Africa. It grows to a length of 35 cm and lives usually in small groups. Another species living in the Indian Ocean and Red Sea is *S. prashadi*. It is usually found among marine

***Sepia latimanus* at the Waikiki Aquarium (Honolulu, Hawaii).**

Sepia trygonina **during courtship in the Red Sea.** **Photos: Dr. T. Paulus**

Sepia apama **- seen here at the Great Barrier Reef, can grow up to 50 cm in length and reach a weight of 5 kg.** **Photo: M. Cufer**

grasses or corals in shallow water down to depths of about 40 m. In males one arm is hectocotylized, i.e. modified to transfer spermatophores to females during reproduction. *S. prashadi* is a solitary cuttlefish and grows to a length of only 14 cm (ERHARDT & MOOSLEITNER, 1995).

Order Teuthoidea -Squids

The squids (or calamari) of order Teuthoidea are also ten-armed cephalopods. Of the six families in this order we will only deal with family Loliginidae. Among other species, this family also contains *Sepioteuthis lessonianai* that reaches a maximum size of 35 cm and is often seen in

Sepia prashadi **over sandy bottom and in a sea grass bed.** **Photos: Dr. T. Paulus**

Left: *Sepia* species at Madang, Papua-New Guinea (photo: L. Newman and A. Flowers). Right: *Sepioteuthis lessonikanai*, a squid, from the Red Sea (photo: Dr. T. Paulus).

large schools on coral reefs. It is widely distributed throughout the Indo-Pacific. Its relative in the Caribbean Sea is *S. sepioidea* which inhabits shallow water on coral reef and areas with marine grasses, where is can occur in schools of up to 50 specimens. This is the only ten-armed cephalopod common in this area on coral reefs (HUMANN, 1994). According to ERHARDT & MOOSLEITNER (1995) it has been bred in captivity. Here we also find *Dorytheuthis plei*, a pelagic species that also breeds in the same area, although adult specimens are normally not found among corals. Clusters of eggs and juveniles of *D. plei* can be found in reef areas of the western Atlantic.

Family Loliginidae also includes genus *Loligo*, where many species are beautifully coloured. One such example is *Loligo opalescens*, which occurs along the west-coast of America and Mexico. It is traded in America on a large scale.

Order Octopoda - Octopuses and Argonauts

The members of this order are eight-armed cephalopods. From among the four families we will deal only with the octopuses of Family Octopodidae and the argonauts of family Argonautidae.

Family Octopodidae - Octopuses

The dominating genus in this family is *Octopus*, with an unknown number of species and rather variable sizes. The smallest measure barely a few centimetres; however, the Indo-Pacific *Octopus cyaneus* grows to a length of 3 m. It is characterized by the presence of large eyespots between the eyes, and a colouration that is usually dark lilac or nearly black. Octopuses tend to stay in crevasses and caves along the reef. Yet, often they can also be seen in the open on the bottom, where they move about with elegance and litheness. *Octopus dolfleni* is the world's largest octopus with a mantle length of 60 cm and a total length of more than five metres. It lives in moderate- or deep waters in the northern Pacific Ocean, where it is of great commercial importance. The species is commonly on display in public aquaria in Japan and North America.

Octopuses too can flee quickly with the aid of a jet propulsion system, when they become frightened. Normally octopuses are nocturnal and shy creature. They do not possess an internal shell, but they have an enormous capability to change colouration and discharge copious volumes of ink for camouflage purposes.

Octopus rugosus occurs in the Indo-Pacific. It is particularly abundant in the Red Sea and the Indian Ocean. Like all other species in this genus it feeds on other invertebrates, such as clams, snails and crustaceans, but it is also capable of catching benthic fishes. The maximum total body length is 50 cm. The common octopus, *Octopus vulgaris* has a cosmopolitan distribution; it occurs in subtropical as well as in tropical seas. It is one of the most common species of this genus and lives in shallow water, down to depths of about 200 m. It is found on sand as well as rocky bottom. For details on the aquarium care of this species, refer to OTT (1989), SPIES (1983) and TOMEY (1988).

The large *Octopus macropus* has a circum-tropical distribution. It has longish or oval-shaped white spots against a brown-grey background colour. When it feels threatened it turns blood red with shiny white spots. In addition, the mantle is covered by large wart-like papillae. The maximum size of this octopus is about 50 cm; it lives mostly on a sand bottom or among coral fragments on coral reefs.

In the Caribbean Sea there are several species, including *Octopus filosus* (synonym *O. hummelinki*), which is particularly prevalent around Haiti and grows to a size of 25 cm. It can be recognised by an eyespot that forms a blue ring directly underneath the eye. A very attractive species is

Octopus briareus, the most common octopus in the western Atlantic. It reaches a maximum size of 50 cm and often hides in depressions or in tubular sponges that are also rather numerous in the same geographical region. Sand- and mud bottom substrate in the western Atlantic is also inhabited by the small, barely 10 cm long *Octopus joubini*, and the smallest species in this region.

Also included in family Octopodidae are the infamous blue-ringed octopuses of genus *Hapalochlaena*. **Beware of these octopuses. One bite from any of them can have fatal consequences!** Anecdotes about venomous organisms on coral reefs are often unduly dramatised, however, there is no doubt about these creatures: **Blue-ringed octopuses are highly venomous. These are animals one has to be extremely cautious of!**

There are at least two different species in genus *Hapalochlaena*. Both have a colour pattern consisting of a yellow base colour covered by blue rings. Since blue-ringed octopuses are – as exotic aquarium animals – in demand and are available in the aquarium trade, they are discussed in some detail in a Fact Sheet on page 207.

Octopuses occasionally occur in the trade. We know of people who successfully have kept juvenile specimens or small species of octopuses in community tank where they co-existed with invertebrates rather well. However, they grow fast and eventually they had to be removed. Octopuses' habit to catch fishes as well as to release ink as protection all in all make it necessary to give these beautiful invertebrates a special aquarium. A special octopus-aquarium, which will be a real attraction providing the possibility for studying the biology of the molluscs in close up, should have several holes and crevices serving as hiding places and should not be too small. Proper food, such as crabs, shrimps and small fishes must be provided.

Octopuses are frequently encountered on reefs – like this specimen off Kanifinolhu Island (Maldives), which fled when we approached.

Family Argonautidae - Paper Nautiluses, Argonauts

In conclusion of the chapter on Phylum Mollusca a few remarks about Argonauts, also known as paper nautiluses. They belong to genus *Argonauta* and are pelagic, perpetually swimming cephalopods that can be found in all seas. There are several species, and *Argonauta argo* is the most common one. In general, argonauts are not often seen. So far no attempts have been made to keep these animals in aquariums. However, in respect to their reproduction they are rather special and highly interesting animals. Two of a female's arms have been modified in to a type of sail or membrane at their ends. During the reproductive process, this part of the arm excretes a thing, very delicate and beautiful shell, formed like a bivalve shell that is paper-thin. During reproduction the female carries this shell around. It serves as a hiding place for the female as well as a brood chamber for juvenile argonauts. The male is much smaller and does not build a paper shell, but occasionally one can find a male and female together in the shell of the latter. In this manner, argonauts return, for a brief segment of their life cycle to their ancestral mode of life with an external shell – a way of life that nowadays remains only in *Nautilus* species.

Hapalochlaena lunulata **and** *H. maculosa*
Blue-ringed Octopus

Distribution: *Hapalochlaena lunulata* occurs only in the tropical Indo-Pacific (Philippines, Indonesia, Melanesia, northern and western Australia), while *H. maculosa* occurs in the tropical as well as sub-tropical Indo-Pacific; from South-Eastern Australia to the Gulf of Aden.
Size: Maximum length 10 cm.
Reproduction: Reaches sexual maturity in one year. The eggs of *H. lunulata* (3-5 mm) are deposited on a substrate. The eggs of *H. maculosa* (7–8 mm) are being carried by the female among the arms for a period of two months until they hatch; during this period the female does not feed. Components of the venom are transferred to the eggs for their protection.
Food: Both species feed on crabs and other molluscs.
Description: Both species are hard to distinguish from each other; base colouration is brown or greenish-yellow. *H. lunulata* has blue-rings only on its head and abdominal sac; ink sac visibly reduced. *H. maculosa* has sometimes inter-twined blue lines on its head, abdominal sac and arms (instead of blue rings); ink sac well developed. **Both Blue-ringed Octopus species are extremely venomous. A bite can be fatal to humans**. There are already reports about fatalities. The venom consists of two components: maculotoxin and palotoxin. A fraction of a gram of this venom can be deadly to humans as it paralyses muscles and nerves.
Aquarium requirements: Although Blue-Ringed Octopuses are often available in the aquarium trade, they should not be kept! These are distinctly aggressive animals and will readily attack the hand of an aquarist, which might be fatal! Please avoid these animals!

GA: -; **SE:** +/-; **CO:** –

Photos:
Top left: *Hapalochlaena lunulata* (?) on a reef. Photo: L. Newman and A. Flowers
Top right: *Hapalochlaena maculosa* in New South Wales (Australia). Photo: M. Cufer
Bottom: *Hapalochlaena maculosa* carrying eggs between the arms. Photo: M. Cufer

Chapter 3:

Echinodermata

The spiny-skinned animals of phylum Echinodermata include feather stars, sea lilies, sea stars, brittle stars, sea urchins and sea cucumbers. Invariably, these are colourful and conspicuous inhabitants of coral reefs. As detritus feeders, algae grazers or predators, they play a significant role in a reef's ecosystem. Moreover, many echinoderms hold important key positions, so that any fluctuations in their population numbers have direct, flow-on effects on to the species composition of a reef. Many echinoderms make excellent aquarium inhabitants, which are not only very attractive but also quite useful. Yet, on the other hand there are also species that can cause problems in an aquarium because either they are sensitive and delicate to keep or they may be incompatible with other aquarium inhabitants. In essence then, echinoderms are organisms that are not inconse-

Echinoderms - or spiny-skinned animals as they are also called - are among the most conspicuous inhabitants on a coral reef. Apart from the fact that many of them are extraordinarily colourful and amazingly beautiful, they also play an important role in the ecosystem by feeding on detritus and algae, as well as being carnivorous. The photos selected here show two typical echinoderms on the Great Barrier Reef.

Top: The bright Red Indian Sea Star, *Fromia indica*, seen here wandering - like a "moon walker" - across a large *Porites* colony.

Photo: M.Cufer

To the left: *Thelenota ananas*, is an impressive detritus-feeding sea cucumber that can reach a length or more than 50 cm.

Photo: E. Svensen

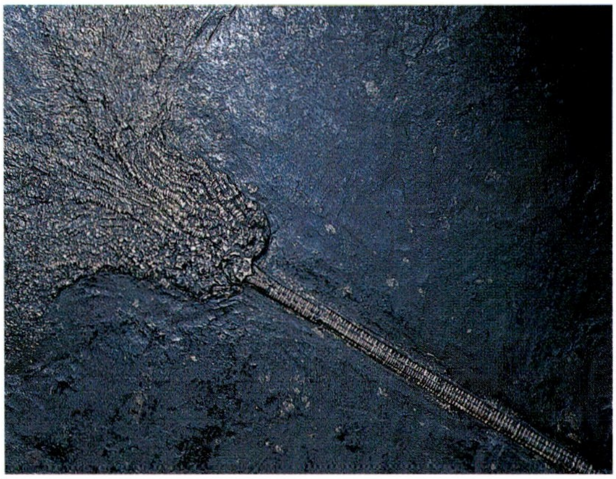

Spiny-skinned animals evolved more than 530 million years ago, during the Cambrian Period, however, fossil echinoderms from that time period are extremely rare. Yet, there are many attractive and well-reserved fossils from younger geological time periods, e.g. from the Jurassic- and the Cretaceous Periods, like this sea lily (left) and a brittle star (right).

quential to marine aquarists. With continuously growing efforts to establish more complete habitats in our aquariums, it seems safe to predict that the significance of echinoderms in the coral reef aquarium hobby will further increase in the future.

From a physiological perspective, the Echinodermata are among the singularly most peculiar animal groups in existence. The origin of this group of exclusively marine animals dates back 530 million years ago to the Cambrian Period. Early fossils of this animal group are very rare, however, there are well-kept finds of sea lilies (class Crinoidea) from the Silurian Period (438 to 408 million years ago).

In the course of time, more than 6,000 species have evolved within this phylum that are now, distributed over six classes:

- Class **Crinoidea**, feather stars, sea lilies
- Class **Asteroidea**, sea stars
- Class **Concentriclycloidea**, deep-sea urchins
- Class **Ophiuroidea**, brittle stars, serpent stars
- Class **Echinoidea**, sea urchins, sand dollars
- Class **Holothuroidea**, sea cucumbers

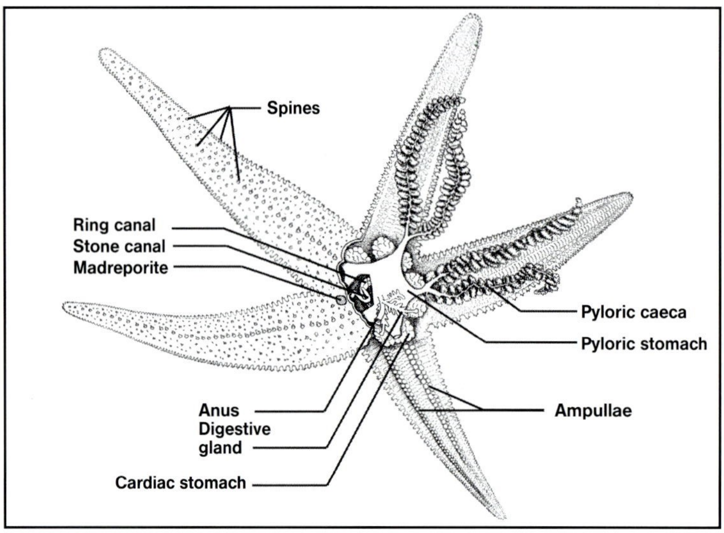

Anatomy of a sea star viewed from above (= aboral side). The arms are shown here at different stages of dissection.

Drawing after FREEMAN & BRACEGIRDLE (1971)

From these six classes only five of them of them are of interest to aquarists. The sixth and most recently established class Concentricycloidea contains deep-sea urchins that were not discovered until the late 1980's, off New Zealand. So far only two of these mysterious animals have been described, one Pacific and one Atlantic species, and both are included in genus *Xyloplax*.

Echinoderms are largely marine animals. There are very few echinoderm species that have a limited adaptation to brackish water; however, no species occur in freshwater.

General Biology

The subsequent treatment of the general biology of the Echinodermata is based on numerous sources. The most important ones are

HENDLER et al. (1995), RUPERT & BARNES (1980) and SEMB-JOHANSSON (1988).

Body Symmetry

The most conspicuous characteristic of echinoderms is their pentamerous radial symmetry. That means the body of an echinoderm specimen can be divided in to five equal parts grouped around a central axis. This is particularly obvious in many sea stars and brittle stars, where most species have five arms radiating from a central oral disc.

Close examination will reveal this characteristic also to be present in the majority of other echinoderm species, even though there may well be multiples of the five-arm arrangement (10, 15, 20 arms, etc.), as in many sea stars, brittle stars and feather stars. Alternately, there may be five rows of tube feet, as in many sea cucumbers or a shell structure made up of five segments, more or less hidden by spines and other outgrowths, as in sea urchins.

Forms of radial symmetry can also be found in some of the other well-known reef animals, e.g. in sponges and cnidarians; however, the symmetry in echinoderms is of a totally different origin and is not indicative of a relationship to other groups of animals. The symmetrical body shape of echinoderms has arisen secondarily from a bilateral base form, whereby it is assumed that this feature is shared with the Chordata. With that in mind, one could claim – somewhat provocatively – that echinoderms kept in a particular aquarium are more closely related to the aquarist than to corals.

The bilaterally symmetrical origin of echinoderms is reflected in the shape of the free-swimming larvae. One can only speculate as to why the pentamerous symme-

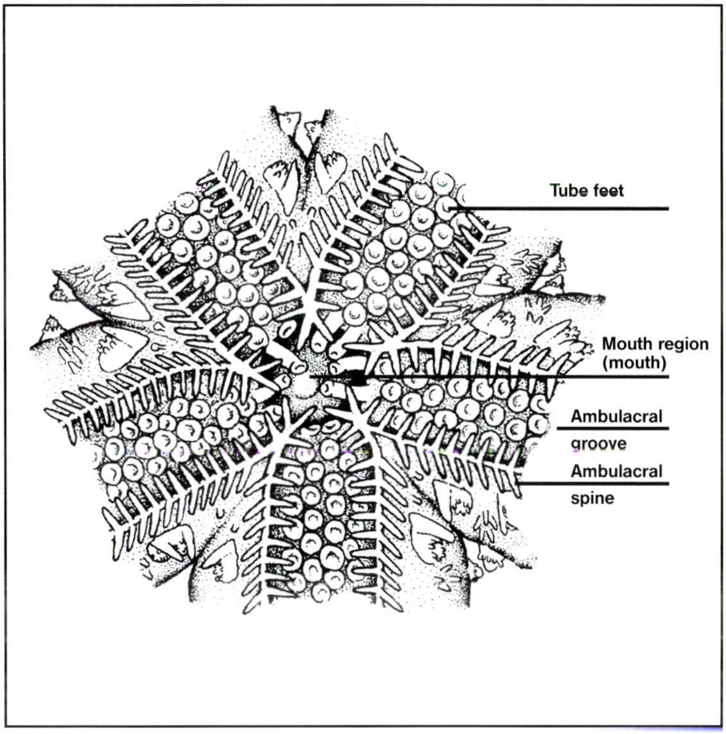

Oral side of a sea star. Drawing after BARNES (1980)

A surprising and rather significant feature of spiny-skinned animals (echinoderms) is their pentamerous symmetry. That means, in most species the body is clearly arranged in five segments around a central axis. This symmetrical body plan is intuitively obvious even to laymen in five-armed sea stars, such as this Blue Sea Star, *Linckia laevigata* (left; seen here at Heron Island, Great Barrier Reef; photo: L. Newman and A. Flowers), as well as in this Parson's Hat Sea Urchin, *Tripneustes gratilla* (right; photo: E. Svensen) seen here in the Red Sea.

Echinoderms are typically oriented in an upright anterior - posterior position (whereby the mouth on the oral side is pointing downward and the anus on the aboral side is directed upward), or they are crawling - in a lateral orientation - on one side, with the mouth at the front end and the anus at the hind end. *Diadema setosum* (left) is an example of echinoderms that are oriented in an anterior - posterior fashion. Its anus is clearly visible as a colourful vesicle on top of the body. On the other hand, the Pineapple Sea Cucumber, *Thelenota ananas* (right) is an example of a laterally oriented echinoderm.
Photos: C. Bryce

try has developed. One possible explanation could be that this creates a stronger skeleton and a more solid body structure.

Because of the pentamerous radial symmetry a protruding head section is absent in most adult echinoderms. Consequently, one usually refers to the front and back of these animals as the oral- (with mouth) and aboral-side (without mouth). Most echinoderms orient themselves upright anterior-posterior, whereby the mouth of the animal is facing downward and the intestinal opening upward. However, sea cucumbers display a lateral orientation and crawl in the direction of its long axis, with the mouth pointing forward (in the direction of movement). In feather stars and sea lilies, the mouth as well as the anus are pointing upward. Therefore, in part of the species there is, besides the radial symmetry, also a weak lateral symmetry as a form of secondary regression, e.g. in the more or less distinct "foot sole" that facilitates crawling locomotion.

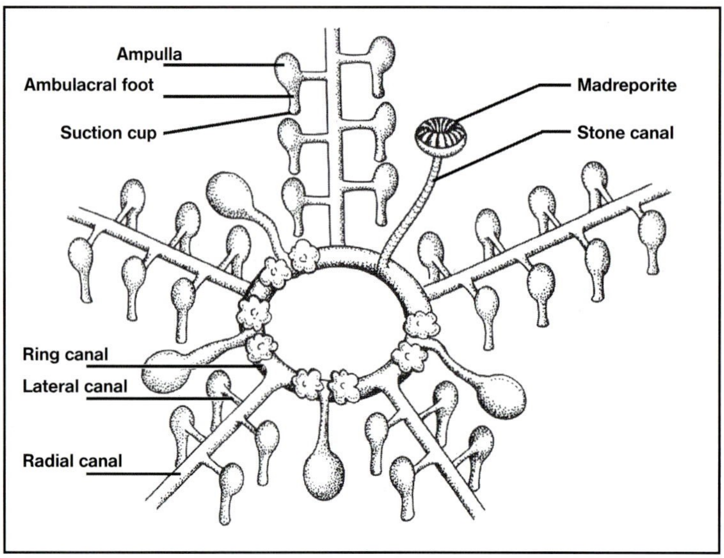

Schematic presentation of a water vascular system of a sea star.
Drawing after BARNES (1980)

Water Vascular System

Echinoderms have a highly specialised water vascular system (water canal system), referred to as the ambulacral system. It is a structural feature unique within the Animal Kingdom. In the (phylogenetically) older echinoderms, it is assumed that this system serves primarily for the intake and transport of food. In addition, the water vascular system contributes to the transport of nutrients, waste products and gases; however, gradually the system developed more in the direction of a hydrostatic pumping system. It facilitates the animal's ability to move, i.e. when using the thin-walled tube feet (podia). These tubular outer extensions (referred to by some authors as tubular- or suction feet) of the water vascular system, of which there can be up to several thousand on a single specimen, penetrate the body wall and the skeleton in areas referred to as ambulacral grooves or –zones.

The entire water vascular system opens to the outside via a special sieve plate, the madreporite (madreporite plate), which leads downward through the stone canal in to the circular canal. In most species, the circular canal is positioned – ring-like – around the mouth. From there individual radial canals lead to individual body parts. Hydrostatic pressure, required for moving the suction feet, is created by contraction of bubble-shaped ampullae.

In sea urchins and sea stars, the madreporite is located on the upper (aboral) side of the animal close to the intestinal opening, while it is on the under (oral) side in brittle stars. The madreporite in sea cucumbers is located inside the animal, and it is absent in sea lilies and feather stars. Instead, the latter have a large number (500 – 1000) microscopically small, ciliated vesicles that lead directly in to the water vascular system to the outside. Both types of opening of the water canal serve primarily to regulate pressure. The water volume, which flows normally through the madreporite, is very small, whereby the fluid in the canals contains – apart from seawater – cells and organic compounds. The cells are principally amoeboid coelomocytes, used for excretion as well as for wound healing and repairs, that is, the regeneration of damaged and lost body parts. Gas exchanges takes place - in part – directly through the thin

Oral side of the Eastern Pacific Nail Sea Star, *Mithrodia bradleyi*, seen here attached to the front glass of an aquarium. The ambulacral grooves with their suction feet are clearly visible.

Photo: S. W. Michael

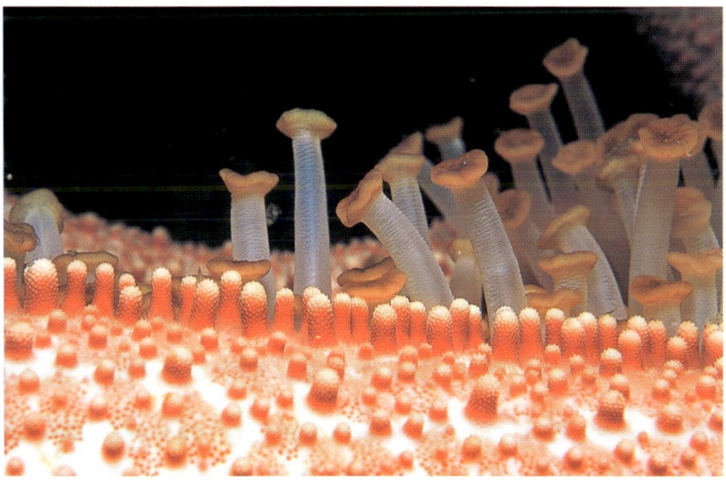

Left: Close-up view of a sea star arm, with the suction feet clearly visible (photo: L. Newman and A. Flowers). Right: Photo-micrograph of a suction foot from a sea star of genus *Asterias* under polarizing light (photo: R. Brons).

The Knoddy Sea Star, *Fromia nodosa*, is one of the most conspicuously coloured sea stars of the Indo-Pacific. In this specimen the madreporite - the external entrance to the water vascular system - can be seen here clearly as an orange-coloured disc on the left side of the aboral side, between two arms.
Photo: Dr. I. Illich

epithelium (outer cell layer of the skin) of the tube feet.

Pedicellaria

In addition to tube feet, spines and warts, all sea urchins and most of the sea stars are also equipped with special outer organs, commonly referred to as pedicellaria. These are specifically developed jaw-like appendages, which – when viewed under a microscope – look like miniature pliers or sugar tongues. In sea urchins the pedicillariase "jaws" are typically located at the end of a long, flexible arm or stem, while they are located directly on the skin in some sea stars.

The first scientist who revealed the presence of the jaw-like pedicillariathese organs considered them to be parasites as known from other animals (for example, anchor worms, *Lernaea* and other specialised crustaceans that are parasitic on fishes), so that he described them initially as parasites on echinoderms. However, in reality they are very important structures for the survival of these echinoderm animals.

Primarily pedicellaria serve to keep the surface of the echinoderm animal clean, and to a lesser degree they have also a defensive function. The "jaws" are opened and closed with the aid of muscles, controlled by radial nerves and specific sensory organs. If an intruder, e.g. the larva of some other animal in search of a settlement site, crawls over the outside of the echinoderm, the touch stimulus triggers sensory receptors, which cause the nearest pedicellaria to react quickly and to capture the intruder.

It has been proven that there are various main types of pedicellaria (see photo-micrographs on page 215) for different tasks, however, the various functions and modes of operation of these organs are far from being fully explained.

Skeleton

Ninety-five percent of the echinoderm skeleton consists of calcium carbonate. Its structure is composed of calcium sclerites (calcium ossicles), which are more or less joined together. In some species, such as brittle stars and most sea stars, adjoining skeletal parts are moveable, so that the animal is fairly articulated and can move readily, in spite of the fact that it has a solid, inter-connected and internal skeleton. The skeleton of (most) echinoderms is sufficiently flexible to bend around projecting objects or to crawl in to tight crevasses or holes. However, in sea urchins the skeleton sections are firmly joined against each other to form a solid "carapace", whereby the individual structural sections are visible on the outside. On the other hand, the skeleton of sea cucumbers is essentially reduced to a few – sometimes only microscopically small – spicules, encapsulated within the skin. For that reason, many sea cucumbers are rather soft and shapeless.

Independent of the degree of firmness of the skeleton, most echinoderms are also equipped with outer spines or sharp, pointed warts, that are visible and/or can be felt through the skin, a fact documented by their name: Echinodermata – spiny-skinned animals.

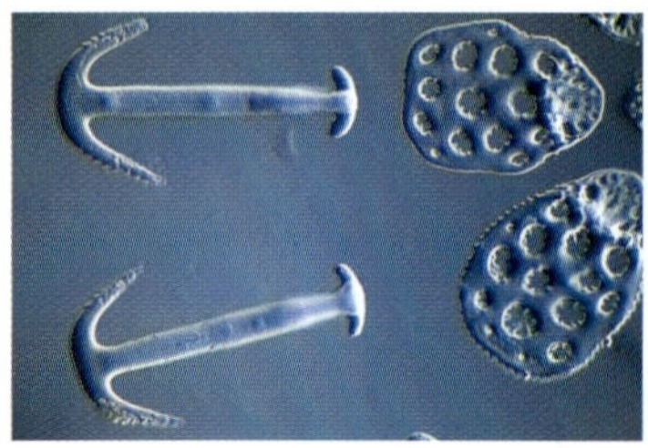

Spicules of *Synapta* sp.
Microphoto: R. Brons

Regenerative Capacity

Another remarkable characteristic of the echinoderms is their ability to replace lost body parts. Sea stars are good examples of this regenerative ability. It is well known that many species are able to regenerate damaged or lost arms. In many cases even internal organs can be completely replaced. In fact, there are also examples where arms that had been torn off, re-develop the remainder of the body, including oral disk and the other arms.

Nervous System

Echinoderms do not possess a central aggregation of nerve cells that could be comparable to a simple brain. Apart from tactile (touch-sensitive) receptors and chemo-receptors (taste- and smell-sensitive), these animals are only modestly equipped with sensory organs. Tactile- and chemoreceptors are distributed over their entire body surface. In addition, there are also a few simple eyes (essentially only light – dark sensitive) located at the tips of the arms of sea stars, some equilibrium organs in a few sea cucumbers and chemical sensory organs in the pedicellaria of sea urchins.

Nerve communication as such takes place via a rather simply laid out nervous system that extends essentially parallel to the water vascular system. It consists of radial nerve strands which lead to individual parts of the animal and which join up in a central nerve ring around the mouth (circum-oral nerve ring), so that the activities of individual arms can be coordinated.

A well-developed layer of skin covers the body surface in most echinoderms. In the basal layer of this epithelium there is a nerve network that controls the outer organs, such as spines, pedicellaria, etc. This network of nerves is connected to the central nerve stem in each arm, and so facilitates coordination between nerve impulses perceived at the skin surface and the subsequent response of muscles and other organs of the animal.

Locomotion

The degree of mobility is highly variable within this group of animals. With the aid of their suction feet or spines, most echinoderms can only move slowly and with great effort. Individual species are capable of moving about by using peristaltic bending movements and through twisting their arms, whereby some can reach quite significant speeds. The vast majo-

Pedicellaria are characteristic for all sea urchins and for sea stars of the order Valvatida and Forcipulatida. It is assumed that these small, tong-like appendages serve principally as a defence against small organisms or larvae that could potentially settle on the body surface, and for cleaning purposes.

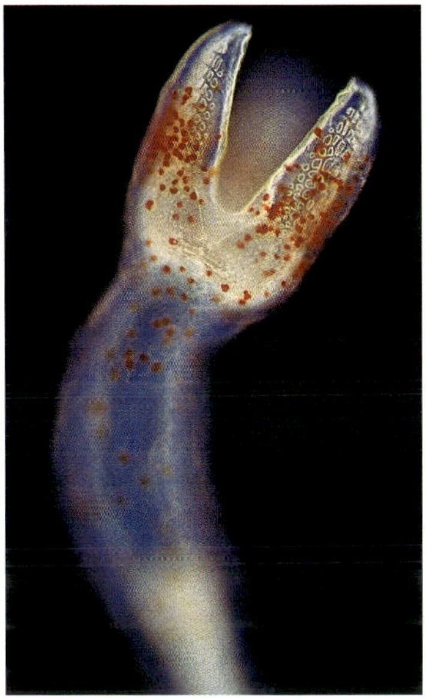

Left: Stalked pedicellaria of the sea urchin, *Psammechinus miliaris* suitable for biting, which can be used for defence as well as a device for cleaning the body surface. When the outside of a pedicellaria is touched it will open up, and when it is then touched on the inside, it snaps shut. Right: Pedicellaria without a stem or stalk of a sea star (genus *Asterias*). Many sea stars have pedicellaria without stems, while those of sea urchins are always with stems.

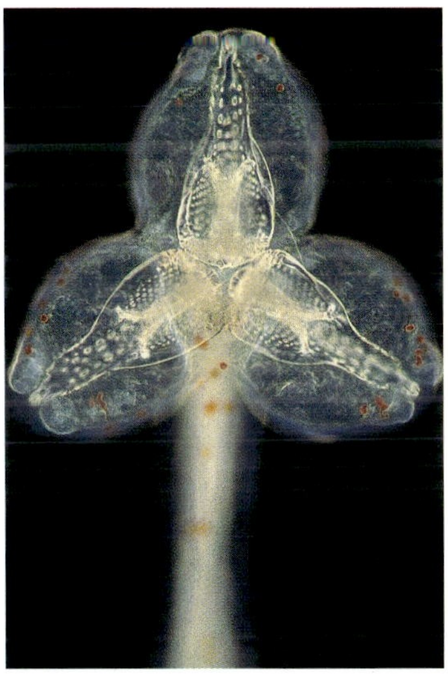

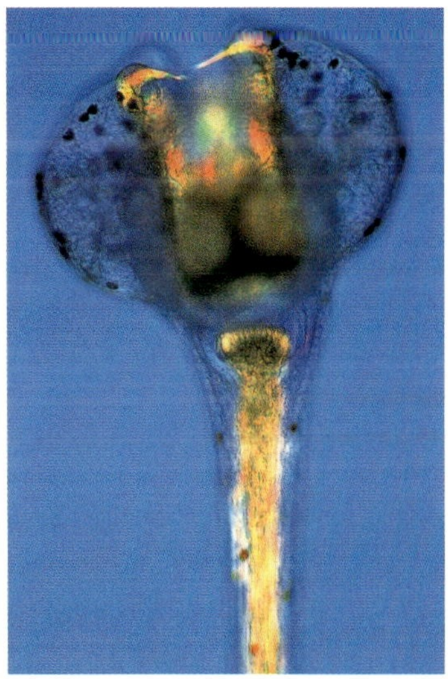

Two venomous pedicellaria. The clamps (jaws) are surrounded by venom-filled vesicles that are connected to the teeth at the tip of the jaws via canals. The venom has a quick-acting, paralysing effect on small animals and it is also capable of chasing away large animals.

Four microphotos shot in polarizing light: R. Brons

The body surface of this Large Pincushion Star, *Culcita novaeguineae* (seen here at Heron Island, Great Barrier Reef) feels rough and warty, caused by calcareous spines and tubercles.

Photos: L. Newman and A. Flowers

Many spiny-skinned animals - especially sea stars - have enormous regenerative capabilities. An arm or even damaged sections of the central disc are regenerated by many species. This *Linckia multifora* (seen here at the Fiji Islands) will quickly regenerate its severed arm.

Photo: S. W. Michael

rity of species are exclusively benthic (live on the bottom), but some (e.g. feather stars) have the ability to swim freely by twisting their hair-like arms.

Suction feet and the musculature are controlled by radial nerves, which are also responsible for the coordination of movement and therefore also for the direction of movement. A sea star moves either in the direction of one arm or follows a direct line indicated by two opposing arms. If the sea star is to move in the opposite direction, it can either reverse its direction or turn around. It appears that in sea stars and sea urchins the angle between the two integuments where the madreporite is located is generally pointed in the direction of

The types of locomotion used by echinoderms are highly variable. Most sea stars (left) and sea cucumbers (here *Bohadschia argus* on the right) move slowly and rather laboriously with the aid of their suction feet and spines.

Photos: E. Svensen

Serpent- or brittle stars (seen here is *Ophiothrix purpurea*; left) accomplish locomotion by means of peristaltic movements of their arms. Feather stars, seen here is a species from family Comasteridae, (right; photo: L. Newman and A. Flowers) can reach access to difficult sites by means of swimming.

movement. This could indicate that the madreporite has a certain sensory function.

Gravity is also perceived by echinoderms. If they are flipped over, they generally return to their former position. The presence of specific balancing (equilibrium) organs has also been shown for sea urchins and some sea cucumbers.

Defensive Strategies

Their regenerative ability has enabled echinoderms to develop specific defensive tactics. An example of this is the ability of sea cucumbers to expel sticky Cuvierien Tubules from their abdominal cavity to scare off an enemy. Once these organs have been ejected they are lost, but there is a gradual re-development for subsequent (re-) use.

The production of a potent poison is another, very important defence strategy in sea cucumbers. Many sea cucumbers discharge the poisonous Holothurin, a triterpene-glycoside, which can have catastrophic consequences for life in a coral reef aquarium (see pages 373-375). This poison occurs mainly in the Cuvierien Tubules, in the skin, the intestine and the respiratory organs. Holothurin creates a soap-like effect in water by eliminating the surface tension of water. It is assumed that this has a direct toxic effect on fishes by damaging their sensitive gill epithelium and so inhibit the uptake of oxygen.

Some sea stars also contain toxic substances, which have probably the same task as Holothurin in sea cucumbers, i.e. to fend off enemies or predators, respectively. Here it is a steroid-glycoside known as Thornasteroside, which is stored in the skin. Such potent venoms are found primarily in sea stars that have an abundance of spines. In addition, feather stars and brittle stars have yet other toxic substances, such as sulphuric acid esters that repel predators.

Many sea urchins (e.g. *Diadema* spp.) have long, sharp spines that can inject neurotoxins when touched. These are very painful to human and probably also to fishes. Consequently, sea urchin predators like triggerfishes have developed a special capture technique. They tend to turn over their prey so that direct contact with the sea urchin's spines is largely avoided. *Asthenosoma* sp. has special spines fitted with poison glands, and yet others (e.g. *Toxopneustes* spp. and *Tripneustes* spp.) have specialised pedicellaria equipped with poison sacs.

Apart from experiences with sea cucumbers, there are no records of such problems caused in aquariums by toxic substances from other echinoderms. Yet, some species, especially certain sea urchins, are so dangerously toxic, that aquarists should exercise extreme caution. Those species that are known to be highly toxic and which could cause problems when kept in an aquarium, will be discussed in relevant sections below or in particular species Fact Sheets. Beyond that, we would like to draw the reader's attention to general literature on toxic or venomous animals, e.g. MEBS (1989 and 1992) and MARSH & SLACK-SMITH (1986).

Food and Feeding

Echinoderms have developed rather variable feeding strategies. There are those that feed on algae and detritus, as well as omnivores and suspension feeders, and also predators. Specifically of great interest for coral reef aquarium maintenance are the detritus-feeding echinoderms. Similarly, some of the algae-grazing species can also be very useful in a coral reef aquarium. Consequently, these two groups are being given particular attention in the discussion below.

Some of the echinoderms digest their food outside the body by everting their stomach out through the mouth and so engulf their prey. They then release enzymes that dissolve the prey so that it is gradually absorbed by the sea star's digestive system.

The expulsion of Cuvierian Tubules is a particularly well-known defensive weapon in certain sea cucumbers. Laymen tend to associate this behaviour often generally with all sea cucumbers, but it is confined to species of family Holothuriidae. It is particularly common in the Eye-spot Sea Cucumber, *Bohadschia argus.* Normally, these tubes are attached to respiratory organs and hidden on the inside of the body. If such a sea cucumber is being irritated or attacked, the anus is directed to the perceived intruder or attacker and the body is contracted. This then expels the tubules out through the anus. The tubules are adhesive and tend to trap the aggressor in a net made of sticky threads. This will totally immobilize small animals, rendering them helpless, thus causing them to die while the sea cucumber crawls away and regenerates its Cuvierian Tubules. Many other sea cucumbers display different defence strategies, which can easily be mistaken for the expulsion of Cuvierian Tubules: that is, they eject their respiratory organs, their digestive tract and/or their reproductive organs.

Photo above: S. W. Michael; photo below: L. Newman and A. Flowers

Sea urchins are true coelomates with an advanced digestive system; however, they do not posses specific excretory organs.

Reproduction

Echinoderms have very simple reproductive organs. There is no copulation in any form. Instead, external fertilization takes place, whereby eggs and sperm cells are released in to open water. Separate sexes is the norm for the majority of species, but there are also hermaphrodites that start out as being males and then end their lives as females, just as we know it from many crustaceans.

Echinoderm larvae usually undergo many successive planktonic stages, before they change to nearly fully adult structures during a distinct and anatomically complicated metamorphosis, and then settle on the bottom. These young benthic echinoderms generally resemble miniature copies of the adult form, until they gradually grow to sexually mature specimens.

In some species the regenerative ability represents a type of asexual reproduction through the replacement of lost or shed body parts or through regular cleavage.

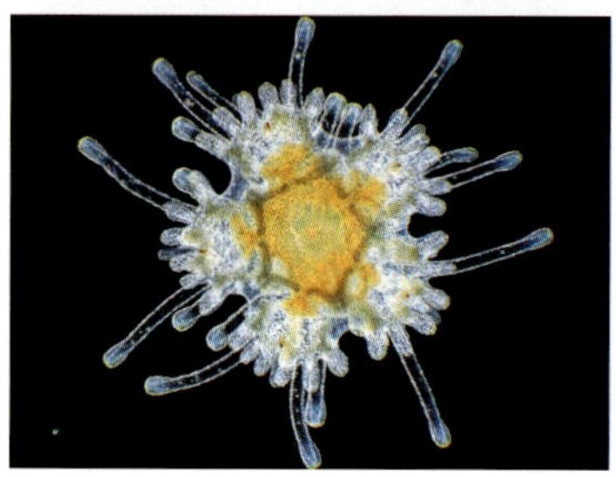

Top: Planktonic brittle star larva (ophiopluteus larva). Bottom: Post larva of *Astgerias rubens* shortly after metamorphosis. Photos: R. Brons

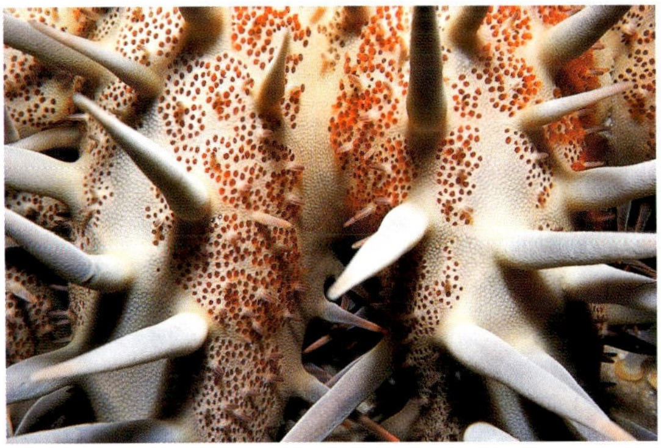

Somewhat smaller venom spines are hidden among the long spines of the Crown-Of-Thorns Sea Star, *Acanthaster planci* (left), which contain the rather potent Thornastroside (photo: L. Newman and A. Flowers). Right: The Tam-O'Shanter Sea Urchin, *Asthenosoma ijimai* has short, but very sharp spines embedded in venom glands (photo: C. Bryce).

The sea urchins *Tripneustes gratilla* (left) and *Toxopneustes pileolus* (right) have venom-carrying pedicellaria, which are so long in *T. pileolus* that they almost cover up the spines. Photos: Dr. I. Illich

Left: *Toxopneustes roseus* is another sea urchin with venomous pedicellaria. The exact composition of the sea urchin venom is not known. In humans it tends to cause immediate pain, swelling and reddening of the puncture site, which can last for several hours. In severe cases there can be extreme dizziness and respiratory complications, which necessitates treatment with anti-histamines. Right: The active substance Holothurin occurs in the body wall (viscera) of many sea cucumbers. Its substantial toxicity has lead to mortalities among fishes, for instances, when it was released by the attractive Sea Apple, *Pseudocolochirus violaceus* during spawning and the gametes were eaten by fishes. Photos: S. W. Michael

Left: This sea star is feeding on a sponge by everting its stomach over the prey, digesting it outside its body (photo: K. Kates). Right: Sea Apples of genus *Pseudocolochirus* are typical suspension feeders. This aquarium specimen has its tentacles spread out in order to capture *Artemia* nauplii.

Left: This feather star (presumably a species from family Comasteridae) collects plankton at night (photo: I. Erga). Right: The diurnal sea cucumber, *Bohadschia graeffei* uses its tentacles to search the bottom for organic material.

Some echinoderms are very popular among aquarists. Especially suitable are certain brittle stars, such as *Ophioderma squamosissimum* (left) from the Caribbean. Although its total spread can be up to 40 cm, its central disk has a diameter of only 4 cm. Its total body mass is only small. Right: Aquarists often consider sea urchins to be effective algae eaters, but in part they can be too efficient for small aquariums. This habitat photo of the Caribbean *Echinometra lucunter* illustrates how "polished" the reef substrate can be after this sea urchin has grazed over it. In an aquarium, this sea urchin could totally eliminate algae from most of the tank decoration.

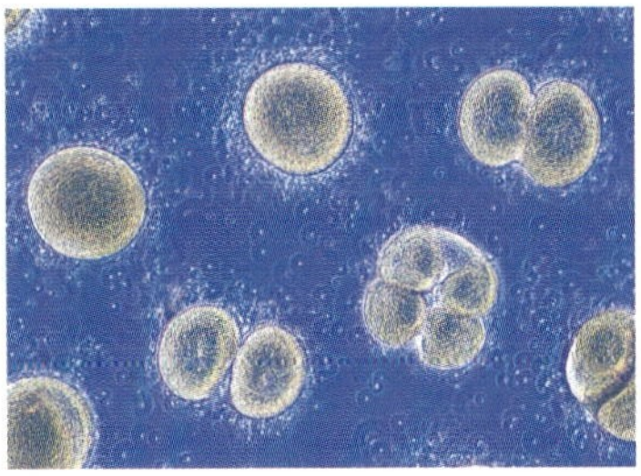 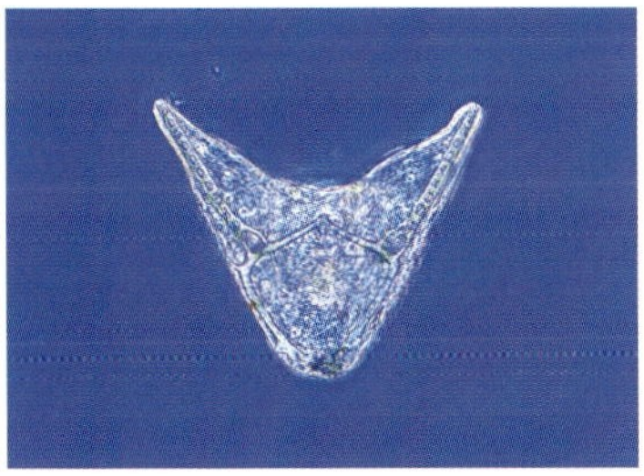 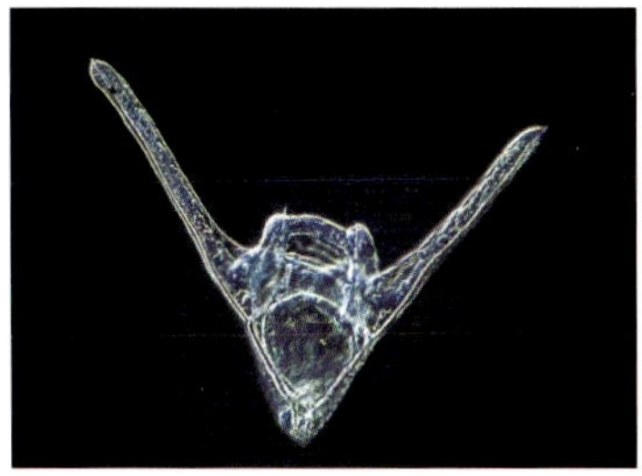

Embryonic development of a sea urchin egg: unfertilised and fertilized (first and second cell division stage) of oozytes, surrounded by many spermatozoa (left), early echinopluteus larva (centre) and pelagic echinopluteus larva.
Microphotos: R. Brons

Class Crinoidea - Sea Lilies and Feather Stars

Sea lilies and feather stars make up the most primitive group of echinoderms. The class has a long fossil history. Sea lilies flourished during the Paleozoic era and were common animals of shallow reefs. 400 million years ago, these animals, as sessile sea lilies were the characteristic animals along the reef bottom, and there they were quite numerous. So far more than 5,000 fossil species are known (MARSHALL, 1979). Today, feather stars with about 600 species are predominant in this class. They are found typically in shallow water of tropical seas, that is, they are distinct coral reef organisms. On the other hand, sea lilies are, however, rare now occurring mainly in very deep water, and so only a few people will have ever seen any of the about 100 species alive.

In terms of their body structure and mode of life, crinoids are distinctly set apart from the other echinoderms. Yet, the relationship manifests itself in the presence of the water vascular system, the pentamerous symmetry and the skeletal structure made of calcium carbonate sclerites (calcium ossicles).

Crinoids have an oral disk, referred to as theca or calyx (cup), which is either supported and held upright by a stalk or it is equipped with a ring of structured cirri that serves as a type of grasping feet (feather stars). In addition, all species have 5 to 200 or more arms (always in multiples of five) that have along their sides jointed appendages, so-called pinnulae. These give the arms an appearance reminiscent of a

Sea Lilies are relicts from ancient times and today only about 100 deep-water species exist. This approximately 45 cm tall animal, possibly *Endoxocrinus parrae*, was collected by submersible from 300 m depth off Salt Cay, the Bahamas.
Photo: Dr. R. Shimek

Only the feather stars of order Comatulida occur on coral reefs; in fact, frequently they are rather numerous. Top: Feather star at Batala, Maldives (photo: Dr. I. Illich); bottom: Feather stars at Manado Tua, Indonesia (photo: R. Hansen).

bird's feather. Finger-like suction feet on the arms capture plankton, which is transported along the arms via mucus strands located in ciliated depressions (ambulacral or food grooves) to the mouth. In those species with large numbers of arms, the combined length of the ambulacral (food) grooves can be more than 100 m (ALLEN & STEENE, 1994).

One of the most extraordinary structural features that distinguishes crinoids from the other echinoderms is the fact, that the mouth as well as the anus are located on the upper side of the calyx. This deviating location of the mouth is probably related to the behavioural mode to filter food particles from the water rather then picking them up off the bottom. The calyx as well as the arms are strongly armoured with densely arranged calcium plates (ossicles), but there are no spines or pedicellaria. The arms break off very easily through usage (e.g. in an aquarium) as well as during predator attacks. However, under favourable conditions regeneration is very rapid.

The class Crinoidea is made up of four orders, of which the first three consist of sedentary or sessile (firmly attached) species, i.e. sea lilies of the order Isocrinidae, Millecrinida and Cyrtocrinida. The systematic overview on page 228 is confined to order Comatulida (feather stars), which is the only one of significance from an aquarist's perspective.

Orders Isocrinida, Millecrinidida, Cyrtocrinida - Sea Lilies

These orders encompass the sea lilies that are constantly attached to some sort of substrate as juvenile as well as adult, either via a stalk (which can be segmented or unsegmented) or with an enlarged foot base or with root-like structures (see drawing to the right).

Recent (currently living) sea lilies can reach a length of up to 50 cm, yet, they are virtual dwarfs compared to their fossil ancestors. Some of them had stalk length of 20 m! It is assumed that those enormous stalk lengths had developed to reach as high as possible in order to be first in line to obtain food descending towards the bottom (VALENTINE, 1973). For suspension feeders that is an intuitive and sensible strategy. Apparently recent forms, which often spread their arms like a collecting box, are still utilizing this strategy. When there are strong water currents, sea lilies point their arms against the flow, so that they are spread out like a filter fan.

Isocrinida species often live in the upper part of the continental shelf, at depths from 100 to 1,000 m, but many other sea lilies live substantially deeper, down to about 6,000 m, and therefore some of them have a rather cosmopolitan distribution. In certain localities sea lilies occur in extremely dense aggregation. Substrate samples taken at depth from 550 to 700 m off Sand Key, Florida, brought so many *Rhizochrinus* specimens to the surface, that Alexander Agassiz, in his account, suggested the bottom scraper must have passed through an underwater forest of sea lilies (MARSHALL, 1979).

Sea lilies are not known to occur on coral reefs, and there are also no published reports available on their care under aquarium conditions.

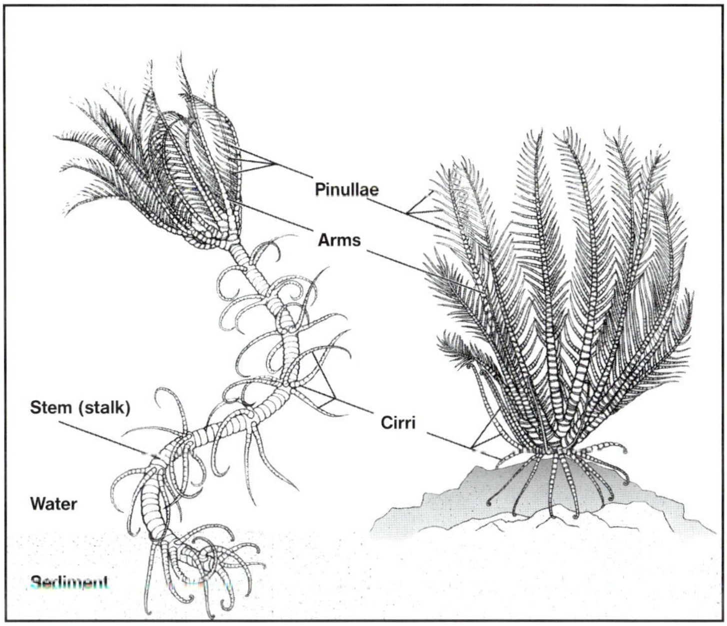

Fundamental body structure of crinoids: sea lilly of order Isocrinida (left) and feather star of order Comatulida (right).
Drawing after GEORGE & GEORGE (1979)

Order Comatulida
Feather Stars

Feather stars of order Comatulida are often described as being the most beautiful and conspicuous of all echinoderms. They are such typical organisms in tropical seas that they are well known to most divers and aquarists. These conspicuous "feather dusters", which can be seen either swimming all over the reef or attached to its structures, simply cannot be overlooked. They occur in all colours of the rainbow: uniformly coloured or with patterns of white, yellow, orange, red, lilac, green, brown and black. In fact, the magnificently coloured tropical shallow water forms are so spectacular, so it is easy to forgot that

Close-up view of feather star *Dichrometra flagellata* that illustrates the arrangement of the arms around the central disc (theca or calyx).
Photo: Dr. I. Illich

Close-up views of arms from various feather stars that underline the wide variability in shape and colour within the order.
Photo (top): S. W. Michael; photo (centre left): Dr. I. Illich

With the aid of the cirri, which function as grasping feet, feather stars can cling to substrate.
Photo: J. Cairns-Michael

Juvenile feather stars at Bintan Island, Indonesia (left; photo: C. Bryce) and at Heron Island, Great Barrier Reef (right; photo: L. Newman and A. Flowers).

within the same order there are also coldwater- deep-sea species. Feather stars occur practically in all seas, from the Arctic to the Antarctic.

Juvenile feather stars live like sea lilies, i.e. they are sessile (by the way, frequent references to this stage as the sea lily stage are misleading). With advancing age they become detached and move about freely. From then on they attach themselves temporarily to some substrate with the aid of special cirri that function as grasping feet (see photo on page 224, bottom). With these, feather stars can firmly attach themselves to a substrate at a suitable location, where current conditions and food availability meet their requirements. The animals can move about partially by crawling or walking with the aid of their cirri and – as is applicable to many species – by executing graceful swimming movements with their hair- or feather-like arms. Consequently, many species are amazingly mobile, yet at relatively slow speeds.

The body of feather stars is of a bony, brittle texture. The skeleton is covered by a thin layer of skin and a seemingly rather minimal component of soft body parts. The oral disk or calyx has a diameter of 3 to 50 mm, and the arms can reach a length of 50 to 550 mm. The more delicate species have only five arms; however, it is more common to find 10 to 20 arms. But there are also species that have five arms as juveniles, and then regularly shed arms and replace each lost arms by regenerating two new ones, so that the final number of arms can become quite large.

The mouth is located on the upper side of the calyx and forms the meeting place where the ambulacral grooves that transport the food collected by the arms, come together. In family Comasteridae the mouth is located along the perimeter of the calyx, in all other families it is in the centre.

Feather stars have available to them a rather limited food selection. Their diet consists exclusively of small planktonic organisms (including foraminiferans, crustaceans, invertebrate larvae and phytoplankton), as well as of organic particles suspended in the water. The preferred particle size is 0.05 to 0.4 mm in diameter (HENDLER et al., 1995). Feather stars are passive suspension feeders that position themselves at a suitable site on a reef and then spread their arms like a fan in the water. Shape and position of arms

The number of arms in feather stars varies from one species to the next, and it can also vary within the development of a particular specimen. **Photos: E. Svensen (left) and S. W. Michael (right)**

Feather stars often act as hosts for many commensal organisms, such as this Pistol Shrimp *Synalpheus stimpsonii*, ... **Photo: C. Bryce**

... the Feather Star Jumping Shrimp, *Allogalathea elegans*, and ... **Photo: E. Svensen**

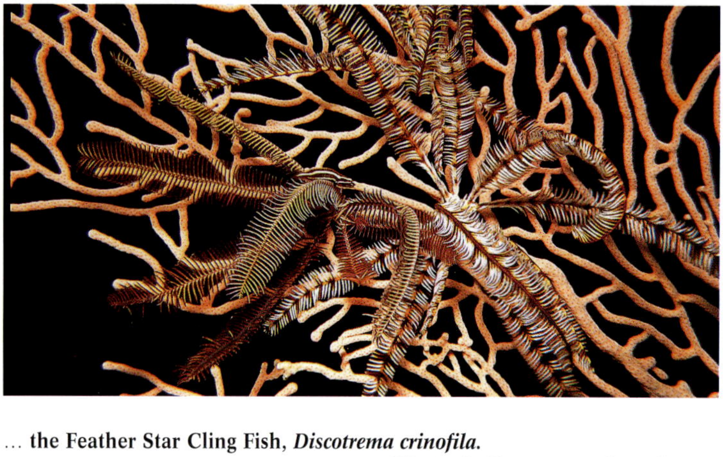

... the Feather Star Cling Fish, *Discotrema crinofila*. **Photo: L. Newman and A. Flowers**

is variable, depending upon what is most effective under the prevailing local conditions. A large segment of the species is primarily nocturnal and spends the day in caves and rocky crevasses. Yet, it is not uncommon to find feather stars also during the day. At dusk they emerge from their hiding places to settle on protruding rocky areas, on sponges or on corals.

Being such conspicuous organisms, some feather stars are prey for certain fishes; however, by and large they are left in peace by other animals. This is probably due to the fact that feather stars carry toxic substances in their skin, which – no doubt- are effective deterrents. The intensive colouration of many species may serve as a warning signal to other animals in the same habitat, that they are inedible.

Because of their rather sedentary way of life, feather stars often act as host for a large number of commensal organisms. It is conceivable that the advantage of such symbiotic relationship may lie in the fact that the commensal organisms are protected by the feather stars against predatory fishes – the inedibility of their host affords them a save heaven. Moreover, the body structure of feather stars is varied in shape and offers so many anchor- and hiding places that several commensal organisms feel secure among the arms and pinnulae of these echinoderms.

In addition, the mucus bands, which transport plankton and other particles from the arms to the mouth are virtual collecting bands of miscellany food items for many other organisms. There they find a nearly inexhaustible supply of easily accessible food, simply by settling on the mucus bands and commence feeding. Some feather stars have practically enclosed ambulacral grooves through which food is transported – possibly a direct counter-measure against the scavenging attempts by their symbionts.

Up to a dozen commensal organisms can inhabit a single feather star, and in extreme cases there may be many more. ZANN (1980) reports to have found three dozen small bristle worms of two species from class Myzostomaria, one large scale worm of family Polynoidae, one dozen small copepods, two species of shrimp, one porcelain crab (Porcellanidae) and three clingfish (Order Gobiesociformes) – and all of them on a single feather star! Various specimens of *Heterometra savignii* have been observed to carry 18 different species of commensals (VINE. 1986). The colouration of the commensals is often a very close imitation of that of the host's markings. Anyone who is looking for further information about the incredible symbiotic relationships between feather stars, worms, crustaceans, snails, brittle stars, fishes and other organisms is advised to consult the comprehensive publication by ZANN (1980).

Although male and female specimens of feather stars seem morphologically rather similar, they do have separate sexes. The reproductive organs are located in each arm at the pinnulae near the calyx. In most species the spawning season last for about one or two months during spring or summer. Mass spawning is also known to occur, as for instance in *Oxycomanthus japonicus* (HENDLER et al., 1995). During such event, masses of individual specimens take part in a simultaneously triggered spawning, which lasts for an hour and occurs only once a year. As observed in corals, the precise timing of spawning among feather stars appears to be related to the lunar cycle. Each female can release up to 2 million eggs and disseminates these with frantic fanning of the arms. A few days after fertilization, an embryo hatches as a dolio (dolilaria) larva. It is characterised by the presence of several ciliated band around the body (cilia rings). A mouth is absent in this larva, which either does not feed at all or nutrients are absorbed directly from the water by means of diffusion. Following a few days of free-swimming, these larvae settle on suitable substrate and grow in to the "sea lily stage", whereby they remain firmly attached – via a stalk – to a substrate. By way of metamorphosis – lasting for at least three months – they develop a mouth, anus, arms and cirri. At that point they break off their stalk and live in a free-swimming, adult stage.

"Sea lily stage" of a feather star. Upon completion of the planktonic stage the larva settles on some substrate, where it undergoes a step-by-step metamorphosis to a juvenile feather star. Drawing after GOSLINGER et al. (1996)

A very small number of feather stars practice brood care, whereby the released eggs are attached to the genital pinnulae or are stored in a "brood pouch".

Feather stars are very delicate and fragile organisms that are easily damaged during capture and transport. Therefore, it may be difficult to collect these animals for taxonomic research or for identification purposes, and even harder yet to transfer them in to an aquarium. Some species have such typical colour markings that they are relatively easy to identify from photographs; however, the majority of the species can be highly variable in colouration and markings. For instance, the Indo-Pacific species, *Comanthina schlegeli* occurs in all shades from yellow to orange and brown to black; multi-coloured specimens occur frequently. For species from the central Pacific we recommend the identification aid listed in GUILLE et al. (1986), and for species from the Caribbean region the reader's attention is drawn to HENDLER et al. (1995). Family identification can be facilitate to some degree on the basis of MATHER & BENNETT (1993).

Coral reef habitats often reveal a large diversity and density of feather stars. For instance, twelve or thirteen species are rather frequent in the Red Sea at a depth of 1 to 25 m (VINE, 1986), while a large, but unknown number of species exist in the Indo-Pacific. In the Caribbean region there are about eight species at depths easily accessible by divers (HENDLER et al., 1995). Feather stars are commonly spotted on almost every reef, especially during the night along the reef edge or reef slope. Under certain circumstances, they can occur in dense aggregation of 50 to 100 specimens per square metre.

The aquarium trade sporadically imports feather stars; however, generally it is very difficult, if not even impossible, to keep them alive in an aquarium for any length of time. So far we are not aware of any documented incidents where an aquarist has been successful in keeping feather stars for more than a few months at best. During the first few days or even weeks, feather stars can be very attractive in an aquarium, but after that parts of the arms start breaking off. The regenerative capabilities of these animals, developed to highest perfection in the wild, does not seem to function under aquarium conditions. Eventually, all that remains is the calyx and a few bare stumps were arms used to be, which then leads to the demise of the animal. With that in mind we are of the opinion, that these animals should currently not be part of an aquarium shop's inventory.

Aquarists, who nevertheless

Systematic Overview of Order Comatulida

after HENDLER et al. (1995); GUILLE et al. (1986); GEORGE & GEORGE (1979); CLARK & ROWE (1971)

Family Comasteridae
Capillaster, Comactinia, Comantheria, Comanthina, Comanthus, Comaster, Comatella, Comatula, Comissia, Nemaster, Oxycomanthus

Family Colobometridae
Analcidrometra, Basilometra, Cenometra, Colobometra, Cyllometra, Decametra, Iconometra, Oligometra, Oligometrides, Petasometra, Pontiometra

Family Eudiocrinidae*
Eudiocrinus

Family Himerometridae
Amphimetra, Craspedometra, Heterometra, Himerometra

Family Mariametridae
Dichrometra, Lamprometra, Liparometra, Oxymetra, Stephanometra

Family Zygometridae*
Cataptometra, Zygometra

Family Aporometridae*

Family Notocrinidae*

Family Asterometridae*
Pterometra

Family Calometridae*
Reometra

Family Charitometridae*

Family Ptilometridae*
Ptilometra

Family Thalassometridae*

Family Tropiometridae
Tropiometra

Family Antedonidae
Antedon, Ctenantedon, Dorometra, Euantedon, Florometra, Mastigometra, Toxometra

Family Apocrinidae*

Family Atelecrinidae*

Family Pentametrocrinidae*

(* not dealt with here)

are determined to keep feather stars – may be to resolve the husbandry problems associated with these animals – need to have a great deal of dedication as well as provide alternate, newly developed husbandry ideas. Most likely the greatest problem will be finding the correct food, possibly in conjunction with improved feeding methods. But there may also be other contributing factors that play an import role here, such as water motion. The types of food that have so far been used with variable success include: yeast, *Calanus*, fish roe, *Artemia* nauplii, *Brachionus*, *Euplotes*, floating suspensions, pulverized substitute foods (GROSSKOPF, 1994) and "mussel milk" (minutely grinded meat of mussels; sperm; LANGE & KAISER, 1991). Live phytoplankton has been used as food for other delicate and difficult organisms, such as soft corals from the genus *Dendronephthya*, and it might be worth the effort to try feeding feather stars with this food source in captivity. It would probably be advantageous, to use feather star specimens that are as small as possibly for this sort of experimental work under aquarium conditions.

A selection of the largest and most important families is discussed in detail below. Beyond that, we refer the reader to the systematic overview provided on the left side of this page.

Family Comasteridae

This family is distinguished from the other feather star families in that the mouth is located over to one side of the calyx and not at the centre. The pinnulae, located towards the centre of the body are very tensile and some of the outer segments are modified in to a comb-like structure.

The general most frequently encountered in the Indo-Pacific are *Capillaster*, *Comissia*, *Comanthina*, *Comanthus*, *Comaster* and *Oxycomanthus*. In the Caribbean region genera *Comactinia* and *Nemaster* are the most common ones. Within this family the

Feather stars of family Comasteridae

Two *Comanthina* species. Photos: L. Newman und A. Flowers

Comanthina nobilis.

Comanthina cf. *andax.*

Comanthina briareus. Photo: C. Bryce

Comanthus sp. Photo: Dr. I. Illich

Feather stars of family Comasteridae

Comanthus sp. Photo: S. W. Michael

Comatula spp.

Comaster multifidus.

Oxycomanthus bennettii. Photo: Dr. I. Illich

Two additional colour varieties of *Oxycomanthus bennettii.*

Comanthina schlegelii. Photo: C. Bryce

Comatula sp. Photo: L. Newman and A. Flowers

The Beautiful Feather Star, *Cenometra bella* in Indonesia.

The Robust Feather Star, *Himerometra robuststipinna* is probably the most-often imported feather star. That (together with its species name) would suggest that it is less delicate and fragile than the other species. However, even with our latest aquarium technology and up-to-date husbandry knowledge, it is nearly impossible to keep this species alive for an extended period of time. Feather stars should only be purchased once appropriate aquarium husbandry knowledge and experience has been acquired.
Photo: C. Bryce

number of arms varies from ten (e.g. *Comaster multifidus* and *Comactinia echinoptera*) up to 130 or more (*Comanthia schlegeli*), and maximum size of adult specimens varies from 10 cm to more than 40 cm in diameter, while the majority measures 20 to 25 cm. Most species are encountered in very shallow water of 1 – 2 m depth, but they are most common at depths from 10 to 40 m.

Family Colobometridae

Many feather stars of family Colobometridae are typically found in deeper water, at depths below 25 m. In contrast to shallow water species, these feather stars display distinct behavioural changes in a day-night rhythm. Specimen can often be found for month located at the same site and with their arms extended in the food-gathering mode. Yet, they can also be attached to gorgonians, spiny corals or sponges.

Genera that occur in the Indo-Pacific region include *Cenometra*, *Colobometra*, *Decametra*, *Oligometra* and *Petasometra*. In the Caribbean region we find the small *Analcidometra armata* that lives commonly on true gorgonians of genus *Pseudopterogorgia* and on whip-coral of genus *Ellisella*. Colouration of these feather stars is – as a camouflage measure – largely adapted to that of the host. The largest colobometrids measure 25 to 30 cm in diameter.

Family Himerometridae

This family includes the shallow water species *Heterometra savignyi*, which is one of the most frequent feather stars in the Red Sea; however, it also occurs in parts of the Indo-Pacific. The species has 20 arms, and has a diameter of 15 to 20 cm. It lives in shallow water, normally down to a depth of about 12 m, and is usually associated with the fire coral, *Millepora dichotoma*. Normally *H. savignyi* is nocturnal, however, in deeper water (below 6 m depth) this feather star can also

Colobometra perspinosa at Bintan Island, Indonesia. Photo: C. Bryce

Decametra sp. (?) at One Tree Island, Great Barrier Reef. Photo: L. Newman and A. Flowers

Heterometra savignyi at Aqaba, Red Sea. Photo: Dr. I. Illich

Feather Stars of family Mariametridae

Dichrometra flagellata.

Lamprometra palmata. Photo: L. Newman and A. Flowers

Lamprometra klunzingeri. Photo: J. Cairns-Michael

Stephanometra spicata.

Stephanometra cf. *indica.* Photo: S. W. Michael

Stephanometra sp. Photo: L. Newman and A. Flowers

be seen feeding during the day.

The genera *Amphimetra*, *Craspedometra* and *Himenometra* are also Indo-Pacific representatives of this family.

Family Mariametridae

All representatives of this Pacific family have more than ten arms and reach sizes of 15 to 40 cm in diameter. The genera *Dichrometra* and *Lamprometra* are particular common. The arms are marked with white cross bands, off set against a darker background colouration, e.g. yellow, red or brown. Many species live in extremely shallow water (less than 2 meters). Normally, they are distinctly nocturnal and remain in hiding until about one hour before sunset. Then they seem to take up a fixed position, often on *Millepora* species.

***Antedon* sp. (seen here in the Red Sea) is a small, rather delicate feather star.**
Photo: E. Svensen

Family Antedonidae

The species of this family are rather delicately built feather stars with ten arms. The maximal size is usually about 10 cm, rarely 20 cm. Many species live in deep water caves, and often they occur epizoic on soft corals of genus *Dendronephthya*. Genera *Antedon* and *Dorometra* are particularly frequent in the Indo-Pacific, while *Ctenantedon kinziei* lives in the Caribbean region. A particular behaviour pattern of this little feather star is the display of a distinct flight reaction, if disturbed by predators, for instance. With elegant, wave-like motions of the arms these feather stars swim out of the danger zone, and if need be, they can remain floating for several minutes.

Class Asteroidea - Sea Stars

Sea stars are probably the best- and most widely known group of echinoderms. In part this is due to their frequent occurrence in marine habitats all over the world, on the other hand it is fact that their distinct star-shape makes them conspicuous and unusual creatures. Their radial shape is in many ways a symbol of the sea and the life forms in it.

The coral reefs of the Indo-Pacific support an abundance of sea star species, in virtually all-accessible, ecological niches, such as the inter-tidal zone, sand-, sea grass- and algae zones. Moreover, they live at all depths among rocks and corals. For the Red Sea alone, VINE (1986) describes 29 species. GUILLE et al. (1986) describe 54 species as being known from the reefs of New Calidonia. CLARK & ROWE

Sea stars occur in a variety of shapes and colours (seen here are two different varieties of *Pentagonaster duebenei* off Western Australia).
Photo: C. Bryce

Systematic Overview of Class Asteroidea

after HENDLER et al. (1995); GUILLE et al. (1986); GEORGE & GEORGE (1979); CLARK & ROWE (1971)

Order Paxillosida
- Family Ctenodiscididae*
  - *Ctenodiscus*
- Family Luidiidae
  - *Luidia*
- Family Astropectinidae
  - *Astropecten, Craspidaster*
- Family Radiasteridae*
  - *Radiaster*

Order Valvatida
- Family Sphaerasteridae
  - *Podosphaeraster*
- Family Chaetasteridae*
  - *Chaetaster*
- Family Archasteridae
  - *Archaster*
- Family Goniasteridae
  - *Anthenea, Anthenoides, Calliaster, Ceramaster, Goniodicaster, Gymnanthena, Iconaster, Mediaster, Monochaster, Ogmaster, Pentagonaster, Pseudogoniodicaster, Pseudoreaster, Siraster, Stellaster, Stellasteropsis, Styphlaster, Tosia*
- Family Asterodiscididae
  - *Asteriodiscides, Asteriodiscus*
- Family Oreasteridae
  - *Choriaster, Culcita, Halityle, Nidorellia, Oreaster, Pentaceraster, Pentaster, Poraster, Protoreaster, Rosaster*
- Family Ophidiasteridae
  - *Bunaster, Celerina, Ceratonardoa, Cistina, Copidaster, Dactylosaster, Ferdina, Fromia, Gomophia, Hacelia, Heteronardoa, Leiaster, Linckia, Nardoa, Neoferdina, Ophidiaster, Pharia, Phataria, Tamaria*
- Family Asterinidae
  - *Anseropoda, Asterina, Diasterina, Nepanthia, Patiriella, Tegulaster*
- Family Asteropsidae*
  - *Asteropsis, Poraniella*
- Family Acanthasteridae
  - *Acanthaster*
- Family Mithrodiidae
  - *Mithrodia, Thromidia*

Order Velatida
- Family Myxasteridae
  - *Myxaster*
- Family Solasteridae
  - *Crossater, Heterozonias, Lophaster, Seriaster, Solaster*
- Family Pterasteridae
  - *Euretaster, Diplopteraster, Hymenodiscus, Pteraster, Retaster*

Order Spinolusida
- Family Echinasteridae
  - *Cistina, Echinaster, Henricia, Plectaster*

Order Forcipulatida
- Family Heliasteridae
  - *Heliaster*
- Family Labidiasteridae
  - *Coronaster, Rathbunaster, Labidiaster, Plazaster*
- Family Asteriidae
  - *Asterias, Coscinasterias, Evasterias, Marthasterias, Pisaster, Pycnopodia, Stylasterias, Uniophora*

Order Brisingida

(* not dealt with here)

(1971) list a total of 231 species in 59 genera and 17 families from the extreme shallow water zones (0 to 20 m) of the entire Indo-Pacific region. By the comparison, the diversity of sea stars in the Caribbean region is rather sparse. According to HENDLER et al. (1995) 18 species are known to occur in shallow water (down to a depth of 46 m). However, if one includes the deeper water regions, the total species number increases significantly to, i.e. to 160 species, which occur down to a depth of 3,660 m. The greatest depth where sea stars have been found is about 9,100 m.

Sea stars are also common in temperate and cold waters. HEYWARD & RYLAND (1996) list 12 species in seven families from the North-West Europe. In these waters "the Common Starfish", *Asterias rubens*, is indeed the most common species, which exists in millions in the shallow waters along the coast of Norway.

More than 1800 sea star species are described. They are distributed over easily 100 genera and about 30 families, contained in six orders. In shape they vary from miniature types (smaller than 1 cm) to real giants with an arm span of more than 1 m. The systematic overview given on the left side of this page contains only a selection of those families of interest to the aquarium hobby, together with a few representative genera.

As in all other Echinodermata, the body structure of sea stars conforms to a pattern of pentamerous symmetry. Typically they have five arms surrounding a central oral disk. Although five arms are the most frequently encountered structure, the number of arms can vary, not only between species but also within a particular species. Among those species with the largest number of arms we can readily find up to 50 arms. There are relatively many species that have from seven to fourteen arms, presenting a slight departure from the customary symmetry of echinoderms. In contrast to the following class (brittle stars), the arms of sea stars are not sharply delineated from the oral disc, but instead flow more or less in to each other. In some instances, they are relatively long in comparison to the oral disc; in others they are rather short. In some species, as for instance in pin cushion sea stars of genus *Culcita* (family Oreasteridae) the arms are so short, that they are hard to distinguish as individual arms or possibly not even at all.

The underside of each arm has a distinct cleft, the ambulacral groove. Along either side of the ambulacral groove there are suction feet arranged in a row (ambulacral feet). If need to be, the suction feet can be directed in to the groove for protection, which is then closed off and covered with special spines. The ambulacral groove of each arm leads in to the mouth of the sea star. The mouth is located in the centre of the oral disc, and so the entire disc area surrounding the mouth together with the respective underside of the arms is referred to as the oral disc. Correspondingly, the opposite (upper) side is called the aboral disc. Most sea stars have an anus (usually rather small) located centrally on the aboral side. In some of the sea stars such anus is absent, and so these sea stars also lack an intestine, i.e. food intake and the elimination of waste products takes place via the mouth opening.

The Horned Sea Star *Protoreaster nodosus* (photographed in Indonesia) is a perfect example of the pentamerous (five-armed) symmetry found in most sea stars. Only a few deviate from this pattern, e.g. with seven or fourteen arms.

The nervous system of sea stars is rather simple. The sensory organs are essentially confined to individual tactile receptors and chemoreceptors. Such epidermal sensory cells are located in large numbers in the skin, specifically on the suction feet as well as along the sides of the ambulacral grooves. Here, aggregations of up to 70,000 sensory cells per square millimetre have been found in some species (as cited by BARNES, 1980). At the tip of each arm, on the oral side, there is an eyespot, composed of 80 to 200 pigmented ocelli. They form an optical knot or –cushion, covered by a type of lens. This lens is formed by thickened, surrounding skin. Each ocellus consists of epidermal cells that contain red photo-labile pigment grains, i.e. pigment grains that can change their chemical composition under the influence of light.

The skeleton of a sea star consists of small, bone-like calcareous sclerites, developed as

The degree of separation between the arms of sea stars varies. For instance, *Nardoa galatheae* (left) has distinctly, separated long arms and only a small central disc, while the arms in *Culcita schmideliana* (right) are shot and barely separated. The arms in this juvenile specimen are still reasonably well visible, but they become increasingly less distinct with advancing growth.

The sea star, *Choriaster grannulata* (seen here at the Maldives) reaches a maximum spread of about 40 cm. It lives on reef slopes towards the sea and land, on live corals or on coral rocks.

rods, crosses or platelets. They are arranged in such a way that they form a net or grate, which is being held together by connective tissue. This grate is structured, very flexible and tensile. This gives most sea stars incredible mobility and litheness, in spite of their hard and solid structure. In most species the body wall is sufficiently strong to withstand destructive wave action against ragged cliff faces. At the same time they can overcome obstacles with the greatest of ease or bend their arms dexterously in to all conceivable angles. Moreover, groups of circular and longitudinal muscles enable sea stars to modify the shape of an arm.

The body surface of sea stars carries large numbers of microscopically small, transparent papillae. These are small protrusions from the body cavity that lead through microscopic pores in to the skeleton. These thin-skinned protrusions carry cilia on their inside and outside, which act in conjunction with the suction feet

The Reticulated Pillow Sea Star, *Oreaster reticulatus* (seen here at the Bahama Islands) lives principally on sandy bottom and in shallow water in the Caribbean, from South Carolina to Brazil and also at the Cape Verde Islands. It reaches a maximum spread of up to 50 cm. Photo: S. W. Michael

as a surface for gas exchange. In some species the papillae give the animal a pelt-like surface appearance. In addition, most sea stars also possess a large number of pedicellaria. These are required to prevent other organisms, such as sponges or bryozoans, to settle on them.

Among the sea stars there are algae-eaters as well as detritus feeders, but also virtual omnivores, scavengers and outright predators. Detritus feeders live mostly off the nutrient-rich film or organic material covering rocks, gravel and sand bottom, whereby they also take in algae and tiny organisms that live on or within this film layer. One of the best-known predatory species is the Crown-of-Thorns Sea Star, *Acanthaster planci*, infamous for its voracious appetite for stony corals. In addition, there are also a number of other species that feed on corals, sponges, bryozoans, tunicates, snails, clams, crustaceans or polychaete worms. A few prey on other sea stars, and some are even capable of overpowering fish.

Sea stars that have become specialised to feed on clams are a well-known example of a highly developed strategy for gathering food. The sea star uses its suction feet in order to exert a constant tension (pull) on the valves (shells) of a clam, until the adductor muscle of the clam becomes so exhausted that it gradually slackens off and the valves open up. Then the sea star evertes its stomach through its mouth and inserts it between the valves. It matters little if the valves are not completely open, because all that is needed is a gap of about 0.1 mm (!) in order to manipulate the stomach through the valves. As soon as the stomach is inside the shell it commences to digest the clam's body. Once the meal has been digested the sea star withdraws its stomach back into its own body. Similar methods of stomach eversion are also deployed by many other predatory sea stars.

Some sea stars are not particularly selective in their food requirements, even though they may prefer one or the other type of special food (pin cushion stars of genus *Culcita*, which prefer stony corals, can also be omnivores if required by prevailing circumstances). On the other hand, there are sea stars, which are distinct food specialists and starve to death if their customary food is not available.

Sea stars have a very high capacity to regenerate lost body parts that extends all the way to a complete replacement of large body parts. For most species that means, that as long as one-fifth of the oral disc still remains intact, the remainder of the body can be redeveloped. If the remaining part of the oral disc includes the madreporite (keeping in mind that there is only a single madreporite per animal), a sea star can be totally regenerated even from less than one-fifth of an oral disc; however, under such circumstances the regenerative process is not quick and it can take a year or more for the entire animal to be reconstructed. Yet, even in the meantime sea stars retain proper body functions.

Some species utilise their regenerative capability regularly for the purpose of asexual reproduction, whereby a specimen splits in to two parts right through the middle of the oral disc. After that, both parts gradually develop the respective, missing part. In genus *Linckia* (among others) that is also known as an aquarium inhabitant, we find the unique characteristic to shed individual arms without losing any part of the oral disc. The shed arm regenerates a new oral disc, as well as all other (missing) arms. Specimens that are in the process of developing a new body from a single arm, are being referred to as "comets" (see photo series below).

Most sea stars have separate sexes. Those few hermaphrodites that exist among sea stars are protandrous, i.e. small juvenile specimens are males, however, with increasing size and age they change in to females. The presence of ten gonads is typical for

The ability to regenerate lost or damaged body parts is in some sea stars extremely well developed. Particularly among *Linckia* species one comes across so-called "comets" rather frequently; these are new individuals as they are being developed from a cast-off arm. Theoretically, five new specimens can develop from what was originally a single specimen. This series of photos shows comets of *Linckia laevigata* at various stages of regeneration (below left) and of *L. multifora* (top and bottom right). **Photo top left and bottom right: L. Newman and A. Flowers Photo top right: Dr. I. Illich; photo bottom left: J. Cairns-Michael**

most adult sea stars, i.e. two per arm. The gonads can resemble a cluster of grapes. Just prior to spawning they can be very large, so that that they fill out large sections of the hollow spaces in the arms. But outside the breeding season they shrink substantially.

Spawning normally takes place by releasing the sex products directly in to open water, where fertilisation takes place. As soon as eggs and sperms enter the water, it tends to stimulate other specimens also to spawn, which is referred to as synchronous spawning. Normally spawning takes place only once a year. Large females of some species can release several hundred thousand eggs or even up to several million! In most species the eggs and subsequent larvae are planktonic, but there are also a few species that engage in brood care and then carry their progeny around with them or guard them by placing their body over the egg masses. In brood-caring species the embryos develop directly in to miniature copies of adults, i.e. without going though any intermediate larval stages.

Many sea stars are suitable for aquariums. They are beautiful and interesting, as well as being hardy aquarium specimens. There are, however, also species that are not so suitable for the reef aquarium either because they are predators or require a specific food source to survive. Since the differences among genera and families are often rather significant, details about their respective aquarium maintenance are being elaborated on in some detail in the systematic treatment of – what are from an aquarist's perspective – the most important sea star families.

Order Paxillosida

The species in this order normally have five arms, however, genus *Luidia* (family Luidiidae) deviates from this scheme. Many species in this particular family have a larger number of arms. All paxillosidans have one or two rows of well-developed and easily recognisable marginal plates, which border around the arms.

The majority of paxillosidans are burrowing species that search for food in sand- or mud bottom. Molluscs (clams and gastro-

An Egyptian Sea Star, *Gomophia egyptiaca* with six arms (see here at Christmas Island). The reason for this configuration is that four arms are being regenerated from three that were lost earlier – a typical anomaly in many sea star species. Photo: P. Lange

pods) – swallowed whole – are the normal prey organisms of these sea stars. Suction feet are of little use for locomotion in this type of substrate. Consequently, in contrast to most other sea stars, suction feet are absent in paxillosidans.

In terms of their colours, most species are rather uniformly brown with a more or less distinct, marbled pattern. Therefore, they are largely inconspicuous to collectors for the aquarium trade, so that they are virtually never available in aquarium shops. Below is a discussion of the two most important families.

Family Luidiidae

This family is monotypic, with *Ludia* as its only genus. *Luidia* species are generally rather attractively marked sea stars, usually with a marbled pattern in shades of delicate light- or dark brown. The arms are bordered by arrow of marginal plates, equipped with thin, sharp spines. Small spines cover the surface, which tends to give a velvet-like appearance to these sea stars. The number of arms varies from five to fifteen.

Most species found in coral reefs occur in relatively shallow water, from a few metres down to a depth of about 50 m; however, a few species can be found in substantially deeper water. The typical habitat of these sea stars is a lose bottom substrate, where they can burrow in search of food. The majority of species are predators, but they also scavenge and feed on organic detritus, as well as on the small organisms it contains, e.g. molluscs, crustaceans and brittle stars. Many species can also handle rather large prey organisms. For instance, It has been observed that specimens of the Caribbean *Luidia calthrata* somehow managed to swallow "sand dollars" (= sea urchins of order Clypeasteridae, see page 358-367). These are prey organisms, which are significantly larger (in diameter) than the oral disc of the sea stars that swallowed them, so that their oral disc became over-extended and was subsequently torn (Walls, 1982)! Yet, one can assume that in light of the extremely high regenerative ability of the animal, this would only be a temporary problem.

Luida alternata **from the Caribbean region is a typical representative of family Luidiidae. Drawing from a photo in Hendler et al. (1995)**

As in many other sea stars, symbiotic organisms also are frequently encountered in *Luidia* species. For instance, Caribbean species are apparently the host animals for the dark brown bristle worm, *Podarke obscura*, which grows to a length of up to 2.5 cm. It lives among the suction feet in the ambulacral grooves (Ruppert & Fox, 1988).

Luidia species occur in all tropical and subtropical seas. *Luidia alternata*, *L. clathrata* and *L. senegalensis* are known from the Caribbean region. The two former species have five arms, but the latter has nine arms. The maximal size of all three species is 30 to 40 cm, however, more commonly one encounters substantially smaller specimens. From among the Indo-Pacific species, *L. maculata* (seven to nine arms), *L. magnifica* (nine or ten arms) and *L. savignyi* (six or seven arms) must be mentioned, with maximal sizes of 20 to 30 cm.

Family Astropectinidae

It is estimated that this family encompasses 25 valid genera; however, the only one common on coral reefs is genus *Astropecten*.

All comb sea stars of genus *Astropecten* have five arms. In addition, they are characterised by two rows of distinctly marked (supero- and infero-) marginal plates that surround the arms. On each plate of the lower (infero-marginal) row there are two flat, sharp spines, sticking out at right angle to the arm.

The multi-spined comb sea star, *Astropecten polyacanthus* from the Indo-Pacific region is an extremely "spinous" species, with spines on the marginal plates that are nearly a centimetre long. This makes it easy to identify this species, which has a maximal size of about 17 cm. The normal colouration of the multi-spined comb sea

The Many-spined Comb Sea Star, *Astropecten polyacanthus* is an extremely spinous species, widely distributed in the tropical- and subtropical Indo-Pacific: The photo on the left shows a specimen off Western Australia (photo: C. Bryce), and the one on the right a specimen off Sydney, Eastern Australia (photo: M. Cufer).

star is a pinkish light brown to a marbled brown pattern, with contrasting lighter or darker marginal plates and white spines.

Astropecten polyacanthus has a wide distribution. It occurs in the Red Sea, southward along the East African coast to South Africa. Its eastward distribution extends to Hawaii. Just like other paxillosidans, it too is tied to a soft sand bottom and it lives at depths from 1 to 40 m. The typical diet of this sea star consists of small clams, which it digs up from the sand. So far we are not aware that this species has ever been imported commercially. However, we assume that *Astropecten polyacanthus* can be kept in aquariums with a sand bottom (at least 10 cm deep), in conjunction with the "Jaubert-System" (see Volume 1, page 256), provided sufficient food is available in the form of small clams.

The Caribbean region is inhabited by two species that are linked indirectly to coral reefs: *Astropecten duplicatus* and *A. articulatus*. The latter can easily be identified on the basis of its colouration. Its aboral side is brown, blue or lilac and contrasting yellow or orange-coloured marginal plates conspicuously border it. The oral side is white or beige-coloured. *A. articulatus* reaches a diameter of 15 to 20 cm and lives on a soft, sandy bottom, from the inter-tidal zone down to a depth in excess of 150 m, in most areas of the Caribbean. At dusk it commences to search for food, whereby it feeds predominantly on molluscs, small crustaceans and echinoderms. WELLS et al. (1961) identified 91 species of prey organisms from the stomach content of 124 specimens.

Astropecten duplicatus obtains a diameter of up to 20 cm. It too is common throughout the Caribbean region, i.e. from North Carolina southward to Brazil. The depth distribution extends from 1 m down to 500 m, but it is most commonly encountered between 5 m and 20 m. *A. duplicatus* varies substantially in its colouration: from a simple grey or brown to various marbled patterns, and pink to reddish brown marginal plates.

Astropecten articulatus **from the Caribbean region (seen here off Puerto Rico) can easily be identified on the basis of its conspicuous and contrasting colouration. The photo on the right shows this species as it is commonly seen in its habitat: partially buried in soft sand bottom.**

Spines located on the marginal plates are generally white. This is a burying comb sea star that searches predominantly for small molluscs in soft bottom substrates. It is able to bury itself in sediments, by executing slow turning forward- and reverse movements. With that it can cover itself completely with sediments within two minutes (MILLER, as cited in HENDLER et al., 1995).

Astropecten armatus lives along the American Pacific coast, from southern California to Peru, where it grows to a size of up to 35 cm in diameter. It is similar in appearance to *A. articulatus*, and feeds on benthic invertebrates, primarily on gastropods, sea feather and sea urchins, but will also scavenge on dead organisms (MEINKOTH, 1994).

This *Linckia laevigata* specimen shows damage incurred as a typical consequence of incorrect handling of sea stars. This includes introducing new specimens in to an aquarium without proper acclimatisation, or exposure to air, which can destroy the water vascular system. It is important to keep in mind that such injuries often do not heal under aquarium conditions, in spite of the fact that sea stars have large regenerative capabilities.

Order Valvatida

With very few exceptions, the sea stars in this order have five arms. Although arm length can be variable, most valvatids show the characteristic (pentamerous) star shape, which – to lay people – is synonymous with typical sea stars. Yet, beyond that there is much diversity, for example, in terms of size (from miniature forms to virtual giants). Some *Asterina* spp. (family Asterinidae) are smaller than 10 mm as adults, while *Thormidia* spp. (family Mithrodiidae) include some of the largest sea stars known, with arm spans of up to 75 cm and an arm thickness of 12 cm and more.

Similarly, the diet of the Valvatida also encompasses a wide range. A large number of them are microphage omnivores or detritus feeders, that pick up small organisms, algae and/or organic detritus off the bottom, while other species are more or less food specialists and predators, feeding on various invertebrates, such as crustaceans, molluscs, corals and echinoderms. It has also been documented that certain large and aggressive species are even capable of capturing fishes.

The waste majority by far of sea stars traded commercially belong to the Valvatida. Even though not all species are suitable for the average aquarium, they are – nevertheless – by-and-large rather hardy and will live a long time. However, a prerequisite for success with sea stars is proper transport and correct acclimatization. In general, it is important that sea stars are being transported in a sufficiently large water volume, so that they remain totally submersed at all time. As soon as they come in to contact with air, there is the risk of damage to the water vascular system by air entering the canals and blocking them to a point where they can burst. It appears that the sensitivity is variable among different species; however, the larger sea star specimens are the greater exposure to this problem. On the other hand, this is not a reason for not buying a suitable specimen if an opportunity presents itself. Similarly, it is also very important to acclimate a newly acquired specimen very slowly to new water conditions, preferably over a period of several hours (ideally for 24 hours) in a special acclimatisation tank (see Volume 1, page 165) specifically on these particular problems. In spite of their excellent regenerative capabilities, one cannot expect that this is sufficiently effective under aquarium conditions, in order to be able to repair severe internal injuries.

Family Archasteridae

Family Archasteridae is monotypic, with its only genus *Archaster*, which occurs in the Indo-Pacific region. These are burying sea stars that have superficially a strong resemblance to *Astropecten* species, but *Archaster* spe-

Iconaster longimanus **has large marginal plates along its arms and small plates on the central disc, which makes it easy to identify this species. Left: Specimen in its natural habitat at Bintan Island, Indonesia (photo: C. Bryce). Right: Specimen in an aquarium (photo: S. W. Michael).**

cies have tube feet. An unusual spawning behaviour has been reported for this genus, generally referred to as "speudo-copulation". Spawning takes place in a manner whereby specimens of different sexes crawl on top of each other and then release eggs and sperms in to the surrounding water. It is assumed, that this type of behaviour leads to an enhanced fertilization rate; of course, this does not involve internal fertilization.

Family Goniasteridae

Currently, this family contains more than 50 valid genera. All species have five arms, yet in terms of their morphology and geographical distribution they represent a variable group.

The species of this family are distributed throughout all seas and occur at variable depths. Various regions of the North Atlantic are inhabited by genera *Ceramaster* and *Hippasteria*. In subtropical temperate waters around southern Australia and New Zealand there is a large number of magnificently coloured species from genera *Nectria* and *Pentagonaster*, and others. Some of these occur in rather warm water of 20 to 22 °C, but never in the vicinity of coral reefs.

Coral reefs of the Indo-Pacific are inhabited by some species that are so frequent in rather shallow water, that divers can observe them. Most of them belong to genera *Iconaster*, *Monachaster*, *Omgaster*, *Stellaster*, *Stellasteropis* and *Tosia*. Only one particularly characteristic species will be discussed here in some detail.

Iconaster longimanus, with a diameter of about 10 cm, is rather small, yet it is unbelievably beautiful. Along its arms there are distinctly coloured marginal plates and less contrasting plates at the centre of the oral disc. The upper marginal plates are so large that they actually come together in the middle of the narrow, thin arms, which gives this species an unmistakable appearance. The dominant colour ranges from grey via beige all the way to reddish brown, in a distinctive, offset pattern.

Iconaster longimanus lives along reef slopes at depths from 35 to 85 m, and so it is not often encountered. Normally it can be found on sand- or gravel bottom. The geographic distribution extends from the Arabian Peninsula to Australia, Indonesia and to the Philippines. Its food requirements are not too well known; however, it is assumed that *I. longimanus* feeds – like the other species in this family – on sessile organisms, such as sponges, moss animals and tunicates. Although this sea star has been available in the aquarium trade, nothing is known about its aquarium maintenance and longevity. But, it stands to reason that providing it with a species-correct diet would be problematic for most aquarists.

Family Asterodiscidae

Some authors are placing the genera *Asterodiscides* and *Asterodiscus* in to the families Goniasteridae and Oreasteridae, respectively. But in more recent times they are generally being included in the separate family Asterodiscidae. These species are widely distributed throughout the tropical and subtropical regions of the Indo-Pacific and the eastern Pacific, yet they are never in abundance.

A rather decorative sea star in this family is *Asterodiscus helenotus*, which occurs from southeastern Australia to Japan. With a maximal diameter of easily 30 cm and very intensive red to yellowish colour markings, it is a rather conspicuous species. A thick skin, densely covered with large papillae, covers the upper surface. This species feeds mainly on sessile, small and large invertebrates.

Family Oreasteridae

Currently, this family contains 19 genera. Except for the two Atlantic species of genus *Oreaster*, all other species are native to the

Family Goniasteridae contains some rather beautiful and attractively marked species, that are endemic to the temperate and cold waters of South Australia, Tasmania and New Zealand Although they are not coral reef animals, they are shown here as reference information for SCUBA-diving aquarists. Top: *Nectria multispina*; Bottom: *Pentagonaster duebeni.* Both specimens were photographed off New South Wales (Australia). Photos: S. W. Michael

Asterodiscides helenotus **(seen here off Ulladulla, New South Wales, Australia) is never found in abundance, but it has a wide distribution. It is a predatory sea star, that has not yet been imported for the aquarium trade.**

Photo: S. W. Michael

Indo-Pacific or the eastern Pacific, respectively. Generally these species are large (mostly 30 – 50 cm as adults) and well-armoured sea stars, which feed principally as microphages or micro-herbivores and sometimes they are scavengers. However, a few are highly selective feeders (e.g. many *Culcita* spp.) that prefer the polyps of stony corals. Most species evert their stomach through the mouth and then digest the food outside their own body.

Many oreasterid sea stars are rather decorative, equipped with heavy plates, spines and wards that form distinct patterns. This is often combined with magnificently contrasting colours, so that even dead specimens are still rather conspicuous. Because of their beautiful appearance many species are in demand from curiosity shops. Dead, dried out sea stars have been, (and still are being) sold world-wide in souvenir- and hobby shops. In fact, often these dead "carcasses" are even painted with bright colours. One could shrug this off with: "It's all a matter of taste", were it not for the enormous magnitude of the trade in certain sea star species. And yet, it is the aquarium market that is being attacked for its comparatively modest use of these natural resources. It may well be time for thoughtful reflections when one sees row upon row of boxes with dried sea stars in the souvenir shops of certain tourist destinations. A case in point is the well-known Reticulated Cushion Star, *Oreaster reticulatus* that used to be common throughout the Caribbean region. Now it has become very rare over most of its former geographical range.

From the perspective of a coral reef aquarist, the most interesting sea star species are found in the genera *Choriaster*, *Culcita*, *Oreaster*, *Pentaceraster* and *Protoreaster*. A selection of these are presented and discussed in detail in Fact Sheets, from page 248 onward. Many species are easy to keep in an aquarium, provided they are given a variable diet, consisting of clams, snails, raw fish flesh and dry food in tablet form or as pellets. However, they

Sea stars of family Oreasteridae - here the Horned Sea Star, *Protoreaster nodosus* on a sand bottom at Bunaken, Indonesia - are heavily armoured. Most species are opportunistic feeders that can readily adjust their normal diet, should this be required. Therefore, in an aquarium they should not be kept together with most other invertebrates.

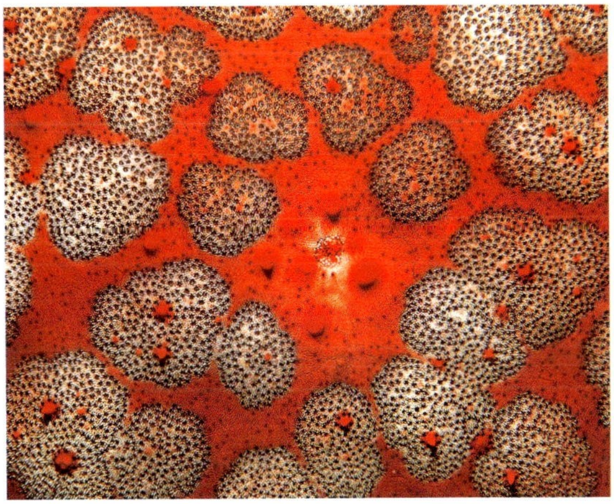

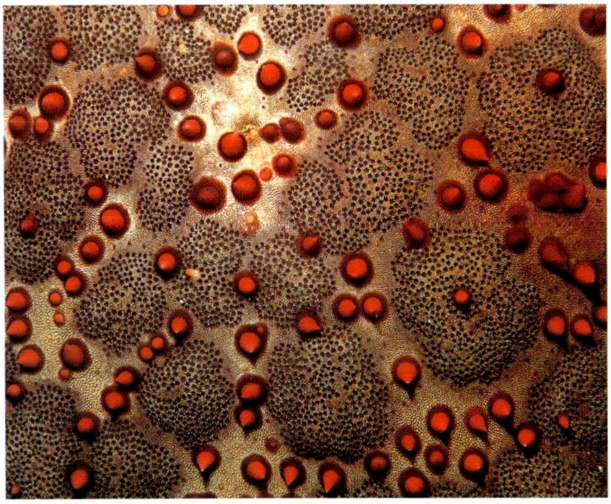

The aboral side in many species of family Oreasteridae is very attractively marked with plates, spines, tubercles and warts. These two close-up photos show two attractively marked *Culcita schmideliana* specimens at the Maldives. Please note the red spines and grey porous patches. **Photo: J. Cairns-Michael**

are **not** suitable as companion species for other invertebrates. Most of the latter, including sponges, cnidarians, worms, molluscs and echinoderms can easily become prey when preferred food sources become scarce.

Family Ophidiasteridae

This family of sea stars, with its 30 genera, is abundantly represented in all tropical seas. Generally speaking, these sea stars are among those best known to aquarists, since the majority of them are very frequently imported. In fact, the most suitable aquarium species belong to this family. Particularly, *Fromia*- and *Linckia* species are commonly available in the aquarium trade and regularly discussed in the aquarium literature. However, other genera are also frequently represented in aquarium shops.

By and large this family Ophidiasteridae contains rather slender, partially even streamlined species. They have five long, relatively smooth and rather evenly shaped, stout arms that radiate from a small oral disc. There can also be specimens with a deviating number of arms (less, but also more). Yet, these are rarely genetic variations, but more likely defects that occurred due to mechanical damage. In a number of species the oral disc is so small, that it is only about the size of the central area from where the arms radiate outward. Many species have impressive colouration, which is an important reason for their regular appearance in the aquarium trade. The majority of species, together with their greatest diversity can be found in the tropical Indo-Pacific. But there are also species endemic to the

The Western Pincushion Star, *Culcita schmideliana*, on coral substrate in the Maldives (for further details refer to Fact Sheet on page 251).

Choriaster granulatus

Distribution: Occurs widely throughout the Indo-Pacific; from the Red Sea to the South Pacific.
Size: Up to 40 cm, in diameter, but usually smaller.
Reproduction: Separate sexes; free spawner; details about asexual reproduction unknown.
Food: Feeds mainly on detritus and dead organisms; however, this sea star may also take small, sessile invertebrates, including (but not limited to) sponges and cnidarians if given the opportunity. Yet, under aquarium conditions it will accept most substitute foods, such as clam-, musselmeat and parts of raw fish, as well as dry food in tablet- or pellet form.
Description: Characteristically, the entire body surface is covered by a smooth, thick skin that is interrupted only by rather small ambulacral spines and warty patches. Colouration variable, from whitish, yellow, pink to orange or brownish; the warty patches are orange or red. The arms are short, but well developed, with rounded off tips. Normally lives at depths from 5 to 40 m, on hard substrate and/or sandy bottom. Frequently used as host by commensal Pearl Fishes of family Carapidae.
Aquarium requirements: Remarkably durable, however, like most sea stars, this species is rather sensitive to adverse transport- and acclimatization conditions. Should be kept preferably in very large aquariums that are decorated with calcareous rock or live rocks with dense epiphytic growth, but also contains sandy patches. Because of its feeding behaviour it can only be kept together with fishes and large, active invertebrates, such as crabs and shrimps. In a coral community aquarium the starfish will soon destroy many of the invertebrates.

GA: +/-; **SE:** +/-; **CO:** –

Photos:
Several colour varieties of *Choriaster granulatus*, from the Red Sea (top: photo: E. Svensen), from the Maldives (centre; photo: S. W. Michael) and from Palau (bottom; photo: Dr. I. Illich).

Culcita novaeguineae
Eastern Cushion Star

Distribution: Eastern Indian Ocean and Pacific Ocean; from Bay of Bengal to Hawaii.
Size: Up to 30 cm in diameter.
Reproduction: Separate sexes; free spawner; details about asexual reproduction unknown.
Food: Prefers to feed on the polyps of stony corals (*Acropora* spp. and *Porites* spp.), however, will also take slow-moving and other sessile invertebrates, as well as apparently also detritus and micro-organisms. Will readily accept most substitute foods, such as clam and mussel meat, pieces of raw fish and also dry food.
Description: Adult individuals can be recognised like the other 3-4 known adult *Culcita* species by the typical pillow shape. The five arms are extremely short and barely noticeable. Juvenile specimens are more star-shaped with distinctly visible arms. This is the most widely distributed species of its genus and extremely variable in colouration. Lives on platform-, barrier- and fringing reefs, at depths from 1 to 30 m. The delineating features between *C. novaeguineae* and *C. schmideliana* (see page 251) are not de-

continuation on next page.

Photos:
Top: Aboral- (left) and oral side (right) of a cushion star on the Great Barrier Reef. Photos: L. Newman and A. Flowers
Centre: Aquarium specimen of *Culcita novaeguineae.*
Bottom: Eastern Cushion Star at Bunaken, Indonesia.

(continuation from page 249)

finitive. *C. schmideliana* is not supposed to have spines on its porous areas, while in *C. novaeguineae* there are spines in these areas. Irrespective of this, there are also transitional forms. The most reliable feature for identification appears to be the respective geographic distribution of both species.

Aquarium requirements: Remarkably durable, however, like most sea stars, this species is rather sensitive to adverse transport- and acclimatization conditions. Should be kept preferably in very large aquariums that are decorated with calcareous rock or live rocks. Because of its feeding behaviour it can only be kept together with fishes and large, active invertebrates such as crabs and shrimps.

GA: +/-; **SE:** +/-; **CO:** –

Photos:
Specimens of the Eastern Cushion Star, *Culcita novaeguineae* at Bintan Island, Indonesia (top left; photo: C. Bryce), at Palau (top right; photo: Dr. I. Illich) and at Madang, Papua-New Guinea (bottom; photo: L. Newman and A. Flowers).

Culcita schmideliana
Western Cushion Star

Distribution: Indian Ocean; from East Africa, Madagascar, Reunion Island, Mauritius and Seychelles to Western Australia.
Size: Up to 25 cm in diameter
Reproduction: Separate sexes; free spawner; details about asexual reproduction unknown.
Food: Prefers to feed on the polyps of stony corals (*Acropora* spp. and *Porites* spp.), however, will also take slow-moving and other sessile invertebrates, as well as apparently also detritus and micro-organisms. Will readily accept most substitute foods, such as clam and mussel meat, pieces of raw fish and also dry food.
Description: Highly variable in colouration. Adult specimens can be recognised by their typical cushion shape. The arms are barely recognisable. Juveniles are more star-shaped, with larger lateral plates. Therefore, they have been described as independent species in various genera. The delineating characteristics between *C. schmideliana* and *C. novaeguineae* are not definitive (see page 249). Therefore, the respective geographical distribution of each species appears to be the most useful feature for identification. The related Red Sea Cushion Star, *C. coriacea* is endemic to the Red Sea.
Aquarium requirements: Remarkably durable, however, like most sea stars, it is rather sensitive to adverse conditions during transport and acclimatisation. Should be kept preferably in a very large tank decorated with calcareous- and live rocks. Because of its feeding behaviour it can only be kept together with fishes and large, active invertebrates. Will also attack cnidarians with potent stinging cells as well as other sea stars, including sibling specimens.

GA: +/-; **SE:** +/-; **CO:** –

Photos:
Specimens of the Western Cushion Star, *Culcita schmideliana*, in an aquarium (top; photo: S. W. Michael), at the Amirante Islands, west of the Seychelles (centre; photo: Dr. I. Illich), and off Western Australia (bottom; photo: C. Bryce).

Oreaster reticulatus
Reticulated Cushion Star

Distribution: In the Caribbean Region and tropical Atlantic; from North Carolina and the Bermuda Islands, southward to Brazil and eastward to the Cape Verde Islands.
Size: Up to 50 cm in diameter.
Reproduction: Separate sexes; free spawner; details about asexual reproduction unknown.
Food: Feed principally on micro-organisms and organic detritus on sand bottom, among sea grass (SCHEIBLING, 1982). To some degree it also feeds on algae, sponges and – under certain circumstances – also other echinoderms (e.g. *Tripneustes* spp., *Meoma* spp.) and also sibling specimens; in essence, an omnivore that is also an aggressive feeder. In order to feed it everts its stomach through the mouth and digests its prey outside of its own body.
Description: Lives in rather shallow (1 to 37 m deep) and calm water of reef flats, lagoons and in mangrove channels, on sea grass covered sand bottom (HENDLER et al., 1995). Very robust species with massive central disk that appears as if inflated, with five, relatively narrow-tipped arms. The aboral side is equipped with thick, heavy plates, which form a reticulated (net-like) pattern. These plates have numerous, protruding tubercles. Colouration consists of bright red or yellow colour tones, off set by the contrasting pattern of reticulated plates. Because of its attractiveness this sea star is in strong demand by the souvenir- and curiosity trade.
Aquarium requirements: Remarkably durable, however, like most sea stars, it is rather sensitive to adverse conditions during transport and acclimatisation. Should be kept preferably in a very large tank with a large sandy area. Because of its feeding behaviour it cannot be kept together with other invertebrates.

GA: +/-; **SE:** +/-; **CO:** –

Photo: Reticulated Cushion Star in its typical habitat in Puerto Rico.

***Pentaceraster* cf. *cumingi*,**
(formerly named *Oreaster occidentalis*)
Panamanian Horned Sea Star

Distribution: Eastern Pacific; Gulf of California (Sea of Cortez) southward to Peru and the Galapagos Islands, westward to Hawaii.
Size: Up to 20 cm in diameter.
Reproduction: Free spawner; no details available on asexual reproduction.
Food: Feeds on micro-organisms, algae and organic detritus; however, it is possible that it is omnivorous. Will accept most substitute foods in an aquarium.
Description: Lives on the rocky substrate on platform reefs, from the inter-tidal zone down a depth of 180 m. Very attractive and robust sea star that has a superficial resemblance to *Oreaster reticulatus* (see page xxx). Hard, swollen body wall (as if inflated), with stable, immovable spines that protrude like nails from the aboral side. Some colour variation, normally red or with conspicuous red markings.
Aquarium requirements: This should really be a rather hardy species; however, so far there have only been a few reports about its aquarium husbandry. Distinctly sensitive to adverse transport and acclimatisation conditions. Aggressive feeder and a potential predator of many other aquarium animals. Must not be kept together with other invertebrates.

GA: +/-; **SE:** +/-; **CO:** –

Photo:
Panamanian Horned Sea Star, *Pentaceraster* cf. *cumingi* in an aquarium. Photo: S. W. Michael

Pentaceraster mammilatus
Green Horned Sea Star

Distribution: Indian Ocean and Red Sea.
Size: Up to 40 cm in diameter.
Reproduction: Free spawner; no details available on asexual reproduction.
Food: Feeds on small molluscs and other benthic invertebrates, probably also on certain algae, organic detritus and dead organisms. Under aquarium conditions it will accept most substitute foods, such as clam- and mussel meat, crustaceans and fishes as well as dry food tablets.
Description: Lives on soft sand bottom at depths from 2 to 15 m. This is a frequently imported sea star, attractively coloured and with a knobbly body surface. Colouration is green to greenish-blue with yellow or orange-coloured patterns. Has already been extensively kept in aquariums. Some aquarists have reported on its aggressive feeding behaviour in respect to most other invertebrates, including other sea stars. However, since this does not seem to occur in every case, other (individual or environmentally related) factor may have triggered such predatory behaviour.
Aquarium requirements: Hardy aquarium inhabitant, but sensitive to adverse conditions during transport and subsequent acclimatization. It is an aggressive feeder and potential predator of many other aquarium inhabitants. If possible, should best not be kept together with corals and other delicate invertebrates as might very well prey on these in captivity.

GA: +/-; **SE:** +/-; **CO:** –

Photo:
Green Horned Sea Star, *Pentaceraster mammilatus* in a display aquarium at the Bochum Tierpark (in the background the Sea Star, *Protoreaster lincki*; see Fact Sheet on page 256).

Pentaceraster tuberculatus
Grey Horned Sea Star

Distribution: Indian Ocean and the Red Sea.
Size: Up to 30 cm in diameter.
Reproduction: Free spawner; no details available on asexual reproduction.
Food: Feeds on small molluscs and other benthic invertebrates; probably certain algae, organic detritus as well as dead organisms probably also make up part of its diet. Under aquarium conditions it will accept most substitute foods, such as clam- and mussel meat, crustaceans and fishes as well as dry food tablets.
Description: Lives on soft sand bottom, at 1 to 80 m depth. Rather attractive sea star, which is easily mistaken for the more frequently imported Green Knobbly Sea Star, *Pentaceraster mammilatus*. Both species show great colour variability, so that colouration is not a reliable distinguishing characteristic; however, the porous patches in *T. tuberculatus* flow more readily in to each other, than those in *P. mammilatus*. Moreover, the latter does not have any distinct upper marginal spines.
Aquarium requirements: Hardy aquarium inhabitant, but sensitive to adverse conditions during transport and subsequent acclimatization. It is an aggressive feeder and potential predator of many other aquarium inhabitants. If possible, should not be kept together with other invertebrates. Best not be kept together with corals and other delicate invertebrates as might very well prey on these in captivity.

GA: +/-; **SE:** +/-; **CO:** –

Photo: Grey Horned Sea Star, *Pentaceraster tuberculatus*, in an aquarium. Photo: S. W. Michael

Protoreaster lincki

Distribution: Occurs widely throughout the Indian Ocean; from East Africa to Western Australia.
Size: Up to 30 cm in diameter.
Reproduction: Free spawner. We have not been able to find reports about any asexual reproduction. Yet, ERHARDT & MOOSLEITNER (1995) are indicating that asexual reproduction can take place, but without providing any reference details.
Food: According to GOSLINER et al. (1996) *Protoreaster lincki* may be feeding on sponges; however, most authors assume it is omnivorous and feeds mainly on material of organic origin, such as detritus, dead organisms, algae, invertebrates of different sizes, and – according to ERHARDT & MOOSLEITNER (1995) – on sleeping fishes. In an aquarium it has shown to be somewhat more selective in respect to substitute foods than those species discussed earlier. This suggests that it is less of an omnivore than had been assumed all along. Yet, this sea star will normally accept mussel and clam meat, shrimps and (raw) fish meat.

Description: Prefers to live in shallow water on a soft and sandy bottom substrate, at depths from 1 to 30 m. Can also be found in sea grass beds, and occasionally on hard substrate. The contrasting and exceedingly attractive grey-red colouration is typical for *Protoreaster lincki*. It is rather frequently imported.
Aquarium requirements: Hardy aquarium inhabitant, but sensitive to adverse conditions during transport and subsequent acclimatization. It is an aggressive feeder and potential predator of many other aquarium inhabitants. If possible, should not be kept together with small, slowly moving and/or sessile invertebrates.

GA: +/-; **SE:** +/-; **CO:** –

Photo: *Protoreaster lincki* in an aquarium.
Photo: S. W. Michael

Atlantic, the Mediterranean Sea and in the eastern Pacific.

The food preference of these sea stars is generally so that they are suitable for a coral reef aquarium. Many species – probably the majority of them – feed on epiphytic growth, microscopic algae and small organisms among them, as well as on organic detritus.

Asexual reproduction occurs particularly in genus *Linckia*, but also in a few other ones, whereby arms are cast off from which "comets" develop (please do also refer to page 239). In many species this occurs frequently before specimens have reached sexual maturity resulting in juvenile specimens often being more frequent than adults. Fully grown specimens are capable of sexual reproduction by releasing eggs and sperm, but this appears to occur rarely (KAPLAN, 1982). The loss and subsequent regeneration of arms is also the reason why individual specimens are rarely perfectly symmetrical. The arms are usually not equally long and are at different stages of regeneration.

The most widely occurring species – and those of greatest interest from an aquarist's perspective – are contained in genera *Fromia*, *Gomophia*, *Leiaster*, *Linckia* and *Nardoa* (please do also refer to the colour plates on pages 258/259 as well as to the Fact Sheets from page 261 onward).

Family Asterinidae

This family contains 18 genera of predominantly very small sea stars that have a worldwide distribution in tropical and subtropical seas. Many of them measure only 1 or 2 cm in diameter; the largest species of genus *Anseropoda* reach sizes of 10 to 15 cm only. Moreover, the majority of these sea stars is not particularly colourful. Consequently, on a coral reef they are easily overlooked, even when they occur in abundance. The number of arms

Linckia guildingi **(seen here at Marie Louise, Amirante Islands in the Seychelles) has a circum-tropical distribution. It grows rather large and feed – much like its relatives – on all edible material found among epiphytic growth, as well as on detritus, microorganisms, algae and small invertebrates.**
Photo: Dr. I. Illich

The Blue Sea Star, *Linckia laevigata* and the New Caledonian Sea Star, *Nardoa novaecaledoniae*, on a reef of the central Great Barrier Reef, near Townsville, Australia. Photo: L. Newman and A. Flowers

Sea Stars of family Ophidiasteridae

Fromia cf. *monilis*. Photo: L. Newman and A. Flowers

Fromia cf. *nodosa*. Photo: S. W. Michael

Fromia cf. *nodosa*. Photo: E. Svensen

Fromia sp. "KA6-ECH-01". Photo: C. Bryce

Fromia sp. "KA6-ECH-02".
Photo: L. Newman and A. Flowers

Fromia sp. "KA6-ECH-03".
Photo: E. Svensen

Sea Stars of family Ophidiasteridae

Gomophia cf. *mammillifera*. Photo: C. Bryce

Neoferdina glyptodisca. Photo: L. Newman and A. Flowers

Tamaria megaloplax. Photo: Dr. I. Illich

Nardoa rosea. Photo: L. Newman and A. Flowers

Linckia sp. "KA6-ECH-04". Photo: L. Newman and A. Flowers

Nardoa sp. "KA6-ECH-05". Photo: L. Newman and A. Flowers

Linckia laevigata (right) and *Laiaster leachi* (left) in an aquarium. Most members of family Ophidiasteridae thrive well in a coral reef aquarium provided they have been carefully transported and properly acclimated. Because of their food preference for epiphytic growth, they are not particularly difficult to feed. Most specimens accept dry food, especially when given in tablet form. Moreover, they cause only little or no damage at all to large invertebrates.

Fromia monilis at Siladen, Northern Sulawesi, Indonesia.
Photo: R. Hansen

is variable. Five arms are more or less the norm, but it is not uncommon to find specimens with six, seven or eight arms. In fact, in a few species there can be up to 15 or 16 arms.

These sea stars are of particular interest to the coral reef aquarium hobby since they are relatively often introduced inadvertently in to our aquariums, together with live rocks or corals (see Fact Sheets on pages 261-280). Once they have become established in an aquarium there can be an almost explosive population growth. That can both be a joy or anguish, depending on the personal preference of the respective aquarist. In any event, such a rapid reproductive rate could be due to asexual reproduction, but it is also conceivable that sexual reproduction can occur under aquarium conditions.

Since taxonomic research on this group of sea stars has so far

Fromia ghardaqana
Ghardaqa Sea Star

Distribution: Endemic to the Red Sea.
Size: Up to 7 cm in diameter.
Reproduction: Free spawner; asexual reproduction by means of division can occur, however, this is not common.
Reproduction: Feeds primarily on periphytonic overgrowth, that is, on microscopic algae and the invertebrate fauna it contains, as well as on detritus. A few specimens may find enough food in a large aquarium, provided there is sufficient microscopic life growing on the decoration. If need be, the diet can be supplemented with mussel and clam meat or with crustaceans as well as with commercially prepared food.
Aquarium requirements: A very spectacular sea star that was first described by the Danish scientist MORTENSEN in 1938. This particular description has often been overlooked, or – respectively – the Ghardaqa Sea Star has often been incorrectly identified in literature on the echinoderms of the Red Sea. This may have been in part due to the fact that this sea star is highly variable in its colouration; that is, *F. ghardaqana* – like most *Fromia* species – varies from one population to the next. But all of them have a scarlet-red base colour; however, the size, number and distribution of the chalky white (sometimes nearly blue-white) spots can be highly differential. The specimen depicted in the photo below belongs to those with the greatest density of spots. But there are also specimens, which have spots only at the tips of their arms. *F. ghardaqana* lives individually among coral rocks in shallow water, down to a depth of about 20 m. It remains in hiding during the day.
Aquarium requirements: Generally, *Fromia* species do not seem to be quite as sensitive to transport conditions, as are other sea star species; however, they are often hard to keep alive in an aquarium, which could possibly be indicative of incorrect feeding. *F. ghardaqana* seems to be the most sensitive sea star species of this genus. It should be kept in an aquarium with an amply supply of microscopic life in order to provide their natural dietary requirements. This sea star normally does not damage other invertebrates, although there is some risk that it might feed on for sponges, bryozoans, etc. Keeping this sea star together with larger predatory fishes, crustaceans and other echinoderms that represent potential predators, must be avoided.

GA: +/-; **SE:** +/-; **CO:** +

Photo:
Ghardaqa Sea Star,Fromia ghardaqana, in its natural habitat in the Red Sea. Photo: J. Cairns-Michael

Fromia indica
Indian Sea Star

Distribution: Occurs widely throughout the Indo-Pacific, from the Maldives to Hawaii.
Size: Can reach a diameter of up to 9 cm, but usually remains smaller.
Reproduction: Free spawner; asexual reproduction through division can occur, but this is rare.
Food: Feeds mainly on periphytonic overgrowth, i.e. on microscopic algae and any invertebrates living in them, as well as on detritus. A large aquarium can provide enough food for a few specimens, if there is sufficient microscopic life present on the aquarium decoration. If need to be, the diet can be supplemented with mussel- or crustacean meat and with commercially prepared food.
Description: Rather easy to identify on the basis of the dense cover of tubercles on the aboral side, which are separated by thin black lines. Base colouration always dark- or brick red. Tips of arms in some populations dark, as seen here in two of the specimens. *Fromia indica* occurs from the inter-tidal zone down a depth of about 25 m, where it is typically found in the vicinity of stony corals. It actively searches for food during the day.
Aquarium requirements: Although the Indian Sea Star tolerates transport conditions rather well, it is not always easy to keep alive in an aquarium for prolonged periods of time. This could be due to an incorrect diet or the lack of suitable food. This sea star should be kept in an aquarium decorated with live rocks containing an abundance of microscopic life forms in order to meet its natural food requirements. Can be kept together with most other invertebrates, although there is always risk for minor loss of invertebrate life when this sea star crawls over sponges, bryozoans, etc. Should not be kept together with larger predatory fishes, crustaceans and other echinoderms that could be potential predators.

GA: +/-; **SE:** +/-; **CO:** +

Photos:
Fromia indica on the Great Barrier Reef (top), at the Maldives (centre; photo: J. Cairns-Michael), and in an aquarium (bottom; photo: T. Luther).

Fromia milleporella
Thousand-dotted Sea Star

Distribution: Occurs widely in the Indo-Pacific, from the Red Sea eastward to the Marshall- and Society Islands.
Size: Up to 7 cm in diameter.
Reproduction: Free spawner; asexual reproduction through division can occur but this is rare.
Food: Feeds mainly on periphytonic overgrowth, i.e. on microscopic algae and any invertebrates living in them, as well as on detritus. A large aquarium can provide enough food for a few specimens, if there is sufficient microscopic life present on the aquarium decoration. If need be, the diet can be supplemented with mussel- or crustacean meat and with commercially prepared food.
Description: Coloured uniformly red with – as indicated by the name – many clearly visible pores distributed over the entire aboral side. Specimens from the Red Sea sometimes with irregularly scattered blue and white dots, which could possibly represent an imitation of *Fromia ghardaqana. F. milleporella* lives in shallow reef sections at depths from 2 to 10 m, and actively searches for food during the day.
Aquarium requirements: Although the Indian Sea Star tolerates transport conditions rather well, it is not always easy to keep it alive in an aquarium for prolonged periods of time. This could be due to an incorrect diet or the lack of suitable food. This sea star should be kept in an aquarium decorated with live rocks containing an abundance of microscopic life forms in order to meet its natural food requirements. Can be kept together with most other invertebrates, although there is always risk for minor loss of invertebrate life involved when this sea star crawls over sponges, bryozoans, etc. Should not be kept together with larger predatory fishes, crustaceans and other echinoderms that represent potential predators.

GA: +/-; **SE:** +/-; **CO:** +

Photos:
Fromia milleporella at Kandooma, Maldives (top; photo: Dr. I. Illich), in an aquarium (centre; photo: S. W. Michael) and a blue-spotted variety in a display tank at Welkes Zoo Center, Lünen, Germany (bottom).

Fromia* cf. *monilis
Pearl Sea Star

Distribution: The documented distribution is restricted to the eastern sections of the Indo-Pacific: from the Indo-Australian Archipelago to the islands of the Central Pacific. Reports of its occurrence in the Indian Ocean can be correct, but it is more likely that these refer to *Fromia nodosa* or to a related and possibly yet-to-described species.
Size: Up to 8 cm in diameter.
Reproduction: Free spawner; asexual reproduction by division is possible, but very rare.
Food: Feeds mainly on periphytonic overgrowth, i. e. on microscopic algae and any invertebrates living in them, as well as on detritus. A large aquarium can provide sufficient food for a few specimens, if there is sufficient microscopic life present on the aquarium decoration. If need be the diet can be supplemented with mussel- or crustacean meat and with commercially prepared food.
Description: Belongs to a conglomerate of similar looking species or geographical and ecological forms, respectively, with a wide spectrum of intra-specific variations (see photos on page 258). This has led to a large number of misidentifications, especially in the aquarium literature. *Fromia* cf. *monilis* is often mistaken for *Fromia* cf. *nodosa* (see Fact Sheet on page 265). *Fromia* cf. *nodosa* has clearly visible, convex carinal plates, which in *Fromia* cf. *monilis* normally do not project beyond the outer skin. *Fromia* cf. *monilis* lives in shallow water of lagoons and canal systems between reefs, but can also be found along steep reef slopes, down to a depth of about 35 m.
Aquarium requirements: Problems observed in aquariums are due to insufficiently available, natural food. Otherwise, this species does rather well in aquariums that have been decorated with live rocks and contains a rich micro-fauna and –flora. Can be kept together with most larger invertebrates, including corals; however, this sea star presents a danger to small invertebrates, especially when they are malnourished. Should not be kept together with larger predatory fishes, crustaceans and other echinoderms that represent potential predators.

GA: +/-; **SE:** +/-; **CO:** +/-

Photos:
Top: *Fromia* cf. *monilis* at Sape, Sumbawa, Indonesia.
Photo: C. Bryce
Bottom: *Fromia* cf. *monilis* in Indonesia.
Photo: S. W. Michael

Fromia* cf. *nodosa
Knotty Sea Star

Distribution: The documented distribution is limited to the western Indian Ocean, including the Maldives and Sri Lanka. However, it is possible that the distribution area is either much larger (see comments in Fact Sheet on *Fromia* cf. *monilis* on page 264) or much smaller (see comments under "Description", below).
Size: Up to 8 cm in diameter.
Reproduction: Free spawner; asexual reproduction by division is possible, but very rare.
Food: Feeds mainly on periphytonic overgrowth, i.e. on microscopic algae and any invertebrates living in them, as well as on detritus. A large aquarium can provide sufficient food for a few specimens, if there is sufficient microscopic life present on the aquarium decoration. If need be the diet can be supplemented with mussel- or crustacean meat and with commercially prepared food.
Description: Belongs to a conglomerate of similar looking species (see comments in Fact Sheet on *Fromia* cf. *monilis* on page 264). The original description was based on a specimen from the Amirante Islands, which represented a rather specific – rarely found – morph. The specimens from the Maldives shown here are identical to the form commonly referred to as *F. nodosa*. MOOSLEITNER (1997) argues that it may be correct to grant this form the status of a valid species or at least that of a subspecies described as *F. nodosa maldivensis*. As long as this massive confusion with the systematic of this species complex has not been resolved by a comprehensive revision, we prefer not to follow MOOSLEITNER'S reasoning.
Aquarium requirements: Problems observed in aquariums are due to insufficiently available, natural food. Otherwise, this species does rather well in aquariums that have been decorated with live rocks and contains a rich micro fauna- and –flora. Can be kept together with most larger invertebrates, including corals; however, the Knotty Sea Star presents a danger to small invertebrates, especially when they are malnourished. Should not be kept together with larger predatory fishes, crustaceans and other echinoderms that represent potential predators.

GA: +/-; **SE:** +/-; **CO:** +/-

Photos:
Fromia cf. *nodosa* at the Maldives.
Photos: Dr. I. Illich (top); J. Cairns-Michael (centre)

Gomophia egyptiaca
Egyptian Sea Star

Distribution: Occurs widely in the Indo-Pacific, from East Africa and the Red Sea, eastward to Australia, New Caledonia, Guam, Palau and the Society Islands.
Size: Up to 16 cm in diameter.
Reproduction: Free spawner.
Food: Feeds mainly on periphytonic overgrowth, i.e. on microscopic algae and any invertebrates among them, as well as on detritus. Will not always accept substitute foods, but various types of dry food should be tried.
Description: Aboral side with clearly visible conical plates carrying a cone-shaped spine. Colouration variable, but normally cream-coloured to reddish-brown with lighter or darker spines. Never in abundance. Normally found on corals and rocks along reef slopes, lagoons and outer reefs, at depths from 5 to 50 m.
Aquarium requirements: Appears to be rather hardy when kept in large aquariums decorated with live rocks and containing a rich micro-fauna and –flora. *G. egyptiaca* tends to starve to death in sterile surroundings. Normally this sea star will not damage large invertebrates, however – as with other sea stars – caution is advisable, especially then when there is a lack of natural suitable food for this species.

GA: +/-; **SE:** +/-; **CO:** +/-

Photos:
Gomophia egyptiaca in Indonesia (top; photo: S. W. Michael) and in the Red Sea (bottom; photo: E. Svensen).

***Gomophia gomophia* (?)**

Distribution: Western to Central Pacific; from the Philippines to Polynesia.
Size: Rarely more than 25 cm in diameter, but according to some authors up to 40 cm.
Reproduction: Free spawner.
Food: Feeds mainly on periphytonic overgrowth, i.e. on microscopic algae and any invertebrates among them, as well as on detritus. Will not always accept substitute foods, but various types of dry food should be tried.
Description: The correct name and identity of this sea star is somewhat uncertain. Some authors are including it in genus *Nardoa*, occasionally as *Nardoa gomophia* (e.g. GUILLE et al., 1986), in other cases as *N. frianti* (e.g. GOSLINER et al., 1996). Our identification is based on a comparison of many different sources, but it may not be necessarily correct. Although its colouration is variable, this sea star is readily identified on the basis of its distinct tubercular warts (which cover the entire aboral surface), together with the cream- to pink base colour and the wide, dark bands on its arms. Lives in shallow water, normally on a hard substrate with coral growth, but can also be found on soft, sandy or muddy bottom. Depth distribution from 1 to 45 m.
Aquarium requirements: When kept in large aquariums decorated with live rocks and containing a rich microfauna and –flora, *Gomophia gomophia* is remarkably durable. However, it tends to starve to death in sterile surroundings. Normally this sea star will not harm large invertebrates, however – as with other sea stars – caution is advisable, especially then when there is a lack of natural suitable food for this species.

GA: +/-; **SE:** +/-; **CO:** +/-

Photos:
Gomophia gomophia in the Philippines (top and centre; photo: P. Schupp) and at Palau (bottom; photo: Dr. I. Illich).

Gomophia watsoni
Watson's Sea Star

Distribution: Restricted to parts of the Indo-Australian Archipelago: Queensland, New Caledonia and Papua-New Guinea.
Size: Up to 12 cm in diameter.
Reproduction: Free spawner.
Food: Feeds mainly on periphytonic overgrowth, i.e. on microscopic algae and any invertebrates among them, as well as on detritus. Will not always accept substitute foods, but various types of dry food should be tried.
Description: Rather variable in terms of its colouration. The two specimens shown here represent two extremes. All colour varieties have a scattering of small spines, cross bands on the arms with light-coloured zones at their tips. Normally encountered on corals and rocks along reef slopes in lagoons and outer reefs, at depths from 10 to 50 m.
Aquarium requirements: So far we do not have any personal aquarium husbandry experience with *Gomophia watsoni*, but we would expect it to be as hardy as are its closest relatives. Therefore, it should be kept in large aquariums decorated with live rocks and containing a rich micro-fauna and –flora. It tends to starve to death in sterile surroundings. One would expect that this sea star will not harm large invertebrates, however – as with other sea stars – caution is advisable, especially then when there is a lack of natural suitable food for this species.

GA: +/-; **SE:** +/-; **CO:** +/-

Photos:
Gomophia watsoni in Papua-New Guinea.
Photos: P. Schupp (top); S. W. Michael (bottom)

Leiaster coriaceus
Picasso Sea Star

Distribution: Occurs widely in the Indo-Pacific, from East Africa and the Red Sea to Japan and the South-Pacific islands.
Size: Usually up to 35 cm in diameter only, but occasionally specimens with a diameter of 46 cm can be found.
Reproduction: Free spawner; asexual reproduction by means of division can occur, but this is not common.
Food: Feeds principally on organic detritus, dead organisms and microscopic invertebrates as well as on algae. Will accept various substitute foods, e.g. dry food in the form of tablets or pellets. This is a very active omnivore that looks for food, even in competition with fishes.
Description: An extremely spectacular sea star with beautiful colouration. The base colour consists of a creamy or brownish yellow. Large bright-red or red-brown spots are arranged in an irregular pattern on the aboral side. Can be found on hard substrate at depths from 5 to 25 m. Has a tendency to hide during the day.
Aquarium requirements: The Picasso Sea Star is frequently imported. It is the hardiest sea star in this family, and remarkably unproblematic in respect to food and harm to other invertebrates; however, the Picasso Sea Star – like all other sea stars – can not be completely trusted, especially since it is rather large. When purchasing this sea star it is imperative to remember that it is sensitive to adverse transport and acclimatization conditions. Any damage sustained this way can have serious consequences, in spite of its well-developed regenerative capabilities. Although this sea star is normally nocturnal, it will quickly learn to search protected areas in the aquarium for food also during the day.

GA: +; **SE:** +/-; **CO:** +/-

Photos:
Leiaster coriaceus in the Philippines (top; photo: P. Schupp), at One Tree Island, Great Barrier Reef (centre; photo: L. Newman and A. Flowers) and in an aquarium (bottom).

Leiaster leachi
Leach's Sea Star

Distribution: Occurs widely in the Indo-Pacific, from East Africa and the Red Sea to Japan and Hawaii.
Size: Reaches a diameter of 26 cm.
Reproduction: Free spawner; asexual reproduction by division possible but not common.
Food: Feeds principally on organic detritus, dead organisms and microscopic invertebrates as well as on algae. Will accept various substitute foods, e.g. dry food in the form of tablets or pellets. This is a very active omnivore that looks for food, even in competition with fishes.
Description: Can be identified without difficulties on the basis of colouration, consisting of an irregularly distributed pattern of bright red and yellow spots. Although *Leiaster leachi* has a large distribution area, it is never found in abundance anywhere. Since it is a nocturnal feeder, it is rarely ever seen during the day. It lives on hart substrate at depths from 5 to 30 m.

Aquarium requirements: Fundamentally unproblematic in terms of food requirements and compatibility with other invertebrates, but apparently it is somewhat less hardy than *Leiaster coriaceus* (see Fact Sheet on page 269). As with most sea stars, there is a slight risk when keeping small and delicate invertebrates in the same tank; nevertheless, it is an excellent aquarium inhabitant. When purchasing this sea star it is important to keep in mind that it is sensitive to improper handling during transport and acclimatization, and any damages sustained by these animals can have serious consequences.

GA: +/-; **SE:** +/-; **CO:** +/-

Photo:
Leiaster leachi in an aquarium.

Leiaster speciosus
Magnificent Sea Star

Distribution: Occurs in the Indo-Pacific, from the Indo-Australian Archipelago eastward to the Marshall- and Fiji Islands and northward to Japan.
Size: Up to 60 cm in diameter.
Reproduction: Free spawner; asexual reproduction by division possible but not common.
Food: Feeds principally on organic detritus, dead organisms and microscopic invertebrates as well as on algae. Will accept various substitute foods, e.g. dry food in the form of tablets or pellets. This is a very active omnivore that looks for food, even in competition with fishes.
Description: A magnificent, intensely purple-red sea star (speciosus, Latin = beautiful, handsome). Can be distinguished from most other bright red sea stars by its long, flexible arms (like those in other *Leiaster* species) and the regularly spaced longitudinal rows of dark spots on the arms. It lives on hard substrate at depths from 10 to 30 m. In search of food primarily at night.

Aquarium requirements: So far we have not yet seen this sea star available in the aquarium trade, however, due to its attractive colouration imports of this species could indeed be economically successful. It is not expected that the aquarium husbandry requirements of *L. speciosus* differ from those of the other *Leiaster* species.

GA: 0+/-; **SE:** 0+/-; **CO:** 0+/-

Photo:
Leiaster speciosus at Guam. Photo: P. Schupp

Linckia guildingi
Guilding's Sea Stars

Distribution: Circum-tropical. Occurs widely in the Indo-Pacific and Caribbean region.
Size: Reaches a diameter of 50 cm.
Reproduction: Free spawner; asexual reproduction by means of division is common.
Food: Feeds mainly on organic detritus, dead organisms and microscopic invertebrates, as well as on algae. Will often not accept substitute foods.

Description: Because of this species' customary asexual reproduction by division there is an increasing number of specimens with an irregular number of arms and different arm lengths. A normal-shaped specimen has five arms, but six or seven arms also occur frequently. Comets, at different developmental stages, are not uncommon. Colouration is variable, however, generally adult specimens are uniformly coloured brownish-yellow, rarely light brown or violet. Juveniles can have a sprinkled pattern, somewhat similar to *Linckia multifora*. Lives at depths from 1 to 35 m, usually in hiding in the proximity of corals or coral rubble, especially as a juvenile. Large specimens from about 30 cm in diameter onward do not behave quite as cryptically. Characteristic habitats of this species are lagoons, less commonly outer reefs.
Aquarium requirements: Rarely ever available from the aquarium trade. Rather hardy aquarium inhabitant once it has become properly adapted to the prevailing conditions. Physical handling and transport appear to be very stressful to this species, so that a deliberate and slow acclimatization is absolutely imperative. In spite of its substantial size it does not harm corals and other large invertebrates. Yet, this sea star should be monitored, since it can knock over insecurely fastened rocks or corals.

GA: +/-; **SE:** +/-; **CO:** +

Photos:
Adult specimens of *Linckia guildingi* at the Maldives (left and bottom left; photo left: S. W. Michael) and at Puerto Rico (bottom right).

Linckia laevigata
Blue Sea Star

Distribution: Occurs widely in the Indo-Pacific, from East Africa eastward to Hawaii and northward to Japan.
Size: Normally reaches a diameter of 30 cm, rarely to 40 cm.
Reproduction: Free spawner; asexual reproduction by division is common.
Food: Feeds principally on organic detritus, dead organisms and microscopic invertebrates, as well as on algae. Substitute foods are often not accepted. Attempts should be made to offer dry food in the form of pellets or tablets.
Description: Specimens living in the shallow water of lagoons, tide pools, reef flats or barrier reefs have a magnificent blue colouration, a very intense, bright colour rarely ever seen in other sea stars. Specimens living in deeper water are more grayish, pink, yellowish or even khaki-coloured. The depth range of this species is from 0 to 25 m, sometimes even down to 60 m. Because of the common asexual reproduction by division specimens with an irregular number of arms and differential arm lengths are not unusual. Many specimens accommodate parasitic snails of the species *Thyca crystalline* (family Eulimidae). These snails live on the oral side of these sea stars, partially embedded in the ambulacral grooves. They have cap-shaped shells with numerous ripples and they use their proboscis to suck body fluid and tissue from their host. In its natural habitat a sea star can be the host for several snails without sustaining any damage; however, in an aquarium this additional stress factor should be manually removed with a pair of forceps or similar implement.
Aquarium requirements: The Blue Sea Star is probably the one most frequently traded, but unfortunately it is not the one most easily to keep. It appears that transport conditions impose an enormous stress, which all too often leads to the demise of specimens. Consequently, it is absolutely imperative to proceed very slowly and with great deliberation during acclimatization. If acclimatization is successful and if the starfish is kept in an aquarium with plenty of turf-algae as well as micro- and macro-organisms, the blue sea star can live for many years. This sea star does not harm corals and other large invertebrates. Must be kept in aquariums decorated with live rocks and with an abundance of microfauna and –flora, in order to meet its food requirements.

GA: +/-; **SE:** -; **CO:** +

Photos:
Linckia laevigata at Bunaken, Indonesia (top; photo: R. Hansen), at Palau (centre; photo: Dr. I. Illich) and at Kimberley, Western Australia (bottom; photo: C. Bryce).

Linckia multifora

Distribution: Occurs widely in the Indo-Pacific, from the Red Sea southward to South Africa, eastward to Hawaii and northward to Japan.
Size: Reaches a diameter of 15 cm.
Reproduction: Free spawner; asexual reproduction by division is rather common. Arms are cast off readily, that quickly form a new oral disk and new arms.
Food: Feeds principally on organic detritus, dead organisms and microscopic invertebrates, as well as algae. Most substitute foods only reluctantly accepted; however, dry food in the form of tablets or pellets is rather eagerly taken.
Description: Because of the common asexual reproduction there are many specimens with a variable number of arms and arm lengths. Comets that have developed from a single arm can be found in various stages of development. Colouration is extremely variable, but its base is usually cream-coloured, bright yellow or orange-brown. The aboral side is covered by irregularly shaped small spots and larger humps in orange, red and/or brown. This sea star commonly inhabits the inter-tidal zone down to a depth of 40 m. It accommodates the parasitic snail *Thyca crystalline* (see Fact Sheet for *Linckia laevigata* on page 273).
Aquarium requirements: Remains substantially smaller than *Linckia laevigata* and so it is also suitable for smaller aquariums. Transport appears to cause considerable stress. Cautious, slow acclimatization is absolutely imperative. Does not harm corals and other large invertebrates. Must be kept in an aquarium decorated with live rocks that contains an abundant micro-fauna and –flora, in order to meet the food requirements of this species.

GA: +/-; **SE:** +/-; **CO:** +

Photos on this page:
Linckia multifora on the Great Barrier Reef (top) and a comet at the Maldives (bottom).

Photos on page 275:
Linckia multifora at the Maldives (top left and right; photos: S. W. Michael), in Papua-New Guinea (centre left and right; photos: L. Newman and A. Flowers), at the Christmas Islands (bottom left; photo: C. Bryce) and at Palau (bottom right; photo: Dr. I. Illich).

Nardoa* cf. *galatheae
Galathea's Sea Star

Distribution: Indo-Pacific; details unknown but possibly with a wide distribution.
Size: Reaches a diameter of 25 cm, but usually remains smaller.
Reproduction: Free spawner; asexual reproduction by division is rare.
Food: Feeds on microscopic organisms, probably also on detritus and dead organisms. Hardly touches any substitute foods.
Description: *Nardoa galatheae*, described in 1864 by the Danish scientist LÜTKEN, belongs to a group of rather similar looking species, where the dark brown base colouration is covered by light brown plates, so that they look like being sprinkled. We believe that the specimen shown here is identical with *Nardoa galatheae*, but it could also be *N. novaecaledoniae*, *N. variolata* or *N. lemonnieri*. This taxonomic confusion is only of theoretical significance, since the ecology, ethology and feeding behaviour should be much the same for all four species mentioned. *Nardoa* cf. *galatheae* is found mainly on hard substrate of platform reefs and along reef slopes, at depths from 1 to 20 m.
Aquarium requirements: Rather frequently available in the aquarium trade. One of the most suitable sea stars for a coral reef aquarium. Normally it thrives very well with a substantial longevity, provided there has been a slow and deliberate acclimatization. Beyond that, it must be kept in an aquarium decorated with live rocks and there needs to be an abundant micro-fauna and –flora in order to meet the food requirements of this sea star.

GA: +; **SE:** +/-; **CO:** +

Photo:
Nardoa cf. *galatheae* in its natural habitat.
Photo: L. Newman and A. Flowers

Neoferdina cumingi
Cuming's Sea Star

Distribution: Occurs in the eastern Indo-Pacific, from the Indo-Australian Archipelago northward to the Philippines and eastward to Polynesia.
Size: Reaches a diameter of 8 cm.
Reproduction: Free spawner; no information available about asexual reproduction by division.
Food: Feeds on microscopic algae and the associated invertebrate fauna. Tends to reject substitute foods; however, should find sufficient food in large aquariums with an abundant micro-fauna and –flora.
Description: An attractive sea star, but with highly variable colouration. The specimen shown here is typical for the Philippines. All colour varieties have distinct wart-like and contrasting lateral- and central plates on their arms. Lives at depths from 5 to 35 m.
Aquarium requirements: Rarely ever imported. Possibly as difficult in its maintenance requirements as *Fromia* species, however, this prediction is unreliable in view of limited aquarium experiences. Should definitely be kept in aquariums with live rocks and a rich micro-fauna and –flora, in order to meet the food requirements of this species.

GA: +/-; **SE:** +/-; **CO:** +

Photos:
Neoferdina cumingi at the Philippines: habitat- (top) and close-up photo of aboral side of the central disc (below).
Photos: P. Schupp

Photo:
Phataria unifascialis in an aquarium. Photo: S. W. Michael

Phataria unifascialis
Brown Sea Star

Distribution: From the Gulf of California (Sea of Cortez) southward to northern Peru and westward to the Galapagos Islands.
Size: Reaches at diameter of 20 cm.
Reproduction: Free spawner; no information available about asexual reproduction by division.
Food: Probably feeds exclusively on plant material.
Description: The most common sea star in the Gulf of California. Occurs from the inter-tidal zone down to a depth of 100 m. Often lives in association with calcareous algae, but it is unknown to what degree (if at all) it feeds on them when other algae species are available. Colouration variable from brown to an orange-reddish brown. Body and arms stronger and less rigid than in other long-armed species of this family.
Aquarium requirements: Rarely ever imported in to Europe. Relatively durable under aquarium conditions, but transport seems to be the greatest stress problem. Therefore, this sea star - like most others - must be acclimated cautiously and slowly. Apparently it does not harm corals and other large invertebrates.

GA: +; **SE:** +/-; **CO:** +

Photo:
Tamaria stria in an aquarium. Photo: S. W. Michael

Tamaria stria
Tamarisk Sea Star

Distribution: From the Gulf of California (Sea of Cortez) southward to Colombia.
Size: Reaches a diameter of 18 cm.
Reproduction: Unknown.
Food: Feeds on epiphytic growth, i.e. microscopic algae and associated invertebrate fauna.
Description: Can be recognized by the deep purple colouration. Lives on rocks and ruble substrate, at depths from 12 to 50 m; never in abundance.
Aquarium requirements: Only limited reports on first-hand experience, but these indicate that this sea star is rather hardy when kept in aquariums with a rich micro-fauna and –flora. Does not appear to harm corals and other large invertebrates. There remains to be the same potential risks as are associated with most sea stars.

GA: +/-; **SE:** +/-; **CO:** +

Asterina burtoni
Burton's Goose-footed Sea Star

Distribution: Occurs widely in the Indo-Pacific; from East Africa and the Red Sea to Hawaii.
Size: Reaches a diameter of 5 cm, but usually smaller.
Reproduction: Details about sexual reproduction are not known. Asexual reproduction by division of the oral disc is common, even in aquariums.
Food: Feeds on microscopic invertebrates, such as molluscs, crustaceans, bryozoans, sponges, etc., as well as on dead organisms, possibly also on algae and detritus.
Description: Normally with five arms, but six or seven arms are not unusual because of division and irregular regeneration. Colouration generally grey spotted, but variable. Lives hidden among rocks and corals, at depths from 5 to 30 m.
Aquarium requirements: *Asterina burtoni* and closely related species, which are difficult to keep apart, are often introduced in to aquariums as "stowaways" attached to live rocks and corals. If such an aquarium is decorated with live rocks and also contains a rich micro-fauna, supplementary feeding is not required. The reproductive rate in an aquarium can be rather high, possibly due to asexual reproduction by means of division. As far as it is known, this species will not harm large invertebrates, however, small ones always represent potential prey. Large fishes or invertebrates will feed on this shy sea star; however, because of its high reproductive rate it is not likely to die out, once it has become established in an aquarium.

GA: +; **SE:** +; **CO:** +

Photo:
Asterina burtoni in the Gulf of Eilat, Red Sea.
Photo: R. Brons

Asterina* cf. *sarasini
Sarasin's Goose-footed Sea Star

Distribution: Indo-Pacific, but details unknown. Probably widely distributed.
Size: Reaches a diameter of 4 cm, but usually smaller.
Reproduction: Details about sexual reproduction unknown. Asexual reproduction by division of oral disc common, even in aquariums.
Food: Feeds on microscopic invertebrates, such as molluscs, crustaceans, bryozoans, sponges, etc., as well as on dead organisms, possibly also on algae and detritus.
Description: Belongs to a species complex that so far has been insufficiently researched taxonomically. It includes several rather similar looking sea stars that are all characterized by a granulated body surface and short arms. Normally this species has 5 arms, however, because of division and irregular regeneration there can also be more than 5 arms. Colouration grey- or green-spotted with a contrasting star pattern on the body disc.

Aquarium requirements: Many *Asterina* species are often introduced in to aquariums as "stowaways" attached to live rocks and corals. If such an aquarium is decorated with live rocks and also contains a rich micro-fauna, supplementary feeding is not required. The reproductive rate in an aquarium can be rather high, possibly due to asexual reproduction by means of division. As far as it is known, this species will not harm large invertebrates, however, small ones always represent potential prey. Large fishes or invertebrates will feed on this shy sea star; however, because of its high reproductive rate it is not likely to die out, once it has become established in an aquarium.

GA: +; **SE:** +; **CO:** +

Photo:
Asterina cf. *sarasini* in Jelly Fish Lake on Palau.
Photo: Dr. I. Illich

been relatively inadequate, a correct identification of genera is difficult, and becomes even more difficult down to species level. It is likely that the vast majority of species that appear in aquariums belongs to genera *Asterina*, *Diasterina* or *Nepanthia*.

Most of the small species of family Asterinidae are probably microphages that feed on small organisms, such as sponges, crustaceans, algae, detritus or they are scavengers. Because of their small maximal sizes, damage inflicted by these sea stars in a coral reef aquarium is not likely to be excessive; at least, that has been our experience.

SPRUNG & DELBEEK (1997) believe it to be conceivable that some species feed on coral polyps. Of course, that cannot be totally excluded, but so far there have not yet been any reports about such feeding behaviour. Basically, all animals introduced in to an aquarium should (initially) be closely monitored, since available knowledge and experienced gained from aquarium maintenance can never be totally conclusive and certain. Consequently, an aquarist needs to be prepared to deal with unpredicted behaviour pattern or unexpected food requirements. For instance, it is not uncommon for animals that are normally herbivorous (be they gastropods, echinoderms, fishes or others), suddenly selected corals with symbiotic zooxanthellae as their preferred food, whenever there is no other plant food available or only in insufficient quantities. There are hardly any animal species that offer 100 % certainty in this respect. And for that reason we need to be impartial towards unknown organisms that are newly introduced in to an aquarium, until we can be certain whether or not they inflict any damage on to other important organisms in the same aquarium. It is only too common for aquarists to busily eradicate organisms that have suddenly appeared in their tanks, when indeed such animals should really be given a chance!

Goose-footed Stars of genus *Anseropoda* (anser, Latin = goose, and pous, genetive podos, Greek = foot) are the largest representatives of family Asterinidae. Shown here is an unidentified species off southwestern Australia.
Photo: C. Bryce

Family Acanthasteridae

Because of the Crown-of-Thorns Sea Star, *Acanthaster planci* – is an acknowledged, serious predator of stony corals – its entire family (monotypic with genus *Acanthaster*) has become infamously notorious. The family is monotypic with genus *Acanthaster* as the only genus. Some authors list two additional species, i.e., *A. ellisii* and *A. brevispinus*. *A. ellisii* has been described from the eastern Pacific (Gulf of California to the Galápagos Islands and Peru), but it should probably be considered as a synonym for *A. planci*. *A. brevispinus* that was described from the Philippines in 1917, has substantially shorter spines, as well as shorter but thicker pedicellaria than *A. planci*. MADSEN (1955) indicated that it too might be a synonym for *A. planci*. In any event, we presume that this family contains only a single species of *A. planci*, even though there is no absolute consensus on this question within expert circles.

But what a species! The Crown-of-Thorns Sea Star, *Acanthaster planci* is beyond doubt the sea star most talked (and written) about in recent times. There have been hundreds of publications on this species and its effect on coral reef habitats since the 1960's. For instance, the journal "Coral Reefs" has come out with two special editions about the *Acanthaster* problem (WILKINSON, 1990; WILKINSONI & MACINTYRE, 1992).

In short, *Acanthaster planci* is a sea star that feeds almost exclusively on stony coral polyps. During the 1960's the populations of this sea star on the Great Barrier Reef increased to such a degree that there were serious concerns that large sections of the Reef would be virtually destroyed. This massive increase in the numbers of *A. planci* triggered – what was probably – the very first global debate on environmental destruction possible caused by man. This debate intensified to such a degree, that in a Letter to the Editor by a reader of one of the Australian national newspapers, the question was raised: "How will we be able to look our children in the eyes when they ask: What was the Great Barrier Reef?" (as cited from WILKINSON, 1990).

The Crown-Of-Thorns Sea Star, *Acanthaster planci*, – seen here on an Indonesia coral reef - is one of the most efficient predators of stony corals. Its massive occurrence in many parts of the Indo-Pacific during the 1960's caused much excitement and triggered lengthy debates on its causes.

The Crown-Of-Thorns Sea Star, *Acanthaster planci*, feeding on an *Acropora* colony on the Ngaloa Barrier Reef at Fiji. Clearly visible is the white area where the coral polyps have already been eaten away.

Photo: S. W. Michael

The Crown-of-Thorns Sea Star, *Acanthaster planci* was first described in 1705 by the biologist RUMPHIUS and named by LINNÉ in 1758. Thus, it has been known for quite a long time, yet, under normal circumstances it is rarely ever seen on a reef. But sometimes nature proceeds along strange pathways. The first report of mass occurrences of this sea star dates back to the end of the 1950's, from Japan and in 1962 from the Great Barrier Reef. Soon thereafter similar mass occurrences were found in Micronesia, at the Ryukyu Island, Samoa and the Society Islands. With some deviations, the population densities of this sea star remained high until well in to the 1980's. While the normal population density ranges from one specimen per 6 km^2 to one specimen per 20 km^2, during that particular period the density rose to one specimen per 20.6 m^2!

With such a massive population explosion of a significant coral predator, damages to the Reef soon became obvious. In some areas there was a decline in live coral stands from 78% down to 2% within a period of only six months (MORAN in WILKINSON, 1990). One counter measure in the fight against the destructive effects of the Crown-of-Thorns Sea Star was a mass killing, using such methods injecting copper sulphate. Around Samoa about 490,000 sea stars were killed this way within a short period of time, and around the Ryukyu Island about 13 million specimens! Nevertheless, soon this turned out to be a hopeless battle. The reproductive rate of *Acanthaster planci* is high, and ocean currents disseminate the larvae widely.

What can be cause of such mass occurrence? There were (and still are) many theories. It goes without saying that many people held human intervention, such as habitat destruction, water pollution and overexploitation of the predators of *Acanthaster planci*, as being responsible. Yet, it is important to realise that such mass occurrences do not only occur in *A. planci*, but also among many other aquatic- and terrestrial animals. We only have to recall the periodic mass occurrences of rodents in Europe. Moreover, for *A. planci* there is also anecdotal evidence that such population explosions have already occurred previously, for instance, at the beginning on last Century, around Samoa (MORAN in WILKINSON, 1990). Similarly, geological investigations by WALBRAN et al. (1989) have revealed from fossil finds, that there have been cause-and-effect relationships. Presumably, we are not very likely to ever obtain a definitive explanation. What is important for us is the observation that overall the *Acanthaster planci* populations tend to return to normal levels after a few years and that the coral biotopes then recover. In essence then, the mass occurrence of *A. planci* was not a sign that our day of reckoning has come!

From a biological perspective, the Crown-of-Thorns Sea Star, *Acanthaster planci* is rather inte- they turn in to brachiolaria larvae that have a complicated body structure. In essence this can be construed as double specimen, whereby the anterior section makes up the swimming and feeding component, and the posterior section accommodates the sea star embryo. When such larva is then carried to a coral reef by water currents and drops to the bottom, metamorphosis sets in and the larvae develops in to a juvenile, free-living sea star. It looks exactly like an adult, however, its

Even the most aggressive predators are themselves subject to predation. This rare photo shows two Harlequin Shrimps, *Hymenocera picta* feeding on a juvenile The Crown-Of-Thorns Sea Star, *Acanthaster planci*, at Madang, Papua-New Guinea.
Photo: L. Newman and A. Flowers

mass occurrences of Crown-of-Thorns Sea Stars on the Great Barrier Reef at certain intervals over the last 7,000 years. Although it cannot be excluded that man over the years has had an impact on the development of such population explosions (man has influenced nature in many ways, irrespective of his original intent), we need to appreciate that there are multiple and complex resting (ZANN et al., 1987). The sexes are separate and the sex products are released in the open water. A single female can shed 4 to 65 million eggs during a single spawning. Fertilised eggs measure only 0.2 mm in diameter. After a development period of 24 hours the eggs give rise two pelagic, transparent bipinnaria larvae that feed on microscopically small algae. After several weeks diameter is initially only half a millimetre and it has 5 arms. It is actually so small that it cannot feed on corals (in fact, it is probably more likely to be eaten by corals). For about six months the little Crown-of-Thorns Sea Star feeds largely on diatoms, and may be also on detritus, and as it grows it develops more arms. Once it has reached a diameter of 1 cm it has already ten arms and then commences to feed on coral polyps.

At an age of two years and with a diameter of up to 20 cm, the Crown-of-Thorns Sea Star has become a really effective coral feeder. The stomach is everted over a coral colony, enzymes dissolve the live tissue of the coral an immediately digest it *in-situ*; what remains is only the chalky white, dead coral skeleton. *A. planci* prefers rapidly growing coral species, such as *Acropora* spp. and *Montipora* spp. (probably one of nature's deliberate schemes to minimize any danger of a lack of food). In laboratory experiments it has been shown that adult specimens of *A. planci* can also change over to other food if the need arises, including soft corals, gorgonians and other cnidarians. In fact, it can also go completely without food for six to nine months.

The Crown-of-Thorns Sea Star, in turn, has its own predators. Its larvae and juveniles are eaten by many different organisms, e.g. by corals, fishes, coral crabs and shrimps. Even adults are eaten by Triton snails, harlequin shrimps, as well as by large puffers of genus *Arothron*. Even when these predators feed on *A. planci* during a population explosion, it has little – if any – effect on overall population numbers of the sea star. Far for dangerous for the sea star is the spread of certain epidemic diseases, where – for instance – a mortality of 99 % has been observed among juveniles.

Crown-of-Thorns Sea Star measures about 50 cm in diameter when fully grown, and specimen with record sizes of up to 80 cm have also been seen. Adult specimens have 15 to 20 arms. It is an extraordinarily attractive and conspicuous sea star with variable colour markings in grey, green, blue, lilac, yellow and brown and/or red. Eggs and larvae, as well as juveniles and adults contain toxic substances (proteinaceous saponins = glucosides). Adult specimens can transfer these toxins to humans by means of sharp spines. Such a puncture wound is painful and can cause swelling, itching, nausea and vomiting.

Although Crown-of-Thorns Sea Stars can probably be kept in very large aquariums, it is not among the most suitable aquarium animals. Even when an aquarium specimen accepts a limited amount of substitute food, a dietary supplement of stony corals is a prerequisite for its successful aquarium maintenance.

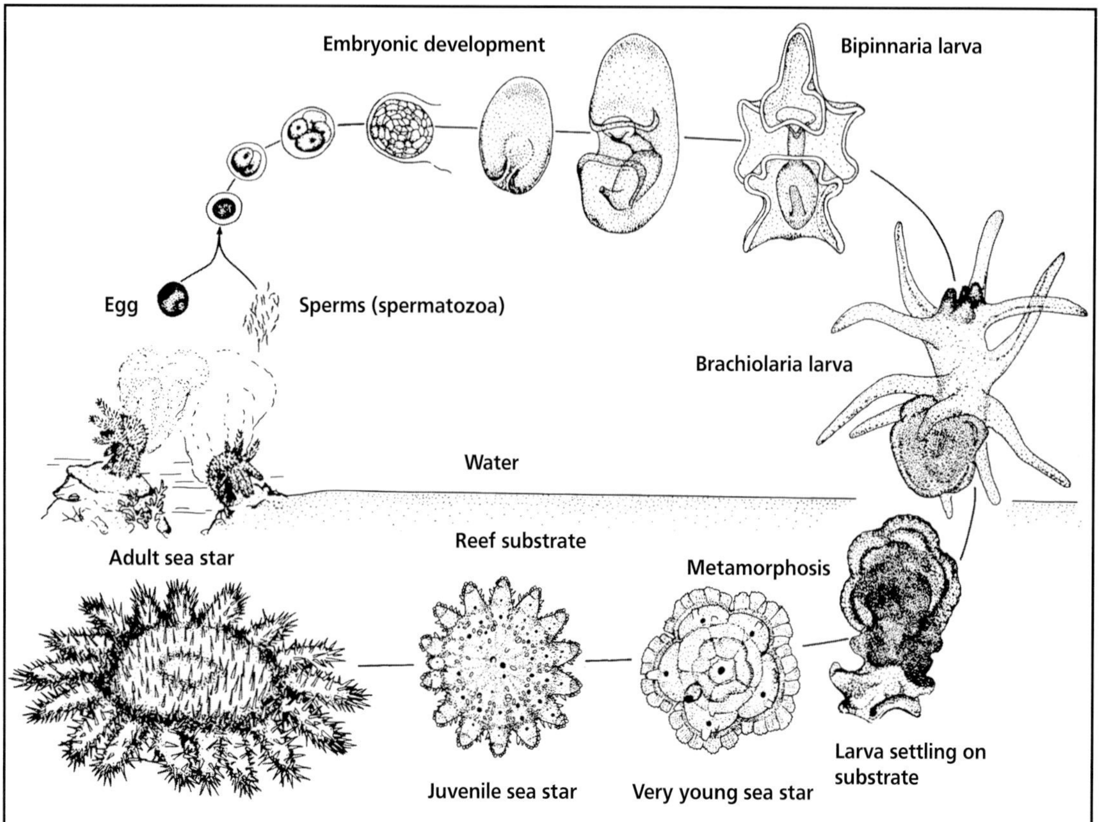

Simplified schematic representation of the life cycle of the Crown-Of-Thorns Sea Star, *Acanthaster planci*.

Drawing after ZANN et al. (1987)

Examples of colour diversity of the Crown-Of-Thorns Sea Star, *Acanthaster planci*

Acanthaster planci at the Maldives (photo: I. Erga), ...

... at the Great Barrier Reef (photo: M. Cufer), ...

... at Madang, Papua-New Guinea (photo: L. Newman and A. Flowers), ...

... at Chuuk, Micronesia (photo: P. Schupp), ...

... at Kimberley, Western Australia (photo: C. Bryce), ...

... in the Red Sea (photo: E. Svensen).

The Brown-tipped Sea Star, *Thromidia catalai* (seen here in Papua-New Guinea) reaches a maximum diameter of 60 cm, and is certainly too large and aggressive for aquarium maintenance. Photo: S. W. Michael

Family Mithrodiidae

With genera *Mithrodia* and *Thromidia*, this family contains only two genera which are both common throughout the Indo-Pacific and in the eastern Pacific. Genus *Thromidia* includes the largest sea stars in the oceans, with a maximum, measured arm spread of 75 cm and an arm thickness of 12 cm.

All species are predominantly epiphytic feeders, removing the organic layer off hard substrates. However, fully-grown specimens will also take substantial amounts of larger sessile animals, such as sponges, moss animals, tunicates and probably also cnidarians. Under aquarium conditions, these sea stars readily accept substitute food items. Therefore, they could be attractive choices for a special aquarium, possibly even in combination with fishes. However, joint maintenance with other invertebrates is inadvisable. Sizes of 30 to 40 cm are not uncommon among adults. *Mithrodia clavigera* is a common species in the Indo-Pacific and the Red Sea. This and a species from the Eastern Pacific are being presented in Fact Sheets from page 287 onward.

Order Velatida

This order, with its families Myxasteridae, Solasteridae, and Pterasteridea, contains primarily cold- and deep-water species. In general, they are distinct predators that do not hesitate to feed on sibling specimens, as well as on a varied selection of other invertebrates. It can be said, that they are not particularly selective feeders, which will feed on whatever is available at the time.

The family Solasteridae contains – among others species — *Crossaster papposus* and *Solaster endeca*, the so well-know sun sea stars from northern European seas. Their average adult diameter is 25 to 30 cm, i.e. they are not particularly large, yet they are aggressive predators. There are also many other species that occur in the cold seas of high northern and southern latitudes of the Atlantic- and Pacific Ocean.

There are only very few tropical species in this order. Moreover, most of them live in deep water, so that they are normally not seen. We are not aware that they have ever been imported for the aquarium trade. Also, because of their extreme food selectivity, these sea stars are not really suitably for aquarium maintenance.

Order Spinulosida

For ages there have been different understandings of the higher taxonomy of sea stars, thus giving rise to questions much as they are being discussed relative to other groups of animals. For example, the order Spinulosida is often considered in much broader and more comprehensive perspective, than depicted by us here in this particular context. So for instance ETHARDT & MOOSLEITNER (1995) – among other authors – follow the older practice of placing several families in order Valvatida, that we have included in order Spinulosida. However, exactly like HENDLER et al. (1995) we follow CLARK & DOWNEY (1962), who have delineated order Spinulosida to contain only family Echinasteridae.

Mithrodia bradleyi
Eastern Pacific Nail Sea Star

Distribution: From the Gulf of California (Sea of Cortez) to the Galápagos Islands.
Size: Reaches a diameter of 35 cm.
Reproduction: Free spawner; details about asexual reproduction by division unknown.
Food: Feeds on periphytonic overgrowth, i.e. microscopic algae and associated invertebrates, as well as on detritus. Large specimens have a tendency to "graze" on sponges, bryozoans, tunicates and possibly also on corals. Will accept most substitute foods, including dry food in tablet form.
Description: Easy to identify on the basis of its bright red colouration and the "nail" covered arms with a reticulated pattern of inter-connected plates. The nail-armored arms are a characteristic that this species shares with the somewhat less colourful *Mithrodia clavigera* (see Fact Sheet on page 288) from the Indo-Pacific, and for which it is possibly a synonym. Lives on rocky substrate, from the intertidal zone down to a depth of 50 m; it is primarily nocturnal.
Aquarium requirements: Rarely ever imported in to Europe. Hardy aquarium inhabitant, once it has adapted to aquarium conditions. Transport appears to be the largest stress factor, and it takes a long time to become fully adapted; will quickly cast off its arms when stressed. Therefore, acclimatization needs to take place very slowly and deliberately. As with other echinoderms, it must never be lifted out of water. Under no circumstances it is advisable to keep this sea star together with colonial invertebrates; however, it is a good choice for a fish aquarium or in an aquarium together with other large, fairly mobile invertebrate.

GA: +; **SE:** +/-; **CO:** –

Photo:
Mithrodia bradleyi in an aquarium. Photo: S. W. Michael

Mithrodia clavigera
Indo-Pacific Nail Sea Star

Distribution: From the Red Sea and East Africa to the central Pacific, possibly up to Hawaii.
Size: Reaches a diameter of 40 cm.
Reproduction: Free spawner; details about asexual reproduction by division unknown.
Food: Feeds on periphytonic overgrowth, i. e. microscopic algae and associated invertebrates, as well as on detritus. Large specimens have a tendency to "graze" on sponges, bryozoans, tunicates and possibly also on corals. Will accept most substitute foods, including dry food in tablet form.
Description: A nocturnal sea star that lives on rocky substrate at a depth range from 5 to 50 m. Has a network of inter-connected plates. Base colouration is beige- to brown-spotted with irregular dark brown bands around the arms. Around Hawaii there is, what is clearly, a transitional form between *Mithrodia clavigera* and *M. bradleyi* (see Fact Sheet on page 287); however, this could also be a variety of one or the other species, or all three forms could belong to *M. clavigera*.
Aquarium requirements: A rather frequently imported species, which is hardy once it has adapted to aquarium conditions. Transport appears to be the largest stress factor, and it takes a long time to become fully adapted; will quickly cast off its arms when stressed. Therefore, acclimatization needs to take place very slowly and deliberately. As with other echinoderms, it must never be lifted out of water. Under no circumstances is it advisable to keep this sea star together with colonial invertebrates; however, it is a good choice for a fish aquarium or in an aquarium together with other large, fairly mobile invertebrate.

GA: +; **SE:** +/-; **CO:** –

Photos:
Mithrodia clavigera in Papua-New Guinea: habitat- (top) and close-up view of the aboral side of the body disc (bottom). Photos: P. Schupp

Thromidia catalai
Brown-tipped Nail Sea Star

Distribution: From the Indo-Australian Archipelago northward to Japan and eastward to Hawaii.
Size: Reaches a diameter of 60 cm.
Reproduction: Free spawner; no information available about asexual reproduction by division.
Food: Feeds on periphytonic overgrowth, i.e. microscopic algae and associated invertebrates, as well as on detritus. Large specimens have a tendency to "graze" on sponges, bryozoans, tunicates and possibly also on corals. Will accept most substitute foods, including dry food in tablet form.
Description: Very impressive-looking sea star, especially in view of its maximum size. Easy to identify by its uniformly whitish-beige or pink-creamy colouration, together with reddish-brown or orange-coloured tips on the arms. In specimens from around Hawaii the arms are completely orange-brown, and only the body disk is whitish. In spite of its size and conspicuous colouration was only scientifically described as late as 1977. Lives generally on rocky substrate, but occasionally it can also been seen on sand bottom, at depths from 10 to 45 m; it is a diurnal sea star.
Aquarium requirements: Possibly a durable aquarium inhabitant; however, we have had no experience at all with this species. Similarly the relevant aquarium literature is also devoid of first-hand accounts with it. Under no circumstances should this sea star be kept together with colonial invertebrates; however, it could be a good choice for a fish aquarium or for an aquarium with other large, highly mobile invertebrates.

GA: 0; **SE**: 0; **CO**: –

Photo:
Thromidia catalai in Papua-New Guinea. Photo: P. Schupp

The European Sun Sea Star, *Solaster endeca* is a typical representative of order Velatida, which contains mainly cold- and deepwater species. Those few tropical species also carried in this order are irrelevant for the aquarium hobby.
Photo from the coast of Norway: E. Svensen

Family Echinasteridae

This family is probably best known for its genus *Henricia*. In particular, the Arctic species complex of this genus is one of the most frequently discussed, systematic problems among the Echinodermata, and – no doubt – as such it has already given many population researchers quite a few grey hairs. *Henricia* is also represented in the southern hemisphere, for instance with the decorative *Henricia ornata* that occurs in the subtropical waters around southern Australia; however, species that live on coral reefs are not known. The Mosaic Sea Star, *Plecaster decanus* is another unusually attractive representative of this family that occurs around

***Plecaster decanus* is an attractive representative of family Echinasteridae, from the subtropical- and temperate waters around the southern half of Australia.**
Photo: M. Cufer

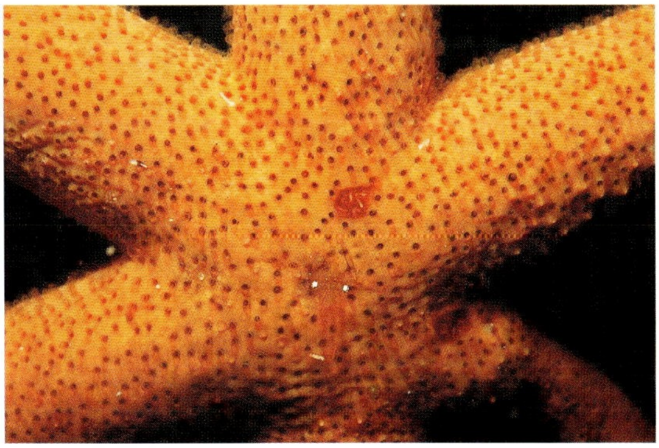

Left: The Red-spined Sea Star, *Echinaster echinophorus* (seen here in an aquarium) is occasionally imported for the aquarium hobby. It is a hardy aquarium inhabitant. Although it feeds usually on detritus and microscopically small invertebrates, it can also attack large sessile invertebrates in an aquarium. Right: Close-up photo of an *Echinaster* species that shows the characteristic, large skin pore (photo: J. Cairns-Michael). Please note the small commensal shrimp (genus *Periclimenes*) in the lower centre of this photo.

southern Australia, Tasmania and New Zealand.

Considering tropical seas and their coral reefs, then *Echinaster* is the most important genus in this order. It occurs in the Indo-Pacific as well as in the tropical Atlantic. This genus includes several very beautiful species, which are regularly offered by the trade for aquarium maintenance.

Large, distinct pores that give a rather characteristic appearance to these sea stars cover the skin of *Echinaster* species. Some species are also very spiny, where the arms are covered with strong, conical and thorn-like (1.5 to 3 mm long) spines. As a rule, these spines are arranged in rows along the upper side of the arms. Most of the species are small to medium size sea stars with maximal diameters from 7 to 30 cm. Many of them are coloured red, orange or yellowish. Overall, *Echinaster* species are harmless detritus feeders or they feed on tiny organisms and the organic layer over the substrate. Therefore, they are generally quite suitable for aquarium maintenance; however, there are a few exceptions, especially as far as the large species are concerned.

According to HENDLER et al. (1995), the Caribbean region is inhabited by *Echinaster echinophoirus*, *E. graminicola*, *E. guyanensis*, *E. paucispinis*, *E. sentus*, *E. serpentarius* and *E. spinulosus*. For a distribution in the Pacific, CLARK & ROWE (1971) list the coral reef species *E. callosus*, *E. luzonicus*, *E. purpureus*, *E. superbus* and *E. versicolor*. A few of these species are presented in Fact Sheets from page 292 onward.

Order Forcipulatida

In part this order contains some rather extreme species, of which only very few are of interest to the coral reef aquarium hobby. Nevertheless, we would like to devote some space to them to underline the enormous variability among the Echinodermata.

With 12 to 35 arms, most Forcipulatida are extremely multi-armed. An exception that confirms the rule can be found in family Asteriidae, where most species conform to the standard pattern with five arms. Beyond that, the asteriids contain many of the common and for Europeans so

***Henricia ornata* in New South Wales, Australia.** **Photo: K. Kates**

Echinaster callosus
Lumpy Sea Star

Distribution: Occurs throughout the Western Indo-Pacific, from East Africa to the Indo-Australian Archipelago.
Size: Reaches a diameter of 26 cm.
Reproduction: Free spawner; asexual reproduction unknown.
Food: Feeds primarily on detritus and microscopic invertebrates, possibly also on algae. Will also attack large invertebrates. Often does not accept substitute foods.
Description: Can be recognized by characteristic rows of round tubercles along the arms. The tubercles are fused and form ring-like bands around the arms, from about the outer half of the arms and up to their tips. Base colouration is variable and obviously depends on the habitat occupied by particular specimens. According to GUILLE et al. (1986), specimens that live on soft or sandy substrate are yellowish-beige with brown arm rings, while those that live among corals are coloured bright yellow and have reddish arm rings. Occurs at depths from 5 to 20 m.
Aquarium requirements: An attractive aquarium inhabitant that is fairly durable. Must be kept in an aquarium decorated with live rocks, and with a rich micro-fauna and –flora, in order to assure an adequate food supply for this sea star. Cannot be trusted in respect to colonial invertebrates, such as sponges and corals.

GA: +/-; **SE:** +/-; **CO:** –

Photo:
Echinaster callosus in the Philippines. Photo: P. Schupp

Echinaster echinophorus
Thorny Sea Star

Distribution: Caribbean Region; from the Florida Keys southward to Brazil; sometimes rather frequent in it southern parts of the distribution area.
Size: Reaches a diameter of 7 cm.
Reproduction: Free spawner; asexual reproduction unknown.
Food: Feeds primarily on detritus and microscopic invertebrates, possibly also on algae. Will often not accept substitute foods.
Description: A magnificently coloured, small sea star. Can be identified by its bright red or purplish-red colouration, combined with rows of thick, thorny spines. Occurs mainly in shallow water, but it has also been found at a depth of 55 m. Lives in the proximity of coral rubble and rocks, and it is occasionally seen in mangrove areas.

Aquarium requirements: A beautiful aquarium inhabitant, which is fairly hardy once it has adapted to aquarium conditions. The species is commonly seen in the trade. It is absolutely essential to keep this sea star in an aquarium decorated with live rocks and that contains a rich micro-fauna and –flora, in order to meet its dietary requirements. Does not appear to present any problems in the company of corals or other larger invertebrates, however, it cannot be totally trusted in respect to encrusting sponges and bryozoans.

GA: +/-; **SE:** +/-; **CO:** +/-

Photo:
Echinaster echinophorus in an aquarium.
Photo: S. W. Michael

Echinaster luzonicus
Luzon Sea Star

Distribution: Eastern part of the Indo-Pacific; from the Indo-Australian Archipelago northward to Japan and eastward to the Society Islands.
Size: Reaches a diameter of 22 cm.
Reproduction: Free spawner; asexual reproduction by division takes place.
Food: Feeds primarily on detritus and microscopic invertebrates, possibly also on algae. Often will no accept substitute foods.
Description: Frequently occurring sea star, on sandy as well as hard substrate, at depths from 3 to 30 m. With highly variable colouration that can range from yellow to brown and red. Always with distinctly visible pores in the skin. Often the host for commensal ctenophore *Ctenoplana astericola* that lives on its aboral side.

Aquarium requirements: A beautiful aquarium inhabitant, which is fairly hardy once it has adapted to aquarium conditions. It is absolutely essential to keep this sea star in an aquarium decorated with live rocks that contains a rich micro-fauna and –flora, in order to meet its dietary requirements. Does not appear to harm corals or other larger invertebrates, however, it cannot be totally trusted in respect to encrusting sponges and bryozoans.

GA: +/-; **SE:** +/-; **CO:** +/-

Photo:
Echinaster luzonicus with commensal ctenophores, in Indonesia. Photo: S. W. Michael

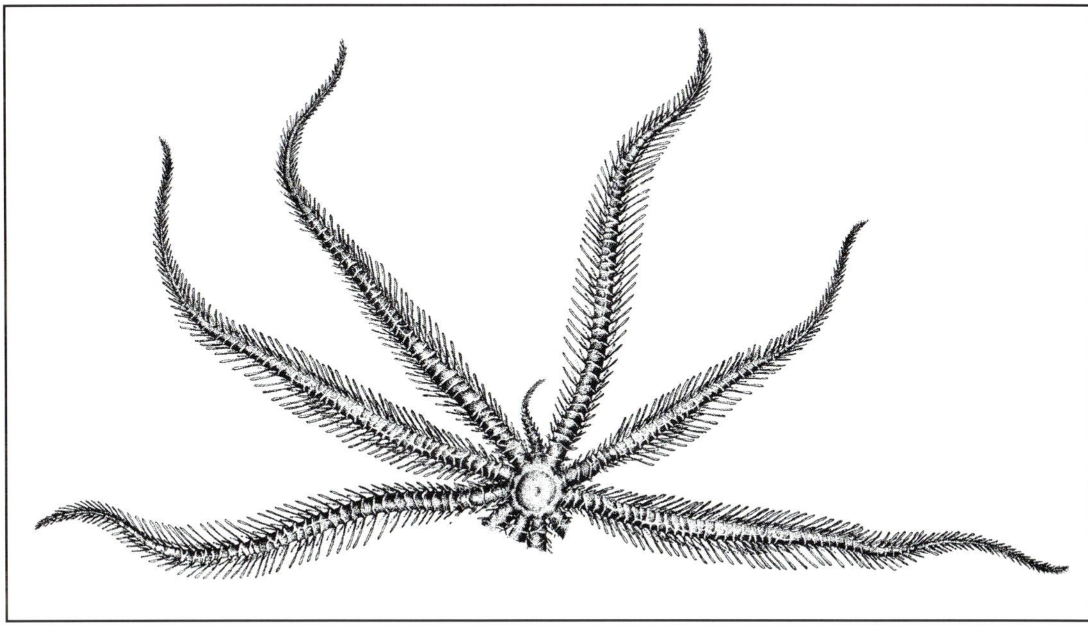

Deep-sea Sea Stars of order Brisingioda (here a *Brisinga* species) are suspension feeders and can easily be mistaken for serpent stars of class Ophiuroidea. **Drawing after MARSHALL (1979)**

well known sea stars, such as the common sea star, *Asterias rubens*, which is found from Iceland and the White Sea along the coasts of the eastern Atlantic to Senegal in western Africa, but missing in the Mediterranean. After an intense blooming of the planktonic algae *Chrysocromulina polylepis* along the coast of southern Norway back in 1988 – among other causing a number of salmon in fish farms to die – the survival rate of juvenile *Asterians rubens* seem to have increased dramtically. In some places the bottom was literally covered with juvenile common sea stars, showing how numerous certain echinoderms can be found at some locations. Many members of this family are characterised by their considerable size. The largest species of all is the sunflower star, *Pycnopodia helianthoides*, form the Pacific coast of North America. It can obtain a diameter of 90 cm. For the purpose of locomotion, such a large sea star needs to coordinate more than 15,000 tube feet!

Coscinasterias tenuispina is a species from the Mediterranean Sea, which also occurs around Bermuda; however, the latter locality is probably the result of human intervention (deliberate or inadvertent introduction). The reason why this species is so interesting is the fact that it reproduces asexually by casting off arms that develop into somites, that is, in the same manner as described for *Linckia* species (see page 257).

All species in family Labidiasteridae are multiple-armed (12 – 20) and live at moderate depths of the Atlantic and Pacific. With the aid of their arms and pedicellaria, many are capable of capturing rather large crustaceans.

Family Heliasteridae is monotypic, with the eastern Pacific *Heliaster* as the only genus. This group includes some extraordinarily strange sea stars with 20 to 35 arms that are characteristically hidden under a large oral disc. Because of this feature, together with the fact that they also have a rather spinous surface area, they can – at first glance - be easily mistaken for sea urchins.

The green or reddish brown *Heliaster microbanchius* occurs along the west coast of Central America. It reaches a maximal size of 15 cm and feeds on clams and acanthocephalans commonly found in the inter-tidal region.

Order Brisingida

This order – with its families Brisingidae and Freyellidae – contains deepwater sea stars that live at depths of more than 6,000 m. Consequently, they are never encountered on coral reefs. The only reason for mentioning them here is the fact that they are distinctly different from all other sea stars. They are suspension feeders that extend their rather long, whip-like arms (usually 12 to 18) in to the open water and then use their dense pedicellaria cover to capture plankton and other small organisms floating by. Moreover, their arms are sharply delineated from the oral disc, so that they look more like a brittle star than a regular sea star. Natural evolution always finds a way for living-beings to adapt to extreme environmental conditions.

Meaning of some anatomical terms used in conjunction with spiny-skinned animals (Echinodermata)

Aboral side
The side opposite to that of the mouth. Since the mouth in sea stars, brittle stars and sea urchins always points down towards the sea floor, the term **aboral** designates the upper side of the animal.

Ambulacral system
The water vascular system of echinoderm animals that is linked via the madreporite and by means of osmosis through the ambulacral feet (tube feet, suction feed or podia) to the surrounding seawater. It consists of the stone canal and the ring canal, that surrounds the mouth and from which five radial canals are branching out that are located under ambulacral plates.

Ambulacrum
Also referred to as ambulacral field or –row, one of the five radial double rows of calcareous plates (= ambulacralia). Its rows of pores make it possible for the ambulacral feet that extend from the radial canals, to access the environment. The closed rows of plates between the ambulacra are referred to as **interambulacrum**.

Apical system
Apical = at the apex, located at the highest point; also referred to as apical- or central plate, normally a system consisting of ten small plates (five genital plates and fine terminal plates) located at the apex of a sea urchin. **Terminal plates** are often erroneously referred to as ocular plates.

Cilia bands
Cilia-covered bands on spines and coronas of irregular sea urchins.

Corona
The exoskeleton or test (case) of a sea urchin (corona, Latin = Crown); often erroneously referred to as shell.

Fasciole
Band covered by tiny spines and cilia, extending over the corona of some irregular sea urchins. Its task is to remove foreign bodies from the surface of the corona by means of a water current created by the cilia.

Gonopores
Small openings in the genital plates, used for ejecting gametes (eggs or sperms).

Madreporite or Madrepore plate
Sieve plate that closes off the water vascular system (ambulacral system) to the outside, and facilitates the connection to the surrounding seawater. However, in sea cucumbers the madreporite is located inside the body cavity and provides the connection between the ambulacral system and the body fluid.

Oral side
Side where the mouth is located. Since the mouth in sea stars, brittle stars and sea urchins is always directed down towards the sea floor, this term also designates the underside.

Papilla
Wart-like elevation on the surface of organs; protrusion from the epidermis.

Pedicellaria
Tiny, stalked grasping tongs (clamps) on the skin of sea stars and sea urchins; often with venom glands. There are different functional types of pedicellaria; i.e. those that inject a venom, bite, grasp, clean or serve as defence-, grasping-, feeding- or cleaning organs. Several types can occur simultaneously on the surface and among the spines. Since their structure is highly differential, they are also used for species identification in many of the recent sea stars and sea urchins.

Class Ophiuroidea – Brittle Stars, Serpent Stars, Basket Stars

Brittle stars are well known to most coral reef aquarists, although these echinoderms live a rather cryptic life and so they often remain out of sight. From an aquarist's point of view, brittle stars are very useful as detritus feeders and scavengers, and as such they take on an ecologically important role.

The class Ophiuroidea (see systematic overview to the right) encompasses about 2,000 described species. This then makes it the largest class in phylum Echinodermata. The majority of the species live in areas of the deep sea, on soft bottom substrate; however, brittle stars are also quite numerous on coral reefs throughout the entire world. Indeed it appears that there is hardly a cave or crevasse not occupied by one or more brittle stars. It has been calculated for the Caribbean Sea that there is a density of brittle stars of 20 to 40 specimens per square meter, associated with coral reefs, as opposed to densities ten- to hundred-times larger on soft substrates! Consequently, brittle stars are quantitatively among the most important animals on a reef (HENDLER et al., 1995). Yet, within this context it must also be noted that many brittle stars are very small. Although the majority of brittle stars known to aquarists have oral discs with a diameter of a few centimetres and arms that are 10 to 15 cm long, there are also many species that, when fully grown, have a total diameter of not more than 2 to 3 cm. The smallest species measures barely 3 mm as an adult! On the other hand, there are also giants with a diameter of more than 1 m.

Overall, brittle stars have many conspicuous anatomical features in common with sea stars of class Asteroidea. For instance, brittle stars also have arms that radiate outward from an oral disc, with an oral- and aboral side. However, on certain points the differences between both classes are quite clear:

- In comparison to sea stars, the arms of brittle stars are extremely long and thin, and they are distinctly delineated from the oral disc.
- Brittle stars do not possess open ambulacral grooves, since lateral plates always cover them.
- The tube feet (= podia) are of less significance for locomotion; they serve primarily as sensors.
- The madreporite is located on the oral side and not – as in sea star – on the aboral side.

Systematic Overview of Class Ophiuroidea

After HENDLER et al. (1995); GUILLE et al. (1986); GEORGE & GEORGE (1979); CLARK & ROWE (1971)

Order Phrynophiurida

Family Asteronychidae

Family Asteroschematidae
Astrobrachion, Ophiocreas

Family Gorgonocephalidae
Astroboa, Astrochalcis, Astrocladus, Astroglymma, Astrophyton, Asteroporpa, Astrospartus, Gorgonocephalus, Schizostella

Family Euryalidae
Euryale

Order Ophiomyxida

Family Ophiomyxidae
Ophioblenna, Ophiomyxa

Family Ophiocanopidae

Order Ophiurida

Family Ophiuridae
Dictenophiura, Ophiolepis, Ophioplocus, Ophioteichus, Ophiotyloo, Ophiozonella, Ophiura

Family Ophiocomidae
Ophiarthrum, Ophiocoma, Ophiocomella, Ophiocomina, Ophiomastix, Ophiopsila

Family Ophionereididae
Ophionereis

Family Ophiodermatidae
Cryptopelta, Distichophis, Ophiarachna, Ophiarachnella, Ophiochaeta, Ophiochasma, Ophioclastus, Ophioconis, Ophioderma, Ophiodyscrita, Ophiopeza, Ophiopsammus, Ophiostegatus, Pectinura

Family Ophiacanthidae
Ophiacantha, Ophiothauma

Family Hemieuryalidae
Sigsbeia

Family Ophiactidae
Hemipholis, Ophiactis, Ophiodaphne, Ophioasphaera

Family Amphiuridae
Amphilycus, Amphiodia, Amphioplus, Amphipholis, Amphiura, Dougaloplus, Ophiocentrus, Ophiocnida, Ophionephthys, Ophiophragmus, Ophiostigma, Paracrocnida

Family Ophiotrichidae
Gymnolophus, Lissophiothrix, Ophiocnemis, Ophiogymna, Ophiolophus, Ophiomaza, Ophiophthirius, Ophiopsammium, Ophiopteron, Ophiothela, Ophiothrix, Macrophiothrix, Placophiothrix.

The pentamerous symmetry – so typical for Echinodermata – is generally also quite distinct in brittle stars. Most species have five arms, and there are only a few species with six or seven arms, or – alternately – with many arms that have branched in to a strong, thread-like network. The latter has developed in to an extreme in family Gorgonocephalidae, commonly referred to as basket stars. These are characterized by the presence of particularly thin arms, which have developed a vast, basket-like complex of branches.

The arms of brittle stars consist of a row of easily movable segments or "joints" that have a remote similarity to the vertebral column in humans. Each segment is connected with muscles and connective tissue. Below the oral disc there are modified arm shields that form triangular maxillary plates surrounding the mouth. Protective plates that carry lateral spines, used for food intake and defence, cover the arms.

View on to the aboral (above) and oral side (below) of *Ophiarachna incrassata*. **Photos: H. Kastl**

At the tip of each arm there is a final, round terminal plate, which also represents the point of growth. Consequently, the chronologically youngest arm segments are at the outer end of each arm. On the underside of the arms are tube feet (= podia) that are connected to the water vascular system and controlled by it. In contrast to most sea stars, suction cups are absent from the feet since they act primarily as tactile organs. In some species the feet are also used for transporting food to the mouth.

By comparison to other echinoderms, the body cavity (coelom) in brittle stars is substantially reduced. It accommodates mainly the stomach, the gonads and the bursae (see further below). The circulatory- and nervous system is rather simple, and specific sensory organs are absent. Chemo- and photoreceptors react to odorous- and gustatory substances, as well as to light and dark.

Respiration takes place by means of ten internal sacs, the so-called bursae, i.e. one bursa on each side of an arm. They are formed at small indentations (macula) on the oral underside of the disc. This means that the bursae are open to the outside through clefts extending along the underside (oral side) of the arms. The movement of cilia and/or pumping movements through contraction of the oral disc musculature maintain the water exchange inside the bursae so that effective respiration is taking place. In many species, the bursae also play a role in reproduction (see page 302), and in part they also facilitate the elimination of waste products.

The relevant German literature refers to most members of the Ophiuroidea as serpent stars, while "brittle stars" as the term commonly found for these echinoderms in English-language publications. The English designation makes reference to the rather fragile arms of these animals. As part of their defensive strategy, many species cast off part of the

Fundamentally brittle stars are built as sea stars, but are distinguished from the latter most notably by their extremely long arms, and other characteristics. Shown here is an unidentified species of family Ophiuridae.

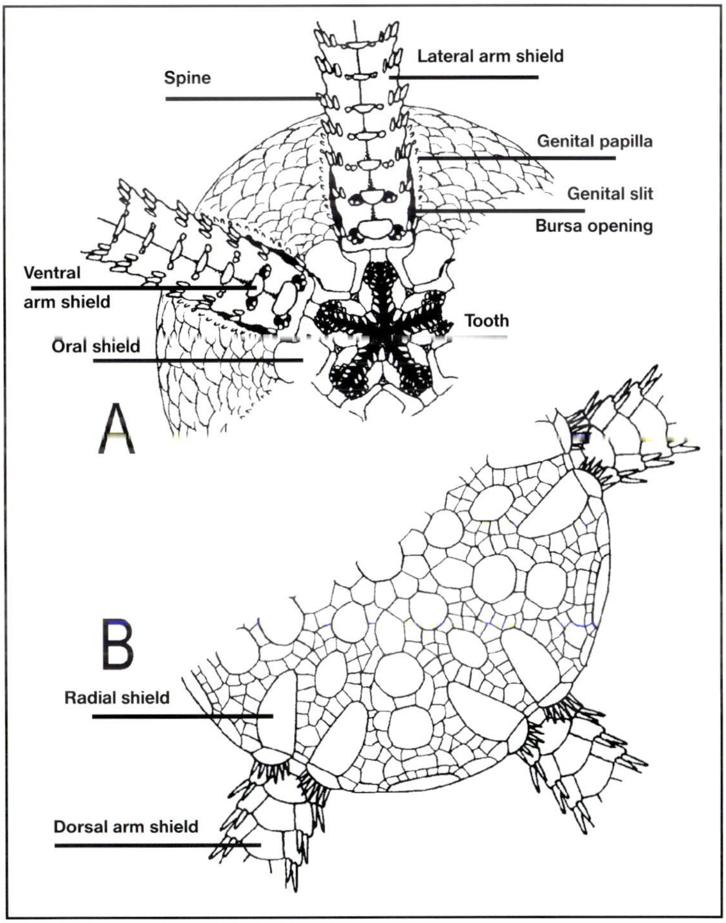

Schematic body structure of a brittle star: A - oral side; B - aboral side. Drawing after CLARK & ROWE (1971)

Brittle stars are extraordinarily mobile and able to slip in to even the narrowest gap. Here we have an unidentified species of family Ophitrichidae hiding inside a sponge.

arms when in danger. As in many other echinoderms, the regenerative capability is so well developed, that lost arms are quickly replaced. There are even extreme examples of self-mutilation, whereby certain species when attacked by a predator, even sacrifice their oral disc (including the internal organs), which is then completely regenerated within the course of two to eight weeks (HENDLER et al., 1995). Other defensive means are based on toxic substances, unpleasant taste, bioluminescence and rapid escape.

The tube feet play only a subservient role in locomotion, or none at all. And yet, brittle stars are among the most mobile of all echinoderms, and because of their extremely flexible arms they can develop considerable speeds. The oral disc is lifted off the substrate, one or two arms are directed forward while one or two are pointing to the back, and the two remaining, lateral arms execute rapid circular movements. This results in the animal being pushed forward in a jerking fashion, by what looks like rowing- or propeller-like movements. The spines maintain the direction of the movement, but there is no evidence that brittle stars give preference to a particular side or arm as their "anterior part". In fact, they can move equally fast in all directions. In addition, many species use their arms to pull themselves forward, by embracing or holding on to suitable objects. The arms can also be used as holdfasts, as for instance, by species that live symbiotically on gorgonians.

Various forms of symbiosis are common among brittle stars. In contrast to many other echinoderms that often serve as host for other species (e.g. crustaceans, molluscs and worms), brittle stars themselves often live on larger species. Quite often brittle stars are found on or in sponges, hydroids, stony corals, soft corals, black corals and gorgonians. Consequently, when these animals are purchased, brittle stars – as stowaways – are often introduced in to our aquariums at the same time.

Among the animals used as host by brittle stars include worms, such as the polynoid bristle worm, *Hololepidella* sp. Brittle stars provide their host with mucus and extract food for them-

Many brittle stars live as commensals or parasites on other animals. Left: Two different colour varieties of *Astrobrachion constrictum* on a black coral (photo: M. Cufer). Right: Tiny *Ophiothela* species on an *Echinogorgia* Gorgonian coral.

selves. There are also a few crustaceans, such as copepods, and less frequently shrimp or gastropods that – as host – can derive benefits from brittle stars.

Brittle stars are uncomfortable in bright light. Therefore, they tend to hide in dark, shaded places, pressing themselves – in a characteristic manner – firmly against the substrate. Eyes are absent in brittle stars; yet photosensitive cells are located in the skeletal plates. Only after the pressure from predators lessens at dusk or nightfall, that is the time brittle stars emerge from their hiding place in order to feed.

Mode of feeding and food selection also varies considerably among brittle stars. Some substances are taken in directly via the skin, by diffusion. However, brittle stars feed mainly on various types of large and small food particles, as well as on other organisms. Some species can wrap their arms around their prey and can even capture mobile organisms, such as small fish, in this manner. Yet it is more common for brittle stars to feed on microscopically small organisms or they are scavengers and detritus feeders.

Suspension feeders filter water, whereby they stir up the surrounding water so that plankton and detritus become attached to mucus strands among the spines on each arm. Particles that have become attached are transported to the mouth by cilia movements and action of the tube feet (= podia). Benthic feeders pick up medium size particle directly with their tube feet. Larger particles, such as dead organisms, are embraced with the arms and are then rolled to the mouth. Some species graze along food sources whereby food is taken in directly with the mouth. Moreover, it is not unusual to find different types of feeding utilized by the same species.

The digestive system is rather simple. An oesophagus leads from the mouth in to a large sac-like stomach that makes the largest volume of the oral disc.

Brittle stars can be extremely numerous on coral reefs; in fact, practically they occupy every cave, crack or crevasse. They play a significant role in the ecosystem of a coral reef as well as in a coral reef aquarium.

Photo: D. Knop

A close inspection of a particular reef biotope quickly reveals that brittle stars live almost everywhere.

Ophiothrix fragilis **lives in the Mediterranean Sea where it is an "open spawner" releasing the gametes in the surrounding water for fertilization outside the body. Here a female specimen is expelling clouds of eggs.**

Photo: H. Steiner

There is no intestine and anal opening. Therefore, the mouth is used for the elimination of indigestible, leftover food.

MORIN (1988) has described how *Ophiarachna incrassata*, a species that is often kept in aquariums (see Fact Sheet on page 317) is able to deceive small fishes that look for a hiding place under its oral disc. For that, the arms are slung around the fish like a cage, so that it can easily be eaten (see drawing on page 303). Many aquarists have reported that this brittle star also captures aquarium fishes. Nevertheless, this does not appear to be a major problem (?), when compared to the number of specimens kept in aquariums worldwide. Similarly, large problems are also being reported in association with brittle star species that feed on other aquarium animals such as corals, bristle worms, sponges, etc. Apart from those few food specialists for whom these particular animals are the principal food, many other aquarium inhabitants can also become their prey if they are sufficiently hungry. On the other hand, if a steady, adequate food supply is assured, for instance by giving deep-frozen foods or dry food, this sort of problem is vastly minimized. With that in mind, aquarists should not be deterred from gathering experiences with these fascinating echinoderms.

The majority of brittle stars has separate sexes but generally they do not exhibit sexual dimorphism. However, there are also hermaphroditic species. The gonads consist of two small sacs that are located adjacent to the inner wall of a bursa, and one or two gonads are contained in one bursa. Once the gonads are mature, they empty out inside the respective bursa. Some species release their sex cells in to open water, with the aid of respiratory movements by the bursa. Free-spawning species disseminate their eggs and sperm in this manner in to the open water where fertilization and subsequent larval development take place.

Interestingly though, brood care of the eggs and larvae are also a fairly common method of reproduction, especially in hermaphroditic species; however, it also occurs in a few species with separate sexes. When brood care is practiced, there is internal fertilization, whereby eggs and larvae develop inside the bursa of the maternal animal. Usually there are only a few eggs or larvae, respectively, in one bursa. The larvae remain inside the bursa until they are sufficiently large to leave the bursa under their own power.

In most brood-caring species there is no contact between mother and progeny. In other words, the entire food supply comes from the egg yolk (ovoviviparity). True viviparity (i.e. the embryo is attached to the wall of the bursa, receiving a food supply in a placenta-like fashion from the female) takes place in very species; one example of this is the cosmopolitan *Amphipholis squamata* of family Amphiuridae.

The embryology of free-spawning species is by-and-large identical to that of sea stars. But the pelagic larva, referred to as ophiopluteus larva (see photo on page 218), is typical for brittle stars. It has four pairs of longish, cilia-covered arms that have been structurally reinforced with calcareous rods. Mistaking it for the brachiolaria larva of sea stars (see drawing on page 284) is virtually impossible. Metamorphosis of the ophiopluteus larva takes place during the continued pelagic stage (i.e. it does not involve a sessile stage). The small brittle star sinks to the bottom and there immediately commence a life like that of its parents. The development periods vary from two to seven weeks. In brood-caring species, development in to benthicfree-swimming juveniles specimens usually take several months. However, details of the embryology of brood-caring species are still largely unknown.

Asexual reproduction by means of division also exists among brittle stars. Certain species divide right across the oral disc,

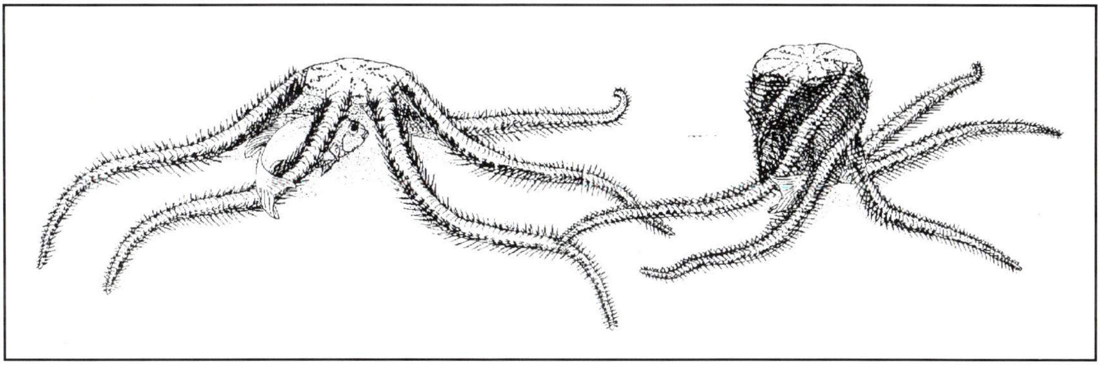

Some brittle stars are able to supplement their menu by capturing fish. It has been our experience that this happens in an aquarium whenever such brittle stars do not receive sufficient food. However, this type of behaviour has also been observed in the sea. About the drawing above: MORIN (1988) described how large *Ophiarachna incrassata* set a trap for small fishes. They create a hiding place with their own body for fish to seek refuge in. Once a fish has entered such "trap" they surround it with their arms thus forming a "barbed wire enclosure". Photos below: We have witnessed a fish being caught by *O. incrassata* in an aquarium, but not in such a clever way. Instead, the brittle star crawled over the unwary fish and then immediately sucked the fish's tail in to its mouthy. In this case the prey was a Pink-and-Blue-Spotted Goby, *Cryptocentrus leptocephalus*, which was probably too large to be killed outright; however, in order to avoid further harm to the fish we decided to release it.

whereby each half then regenerates the missing other half. Once such brittle stars have reached a suitable size, the process of division can repeat itself.

In an aquarium there can be sexual as well asexual reproduction. At protected sites in the aquarium there are quite often large populations of small species (e.g. see Volume 1, page 190), but large species have also bred under captive conditions. For instance, aquarium reproduction of the decorative, blue-green *Ophiarachna incrassata* has been repeatedly reported, by MOUGIN (1981), KASTL (1993) and GIBBS (pers. comm.; see page 315 in this book) and others.

In an aquarium, brittle stars are generally interesting and extremely useful inhabitants. This applies especially to the many small species that often come more or less accidentally in to our aquariums by inhabiting live rocks or that live in sand. We have kept more than 50 specimens of different species in an 850-litre tank without problems. In the sea as well as in aquariums, brittle stars are among the most important detritus feeders and organisms that facilitate the degradation of organic material. They work through the substrate, searching for dead organisms and other leftover organic material that serve as food for them. In this process, they tend to aerate the sand and keep the pores of rocks open, so that organic decay cannot develop there.

Many large and conspicuous species are commonly available in the aquarium trade. By-and-large, they can all be recommended for a coral reef aquarium. In a modern, biologically diverse and technically well-equipped aquarium they are hardy and easy to keep. In fact, under optimal conditions they can live for many years and even breed in an aquarium. Brittle stars are rather active, by comparison to other echinoderms; however, these activities take place primarily at night. As long as there is (natural or artificial) light, they tend to remain in hiding under rocks or behind items of the aquarium decoration. But gradually, they also emerge during the day during the feeding. Some brittle star species will readily take Mysis, by wrapping their arms around the shrimp. Other species prefer smaller food particles. Brittle stars will react immediately by extending their spine-covered arms out from various hiding places, when they are being fed with plankton and other free-swimming food.

We firmly believe that these detritus feeders belong in to every coral reef aquarium; in fact, there should also be a substantial number of them. The common opinion among aquarists, that a few of these echinoderms should be acquired, yet the bulk of available funds are better spend on colourful corals and reef fishes, is regrettable. Even if some of the brittle stars are only grey and appear generally boring (but there are also some magnificently coloured species) they are – because of their food preferences – genuine cleaning specialists in an aquarium. Consequently, they should be numerous as aquarium inhabitants – just as they are on coral reefs. With an average diameter of 15 to 20 cm (including the arms) they do not take up much space. In fact, a ratio of one brittle star per 20 l aquarium volume is quite acceptable, which would add up to 25 specimens in a 500 litre tank. We have kept even larger numbers with good success.

Although a few species can display some predatory tendencies by damaging or even killing other aquarium inhabitants, yet this problem – should it arise – can be substantially minimized by an ample food supply. We should remember that similar problems could also occur among other aquarium inhabitants, if they are not fed properly, so they need to supplement their normal food requirement from other sources. The by far most useful habit of brittle stars in relations with the modern coral reef aquarium is their ability to remove waste products, food remains and detritus from the substratum. By keeping a number of brittle stars in the tank we ensure that much organic remains are removed from the difficult accessible spaces below rocks and behind the decoration. Brittle stars should be kept in every reef aquarium.

The Greedy Brittle Star, *Astrobrachion constrictum* (seen here in New South Wales, Australia) is probably one of the most attractive species of family Asteroschematidae. It lives at medium depths in the Indo-Pacific, where it occurs exclusively on the branches of black corals. It is assumed that the brittle star is totally dependent upon the protein-rich mucus of its host and the plankton it contains. Colouration can be highly variable: from red, orange, black to white or yellow. Photo: M. Cufer

Order Phrynophiurida

Family Asteroschematidae

The two genera *Astrobrachion* and *Ophiocreas* of this family represent some rather primitive brittle stars. They have smooth arms, covered by a thick, soft skin. Beyond that, they are tightly linked to colonial cnidarians, specifically to black corals of order Anthipatharia, where the brittle stars wrap their long, flexible arms closely around the coral branches.

Most likely, these brittle stars are totally dependent upon the protein-rich mucus layer of their coral host and the adhering planktonic organisms, as their main food source. In contrast to the basket stars of family Gorgonocephalidae that use the host merely as a site for better access to capture plankton, the Asteroschematidae are considered to be pure parasites. Up to 40 or 50 of these brittle stars can readily cling to the branches of a large black coral. Because of their particular mode of life and specific food requirements, these brittle stars may only be of rather limited suitability for aquarium maintenance.

Family Gorgonocephalidae - Basket Stars

If one encounters basket stars during a night dive on a coral reef, they seem like one of the most impressive creations found in the sea. They have derived their scientific family name from Greek mythology: Gorgons were the miss-shaped daughters of sea god Phorky and his wife Keto. Gorgons were alleged to have been fear-inducing, dragon-like creatures with golden scales and snakes as hair …that would turn anyone in to stone that looked at them anyway, so the legend has it! When the heavily branched arms are spread out to maximum extent like a parabolic fan (antenna), basket stars are indeed reminiscent of the heads of gorgons. Yet, a comparison to a sieve- or basket-like structure seems more realistic, which then also explains the common English name for these ophiuroids as basket stars. With their arms fully spread out, the largest species can reach a diameter of up to 1.5 m. In their food gathering methodology for capturing plankton they are more like feather stars than brittle- or serpent stars.

During the day – or when a diver shines his light on them – basket stars contract in to an mass tangle of arms and then look more like a pile of decaying ropes than the magnificent animals they were a little while earlier. During the day, many species will leave their host and crawl back in to hiding places; however, with the onset of darkness they will return to their original host.

Basket stars are often associated with gorgonians but also with various species of stony- and soft corals, as well as with fire corals and sponges. As far as is known, the host serves only as favourite food-gathering site and as such it does not play any role in food gathering. With few exceptions, the food of basket stars consists of various planktonic organisms, specifically of copepods and other crustaceans. But it is also known that some of the large species occasionally capture juvenile fish.

The Clinging Brittle Star, *Astrobrachium adhaerens* (seen here at One Tree Island, Great Barrier Reef) is a highly specialised parasite. It should really only be kept by very experienced hobbyists, because the greatest problem is to keep the host alive in an aquarium under the physical load of several parasites. 50 or more specimens of the Clinging Brittle Star can infest a single black coral. Photo: L. Newman and A. Flowers

It is assumed that most species release the gametes in the water where fertilization occurs; however, other reproductive strategies have also been reported, as well as asexual division.

Basket stars are host to a large number of symbiotic organisms, including shrimps of genera *Periclimenes* and *Synalpheus*, copepods, amphipods, molluscs, other brittle stars and bristle worms. A few small organisms, including copepods and bristle worms live on the inside of basket stars, however, most live on the outside among the arms.

There are many genera, and one of the most commonly encountered on coral reefs of the Indo-Pacific is genus *Astroboa*, represented by *Astroboa clavata*, *A. ernae*, *A. granulatus*, *A. globuliferum* and *A. nuda*. The latter is the most common and widely distributed species. It occurs from the Red Sea and East Africa to the Philippines, and southern Japan. At night this basket star usually sits on gorgonians or stony corals, at depths from 10 to

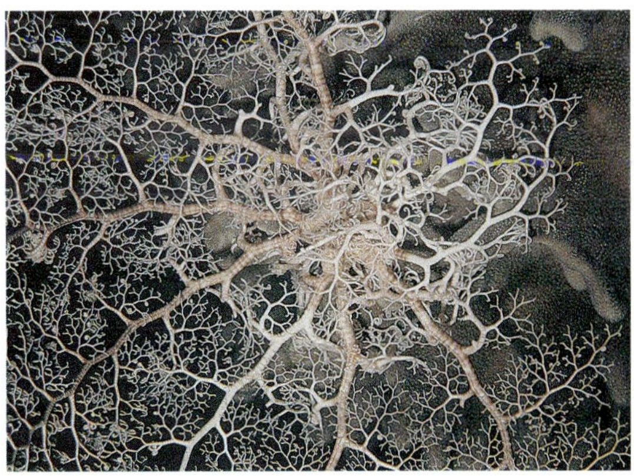

The Naked Basket Star, *Astroboa nuda* (seen here at Papua-New Guinea, at a depth of 10 m) occurs widely in the western Indo-Pacific, from the Red Sea and East Africa to the Indo-Australian Archipelago and the Philippines. This is one of the largest basket stars and often reaches a diameter of more than 1.5 m. Photos: P. Schupp

Basket Star *Astroboa ernae* at the Marmion Marine Park, Western Australia. Photo: C. Bryce

Close-up view of one arm of *Astroboa granulatus*. Photo: E. Svensen

60 m; however, it has also been found as deep as 120 m (Guille et al., 1986). With an oral disc diameter of 10 cm and a maximum arm spread of 50 cm, this is an animal that can hardly be overlooked. Other genera occurring on coral reefs of the Indo-Pacific are *Astrochalcis*, *Astroglymma*, *Astrosierra* and *Astrocladus*.

Astrophyton muricatum is the most frequent species in the Caribbean region. The largest specimens may reach an oral disc diameter of seven cm with an arm spread of 40 to 50 cm; usually though this basket star is not observed that often.

Most basket stars have a variable, brown colouration, but a few

***Astrosierra amblyconus* - seen here in New South Wales, Australia – is a basket star that lives in moderately warm and in subtropical seas, and reaches a maximal diameter of only 15 cm. It is presented here as an example of a basket star with only moderately branched arms. Photo: M. Cufer**

species are characterized by conspicuous, distinctly off set colour markings. For instance, *Conocladus australis*, found in the Indo-Pacific region, has a conspicuous, attractive pattern of black and white spots. The Caribbean *Asteroporpa annulata* is coloured dark brown and has white rings around the arms.

A few basket stars appear sporadically in the aquarium trade. They can indeed be kept in an aquarium, provide the ratio of water volume to animal size is sufficiently large. Yet, they are delicate and very susceptible to transport damages and physical injuries. Their regenerative capabilities, that may save them in the sea, appears to be – like that of feather stars – insufficiently effective under aquarium conditions. It is conceivable, that this may well be related to an inadequate diet. Although basket stars will accept various types of frozen food and plankton substitute, this is not necessarily a nutritious, balanced diet. While some aquarists have reported conditional success with basket stars, we feel compelled to advise against purchasing them; unless, of course, there are special interests in these animals, together with adequate funds for suitable aquarium support technology and appropriate husbandry care, so that they can be kept successfully. Because of their distinctive nocturnal behaviour, basket stars are probably only of limited interest to the average aquarist.

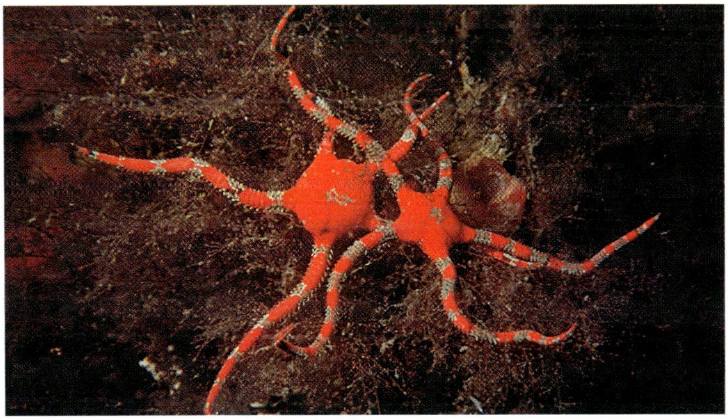

Ophiomyxa australis* - seen here in New South Wales, Australia – is a brittle star of family Ophiomyxidae that is widely distributed in subtropical and tropical areas of the Indo-Pacific. Its basic colouration is rather variable, however, it is always interrupted by red or orange and grey and white bands. It can be found in shallow as well as in deep regions of a reef, where it is active in open areas. Unfortunately, we have not yet encountered this magnificent brittle star in the aquarium trade. Photo: M. Cufer

Order Ophiomyxida

Family Ophiomyxidae

Members of this family are not too often seen neither in the wild nor in the trade. They are characterized by an oral disc with relatively short arms, covered by a soft, thick skin with a slimy surface (Ophiomyxa = slimy snake). Systematically viewed, they are more closely related to basket stars than to all other brittle- or serpent stars, and are therefore, sometimes included in order Phrynophiurida.

In those species, where the feeding behaviour has been closely investigated, it has been found that they actively graze off algae and sponges and live off the detritus gathered up with the aid of their arms. For Caribbean species it has been shown, that they are often hosts to the commensal ostracod *Pontocypria hendleri*, which attaches itself to the arms of these brittle stars (MADDOCKS, 1987).

Family Ophiomyxidae contains brood-caring- as well as open-spawning species. Possibly both types of reproductive strategies occur within the very same species (HENDLER, 1979).

CLARK & ROWE (1971) report on three Indo-Pacific *Ophiomyxa* species: *O. australis*, *O. compacta* and *O. longipeda*. From these, *A. australis* is the most common species with a geographical distribution, which extends from South- and East Africa and parts of the Red Sea (Gulf of Aqaba; VINE, 1986) to the central islands of the Pacific Ocean. This species has a very conspicuous colouration, with red-orange and greyish or whitish markings. It occurs in tropical- as well as subtropical seas; on coral reefs it is often found at depths from 2 to 50 m. The oral disc can reach a diameter of about 2.5 cm, with an arm length of about 10 cm among the largest specimens.

Two species are known from the Caribbean region: *Ophiomyxa flaccida* and *O. antillensis*. The former is highly variable in its colouration, from mainly green, yellow, orange, to red or brown. The arms usually have lighter-coloured cross bands, so that the reddish variants partially resemble the Indo-Pacific *O. australis*. Even more interesting yet is that many juveniles show considerable resemblance to *Ophioderma rubricundum* (family Ophiodermatidae), which inhabits the same habitat. To what extend this similarity is based on camouflage, mimicry or mere chance is not known (HENDLER et al., 1995). *Ophiomyxa flaccida* occurs throughout the zoogeographical region of the Caribbean, from the Bermudas and over most parts of the Caribbean Sea, southward to Brazil and eastward to the St. Paul rocks in the Atlantic. This species is found in all coral reef zones, as well as in sea grass beds and on many variable substrates. The oral disc can have a diameter of 2.5 cm, with an arm length of up to 13 cm.

We have not yet seen Ophyomydidae species in the aquarium trade, however, we assume that they are occasionally available. On the basis of their mode of life and food choices one could postulate that their aquarium care is fairly easy.

Order Ophiurida

Family Ophiuridae

This is the family of brittle stars most commonly represented in the aquarium trade. In fact, primarily species of genus *Ophiolepis* are the ones that are being imported, but *Ophioplocus*- and *Ophiura* species are also regularly available.

Most species in this family are more or less tied to fine-grained sand bottom or other soft substrate, or to small, sediment-filled cavities in the reef structure. Consequently, it is important for the well being of these brittle stars that they are given an aquarium with a small- grained sand bottom. In an aquarium they also tend to frequent open sandy areas as well as shaded sites, such as under overhanging structures or in caves. In contrast to many other brittle stars, the species of this family rarely hide under rocks. In fact, many of them can be seen out in the open throughout the day, and that even on coral reefs. However, they are rarely ever active prior to the onset of darkness.

The arms appear to be more rigid and less brittle than those of most other brittle stars, but their regenerative capabilities also seem to be less developed than those in most of their relatives. When handled carefully during transit they tend to survive transport rather well. Once the have become established in an aquarium they are hardy with substantial longevities.

Most species appear to be sediment feeders. Their diet includes algae, faecal matter as well as various types of organic detritus. Under aquarium conditions they readily accept most substitute food offered, including fresh-frozen plankton and dry food. Many species practice brood care, but there are also species that release the gametes for fertilization in the open water.

There are at least seven genera represented in the Indo-Pacific: *Dictenophiura*, *Ophiolepis*, *Ophioplocus*, *Ophioteichus*, *Ophiotylos*, *Ophiozonella* and *Ophiura*. Singularly the largest and most important genus is *Ophiolepis*, with seven species. From an aquarist's perspective *Ophiolepis superba* (see Fact Sheet on page 309) with its brown colour pattern, has a special position. It is also the most frequently imported species. With its beige-coloured arms, that have a broad dark brown band, and an oral disc with a dark stellar pattern, it can hardly be confused with other brittle stars. Other species observed in the aquarium trade are various *Ophiolepis* spp. and *Ophioplocus imbricatus*.

HENDLER et al. (1995) describe four *Ophiolepis* species from the Caribbean region: *O. elegans*, *O. impressa*, *O. gemma* and *O. paucispina*. From among these four, *O. elegans* is most common available in the aquarium trade. It occurs over the largest part of the Caribbean Sea, and from the Bahamas to French Guinea. The vertical distribution of this brittle star extends from the inter-tidal zone down to a depth of 90 m. It can be very numerous in certain areas. Using a trawl net in Florida Bay, MANNIG (cited by HENDLER et al., 1995) collected more than 1,000 specimens within 5 minutes! The arms and oral disc have a decorative zebra (striped-) pattern, usually in grey and creamy white. With an oral disc diameter of up to

Two frequently imported colour varieties of an *Ophioplocus* species, possibly both of them being *O. imbricatus*, in an aquarium. This brittle star is high variable in its colouration, however, nearly always with a uniformly coloured, brownish body disc. The arms have the same brownish colour, however, 8 to 16 darker bands mark them. It is a typical sediment feeder that feeds principally on organic detritus, faecal matter, dead organisms, etc. In an aquarium it has proven to be an excellent, hardy as well as durable inhabitant that willingly accepts all types of substitute food.

Ophiolepis superba
Marbled Brittle Star

Distribution: Occurs widely from the Red Sea and East Africa eastward to Indonesia, Papua New-Guinea, Australia and New Caledonia.
Size: Central body disc to 3.5 cm in diameter, 12 cm maximum length of arm, but generally shorter.
Reproduction: Possibly ovoviviparous, however, details are unknown.
Food: Sediment feeder; searches principally for food among organic detritus, faecal matter and other organic remnants. Will accept most substitute foods; excellent scavenger.
Description: One of the most attractive and conspicuous brittle stars on the Indo-Pacific. Easy to recognize by its impressive beige to light brown base colouration, together with dark brown or black markings on body disc and arms. Normally encountered under boulders and among coral rubble at depths from 0 to 50 m.

Aquarium requirements: Frequently available from the aquarium trade as imports from various geographical areas. Rapidly growing and durable aquarium inhabitant in properly functioning aquariums with a rich micro-fauna and -flora. We have kept some specimens for more than ten years. Has a tendency to hide during the day under overhanging objects or in caves with a sand bottom, but can be conditioned to emerge during the day when food is offered.

GA: +; **SE:** +; **CO:** +

Photo:
Ophiolepis superba in an aquarium.

2 cm and an arm length up to 6 cm, this brittle star is only of medium size.

Ophiolepis impressa is of about the same size as *O. elegans*, and also has grey-brown and creamy-striped arms. The oral disc has a creamy white marbled pattern with variable shades of brown. *O. gemma* and *O. paucispina* are "dwarf species" with oral discs that have maximal diameters of 6 to 7 mm. The arms are rarely longer than 3 cm. These small brittle stars often live among algae, e.g. *Halimeda*.

Family Ophiocomidae

This family is represented in the Indo-Pacific with about 30 species and in the Caribbean region with 8 species. The Indo-Pacific species are scattered over the genera *Ophiarthrum*, *Ophiocoma*, *Ophiocomella*, *Ophiomastix* and *Ophiopsila*, and those of the Caribbean region over four *Ophiocoma* species, three *Ophiopsdila* species and one *Ophiocomella* species. These are generally rather large brittle stars that often have oral disc diameters of two cm and more. However, genus *Ophiocomella* contains only smaller species that have oral discs with diameters of 5 to 6 mm and arms that are only up to two cm long. The arms of all species are normally covered with numerous erect spines that are approximately as long or longer than the arm diameter.

On coral reefs the members of this family occur on hard and soft substrate in all zones, including sand patches and sea grass- and mangroves zones. Many species hide among rocks and corals, crushed rocks or bury themselves in sand. They occur in deep as well as in shallow water, including very shallow inter-tidal zones. From an ecological point of view, these brittle stars form a rather diverse group that contains many species that are suitable for a coral reef aquarium.

The diet consists typically of organic particles, remnants of plants and animals, which are taken from the bottom, the open water or from the surface. The stomach content of specimens examined in detail consisted mainly of detritus, fleshy algae or sand particles. To some extent various benthic and planktonic organisms are also eaten. Some specimens of the Caribbean *Ophiocoma echinata* seem to have become specialized to prey on the eggs (clutches) of damselfishes (family Pomacentridae; ITZKOWITZ & KOCH, 1991).

As far as is known, all species are open-spawner with a planktonic larval stage, which can lasts two to three months. Large species can produce several hundred thousand eggs! Some species, especially the smaller ones also reproduce by asexual division. For the small Caribbean species *Ophiocomella ophiactopides* (diameter of only 4.5 cm, including the arms, when fully grown), it has been calculated that a single specimen within one year can produce 15 new brittle stars merely through division (MLADENOV et al., 1983).

The genus *Ophiopsila* is of special interest because several species are able to produce bio-luminescence. This ability of live organisms to emit light is known from several groups of organisms, including some bacteria, fungi, cnidarians, worms, insects, crustaceans, molluscs, fishes and also, of course, echinoderms. If one touches the arms of luminescing *Ophiopsila* species during darkness, they will emit light. The significance of the thus ensuing biological "fire works", with sparks and blinking light, presumably serves to scare off predators, such as shrimp and crabs (GROBER, 1989).

Ophiocoma scolopendrina (see Fact Sheet on page 312) lives in shallow water areas of coral reefs in the Indo-Pacific, from the Red Sea and East Africa to Hawaii. Over the entire area of its geographic distribution this is probably the most common brittle star of the inter-tidal zone. It is a territorial species that defends its hiding places vigorously against other brittle stars. This species feeds primarily on organic detritus collected from the surface, among other places. Therefore, the underside of the arms is light-coloured (camouflage), although it is otherwise dark brown to black. Feeding it under aquarium conditions is not difficult. When this brittle star is being kept in a shallow tank and floating food is given (e.g. flake food), one can readily observe their feeding behaviour along the surface.

O. scolopendrina is very suitable and hardy as an aquarium inhabitant. It can be kept together with most other aquarium animals, except for small bristle worm colonies, such as *Bispira* spp. and certain other, small animals that are unable to flee. ERHARDT & MOOSLEITNER (1995) report that there can even be compatibility problems with cleaner shrimp, although we have never had difficulties in that respect. Once again, this is probably a case where an animal can change its feeding behaviour when suitably hungry. Unfortunately, there is no animal that can be kept safely together with all other animals at all times; a rule that applies to brittle stars as well as to other invertebrates animals.

Ophiocoma species are imported from the Indo-Pacific as well as from the Caribbean, however, precise species identification is difficult, especially without access to special literature and reliable details of the collecting site. Among Caribbean species that are apparently most often imported are *O. echinata* and *O. wendtii* (often erroneously identified as *O. riisei*, an Indo-Pacific species; HENDLER et al., 1995). *O. echinata* specimens commonly seen in the aquarium trade, usually have brown marbled arms and a grey-brown oral discs with a large irregular cream-coloured star in the centre. The oral disc can also be uniformly coloured or be sprinkled with brown dots. In its basic colouration *O. wendtii* is often reddish to brick red with a dark brown- or black-lined pattern on the oral disc, as well as dark cross bands around the arms. This species can also have an overall colouration of grey. Both species reach an oral disc diameter of 3 to 3.5 cm and arm lengths of 15 to 18 cm.

The genus *Ophiomastic* contains 14 Indo-Pacific species, and several of these are regularly imported for the aquarium hobby. Most species have large, protruding club-shaped spines among the normal spines. It is probably this particular feature that is the reason why this species is imported so frequently. Beyond that, many species are attractively coloured and have decorative markings, so that they become real "eye-catchers" in an aquarium. Just like the other members of this family, they are relatively easy to keep, which makes them durable and hardy aquarium inhabitants. A few of them are presented in detail on Fact Sheets on pages 313 and 314.

Unidentified brittle star, possibly the Caribbean *Ophiocoma paucigranulata* in an aquarium. Basically, all members of family Ophiocomidae are excellent scavengers.

Family Ophionereididae (= Ophionereidae)

This is a monotypic family with a single genus: *Ophionereis*. There are eleven species living on Indo-Pacific coral reefs and four species in the Caribbean region. These are brittle stars with extremely long arms; adult specimens often have arm lengths that are ten- to fifteen times the diameter of the oral disc. The smallest species, such as *O. dubia* from the Indo-Pacific and the Caribbean *O. olivacea* have oral discs not larger than 5 to 10 mm in diameter and arms that are 4 to 10 cm long. On the other hand though, specimens of *O. fusca* from the Indo-Australian Archipelago and the western Pacific can reach an oral disc diameter of 16 mm and have arms that are 25 cm long. In terms of colouration, there are variable markings and marbled patterns, stripes and bands in shades of grey and brown.

So far we have not yet seen any *Ophionereis* species in aquariums, however, they do probably occur sporadically in the trade. One would imagine that the aquarium care of these brittle stars is rather easy. They feed on small organisms and on detritus, off the bottom. There are species that discharge their sex products in to open water, as well as some that practice brood-care.

Family Ophiodermatidae

From an aquarium hobby viewpoint, this is an important family, since it contains two species that are among the most colourful and frequently imported brittle stars: the olive green *Ophiarachna incrassata* from the Indo-Pacific and the bright red *Ophioderma squamosissimus* from the Caribbean region. Both are discussed in detail on Fact Sheets, pages 318 and 319. Coincidentally, these are also the two largest commercially traded brittle stars, with oral disc diameters of up to 4.5 cm and arms that can be 25 cm or longer, i.e. with a total arm spread (from tip to tip) of 0.5 m or more! Therefore, it is appropriate that the total biomass of these brittle stars is rather modest, even though they are very large. Otherwise there would only be a few aquarists around that had the facilities to keep these attractive animals. In spite of their considerable size, they can readily be kept in a tank with a volume of 200 to 300 liters. They can nestle with ease in to hiding places among the aquarium decoration, and – incredibly – also accommodate their long arms in there.

Both species mentioned above and also most of the other family members can be included among opportunistic omnivores, even though they sometimes prefer dead food (detritus) and tiny live organisms. In fact, when they are sufficiently hungry they can become rather predatory, and that is the reasoning why many aquarists and authors advise against keeping them. Many complain especially about *Ophiarachna incrassata*, which feeds on corals (particularly on soft corals, such as *Xenia* and *Anthelia* species), as well as on fishes. To us, however, this problem is rather insignificant, provided the brittle stars are being given an adequate, variable diet of fresh-, frozen- and dry food.

Ophiocoma scolopendrina
Reef Flat Brittle Star

Distribution: Occurs widely from the Red Sea and East Africa, eastward to Hawaii and northward to Japan.
Size: Disc up to 2 cm in diameter, arms 15 cm long.
Reproduction: Free spawner.
Food: Has become specialized to feed on organic particles present within the surface film (see photo below). Beyond that, it is more or less an omnivorous scavenger taking in various organic materials such as (detritus, dead organisms, small invertebrates, etc.). Will accept most substitute foods, including dry food.
Description: Uniformly brown or nearly black on its aboral side; oral side lighter coloured, in many specimens even white. Arms with protruding spines. This is the most common brittle star of the inter-tidal zone along most Indo-Pacific reefs. Can occur in dense aggregations. It is territorial and defends the crevasse or crack it lives in against other brittle stars. The food gathering behaviour of this brittle star has been studied closely. As long as the water is shallow enough during low tide and while the flood commences, this brittle star extends its long, flexible arms to the surface and collects particle floating by in the surface film. With the aid of cilia and tube feet these particles are then passed on to the mouth. Lives mainly on soft or sandy bottom at a depth of about 1 m.
Aquarium requirements: *Ophiocoma scolopendrina* and the related, similar looking *O. erinaceus*, *O. valenciae* and *O. porrecta* are frequently available from the aquarium trade. All are durable, excellent aquarium inhabitants that normally hide under rocks from where they extend their arms to collect food. The tube feet gather up every little edible particle. Unfortunately, these brittle stars also have a tendency to feed on small and medium-size invertebrates. We have noticed that it is impossible to keep them together successfully with small, colonial tube worms of genus *Bispira*. The tubeworms were being pulled out of their tubes by the brittle stars, which would then haul them back to their hiding place to eat them.

GA: +; **SE:** +; **CO:** +/-

Photos:
Top: *Ophiocoma scolopendrina* on a reef flat in the Gulf of Aqaba, Red Sea. Photo: Dr. I. Illich
Bottom: Specimen gathering food from the surface film of the water. Photo: C. Bryce

Ophiomastix annulosa

Distribution: From the Maldives and Sri Lanka, eastward to Polynesia and north to Japan.
Size: Disc diameter to 3 cm, arms 20 cm long.
Reproduction: Free spawner.
Food: Scavenger that feeds on organic material such as detritus, dead organisms, small invertebrates, etc. Will accept most substitute foods, including dry food.
Description: *O. annulosa*, like most *Ophiomastix* species, has club-shaped spines among the regular spines. A typical feature is the rhomboid pattern (assembled by plates) on the arms. Lives in shallow water, generally on a sand bottom at depths from 5 to 10 m.
Aquarium requirements: A hardy, long-living brittle star that is occasionally imported. Basically peaceful, but may harm small, delicate invertebrates that cannot escape. Corals are normally left untouched.

GA: +; **SE:** +; **CO:** +/-

Photo:
Ophiomastix annulosa in a display aquarium at Aqua-Design, Oldenburg.

Ophiomastix venosa

Distribution: From East Africa eastward to the Philippines.
Size: Disc to 2.5 cm in diameter, arms 17 cm long.
Reproduction: Free spawner.
Food: Scavenger that feed on organic material such as detritus, dead organisms, small invertebrates, etc. Will accept most substitute foods, including dry food.
Description: The narrow, dark line markings on the disc are characteristic. Distinctly recognizable, club-shaped spines among the normally shaped spines. Lives in shallow water, generally on a sand bottom and among algae, at depths down to 20 m.
Aquarium requirements: Probably the most common brittle star of genus *Ophiomastix*; frequently available in the aquarium trade. This is a spectacular aquarium animal that attracts much attention because of the many club-shaped spines on its arms. Rather easy to keep with a substantial longevity. Basically peaceful, but may harm small, delicate invertebrates that cannot escape. Corals are normally left untouched.

GA: +; **SE:** +; **CO:** +/-

Photo:
Ophiomastix venosa in an aquarium.

Ophiomastix variabilis

Distribution: From East Africa eastward to Palau; generally fairly rare, except for the Indo-Australian Archipelago, where it occurs rather frequently.
Size: Disc to 1.6 cm in diameter, arms 12 cm long.
Reproduction: Free spawner.
Food: Scavenger that feeds on organic material such as detritus, dead organisms, small invertebrates, etc. Will accept most substitute foods, including dry food.
Description: As indicated by the species name, colouration is variable, but can be identified by the dark brown, almost black base colour and the light yellow to white markings on disc and arms. Like most *Ophiomastix* species, *O. variabilis* also has club-shaped spines (even though they are small) among the normal spines. Lives in shallow water, generally on sand bottom or at the surface of living corals, at depth from 5 to 10 m, but has also been found as deep as 60 m.

Aquarium requirements: Rarely imported, but it is basically a hardy and peaceful brittle star; however, it may damage delicate, small invertebrates that are unable to escape. Corals are normally not touched.

GA: +; **SE:** +; **CO:** +/-

Photo:
Ophiomastix variabilis at Madang, Papua-New Guinea.
Photo: L. Newman and A. Flowers

Nevertheless, it goes without saying that there is every reason to keep an eye on this problem; clearly such large brittle stars must not be added to an aquarium inhabited by valuable and/or rare animals. Beyond that, it is of course every aquarist's own decision as to what risk he or she is willing to accept. In any event, these brittle stars are hardy and highly fascinating aquarium animals.

At this point it is also interesting to note, that there are many reports especially about *Ophiarachnia incrassata* breeding under aquarium conditions. The first published account was by MOUGIN (1979), from the Aquarium of Nancy, France. Then there is KASTL'S (1993) reports about progeny in his aquarium. In addition, we have been advised by aquarists from around the world about their breeding success with this species. To this day, there is no explanation as to why it reproduces so readily in an aquarium. Neither eggs nor the actual spawning has ever been observed. Therefore, it is also unclear whether this species belongs to the open spawning or to brood-caring brittle stars On the basis of the relatively small number of young that have been observed every time, we would assume that this is a brood-caring species. A free-spawning brittle star of this size would probably produce thousands of young, which would create serious bio-load problems in any closed aquarium system.

Among the reports that we have received about the captive breeding of *Ophiarachna incrassata*, the one from Mitch Gibbs (pers. comm.) from Kentucky, USA, is one of the most comprehensive, and so we like to pass this on to our readers, in a slightly edited version, as follows:

"I will tell you about my experience with *O. incrassata*. On September 17th 1996 I first noticed small red, moving specks in one of my aquaria. I would estimate them to have oral disks at 0.15 mm or less. In about one weeks time they were maybe 0.25 mm. They looked like tiny stars, and at first I misidentified them as *Echinaster luzonicus*, as they are bright orange in colour, and adults were introduced to the system shortly before I noticed the orange specks. Furthermore these tiny youngs did not behave like typical brittle stars. Instead of hiding, they spent the entire day on top of the live rock.

This behavior continued for two months, and they were easy to observe daily (which I did).

On November 7th 1996, when I went to look at them, they were gone. With the aid of a magnifying glass I observed an arm sticking out of a crack very near the site where they had used to stay. By looking at the cracks more closely, I found several more also hiding. When I looked even more closely, I noticed the inner third of the oral disk had now turned maroon. At this time the oral disk diameter was about 1 mm. I realized they couldn't be *E. luzonicus*. I would hypothesize that the orange colour allows them to hide among orange sponges in their early days, possibly even eating sponge as first food. (Authors' remark: This hypothesis is probably correct, since sponges make up an import nutritional element in the diet of several species of family Ophiodermatidae). The ones I had were forced to eat mainly detritus and maybe they, thus, lacked proper nutrition for normal growth.

On December 12th 1996 I first noticed that a tricolour pattern was starting to develop. The outer fith of the arms turned green, followed by a yellow band and maroon. I now suspected the youngs were actually *O. incrassata*.

February 11th 1997 the stars were still tricolour and about 16 mm total diameter. Not all the stars were at the same stage of development. System was nutrient poor and not feed much.

By June 24th 1997, some stars were almost total green in colour, like their parents, and 4-5 cm in total diameter. September 1997 I removed all the rock from the system and found 19 still small stars. At one year of age, they were obviously stunted, as size averaged only 5 cm with some a little larger and some smaller. Two still had a maroon center.

I also had two different customers of mine reporting similar spawning just a few months after the spawning here."

There is yet a large number of other species in family Ophiodermatidae, on Caribbean as well as on Indo-Pacific reefs, and so it is conceivable that some of them appear in the aquarium trade. From the Caribbean region it is only genus *Ophioderma* with its eight species, that is of interest to the aquarium hobby. All of these brittle stars are described in detail in the publication by HENDLER et al. (1995). Apart from *O. squamosissimum*, another species in demand by aquarists and also frequently traded is *O. apressum* (see Fact Sheet on page 318).

The tropical Indo-Pacific contains a vast diversity of brittle star species. However, a distinct lack of popular scientific identification literature makes any species identification very difficult for most aquarists. The following genera are represented: *Cryptopelta* with three species, *Distichopsis* with one species, *Ophiarachna* with five species, *Ophirachnella* with nine species, *Ophiochaeta* with two species, *Ophichasma* with one species, *Ophioclastus* with one species, *Ophioconis* with three species, *Ophioderma* with two species, *Ophiodyscrita* with two species, *Ophiopeza* with three species, *Ophiopsammus* with one species and *Ophiostegatus* with one species. For New Caledonia (and the Indo-Australian Archipelago), GUILLE et al. (1986) are describing eight of the most common species from five genera, which can be of some help to aquarists. CLARK & ROWE (1971) have published scientific identification keys for all species, but without photos and details of colouration.

Captive Breeding of *Ophiarachna incrassata* (photos: M. Gibbs)

Some aquarists have reported the reproduction of the Olive Green Brittle Star, *Ophiorachna incrassata* in their aquariums. Mitch Gibbs is one of the few that have documented the growth periods photographically. Left: At the time of their discovery in September 1996 the young were little more than tiny red starlets, moving openly about on a live rock. Right: In November 1996 their behaviour and colouration changed. They started to hide in cracks and crevasses, and the inner third of the oral disc had turned chestnut-brown.

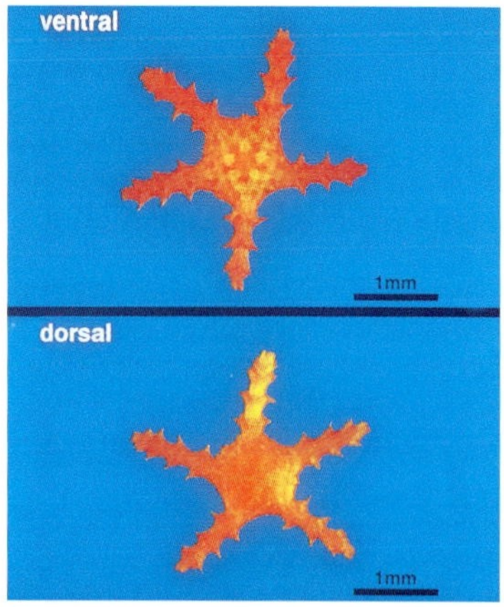

Left: In February 1997 the colouration of the young had taken on a tri-coloured pattern. At that stage their average diameter was 16 mm. Right: At the end of June 1997 some of the young brittle stars had taken on adult colouration and they had reached a diameter of 4 to 5 cm, while the other were still at the tri-coloured stage and their growth had stagnated.

Family Hemiuryalidae

The only reason why this family is of interest to the aquarium hobby is because it contains the Caribbean *Sigbeia conifera*. This brittle star lives in association with lace corals of genus *Stylaster*, in particular with *Stylaster roseus*. *Sigbeia conifera* is a very small brittle star species, with a total diameter not larger than 3.5 cm (oral disc 5 mm, arms 1.5 cm). This brittle star has camouflage colours of red-lilac or pink, which is in adaptation to the colouration of its host. During the day it wraps its arms tightly around the branches of its host coral, but at night it waves its arms through the water in order the collect plankton. This brittle star has to our knowledge not yet been seen in the aquarium trade.

Family Ophiactidae

The brittle stars in this family are generally rather small. The majority of them live as epizoites on sponges or among crushed coral and algae. They do not play any role in the aquarium specimen trade, but it is conceivable that they come in to our aquariums as "stowaways" on live rocks or together with other animals.

The most important genus is *Ophiactis*, with seven species associated with coral reefs in the Caribbean (Hendler et al., 1995) and 18 Indo-Pacific species (Clark & Rowe, 1971). *Ophiactis savignyi* has a circum-tropical distribution, which – in part may have been aided by human activities (e.g. included in marine growth attached to ships' hulls). Other tropical species are contained in the genera *Hemipholis*, *Ophiodaphne* and *Ophiosphaera*. Species identification in this family is generally difficult, since many species have an oral disc diameter of not more than 2 to 3 mm. Moreover, there is a high degree of uncertainty surrounding the taxonomic legitimacy of some species. This problem is further complicated by the fact that many

Ophiarachna incrassata
Olive-green Brittle Star

Distribution: Occurs widely in the Indo-Pacific, from East Africa eastward to the Fiji Islands.
Size: Disc to 5 cm in diameter, arms 25 cm long.
Reproduction: Possibly ovoviviparous. There are numerous reports about reproduction in aquariums, however, details about spawning or brooding are absent. Please refer to the text and to the interesting photos by M. Gibbs on pages 315 and 316.
Food: Omnivorous opportunist that prefers dead organic material, such as detritus and dead organisms and small benthic invertebrates, but is also not deterred by other suitable prey, including small fishes (see page 303) when sufficiently hungry. Will accept most substitute foods.
Description: An attractive, rather large and active brittle star that lives on sand bottom and among algae and sea grasses on shallow sand flats, or among coral rubble, at depths from 5 to 40 m. Adults are easily identified by the intense olive-green colouration together with pearly grey, black-edged markings on the aboral side of the disc. Arms with well-developed spines, which are also coloured grey and black. Juveniles undergo a series of colour changes from orange-red to a tri-coloured pattern.

Aquarium requirements: In spite of its aggressive feeding behaviour, this is one of the most frequently imported brittle star. Some aquarists recommend not keeping this species because of the danger that it might attack other aquarium inhabitants, including fishes. However, we are certain, that the majority of aquarists are satisfied with this brittle star, and we have kept it for years ourselves without observing the brittle star to harm other animals in the tank. There is, however, reasons for not keeping *O. incrassata* with the most delicate forms of sessile invertebrates, such as small tube worms (e. g. *Bispira brunnea*, see volume 3, page 132). It seems certain that it cannot be kept together with all other animals, especially not with small and delicate forms, without any risk at all. It is important that this brittle star receives sufficient food as a lack of natural food supply might very well bring about predatory behaviour. Beyond that it is a hardy, rather active and very beautiful brittle star that can achieve considerable longevity in an aquarium.

GA: +; **SE:** +; **CO:** +/-

Photo:
Ophiarachna incrassata in an aquarium.

Ophioderma* cf. *apressum
Zebra Brittle Star

Distribution: Caribbean region; from Bermuda to Brazil. There are also reports about its occurrence in the eastern Pacific; however, these appear to be questionable (HENDLER et al., 1995).
Size: Disc to 2.5 cm in diameter, arms 12.5 cm long.
Reproduction: Free spawner.
Food: Scavenger that feeds on detritus, faecal material, algae, small benthic invertebrates and other organic material. Will accept most substitute foods.
Description: One of the most frequently occurring brittle stars in the Caribbean region. Colouration extremely variable. Many specimens with colours that are nowhere near those of the specimens shown in the accompanying photos. All colour forms fall within two principal patterns: uniform- and harlequin pattern. The popular name used in this Fact Sheet refers to the harlequin pattern, which – because of its attractiveness – is the one most commonly found in the aquarium trade. The disc of uniformly coloured specimens is grey, green or brown, usually with a "pepper and salt" pattern. The harlequin form occurs in a wide spectrum of mutually contrasting colours: green, white, black or yellow. The arms of all colour varieties with grayish or white bands. Lives in sea grass habitats as well as on reefs among coral rubble, protected by branched or leafy corals, from the inter-tidal zone to the fore reef and along the reef slop down to a depth of 50 m.
Aquarium requirements: One of the most spectacular brittle stars from the Caribbean region that is regularly available in the aquarium trade. Remarkably hardy and easy to care for, with a significant longevity and peaceful towards all other aquarium inhabitants except small, delicate forms that are unable to escape. So far we have not yet seen this brittle star damage any corals.

GA: +; **SE:** +; **CO:** +/-

Photo:
Harlequin form of *Ophioderma* cf. *apressum* with zebra pattern in an aquarium. Photo: S. W. Michael

Ophioderma squamosissimus
Red Caribbean Brittle Star

Distribution: Caribbean region; from the Bahamas to Belize.
Size: Disc to 4.2 cm in diameter, arms 20 cm long.
Reproduction: Possibly ovoviviparous; however, no details available.
Food: According to HENDLER et al. (1995) the only specimen examined had sponge parts in its stomach. Yet, aquarium experiences indicate that this species is – much like *Ophiorachna incrassata* – an opportunistic omnivore. Will accept most substitute foods, including frozen and dried food.
Description: Possibly the largest Caribbean brittle star, as well as one of the most beautiful and most spectacularly coloured ones: scarlet red or vermilion in adult specimens. Juveniles with irregular pale orange-yellow spots on the disc. Lives along reef slopes or on platform reefs among or in the proximity of large corals, e.g. brain corals, at depths from 3 to 85 m.
Aquarium requirements: A magnificent brittle star that is often imported. However, there is a degree of skepticism among many aquarists and authors because of its size and voracious appetite. Can certainly not be kept together with small, delicate invertebrates that are unable to escape. We have had basically only few - if any - problems, provided this brittle star is properly fed. Corals are normally not damaged. Easy-to-care for brittle star that is very active and tends to live for a long time.

GA: +; **SE:**; **CO:** +/-

Photos:
Top: Subadult specimen of *Ophioderma squamosissimus* that still shows remnants of the pale orange-yellow sprinkling on its disc; in an aquarium.
Photo: S. W. Michael
Centre: Adult, intensely red specimen in an aquarium.
Bottom: Close-up photo of an aquarium specimen with collapsed disc.

Most species of family Amphiuridae live in holes bored in to sand bottom, from where they extend a few arms to the outside in order to gather up detritus and organic material. This photo shows an amphiurid species during a night dive in Indonesia.

species have rather widely overlapping distribution areas; at the same time there is also a significant range of variation of many morphological characteristics within a particular species.

Even the largest species rarely reach an oral disc diameter of more than 10 mm. Most species reproduce sexually, releasing the gametes into the water for fertilization, (as open spawner) as well as asexually through division. Examinations of stomach contents have revealed that detritus and tiny organisms contained in it make up the major component in the diet of these brittle stars. For some species it has been shown that the fluid inside the water canals contains haemoglobin-like substances. This probably enhances the ability of these brittle stars to live in an oxygen-deficient environment.

Family Amphiuridae

The majority of species in this family lives buried in sand or other soft bottom substrates. Whereby only the arms are extended in to the surrounding water. Pits established by these brittle stars are reinforced with mucus so that their sides cannot cave in. A unique characteristic of this family is the presence of a pair of papillae at the apex of the maxillary below the lowest tooth. The oral disc is often (but not always) covered with spiny scales.

The size of adult specimens varies significantly from one species to the next; however, amphiurids are predominantly small brittle stars. The smallest species have oral disc diameters of less than 5 mm, while the largest ones have oral disc diameters of 15 mm and arm lengths of 25 to 30 cm.

In spite of the large number of species (80 in the Indo-Pacific and 22 in the Caribbean), we do not know whether amphiurid brittle stars are specifically imported for the aquarium trade. However, it stands to reason that they come in to our aquariums as "stowaways" on live rocks, etc. Tropical genera with species of interest to the aquarium hobby are: *Amphylicus*, *Amphiopdia*, *Amphioplus*, *Amphipholis*, *Amphiura*, *Dougaloplus*, *Ophiocentrus*, *Ophiocnida*, *Ophionephthys*, *Ophiostigma* and *Paracronida*.

We made our first acquaintance with a representative of this family, when we removed sand and detritus from behind the aquarium decoration in one of our tanks. There, in the nutrient-rich substrate, we found hundreds of brittle stars with a total diameter of only about 1 cm. The oral discs measured no more than 2 mm. They were coloured much like the sediments they were living in and we discovered them only after we passed the material through a fine-meshed net. Finding so many brittle stars prompted us to search the entire aquarium thoroughly, and we quickly noticed vast numbers of tiny arms projecting through the substrate in many places. We also found the same or similar species among algae on rocks or on glass sides of the aquarium.

With considerable certainty we can state that this brittle star was *Amphipholis squamata*, a species with an almost worldwide distribution in tropical- and subtropical seas. It also occurs in the Mediterranean Sea and in the North Sea. It has been shown that in tropical seas (in the Caribbean as well as in large parts of the Indo-Pacific) this species lives among corals, on rocky substrate, sand bottom, sea grass beds, mangroves and in the inter-tidal zones of river estuaries. As a rule, *A. squamata* occurs in dense aggregations under rocks, in bottom sediments or among sponges, moss animals, clams and algae. This brittle star displays wide tolerance to variable habitat conditions; it lives in the ocean as well as in brackish water. Frequently, it is found in inter-tidal zones, yet it has also been shown to occur at a depth of 1,330 m (GAGE et al., 1983). It appears that specimens tend to increase in size towards higher (northern) latitudes. Specimens collected along the Norwegian coastline had an oral disc diameter of 5 mm, and arms that were 2 cm long. Yet, in the Caribbean region this brittle star species has only 3 mm large oral discs and arm lengths of 11 mm.

Colouration and markings vary. Generally the colours of *Amphipholis squamata* are rather pale. The aboral side is often white, yellow, orange, light brown or grey, usually with a faint marbled pattern. The arms are light brown, yellowish or white, occasionally with faint dark bands. Like other brittle stars, this species is alleged to emit bioluminescence, however, in an aquarium we have never been able to observe light emission. It is conceivable, that this may be due to insufficiently long observations or that not all populations of this brittle star develop this capability.

Amphipholis squamata feeds on particles suspended in the water, on detritus that has drifted to the bottom, as well as on tiny organisms and on planktonic or benthic algae. ROTHANS & MILLER (1991) have demonstrated that this species is directly involved in the decomposition of fish excrements on coral reefs. For aquarium husbandry purposes one can assume that large numbers of brittle stars make up a useful component in the community of organisms involved in biodegradation.

It has been shown that *Amphipholis squamata* in tropical seas reproduces sexually throughout the year. It practices brood care, whereby the maternal animal carries a small number of young in the bursa for three to seven months. Beyond that, it is a hermaphrodite, capable of fertilizing itself. This way large populations of genetically rather similar specimens can become established quickly. RUMRILL (1982) counted more than 500 specimens per square meter, and among algae samples he even found 500 specimens per litre material collected.

Family Ophiotrichidae

In terms of ability to adapt to many different habitats, this family has been very successful, specifically in the Indo-Pacific, where there is a vast number of species. A particularly large number of species lives as epizoites on soft corals, gorgonians and other invertebrates; however, other reef habitats are utilized. For instance, there are species that live in association with calcareous algae, as well as with crushed coral.

The tropical species of this family are often rather colourful so that they are conspicuous to divers. Especially the Indo-Pacific is home to a very large number of species. In total, 89 species have been described that belong to the orders *Gymnolophus*, *Lissophiothrix*, *Ophiocnemis*, *Ophiogymmna*, *Ophiolophus*, *Ophiomaza*, *Ophiophthirius*, *Ophiosammium*, *Ophiopteron*, *Ophiothela*, *Ophiothrix*, *Macrophhiothrix* and *Placophiothrix*. HENDLER et al. (1995) list five species for the Caribbean region that are part of genus *Ophiothrix*.

These brittle stars, whose identification is very difficult and often requires a microscopic examination, have arms that are often covered by long spines, and there can also be distinct spines on the oral disc. There are small as well as some rather large species. Most are suspension- or detritus feeders, characteristics that make them rather suitable aquarium inhabitants. Large benthic species

***Amphiura* sp. at Monte Bellos, Western Australia. Photo: C. Bryce**

and also small species that live as commensals on gorgonians and other cnidarians are regularly available through the aquarium trade. Moreover, it is not uncommon to introduce representatives of this family inadvertently in to an aquarium when they are attached to live rocks.

One of the most frequently observed (and most commonly photographed) Indo-Pacific species is the red brittle star, *Ophiothrix purpurea*, characterized by a pattern on the oral disc and a dark lilac-colour line along the upper side of each arm. It occurs very often in shaded coral caves and tunnels in the upper regions of the fore reef, where the water is only a few meters deep.

A large population of *Amphiphoslis squamata* had become established in our aquarium. This brittle star had been introduced accidentally, most likely together with live rocks. Left: A group of these brittle stars among filamentous algae. Centre: Two specimens that we removed from a detritus layer. The left specimen shows very clearly the typically rough scaled oral disk (8 X magnification). Right: Specimen under 12 X magnification.

Serpent Stars of genus *Ophiothrix*

Ophiothrix proteus at Port Sudan. Photo: Dr. I. Illich

Ophiothrix cf. *suensonii* at the US Virgin Islands. Photo: S. W. Michael

Ophiothrix sp. at the Maldives.

Ophiothrix cf. *nereidina* in Papua-New Guinea. Photo: P. Schupp

Two varieties of *Ophiothrix* cf. *nereidina* in Papua-New Guinea. Photos: L. Newman and A. Flowers

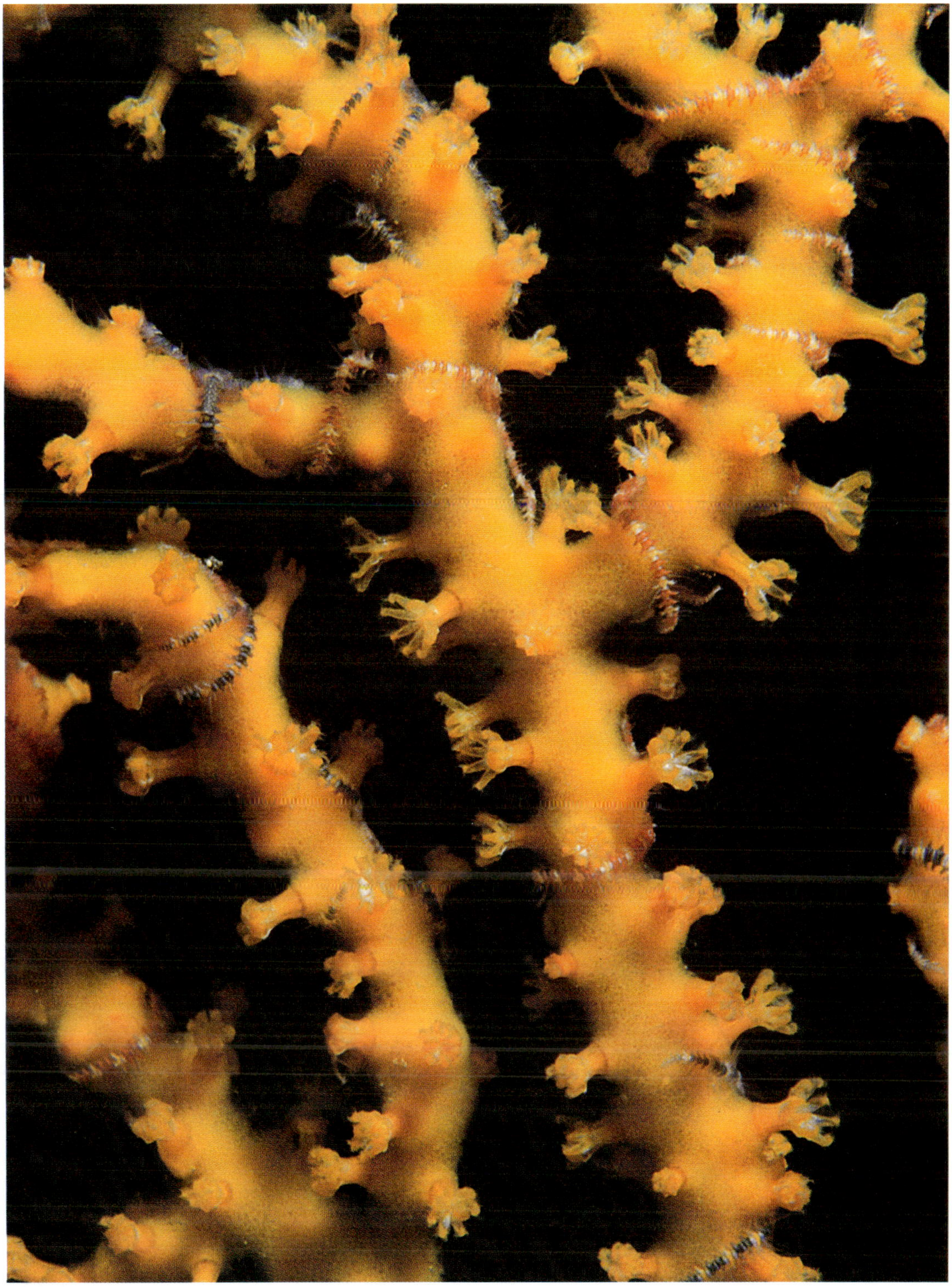

Serpent stars of family Ophiotrichidae usually live as epizooites on soft corals, gorgonian corals and other invertebrates. This beautiful, rather tiny and unidentified species (seen here in the Red Sea) embraces tightly the branches of a gorgonian coral of family Acanthogorgidae. Photo: Dr. I. Illich

Ophiothrix purpurea
Purple Brittle Star

Distribution: Western Indo-Pacific, from the Red Sea and East Africa eastward to the Indo-Australian Archipelago and northward to Japan.
Size: Disc to 1 cm in diameter, arms 12 cm long.
Reproduction: Probably a free spawner, however, details are unknown.
Food: Suspension feeder that filters plankton and organic particles from the water. Will accept various planktonic substitute foods.
Description: An typical brittle star that is not an obligatory commensal on other invertebrates, but instead migrates from one micro-habitat to another, but still frequently seen on sponges, gorgonians, fire corals or other stony corals. Colouration variable, but typical base colours of disc and arms are red or reddish-white, with or without markings. Spines on the arms generally deep red. The arms are marked with narrow, dark – often black – cross bands. Lives at depths from 10 to 60 m.
Aquarium requirements: Only occasionally imported, but appears to be remarkably hardy. Is possibly more difficult to feed than other brittle stars. Peaceful towards other aquarium inhabitants.

GA: +; **SE:** +/-; **CO:** +

Photos:
Ophiothrix purpurea in Indonesia (top and centre; photos: C. Bryce) and on the Great Barrier Reef (bottom; photo: E. Svensen).

Ophiothrix suensonii
Sponge Brittle Star

Distribution: Caribbean region; from Bermuda to Brazil.
Size: Disc to 1.8 cm in diameter, arms 10 cm long.
Reproduction: Free spawner; reproduces continuously throughout the year, permanently but with seasonal peaks. Individuals spawn asynchronously. Females produce up to 47,500 eggs that have a diameter of 0.13 mm. Within ten days the eggs develop in to four-armed ophiopluteus larvae, which have been kept alive for 49 days, but without a successful metamorphosis (HENDLER et al., 1995).
Food: Suspension feeder that filters plankton and organic particles from the water. Will accept various planktonic substitute foods.
Description: Occurs on reefs at depths from 3 to 45 m, but also in mangrove areas and muddy lagoons. Appears to prefer sponges as hosts, but has also been seen on gorgonians and fire corals, and occasionally also on coral rubble. Colouration is variable. Typical base colours of disc and arms are grey, yellow-brown, lavender, pink, orange, red or deep purple. A black, purple-coloured or dark red line extends along the arms; however, this is clearly visible only in pale specimens. The arms are covered by numerous long, glassy spines.
Aquarium requirements: Only rarely imported, but appears to be hardy. May be more difficult to feed than other brittle stars. Peaceful towards other aquarium inhabitants.

GA: +; **SE:** +/-; **CO:** +

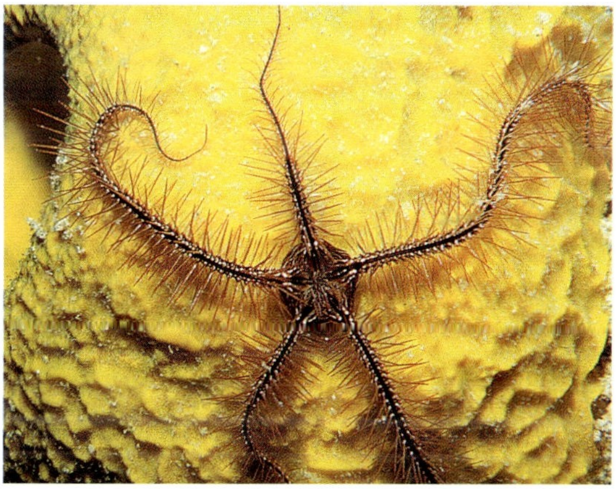

Photos:
Ophiothrix suensonii at Curacao (top; photo: Dr. I. Illich) and at the US Virgin islands (centre and bottom; photos: S. W. Michael).

Ophiothela* cf. *danae
Sling Brittle Star

Distribution: Indo-Pacific; from the Red Sea and East Africa to the South Pacific Islands.
Size: Disc to 0.3 cm in diameter, arms 15 cm long.
Reproduction: Probably free spawner, but details are unknown.
Food: Suspension feeder that filters plankton and organic particles from the water. It is conceivable that it feeds on the same food as that of its host, where it can then steal it. Will accept very small planktonic substitute foods.
Description: Apparently an obligatory commensal on other invertebrates, e.g. branching sponges, soft corals, sea feathers and gorgonians, where large numbers of these brittle stars can be found. It is unknown whether there are certain morphs that live on specific hosts. Normally occurs at depths from 15 to 50 m. Colouration highly variable, but the prevailing colours are light blue, grey, red, orange and yellow. Disc and arms can be coloured differently. Arms often with cross bands. Apparently the host colours are not taken on.
Aquarium requirements: Is occasionally introduced inadvertently in to aquariums when attached to a host, specifically to a gorgonian host. Most specimens that we have seen had been introduced attached to a *Echinogorgia* species. Feeding can be complicated, since this brittle star will take only very small planktonic particles or planktonic substitute foods. If the host animal can be kept alive and induced to grow, then there appear to be no problems with the Sling Brittle Star. Must not be kept together with large fishes and predatory invertebrates, for which this species could be potential prey. Will not harm other animals.

GA: +; **SE:** +/-; **CO:** +/-

Photos:
Ophiothela cf. *danae* on a sea feather of genus *Virgularia* (top; photo: C. Bryce), on an *Echinogorgia* species in a display aquarium of the Löbbecke Museum and Aqua-Zoo, Düsseldorf (centre) and on an *Echinogorgia* species in the aquarium of Kjell Nagy, Norway (bottom).

Class Echinoidea – Sea Urchins, Heart Urchins and Sand Dollars

Sea urchins are spherical, egg-shaped or round and compressed echinoderms that are totally enclosed in a solid shell (test or corona) made up of tightly fitting calcareous plates (= skeletal ossicles). The plates carry tubercles (nodules), which form a ball joint for movable, strong spines. The name "spiny-skinned animals" (Echinoidea = like a hedgehog) is therefore appropriate for these animals with their conspicuously sharp and often very long spines. Other linguistic colloquial equivalents are "Seeigel" (German), "Zeeigel" (Dutch), "Riccio di Mare" (Italian) and "Sjøpinnsvin" (Norwegian).

Sea urchins are numerous on coral reefs. As algae eaters they play an important role in the ecosystem of a reef. Some species are also of significance as scavengers and detritus feeders. Sea urchins also produce calcareous sand as waste products from feeding on calcareous algae. In contrast to brittle stars, sea urchins are easy to spot on a reef. Although most species are active during twilight times or at night, they have a tendency to congregate in large and conspicuous groups. In reefs where the level of nutrient is higher than for the average reef and show a high diversity of algae, sea urchins can be especially numerous.

In total there are about 900 species that occupy most habitats, in tropical, temperate as well as cold seas, from rocks in the inter-tidal zone down to a depth of 6,000 m. Coral reefs and other shallow tropical water habitats in the Caribbean region are inhabited by about 30 species (Hendler et al., 1995). In the same type of habitats in the Indo-Pacific we find 146 species (Clark & Rowe, 1971).

The older, traditional classification system places sea urchins in to two subclasses, on the basis of their characteristic appearance, and considers them as having developed along two different evolutionary pathways: Regularia and Irregularia, i.e. regular and irregular sea urchins. Regular sea urchins display a distinct pentamerous symmetry and are approximately spherical (in some cases slightly oval-shaped), with a mouth located at the centre of the oral side and an anal opening on the opposite, aboral side. The ir-

Sea urchins are abundantly present on reefs. Being herbivores they play a significant role in reef ecology. This photo from Indonesia shows a number of *Echinometra mathaei*, a typical regular sea urchin, which is a very efficient algae eater.

Simplified systematic overview of class Echiniodea

(after HENDLER et al., 1995; GUILLE et al., 1986; GEORGE & GEORGE, 1979; CLARK & ROWE, 1971; and others)

Subclass Perischoechinoidea
Order Cidaroida
Family Cidaridae
Chondrocidaris, Eucidaris, Goniocidaris, Phyllacanthus, Prionocidaris
Family Psychocidaridae
Subclass Euechinoidea
Superorder Diadematacea
Order Pedinoida
Order Echinothuriida
Family Echinothuriidae
Asthenosoma
Order Diadematoida
Family Diadematidae
Astropyga, Centrostephanus, Chaetodiadema, Diadema, Echinothrix, Lissodiadema
Superorder Echinacea
Order Arbaciida
Family Arbaciidae
Arbacia
Order Salenioida
Order Temnopleuroida
Family Temnopleuridae
Mespilia, Mierocyphus, Paratrema, Salmaciella, Salmacis, Temnopleurus, Temnotrema
Family Toxopneustidae
Cyrtechinus, Gymnechinus, Lytechinus, Nudechinus, Pseudoboletia, Sphaerechinus, Toxopneustes, Tripneustes
Order Phymosomatoida
Order Echinoida
Family Echinidae
Echinus, Paracentrotus, Sterechinus
Family Echinometridae
Anthocidaris, Colobocentrotus, Echninometra, Echinostrephus, Heterocentrotus, Zenocentrotus
Family Parasaleniidae
Parasalenia
Family Strongylocentrotidae
Strongylocentrotus
Superorder Gnathostomata
Order Holectypoida
Family Echinoneidae
Echinoneus
Order Clypeasteroida
Family Clypeasteridae
Clypeaster
Family Dendrasteridae
Dendraster
Family Mellitidae
Encope, Leodia, Mellita
Superorder Atelostomata
Order Holasteroida
Order Spatangoida
Family Brissidae
Brissopsis, Brissus, Eupatagus, Meoma, Metalia, Plagiobrissus, Rhynchobrissus
Family Spantagidae
Maretia, Pseudonaretia, Spatangus
Family Loveniidae
Breynia, Echinocardium, Lovenia
Order Cassiduloida
Family Echinolampadidae
Echinolampas
Family Cassidulidae
Cassidulus
Order Neolampadoida

regular sea urchins have a secondarily developed, more or less distinct bilateral symmetry, which means that they have originated from ancestors with the pentamerous symmetry. Most of them have a somewhat oval and often flat, compressed shape along the oral – aboral axis, but they can also be round. The mouth in irregular sea urchins can be positioned anteriorly (at the front end; in oval species) or at the centre (in round species). The anus is always located posteriorly, i.e. opposite the mouth.

It may be practical to retain this arrangement together with the relevant popular-scientific designation and identification of sea urchins. However, modern phylogenetic research has largely abandoned the understanding that this represents two monophyletic, valid taxa. The simplified, systematic overviews presented on the left set up in such a manner whereby the regular sea urchins can be found in the first nine orders, while the last six orders are made up of irregular sea urchins (superorders Gnathostomata and Atelostomata).

For most laymen sea urchins are a rather uniform group of animals, which are normally associated with dried exoskeletons, more or less grotesquely painted and lacquered, offered as lampshades, etc. in souvenir shops. Sometimes sea urchins are viewed merely as gastronomic delicacies. The gonads are used in the Japanese "sushi" and "unishiokara" dishes, or they are served as "Oursin au naturelle" in France. In any event, the best-known sea urchins are the regular, spherical species with long, stiff and sharp spines.

Yet, sea urchins occur in an amazingly vast diversity of shapes, sizes and colourations, and – hopefully – this chapter will help to underline and emphasize this fact. Yet, in spite of this diversity, all sea urchins are characterized by the presence of a hollow exoskeleton, made up of calcareous plates (ossicles) fitted tightly (sutured) together along their edges, to form a solid, strong case. In family Echinothuriidae, however, the plates are not tightly grown together, and so these species have a rather flexible, pliable skeleton.

For an animal with a rigid outer skeleton, a strong musculature would be of little use. Therefore, large internal muscles – like those found in other echinoderms and where they are used for moving the arms and the muscular body – are absent in sea urchins.

The rigidity of the skeleton and its significance for protecting highly vulnerable soft body parts is probably also an important explanation for the lesser-developed regenerative capability in sea urchins than in other echinoderms. Although sea urchins can replace damaged and lost tube feet and spines, however, substantial damage to the corona as such nearly always leads to the demise of the animal.

In regular sea urchins, the mouth is located on the underside and the anus is found – across and above on the aboral side. The anus forms the centre of the apical system, a double ring of five terminal plates and five genital plates (see drawing to the right). In bilaterally symmetrical, irregular sea urchins the anus migrates to the back (posteriorly) during the embryonic stage, so that it is located outside the apical system in adult specimens.

In many regular sea urchins species we find an anal bladder, which forms the terminal section of the intestinal opening. The animal can regulate the size of this bladder by either emptying or filling it with fluid. In groups the size and colour markings of this bladder is an important taxonomic characteristic.

The genital plates of the apical system are penetrated by a gonopore, through which the sex products (eggs and sperms) are discharged in to the open water. In addition, one of the genital pores carries the madreporite, the opening for the water vascular (canal) system. Formerly, the terminal plates used to be erroneously referred to as ocellar plates be-

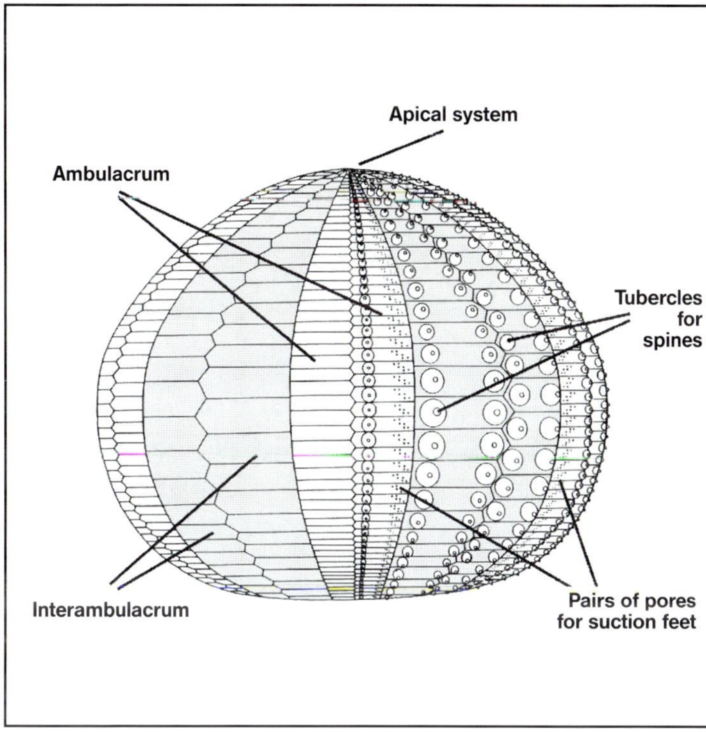

Schematic body structure of corona from a regular sea urchins.
Drawing after CLARK & ROWE (1971)

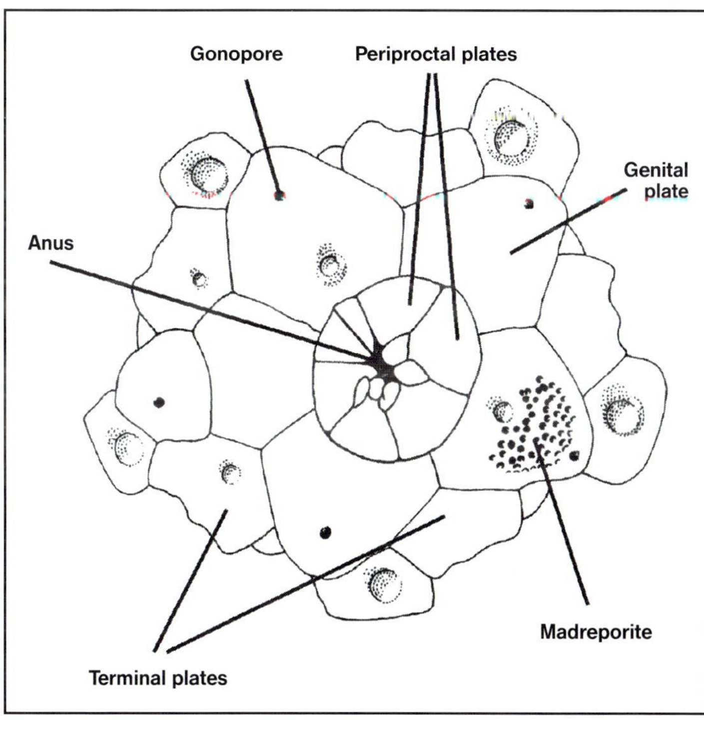

Structure of apical system. Drawing after HENDLER et al. (1995)

Unidentified, irregular sea urchin of order Spatangoida. Irregular sea urchins usually live on soft bottom, where they feed on organic detritus, dead organisms and various microorganisms, as well as small invertebrates.
Photo: P. Schupp

cause it was thought that they function as light-sensitive organs. But later that was found to be incorrect. Actually, they form primarily the growth zones of sea urchins, where – during periods of growth – new plates are added (analogous to the terminal plates at the tips of the arms in sea- and brittle stars). By varying the rate of growth of the remaining plates, sea urchins retain their skeletal shape unchanged.

Regular sea urchins have five distinct ambulacral zones, arranged symmetrically around the animal. Inter-ambulacral zones separate them. The ambulacral zones are analogous to the ambulacral grooves in the arms of sea stars. They have rows of tube feet, which extend from the mouth to the proximity of the anus. The tube feet are used for locomotion as well as for handling materials and perception of the surrounding area. In addition, sea urchins are well equipped with pedicellaria and spines (see photo on page 215).

Most sea urchins have two distinctly different types of spines: large primary spines and small secondary spines. Both are movable from a concave socket at the base that fits over a corresponding tubercle on the test or corona. These ball-and-socket joints are surrounded and controlled by a sheath, a mechanism that gives the spine enormous mobility. On other hand, the spines have a tendon attachment that – if required – keeps the spines rigidly erect As soon as this tendon lock is activated it is easier to break off a spine than to move it. In some sea urchins the spines facilitate locomotion, by moving them and so pushing the animal forward, in part they provide an effective defence. The spines are particularly important as defensive weapon in species that also equipped with venom organs. In yet other species, for instance in burying irregular sea urchins, the spines – in addition to their burying function – also serve as cover and protection against the surrounding sediments.

As in most echinoderms the soft, external parts contribute to the gas exchange process. In addition, regular sea urchins have five gills; it is assumed that these perform the bulk of the respiratory work. The gills are located on the underside, around the mouth opening. They are formed by protrusions from the body wall. The inside of the gills is filled with a fluid and is connected to specific cavities in the body, referred to as peripharyngeal coelom. These are arranged around the jaw apparatus (see below) that assures that the fluid is pumped

Oral- (left) and aboral side (right) of the regular sea urchin *Tripneustes gratilla* at One Tree Island, Great Barrier Reef.
Photos: L. Newman and A. Flowers

in and out through the gills. Irregular sea urchins do not have such gills; instead, they have modified tube feet that facilitate respiration.

A well-known and specific anatomical feature of sea urchins – studied inevitably by every biology student – is the shape of the mouth. In the vast majority of species the mouth is equipped with five chisel-like teeth - and herein again we recognize the pentamerous symmetry. They form a special jaw apparatus, the so-called "Aristotle's Lantern". The comparison to a lantern is based on the presence of five large calcareous plates, called pyramids that are arranged in a ring around the mouth with their tips pointing to the mouth opening. They are sutured together by perpendicular muscle fibres between the plates. The teeth consist of a long, calcareous band that extends along the midline of each pyramid. New tooth material is constantly being produced in a tooth pouch in the upper part of the band. As a tooth is being worn away, new tooth material is gradually pushed down as replacement. Because of heavy wear and tear on a tooth due to the scraping action over hard substrates, this is absolutely essential. Measurements made on *Paracentrotus lividus* have revealed new tooth material being produced a growth rate of 1 to 1.5 mm per week (Barnes, 1980). In addition to teeth, the "Aristotle's Lantern" also consist of a large number of small bones and special muscles. The entire jaw apparatus can be displaced in a lengthwise direction, so that the teeth can be moved in and out of the mouth opening. In the most primitive sea urchins, the teeth – by opening and closing - have little more than a scraping function. Higher developed species, however, can not only use their teeth for scraping, but also for pulling and tearing by being able to move them laterally as well.

Effective scraping teeth are necessary, since most of the tropical shallow water sea urchins are distinctly herbivorous and have a definite preference for algae. In many respects they are therefore nocturnal replacements for diurnal herbivorous fishes, and so to a large measure they are responsible for coral reef aquarium not becoming overgrown with algae. Without the presence of herbivores, rocky surfaces would quickly be covered with algae. Simultaneously with the removal of all sorts of algae, large amounts of coral rock are also scraped off live rocks. With that, sea urchins are one of the most important factors in the erosion of coral reefs, and resulting in the production of coral sand.

The solid teeth, which in some species are strong enough to gnaw through steel (!), leave behind a distinct trail on the sub-

Close-up photo of the oral side of an unidentified, regular sea urchin. The long primary and short secondary spines, attached via round tubercles to the corona are clearly visible. A few flexible tube feet can also be seen among them.

***Heterocentrotus mammilatus* (left) and *Echinometra mathaei* (right) are two algae grazers on shallow reefs of the Indo-Pacific. Without the continuous grazing activities of herbivores these reefs would soon be overgrown by algae.**

Sea urchins are associated with a multitude of symbiotic organisms. Left: The shrimp *Periclimenes colemani* lives exclusively in association with the highly venomous *Asthenosoma varium*, photographed here in Indonesia by J. Cairns-Michael. The shrimp conquers a settlement site on the sea urchin by removing tube feet and spines on that part of the corona. Right: The Yellow-Striped Clingfish, *Diademichthys lineatus* lives in association with a *Diadema* sea urchin, seen here in a display aquarium at the Löbbecke Museum and Aquazoo, Düsseldorf, Germany.

stratum. When these teeth scrape towards the centre of the "Lantern", they leave behind a clear pentagonal star pattern. Such feeding "trails" left behind by sea urchins are even known from fossil finds. It has been calculated that sea urchins in the Caribbean region are responsible for more than 90 % of bio-erosion (HENDLER et al.,1995). For instance, the scraping activity of *Lytechinus variegatus* has lead to a distinct degradation of solid substrate in its habitat.

Echinoderms are important algae grazers and detritus feeders on the coral reefs. The group is commonly found in numbers in shallow rubble zones like here at Rarotonga in the Cook Islands situated in the South Pacific. At this spot, which is less than one square meter, we can see several sea urchins from the genus *Echinometrea*, which have drilled holes in the rocks, the sea cucumbers *Holothuria atra* and *Holothuria leucospilota* are both common, but most obvious is the magnificent blue star fish *Linckia laevigata*.

In several places of the Indo-Pacific (e.g. Togian Islands at Sulawesi and the Andaman Islands off India) we know of characteristic, mushroom-shaped islands which are located in the sea on top of thin columns made of fossil material. BROMLEY (1975) is of the opinion that these geological formations are primarily due to the scraping activities of the small sea urchin *Echinometra mathaei*, working in conjunction with the eroding forces of the sea. *E. mathaei* is particularly active in the degradation of coral rocks, since it not only scrapes off rock material while feeding, but it also bores deep pits in to the rocks where it seeks cover during the day. We also find similar "rock borer" in the Indo-Pacific, *Echinostrephus molaris,* and the eastern Pacific, *Diadema mexicana*. KRÜGER (1998) provides a rather interesting discussion about the phenomenon of echinoid bio-erosion.

By cleaning off areas of substratum, sea urchins provide virtually pristine settlement areas for other animals, for instance, for the planula larvae of corals. Without such "substrate cleaners" many coral organisms would have difficulties finding suitable settlement sites.

Although algae represent the most important food for sea urchins many of them also have tendencies towards omnivorous feeding behaviour. When scraping off the organic surface layer they also take in a wide spectrum of vegetable and animal foods, as well as detritus. This occurs inadvertently during the consumption of algae, but to a certain degree there is also targeted predation, especially when sessile, colonial invertebrates are involved. Some sea urchins even feed on corals, especially when there is a lack of suitable vegetable food. Replacement of normal algae food with animals that contain zooxanthellae is also known from many other herbivores (e.g. Blennies, *Ecsenius* spp. and sea stars.

Most irregular sea urchins feed predominantly on organic particles, detritus, and dead animals as well as on tiny organisms that are found while burying in soft substrate. In fact, the "Aristotle's Lantern" is absent in species of superorder Atelostomata. In these sea urchins modified tube feet dig through the substrate and then transport the food to the mouth. Overall, there are significant differences in the behaviour and ecological role between regular and irregular sea urchins.

All sea urchins have separate sexes, but generally there are no external sexual characteristics. In the nominal form, all have five gonads (ovaries and testicles) that are attached to the inside of plates in the inter-ambulacral zones. Yet, in most irregular sea urchins the particular gonad, located at the inter-ambulacral groove that makes up the posterior end of the urchin, is absent, so that these animals in effect have only four gonads. In most species, eggs and sperms are released in to open water via gonopores in the genital plates, through contraction of muscle layers surrounding the gonads. Fertilization takes place outside the animals in open water. Most shallow water species have annual or monthly reproduction cycles, which are probably controlled by lunar phases.

Many sea urchins – like this long-spined *Diadema setosum* – are extremely efficient algae grazers, and they also maintain this feeding behaviour in an aquarium. Consequently, one must be aware that they can inflict damage on the biological diversity in an aquarium. That is, they will not only eat the ugly filamentous algae, but also the decorative forms such as red calcareous algae, and together with it also the microscopic animal life that is of such importance for the biological stability of a coral reef aquarium.

The larvae of sea urchins – referred to as echinopluteus larvae – have many features in common with sea stars. Some of them feed on plankton, while others derive nutrition from the yolk sac of the egg, and so they do not take in any external food. Complete larval development can take up to four months. While the larva is in its pelagic stage it commences to develop adult characteristics, such as the first skeletal plates, genital and terminal plates. Once the larva has settled to the bottom, metamorphosis will start quickly, often within an hour. The result of this is a small copy of an adult sea urchin that has a diameter of only 1 mm.

There are also a few species – especially Antarctic coldwater forms – that practice brood care. From members of family Cassidulidae it is known that – following external fertilization – they collect the free-swimming larvae and keep them protected among the spines of the maternal animal. A more common occurrence is that brood-caring species transfer their eggs to external depressions in the skeleton. There the young develop – without going through a larval stage (and the need for a metamorphosis) – directly in to small sea urchins. Examples of this type of reproductive behaviour can be found in certain species from several different orders, including Cidaroida, Temnopleuroida, Spatangoida, Cassiduloida and Clypeastroida. In addition, there are other brood-caring species with specific brood pouches. This in turn involves fundamental changes of the body structure. These pouches are directly connected to the gonads and from there to the outside by opening of the genital plates on the aboral side. Such brood chambers can be quite large and take up most of the space inside the body cavity. In those three or four chambers there can be eggs and embryos or juveniles side-by-side in different developmental stages. With a size of 4 to 5 mm at the time of "birth" the young is a miniature copy of the adult sea urchin. This extreme form of brood care has been described by KRÜGER (1997) for deep-water species of the irregular sea urchins in family Urechinidae.

There have been a few unsubstantiated reports about spawning of sea urchins in aquariums, discussing primarily the release of spermatozoa by males; however, we have heard nothing about any juvenile development or the rearing of sea urchins, which is totally in contrast to controlled experiments undertaken in laboratories. The large eggs of many sea urchins are popular research objects in embryology.

Like many other groups of echinoderms, sea urchins also accommodate a diverse selection of small and large symbiotic organisms. Primarily these are crustaceans, such as shrimps (*Allopontonia*, *Periclimenes*, *Stegopontonia* and others) but also crabs (e.g. *Zebrida* and *Echinoecus*) and molluscs (snails and clams). There are also several fish species, which utilize sea urchins (mainly *Diadema* species) for cover and as hiding places. A good example is the relationship between the Razor fishes *Aeoliscus* spp. and the Yellow-Striped Clingfish *Diademichthys lineatus*. Beyond that, we will specifically refer back to the most frequent symbionts of various sea urchins – were appropriate – when discussing particular species in the Fact Sheets below. Those readers, who are looking for a more general overview, are referred to ZANN (1980).

The, for echinoderms, so typical pentamerous symmetry and the many holes and pores in the shell (the latter represent important identifying characteristics) normally become visible only after the spines have been removed. Consequently, there are often considerable difficulties with correctly identifying live specimens. And so, with sea urchins in an aquarium one has invariably little choice but to accept identification down to the level of genus- or even to family only.

Sea urchins are sufficiently well known as potential algae eaters in a coral reef aquarium. In fact, many species are incredibly effective grazers, but (as we have seen above) unfortunately they are also often not particularly selective in their choice of food. Once a sea urchin has crawled over a live rock there is rarely ever anything left behind. All types of algae, including calcareous algae, are completely removed. Even live rocks or other items of the aquarium decoration are being gnawed over so that within a very brief period of time all that is left is a very sterile decoration, totally denuded of the micro- and macro-biological diversity, which characterizes a well-functioning modern aquarium.

It is, however, imperative that we do not totally write off sea urchins as aquarium animals. In very large aquariums or in cases where tanks are being devastated by uncontrollable algal growth, sea urchins can readily be utilized as prudent algae control. For such cases one can use the rule of thumb of approximately one sea urchin per 150 – 200 l aquarium water. But sea urchins must not be used if attractive calcareous algae growth in a tank is to be retained. Of course, on the other hand many algae species are so interesting that a serious, dedicated aquarist can set up special, dedicate algae tanks.

Below we are introducing a limited selection of sea urchins that we consider to be particularly interesting, either as coral reef aquarium inhabitants or simply as topical aquarium organisms.

Subclass Perischoechinoidea

This subclass contains mainly primitive, now extinct sea urchins, known from fossil finds. For instance, *Archaeocidaris* spp. occurred from the Carboniferous to the Permian Period, *Plegiocidaris* spp. during the Jurassic Period and *Hemicidaris* spp. from the Jurassic to the Cretaceous Period. The latter species are relatively frequent among fossil finds. The oldest genus that has been found is *Bothriocodaris*. It originated during the Ordovician Period, i.e. about 450 million years ago. What fascinates us now when studying fossils is the fact that many well-preserved fossil specimens are remarkably similar to their recent relatives. All currently living forms of this subclass belong in to order Cidaroida. It represents the only branch that survived the Cretaceous Period. There are two families: Psychocidaridae and Cidaridae, but only the latter will be discussed.

Family Cidaridae

The species in this family are generally referred to as "lancelet sea urchins", since they are

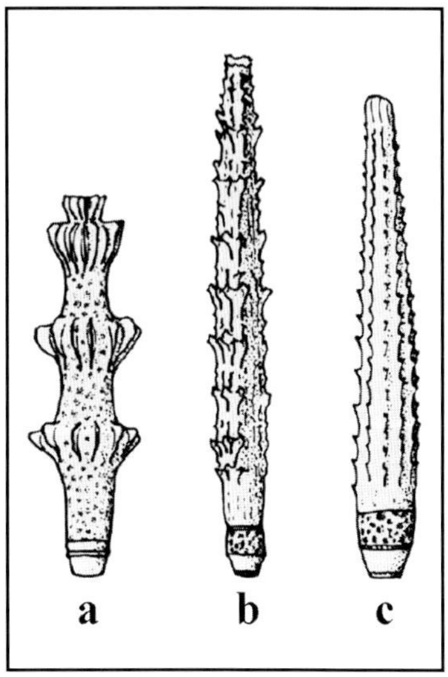

To the right: Examples of the shapes and structures of primary spines of slate pencil urchins: a - *Prionocidaris verticillata*, b and c – *Prionocidaris baculosa*.

Drawing after CLARK & ROWE (1971)

Eucidaris metularia
Small Salter Pencil Urchin

Distribution: Occurs widely in the Indo-Pacific, from the Red Sea and East Africa to Hawaii and the Society Islands, northward to Japan and southward to Australia.
Size: Diameter of corona to 3 cm, total diameter (including spines) 10 cm.
Reproduction: Details unknown.
Food: Feeds principally on algae, but will also take bryozoans, sponges and organic detritus. Will accept substitute foods to some degree, using pieces of molluscs and crustaceans.
Description: Nocturnal; normally solitary along reef margin or along reef slope, at depths from 2 to 40 m. Attractively coloured, spines in older specimens often with epiphytic growth.
Aquarium requirements: Rarely available in the aquarium trade. Remarkably hardy when kept in an aquarium decorated with live rocks and when an ample, diverse microscopic fauna is present. Will feed on most algae species, including red calcareous algae; will not attack large invertebrates.

GA: +/-; **SE:** +; **CO:** +/-

Photo:
Eucidaris metularia in the Red Sea. Photo: R. Brons

Left: *Phyllacanthus imperalis*, photographed during a night dive in Indonesia (photo: R. Hansen), with pencil-thick spines, is a typical representative of family Cidaridae. Right: The subtropical *Goniocidaris tubaria*, seen here at Bussleton, Western Australia (photo: C. Bryce), has extremely thorny spines that are reminiscent of *Acropora* branches.

equipped with strong, lancelet-like primary spines. In many species the spines are as thick as lead pencils. Taking in to account the rigidity of these spines, together with their thickness, the colloquial name for these urchins in the English aquarium literature is "slate pencil urchins". Also, there is no skin covering the surface of the spines, so they tend to be overgrown with algae and various organisms.

In the Caribbean region there is only a single species: *Eucidartius tribuloides*, while there are about 12 species known to occur on reefs in the Indo-Pacific, that are distributed over four genera: *Chondrocidaris*, *Eucidaris*, *Phyllacanthus* and *Prionocidaris*.

Chondrocidaris is a genus primarily characterized by the fact that the spines are rather irregular in their shape. In *Chondrocidaris brevispina* the extremely short and thick spines can be reminiscent of an acorn with stem and shell, while *C. gigantean* has spines that resemble the skeleton of *Acropora* stony corals. Similar shapes can be found in *Goniocidaris tubaria* that inhabits subtropical Australian seas. Several *Prionocidaris* have similarly irregular, *Acropora*-like spines; however, they are far sharper and more pointed than those in the preceding species.

The other two genera mentioned above have rather regular and smooth spines with only a weak indication of a surface structure. *Eucidaris* spp. have a weakly indicated structure of regular rows of tubercles on each spine, while *Phyllacanthus* spp., have longitudinal grooves instead. Both variations should be readily recognizable on photographs of the respective species.

Most species are virtually omnivorous. The main constituent of the diet is algae, but sponges, Bryozoans and other encrusting invertebrates are also often included. In fact, even tunicates, bristle worms and cnidarians are eaten by some species, if the opportunity arises. Under aquarium conditions an exclusive algal diet is insufficient and must be supplemented with fresh or frozen clam meat. Apart from their specific dietary requirements, these urchins are relatively suitable aquarium inhabitants that do fairly well under optimum condition.

Sea urchin imports include frequently *Eucidaris tribuloides* (see Fact Sheet on page 337) from the Caribbean, as well as *E. metularia* and *Phyllacanthus imperialis* (see Fact Sheet on page 338) from the Indo-Pacific.

Subclass Euechoidea

This subclass contains the vast majority of recent (currently living) sea urchins. It encompasses four superorders. Superorders Diadematacea and Echinacea are regular sea urchins that were formerly maintained in Class Regularia. Altogether, these two superorders contain eight orders, of which we will discuss orders Echinothuriida, Diadematoida, Temnopleuroida and Echinoida in some detail. Within these four orders we find the majority of sea urchins that are conspicuous on coral reefs and important to reef ecology. Many of these are also frequently available in the aquarium trade.

Irregular sea urchins (formerly kept in Class Irregularia) belong to the two superorders Gnathostomata and Atelostomata. Since they are unfortunately rarely available in the aquarium trade, as well as being difficult to find on a reef and hard to identify, we will discuss the two superorder and their respective orders together on the following pages.

Eucidaris tribuloides
Caribbean Slate Pencil Urchin

Distribution: Along the east coast of America and throughout the Caribbean region, from North Carolina southward to Rio De Janeiro, Brazil. The subspecies *Eucidaris tribuloides africana* is known to occur off West Africa and the Cape Verde Islands.
Size: Diameter of corona to 4 cm, total diameter (including spines) 13 cm.
Reproduction: Reaches sexual maturity in 2 years. Gametes mature in early spring and summer. Embryos develop from fertilized eggs to metamorphosed youngs in 25 days (McPherson, 1968).
Food: Appears to be omnivorous, although there is a preference for algae and encrusting invertebrates such as bryozoans. Will accept a wide range of substitute foods, including frozen- and dry food.
Description: Because of the typical solid, thick and cylindrical spines cannot be mistaken for any other Caribbean sea urchin species. Primary spines arranged in ten vertical rows and because of the presence of epizootic animals and algae the colouration is highly variable. Spines that have been cleaned often have wide, dirty white or reddish-brown bands. On the reef it lives in narrow cracks and crevasses or among rocks and rubble in lagoon areas from the tidal zone down to a depth of 50 m. Occasionally encountered as deep as 800 m. Host for various more or less parasitic gastropods that live on its oral side.
Aquarium requirements: Does not appear very often in the aquarium trade; imported only occasionally. Hardy when kept in an aquarium decorated with live rocks and with a diverse micro-fauna and -flora. Has reached longevities of up to 5 years in aquariums. Feeds on most algae species, including red calcareous algae, and normally does not attack larger invertebrates. Yet, in the wild specimens have been found with coral fragments in their stomachs.

GA: +/-; **SE:** +; **CO:** +/-

Photos:
Eucidaris tribuloides in an aquarium (top; photo: S. W. Michael) and at Puerto Rico (bottom).

Phyllacanthus imperalis
Imperial Slate Pencil Urchin

Distribution: Indo-Pacific; from the Red Sea and East Africa to the Marshall Islands and to Tonga.
Size: Diameter of corona to 8 cm, total diameter (including spines) 23 cm.
Reproduction: Spawns in large aggregations, releasing the gametes into the surrounding water.
Food: Feeds principally on calcareous algae, but will also take other algae species as well as encrusting invertebrates and molluscs.
Description: Can be readily identified on the basis of its dark corona and spines that are normally overgrown by various epizootic organisms. Cleaned spines are red or with brownish bands. Lives in caves and crevasses on shallow reefs at depths from 0 to 15 m.
Aquarium requirements: Occasionally available in the aquarium trade. Should be kept in an aquarium decorated with live rocks, and with a rich micro-fauna and -flora present. Appears to be rather hardy. Feeds on most algae species including red calcareous algae. Adult specimens can bother larger invertebrates, but corals are usually not harmed.

GA: +/-; **SE:** +; **CO:** +/-

Photos:
Phyllacanthus imperalis in Indonesia (top) and in the Red Sea (bottom; photos: E. Svensen and S. W. Michael).

Asthenosoma varium, seen here in an aquarium, is a highly venomous sea urchin and its stings can cause severe pain and swellings in humans that can last for several hours. In spite of its beauty it should only be selected for aquarium maintenance after due deliberation. Photo: S. W. Michael

Order Echinothuriida

The species in this order are distinguished from all other sea urchins in that the plates of the corona are not tightly fused, and so it feels noticeably flexible. Indo-Pacific coral reefs are inhabited by two interesting representatives of genus *Asthenosoma*: *A. varium* and *A. ijimai*. Both are incredibly beautiful urchins with fluorescent colouration and thin, hollow spines. In essence they are urchins that one would inevitably pick up and examine. But their appearance is deceiving! **These sea urchins are extremely venomous, and their venomous spines can penetrate right through rather thick gloves**.

Each spine is fitted with a simple, large venom gland located at its outer-most tip. When the spine penetrates human skin it simultaneously injects the venom. Usually the tip of the spine breaks off and becomes embedded in the skin, which leads to further, secondary infections. The sting causes immediate severe pain, followed by long-lasting swelling. The most effective treatment consists of keeping the affected body part in hot water of at least 50 °C (as warm as is bearable) until the pain eases. In addition, remaining spine remnants should be surgically removed (Marsh & Smith, 1986).

Asthenosoma varium is by far the most frequently occurring species, with a distribution from the Red Sea to Australia and to the Philippines. There are several colour varieties with purple-red, brown-orange, greenish or yellow-white base colouration. The distinct rows of spines with large bubbles at their tip, and the long, thin and beard-like spines that surround the lower edge of the corona are a typical characteristic of all colour variants. *A. ijimai* is in its distribution confined to the Indo-Australian Archipelago and northward to Japan. This species has uniformly arranged spines, which – in terms of their size – are more even than those of *A. varium*. Both sea urchins are aggressive feeders that will devour anything that comes in to their path, including most of the invertebrates we like to keep in aquariums. Moreover, they move rather rapidly. Adult specimens can reach a diameter (including the spines) of 20 cm (*Asthenosoma ijimai*) or 25 cm (*A. varium*). It goes without saying that these sea urchins can only be kept in a special aquarium. Although they do reasonably well in an aquarium and feeding them is relatively simple (with most types of substitute food), it must be pointed out here that the average aquarist is well advised to stay away from these species.

As a curiosity we should note here that *Kamptosoma abyssale*, which belongs to the same family, is the sea urchins with greatest depth distribution. It has been found at depths from 4,374 m down to 6,235 m. (MIRONOW, 1971). Overall, order Echinothuriida contains many deep-water species, of which some are virtual giants, with corona diameters of 25 to 30 cm.

Order Diadematoida

This order, represented by family Diadematidae, contains what are most likely the most conspicuous sea urchins found anywhere in the sea. There is probably not a single diver or aquarists that has not encountered one or the other *Diadema* species. They occur in all tropical seas, and frequently they inhabit shallow, protected areas, especially around rocks and coral structures, where they seek shade during the day. With their extremely long and needle-sharp spines they are spectacular and beautiful sight. Many species are commonly found in groups, sometimes even in large aggregations. *Diadema* species can occur in densities of more than 15 to 25 specimens per square metre.

During the day, these urchins tend to remain partially in hiding, usually in caves or in reef crevasses. At night they emerge, and then graze the algae off rocks, coral rubble and other hard substrates. Several species show a distinct preference for sea grass beds and mangrove stands. Many of the sea urchins in this family are very active and react rapidly to environmental changes, e.g. disturbances, such as sudden light changes or changes in water movements. Such reactions have been observed in *Diadema*. Species, that they move away rapidly from the affected site or that they take on a defensive posture, whereby most of their spines are pointed in the direction of the source of the disturbance. In fact, the spines represent an effective defence against most enemies; however, it is also not uncommon to see resilient triggerfishes or puffers continuously attacking urchins in spite of suffering severe injuries from the spines.

Humans too are not safe from *Diadema* urchins stings. Although a specific venomous substance has never been isolated from these urchins, the reactions in humans to stings from these urchins, suggest that there must be a substance that causes poisoning. This could possibly involve a neurotoxin, contained in the thin skin that covers the spines of the sea urchins (MARSH & SLACK-SMITH, 1986; MEBS, 1989). The long, needle-thin spines are very fragile and incredible sharp and brittle, so that they can easily penetrate human skin, break off and then become embedded in the skin. Similarly, it is also quite possible for the spines to penetrate a diver's suit or even sneakers. Because of the rough surface area and brittle structure of the spines it is almost impossible to remove them from a flesh wound. The general symptoms after a sting include an immediate, intensive pain that can last from 30 minutes to four hours, which is followed by swelling, often together with inflammatory reactions. Usually such a sting does not cause a severe medical emergency, unless there is an allergic response.

Spine fragments remaining in the wound are usually dissolved and absorbed by the human body without complications. Larger fragments are usually pushed out through the skin surface by a tissue (rejection) reaction. In Indonesia there is a traditional treatment that consists of rubbing the wound with a smooth stone so that the spine fragments are broken in to smaller pieces, which are then absorbed more quickly. Western medicine is more prone to remove the foreign body by surgical means, however, often with less than satisfactory results (MARSH & SLACK-SMITH, 1986).

Genus *Diadema* is represented in the Indo-Pacific with three species, i.e. the rather widely distributed *D. savignyi* and *D. setosum* and with *D. paucispinum* from the South-Pacific islands and from Hawaii. All species are characterized by extremely long spines that can be up to four-times longer than the diameter of the corona. Colouration can be variable, but often they are totally black or black and white, possibly also with a tinge of blue or withy reddish spines. *D. setosum* is clearly the most common species available from the aquarium trade. Normally it can be reliable identified by the orange-red ring around the anal opening. However, some populations and specimens (e.g. from the Red Sea) have a whitish or grey ring instead (for further details refer to Fact Sheet on page 346). *D. savignyi* is being traded less commonly. Normally this species has a uniformly-coloured anal opening, but possibly also with a blue ring. In some specimens the spines may have black and white cross bands; however, these are not definitive identification characteristics. *D. paucispinum* has fewer spines and these are further apart, than in the two other species. In extreme cases, all species can reach a total diameter of 40 to 50 cm, whereby the corona is rarely larger than 9 cm, usually even much smaller.

Diadema antillarum occurs in the Caribbean and the Atlantic, eastward to West Africa. This species is normally also black, but it can have a few white or grey spines. There can also be shades of red or blue, just as in the Indo-Pacific species. Juvenile specimens usually always have spines with black-white cross bands. Fully-grown specimens can reach a corona diameter of 10 cm and a total diameter of 50 cm. The longest of the upper spines can be more than 40 cm long! The aquarium trade also intermittently imports this species.

Diadema mexicanum, a rarely imported sea urchin, occurs in

"Fire works" in an aquarium! Family Diadematidae includes some highly spectacular sea urchins. The extremely long and magnificently coloured spines make this *Diadema* species a virtual feast for the eyes. Photo: T. Luther

the eastern Pacific. It can be clearly distinguished from the preceding species by spines that are substantially shorter relative to the corona diameter.

All *Diadema* species are primarily algae eaters, and as such they are very effective. Juvenile specimens appear to have a preference for calcareous algae, but this tends to diminish with age. This may be due to greater carbonate (calcium) requirements because of rapid growth. On the other hand, with increasing age this species becomes omnivorous, which then substantially increases the risk, that these sea urchins – as they get larger – will prey on sponges, Bryozoans and other small invertebrates.

Diadema species are not the only members of this family. The genera *Astropyga*, *Centrostephanus*, *Chaetodiadema*, *Echinothrix* and *Lissodiadema* are close relatives that live on coral reefs. Many species from these genera are also rather attractive and impressive-looking.

Genus *Astropyga* contains two particularly beautiful sea urchins: the Caribbean *A. magnifica* and the Indo-Pacific *A. radiata*. Both have many anatomical similarities with *Diadema* species, e.g. long and sharp spines. But, in contrast to *Diadema* species by *Astropyga* species have large, naked sections in the interambulacral grooves, so that the spines are joined in "bundles". The base colouration of both species is reddish to orange-pink or golden yellow. The spines are red brown, with yellow or white stripes (for details on *A. radiata* refer to Fact Sheet on page 342).

Echinothrix is a genus endemic to the Indo-Pacific, with two species: *E. calamaris* and *E. diadema*. Both have a wide geographical distribution that extends from East Africa and the Red Sea all the way to Hawaii; however, they are somewhat less frequently encountered on coral reefs than their relatives from genus *Diadema*. There where *Astropyga* species have naked areas on their corona, *Echinothrix* species are equipped with extremely thin, secondary spines, which extend in five distinct bands from the anus to the oral side. The primary spines are substantially stronger and longer, even though they are not as long as those of *Diadema* species.

Both *Echinothrix* species are venomous. The secondary spines carry venom glands at their tips that contain Noradrenalin, which adversely affects the blood pressure of victims. In addition, they also contain a protein that causes considerable pain. When a human is stung, the immediate reactions are burning, stinging pain and swellings. Gradually thereafter the skin turns red or it may be-

Astropyga radiata
Red Diadem Urchin

Distribution: Indo-Pacific; from South- and East Africa to Hawaii.
Size: Diameter of corona to 20 cm, total diameter (including spines) 45 cm.
Reproduction: Details unknown.
Food: Feeds principally on algae, including calcareous algae as well as on encrusting invertebrates and dead organisms. Will accept most substitute foods.
Description: Can be recognized by its attractive red colouration and the long, brownish-banded spines that are grouped in to bundles. Among these spines are grooves with rows of small, light-blue dots. The large anal vesicle is transparent whitish. Lives in sandy or rubble habitats, at depth from 10 to 30 m. Accommodates symbionts that include shrimp fishes, e.g. *Siphamia versicolor* and juvenile *Lutjanus sebae*. *Astropyga magnifica* is a rather similar looking sea urchin from the Caribbean region.
Aquarium requirements: Rarely ever available in the aquarium trade. Appears to be rather hardy in aquariums. Requires sandy areas as well as rocks overgrown with algae. Feeding does not present any problems. Unfortunately caution is advised in terms of compatibility with other invertebrates.

GA: +/-; **SE:** +; **CO:** –

Photo:
Astropyga radiata at Sulawesi.

Photo: J. Cairns-Michael

20 specimens of *Diadema setosum* have gathered on the sandy bottom covered with coral rubble at "Monkey Bay" at the Island of Phi-Phi Don in Thailand. In many Indo-Pacific reef locations this is how the long-spined sea urchins are typically spotted. In some reefs they can be very numerous. Their long spines are efficient weapons against predators as well as against humans.

come severely discoloured. In contrast to stings from *Diadema* urchins, spine remnants remain embedded in the skin and are not absorbed. Instead, they must be surgically removed otherwise there may be serious after effects.

As far as *Echinothrix diadema* is concerned, it is a known fact that this sea urchin often feeds on stony corals, especially on *Porites-*, *Montipora-* and *Echinopora* species. Even in the sea, these sea urchins can inflict serious damage to coral populations. Consequently, one can assume that under the artificial aquarium conditions these sea urchins can cause extreme damage. So far it has not been yet proven conclusively whether *E. calamaris* has the same food preference, yet aquarists must be fully aware of the risks involved; especially so since both species can easily be mistaken for each other, or be incorrectly identified. Beyond that, both species feed also on algae and tiny organisms that live among algae.

Both *Echinothrix* species are occasionally imported, however, they should only be kept by aquarists with special interest in these urchins, and then they must be kept in special aquariums (for further details refer to Fact Sheets on pages 347 and 348).

Order Temnopleuroida

In this order there are two interesting families, Temnopleuroidae and Toxopneustidae, that are found on coral reefs. Both contain many species that are highly characteristic, attractive and colourful. Moreover, they are reasonably often available in the aquarium trade. In general, these sea urchins can be considered as relatively suitable aquarium inhabitants, with algae as their main food requirement, but there are a few exceptions. As with other sea urchins, one always needs to keep in mind that in the event of any food shortages the path from an exclusive herbivorous sea urchin to one that is feeding on a wide range of (available) invertebrates can be rather short indeed! The obvious conclusion is that herbivores are certain algae eaters as long as they have sufficient food available. However, as soon as there is an insufficient algae supply in the aquarium, the urchins will commence searching for substitute food.

Family Temnopleuridae

A characteristic feature that is typical for the majority of members of this family is that the corona bears sculpture-like patterns on its skeletal plates. Often there are also distinct depressions, penetrations or pores at the angles where the plates are fused. In addition, there are often characteristic patterns in the development

Diadema antillarum
Atlantic Diadem Sea Urchin

Distribution: Atlantic and adjacent seas: in the Caribbean region from the Gulf of Mexico and Bermuda Islands to Rio De Janeiro, Brazil; in the eastern Atlantic from the Azores, Madeira, Canary Islands, Cape Verde Islands and the island of Annabón and in the Gulf of Guinea, West Africa.
Size: Diameter of corona to 10 cm, total diameter (including spines) 50 cm. The corona is flattened so that its height is normally less than 50 % of its diameter.
Reproduction: Significance of lunar phases in regard to spawning has been clearly demonstrated. Around the Bermuda Islands during the period of early summer through early winter, spawning is linked to a certain rhythm of lunar phases (ILIFFE & PEARSE, 1982).
Food: Feeds principally on algae, predominantly on turf algae. Young specimens prefer calcareous algae, while large one graze on epipelic- and endolithic algae. Tests have shown that algae grazing by *Diadema antillarum* is of considerable significance to the reef ecology. It was observed, that this sea urchin also feeds on young corals (RYLAARSDAM, 1983) and encrusting anemones (KARLSON, 1983) as well as on sponges and other encrusting invertebrates.

Description: Has extremely long and fragile spines. Typical colour of the corona and spines is black, however, colour variants occur, including specimens with white or silver-grey spines. Juveniles always have spines that are banded black-white, and so can be mistaken for another species. *Diadema antillarum* lives on reefs as well as in sea grass beds and mangrove areas. It prefers calm water and avoid areas where there is wave action. Occurs in large aggregations with a density of 20 or more specimens per square metres. Usually lives at depths from 0 to 50 m, sometimes even down to 400 m. Various crustaceans (shrimps, mysids and copepods) and juvenile fishes live symbiotically with this urchin.
Aquarium requirements: Should be kept in an aquarium decorated with live rocks and where there is a diversity of microscopic life. If it has been properly acclimated it tends to be quite hardy. Feeds on all species of algae, including red calcareous algae. In a small aquarium the decoration is quickly denuded of all growth, giving it a rather sterile appearance. Large specimens cannot really be trusted in terms of compatibility with other invertebrates, although

Photo:
Diademia antillum at Puerto Rico

damage to corals and other cnidarians would be an exception rather than the rule. It is important to remember that this sea urchin can inflict serious injuries to humans, causing intense (neuro-toxic) pain and (secondary) infections.

GA: +/-; **SE:** +/-; **CO:** +/-

Photo:
Diadema antillarum at the Canary Islands.
Photo: E. Svensen

Diadema mexicanum
Eastern Pacific Diadem Sea Urchin

Distribution: Eastern Pacific; Gulf of California (Sea of Cortez), coast of Mexico and at the Galapagos Islands.
Size: Diameter of corona to 8 cm, total diameter (including spines) 25 cm.
Reproduction: Details unknown.
Food: Feeds mainly on algae, including calcareous algae, but also takes encrusting invertebrates such as sponges and bryozoans.
Description: With long and fragile spines that are shorter, overall, than in other species of this genus (rarely longer than the corona diameter). Adults have a virtually black corona and a mixture of black and white spines. The spines in juveniles have black and white bands. Lives in calm, rocky areas, at depths from 0 to 40 m.
Aquarium requirements: Rarely ever imported. Should be kept in aquariums decorated with live rocks and an abundance of microscopic life. When properly acclimated a durable aquarium inhabitant. Feeds on all types of algae. In small tanks the decoration will quickly be denuded of all growth, giving it a rather sterile appearance. Cannot be totally trusted in respect to other invertebrates, but normally will not harm corals. It is important to remember that this sea urchin can inflict serious injuries to humans, causing intense (neuro-toxic) pain and (secondary) infections.

GA: +/-; **SE:** +/-; **CO:** +/-

Photo:
Diadema mexicanum in Mexico. Photo: E. Svensen

Diadema setosum
Diadem Sea Urchin

Distribution: Indo-Pacific; from the Red Sea and East Africa to the South Pacific islands.
Size: Diameter of corona to 9 cm, total diameter (including spines) 40 cm.
Reproduction: Forms large aggregations during spawning seasons, which are determined by lunar cycles.
Food: Feeds mainly on algae, including calcareous algae, but will also take encrusting invertebrates such as sponges and bryozoans.
Description: Has extremely long spines. Adult specimens are generally totally black, but there are also specimens with white spines. In juveniles the spines have a contrasting pattern of black and white bands. There are always five white-blue spots on the aboral side arranged around the anus. The anal vesicle with a characteristic bright orange or red ring, which can sometimes be grey. Lives in sizeable aggregations of 15 to 20 specimens or more per square metre in coral habitats, at depths from 1 to 25 m. Host to a large number of symbiotic organisms: easily identifiable fishes, such as *Aeoliscus* spp., *Diademichthys* spp. and a few apogonids, various shrimps and crabs, as well as a comb jelly of genus *Coeloplana* (a related ctenophore uses the Luzon Sea Urchin, *Echinaster luzonicus* as a symbionts).
Aquarium requirements: Rather frequently available in the aquarium trade. Should be kept in an aquarium decorated with live rocks and where there is an abundance of microscopic life. Very hardy when properly acclimated. Feeds on most algae species. In small tanks the decoration will quickly be denuded of all growth, giving it a rather sterile appearance. Cannot be totally trusted in respect to other invertebrates, but normally does not harm corals. It is important to remember that this sea urchin can inflict serious injuries to humans, causing intense (neuro-toxic) pain and (secondary) infections.

GA: +/-; **SE:** +/-; **CO:** +/-

Photo:
Diadema setosum in an aquarium.

Echinothrix calamaris
Pencil Diadem Sea Urchin

Distribution: Indo-Pacific; from the Red Sea and East Africa to Polynesia and Hawaii.
Size: Diameter of corona to 16 cm, total diameter (including spines) 30 cm.
Reproduction: Details unknown.
Food: Feeds mainly on algae, including calcareous algae, but also dead organisms and other organic waste, also stony corals. Will accept most substitute foods.
Description: Brightly coloured sea urchin. Adult specimens retain the light-dark banded primary spines, as are normally seen only in juveniles of family Diadematidae. In rare instances, very large specimens can be almost totally black. Lives solitary or in small groups only. Can often be found under coral heads and among coral rubble in shallow reef habitats down to a depth of about 40 m. Known as host to symbiotic animals, including fishes such as *Siphamia versicolor* and juveniles of *Lutjanus sebae*.
Aquarium requirements: Rarely available in the aquarium trade, however, a durable aquarium inhabitant. Should be kept in an aquarium decorated with live rocks and where there is a multitude of microscopic life. In small tanks the decoration will quickly be denuded of all growth, giving it a rather sterile appearance. Cannot be totally trusted in respect to other invertebrates, although damage to corals and other cnidarians are rather an exception than the rule. It must be noted here that the closely related *Echinothrix diadema* (a species it can easily be mistaken for) is an aggressive predator of corals. It is also very important to keep in mind that this type of sea urchin can inflict severe injuries to humans, which can cause intensive pain, reddening and swelling. Secondary infection and other after effects arising from spine remnants embedded in the skin are common. The tips of the spines are encased by tissue that contains venom glands. These glands produce Noradrenalin that causes high blood pressure, as well as a type of protein that is assumed to be the cause of the burning pain.

GA: +/-; **SE:** +; **CO:** +/-

Photo:
Echinothrix calamaris in an aquarium.

Echinothrix diadema
Pincushion Diadem Urchin

Distribution: Indo-Pacific; occurs widely from the Red Sea and East Africa to Polynesia and Hawaii.
Size: Diameter of corona to 10 cm, total diameter (including spines) 22 cm.
Reproduction: Details unknown.
Food: Feeds principally on algae and organic detritus. Has proven to be an aggressive predator of stony corals. Appears to prefer *Porites-*, *Montipora-* and *Echinopora* species. Will accept most substitute foods.
Description: Less brightly coloured than the closely related *Echinaster calamaris*. While juveniles are often dark green primary spines with black bands, adults are mostly completely black. Lives among corals at depths from 3 to 10 m.
Aquarium requirements: Rarely seen in the aquarium trade, however, a durable aquarium inhabitant. It is known that it can cause significant, (localized) damage to stony coral populations. Consequently, it must not be kept in a coral reef aquarium; if at all, it should be kept in an aquarium decorated with live rocks and where there is an abundance of microscopic life. It is also very important to keep in mind that this type of sea urchin can inflict severe injuries to humans, which can cause intensive pain, reddening and swelling. Secondary infection and other after effects arising from spine remnants embedded in the skin are common. The tips of the spines are encased by tissue that contains venom glands. These glands produce Noradrenalin that causes high blood pressure, as well as a type of protein that is assumed to be the cause of the burning pain.

GA: +/-; **SE:** +; **CO:** –

Photo:
Echinothrix diadema in the Red Sea. Photo: Dr. I. Illich

Microcyphus rousseaui **(order Temnopleuroida) in the Red Sea.** **Photo: Dr. I. Illich**

of spines, their placement and colour markings. With such conspicuous characteristics, these species are the easiest of all sea urchins to correctly identify.

Beyond that, they are rather affable aquarium inhabitants, using algae as their predominant food. Sponges and other tiny organisms also living on the substrate that is being grazed upon by the urchins, will be included in their diet. Sponges, in particular have frequently been found as part of their stomach content. However, field studies have shown that large invertebrates and corals are not touched. Yet, there are reports to the contrary from aquarists, but here it appears reasonable to assume that this may have been more due to a lack of suitable algal food, rather than being a manifestation of natural feeding behaviour (also see comments above).

Seven genera are known from the Indo-Pacific; these are: *Mespilia*, *Microcyphus*, *Paratrema*, *Salmaciella*, *Salmacis*, *Temnopleurus* and *Temnotrema*, which contain a total of 20 species. Two species – *Mespilia globules* and *Salmacis bicolor* – are very frequently available in the aquarium trade. Therefore, both are discussed at length in Fact Sheets on pages 350 and 351. This family does not occur in the Caribbean region.

Family Toxopneustidae

As already indicated by the name, the members of this family are venomous. Here specialised pedicellaria are carrying venom organs, i.e. not the spines, as in the previously discussed, venomous sea urchins. Most of these sea urchins have rather tiny pedicellaria, which are used in defence and for cleaning off the surface. Typically, they consist of a set of jawbones ("sugar tongs") on top of a flexible stem. Some of the defensive jaws have venom glands and sharp, tusk-like teeth. These occur in many sea urchins, but as a rule they are very small and their venom effect is insignificant, except on tiny organisms. Yet, the venom glands found in the Toxopneustidae are sufficiently large in size and venom efficacy, that they can hurt large animals and even humans.

The purpose of such extremely venomous pedicellaria is primarily to facilitate the capture of small organisms (e.g. crustaceans). It is uncertain, however, to what degree they also serve as defensive weapons against predators. For instance, we know that the main predators of the Caribbean *Lytechinus variegatus* are sea birds, such as seagulls. These birds chop holes in to the urchins or drop them from great heights in order to crack the shell.

We find the most venomous sea urchins and probably the only ones that can be fatal to humans in genus *Toxopneustes*. Two species are known from the Indo-Pacific: *T. pileolus* and *T. maculates*, of which the former is clearly the most dangerous. *T. roseus* lives in the eastern Pacific. The spines in all three species are nearly totally hidden among the large, venomous pedicellaria, of which each can have a diameter of up to 3 mm. When the sensitive cilia on the inside of the jaws of the pedi-

Mespilia globulus
Globular Sea Urchin

Distribution: Central Indo-Pacific; from the Maldives to the Fiji Islands, especially numerous throughout the Indo-Australian Archipelago.
Size: Diameter of corona to 6 cm, total diameter (including spines) 7.5 cm.
Reproduction: Details unknown.
Food: Feeds almost exclusively on algae, possibly also on some encrusting invertebrates, such as sponges; however, the latter constitute only a minor component in their natural diet in the wild.
Description: Very attractive sea urchin, that can be readily identified on the basis of the ten velvet-blue, olive-green or black patches that separate the zones of short and sharp brown spines. The velvety appearance of the spineless areas is created by a cover of small, glandular pedicellaria. Often carries algae, coral fragments or other particles on its spines. Lives in shallow water on coral rubble and on platform reefs, at depth from 1 to 5 m.
Aquarium requirements: Hardy and rather peaceful sea urchin that is considerable better suited for a coral reef aquarium than most other sea urchins. Feeds on virtually all algae, including red calcareous algae. The white patches left over after the sea urchin has grazed, can really make an aquarium look dead and sterile. This is a general risk taken when sea urchins are kept in reef tanks. Therefore, it can create a sterile aquarium decoration when the aquarium is over-crowded. We are, however, not aware of any reports that this urchin damages larger invertebrates.

GA: +; **SE:** +; **CO:** +

Photos:
Mespilia globules in Indonesia (top; photo: C. Bryce) and in an aquarium (bottom).

Salmacis bicolor
Bi-coloured Sea Urchin

Distribution: Indo-Pacific; from the Red Sea and East Africa to the Central Pacific.
Size: Diameter of corona to 10 cm, total diameter (including spines) 13 cm.
Reproduction: Details unknown.
Food: Feeds principally on algae, but also takes some encrusting invertebrates, such as bryozoans and sponges. Will accept most substitute foods, but has a tendency towards food opportunism if there is insufficient regular food.
Description: A fascinatingly beautiful sea urchin, with very thin, sharp spines, banded in red and yellow or in green. Lives normally among corals and on coral boulders at depths from 2 to 50 m.
Aquarium requirements: Rather frequently available in the aquarium trade. Hardy sea urchin that should be kept in an aquarium decorated with live rocks and where there is an abundance of microscopic life present. Cannot be totally trusted together with other invertebrates. Although our experience has shown that it does not harm larger invertebrates (e. g. corals), there are also reports to the contrary. WILKENS (1987) writes, for instance, "...a predator that would rasp mainly on sessile invertebrates, ... and belongs to those few sea urchins, where attacks on even harmless, short-spined relatives and on sea stars ... had been observed..." We believe that this is an example of an opportunistic feeding behaviour, when there is an insufficient supply of suitable food; however, it is enough reason to be particularly cautious with this sea urchin.

GA: +/-; **SE:** +; **CO:** +/-

Photo:
Salmacis bicolor in an aquarium.

The Rose Sea Urchin, *Toxopneustes roseus* belongs to the most dangerous sea urchins. The large, pink-coloured pedicellaria are extremely venomous and can cause severe symptoms in humans, such as collapse, paralysis and respiratory difficulties. This urchin occurs from the Gulf of California to Ecuador and the Galapagos Islands. **Photo: S. W. Michael**

cellaria are stimulated, they snap shut and so force the venom through the tusk-like teeth. Although they are not capable of biting through thick human skin as found on the inside of hands or the underside of feet, the top of the hands and the arms are vulnerable to them. If a human as sustained a sufficient number of bites, the ensuing pain and nausea can be very substantial.

The venom blocks nerve signals and paralyses the musculature. Immediate symptoms include dizziness, breathing difficulties and numbness of lips, tongue and eye lids. Weak paralysis can occur in arms and legs. In serious cases it may become necessary to initiate artificial respiration or – if need be – administer heart massage. General treatment for shock, pain-reducing medication and anti-histamines are recommended as part of the treatment procedures (MARSH & SLACK-SMITH, 1986). Normally, the most severe symptoms diminish after 15 minutes to one hour; however, paralysis can last for several hours. In Japan diving accidents have happened because of *Toxopneustes* bites.

Toxopneustes pileolus, the cause of many venom attacks by sea urchins, is widely distributed throughout the Indo-Pacific, from East Africa to the central Pacific. It occurs from the inter-tidal zone down to depths of 90 m. Therefore, it is not very surprising that this sea urchin appears sporadically in the aquarium trade. Although *T. pileolus* is principally herbivorous (apart from some plankton that gets caught by the pedicellaria) and so it is not particularly problematic when keeping it together with other animals, we consider the normal trade of this species not advisable. For especially interested aquarists, who are familiar with the venom risk, *T. pileolus* is of course a fascinating aquarium inhabitant.

Apart from the intake of animal food caught by these sea urchins with their venomous pedicellaria, algae make up the bulk of the diet of most species. Many prefer rather fleshy algae, such as *Caulerpa* spp. From the small *Lytechinus williamsii* (with a maximum total diameter of 5 cm) from the Caribbean region, it is known that there are populations that have become specialized in feeding on *Acropora cervicornis* (CAYCEDO, 1979). Therefore it cannot be excluded that there are other family members that exhibit similar types of feeding behaviour. However, so far we have not been able to find evidence that this occurs naturally and on a regular basis. Some aquarists have reported that the frequently imported *Tripneustes gratilla* feeds on soft corals, especially on *Xenia* spp; however, we suspect that this is a manifestation of an emergency strategy by the sea urchin. As suitable substitute food in an aquarium we can offer dry food

Lytechinus variegatus **lives in the Caribbean region, at depths from 0 to 50 m. It is a highly variable sea urchin that can reach a total diameter of about 11 cm.** **Photo: S.W. Michael**

Tripneustes gratilla – seen here at El Quseir, Red Sea - is a representative of family Toxopneustidae, and is frequently available in the aquarium trade. It camouflages itself with all sorts of pieces of debris, a type of behaviour found in many species of this family. **Photo: Dr. I. Illich**

that that sinks to the bottom and frozen plankton, and – of course – fresh and frozen algae, or respectively, kelp directly from the sea.

Nearly all species in this family have the habit to cover themselves with pieces of algae, corals, clams, etc. and to cover this debris around with them on the aboral side, attached by means of the tube feet. It stands to reason, that this serves primarily as camouflage, but it could also be a method to provide shade against too much solar radiation. MILLOT (1956) has demonstrated that the Caribbean *Lytechinus variegatus* modified its particular collecting behaviour of such fragments, in response to light and shade conditions in the habitat.

We find the following three species on Caribbean coral reefs: *Lytechinus variegatus*, *L. williamsi* and *Tripneustes ventricosa*. There are at least 17 species in the Indo-Pacific that fall in to genera *Cyrtechinus*, *Gymnechinus*, *Nudechinus*, *Pseudoboletttia*, *Toxopneustes* and *Tripneustes*. From among these species, it is only *Tripneustes gratilla*, of which it can be said that it is frequently available in the aquarium trade. Therefore, we are discussing it in some detail in a Fact Sheet on pages 354/355.

Order Echinoida

This is the last group of regular sea urchins to be discussed here. It includes some of the most common and best-known sea urchins, such as those found frequently along European coastlines of genera *Echinus* and *Paracentrotus* (family Echinidae), as well as genus *Strongylocentrotus* (family Strongylocentrodidae), native to the west-coast of North America. In addition, this order also contains a number of very interesting tropical representatives. We will be discussing the families Echinometridae and Parasaleniidae in some detail.

Family Echinometridae

This family includes round or weakly oval-shaped sea urchins. The corona is smooth, without cavities or depressions in the shell. But that is where the mutual, external features among members of this group come to an end. Within genera *Anthocidaris*, *Colobocentrotus*, *Echinometra*, *Echinostrephus*, *Heterocentrotus* and *Zenocentrotus* we find sea urchins with sometimes rather different appearance. In terms of species diversity, this family is not particularly strongly represented on coral reefs, i.e. there are only twelve species in the Indo-Pacific region and two species on Caribbean coral reefs; however, some of them can be rather abundant in certain areas.

In the Indo-Pacific, *Echinometra mathaei* is probably the most frequently encountered and most numerous sea urchins in shallow water. Its distribution extends from the Red Sea and East Africa to Hawaii. In the eastern Pacific it is replaced by the rather similar *Echinometra vanbrunti*, with an identical mode of life and abundance. *E. mathaei* lives in the inter-tidal zone and on reef flats, down to a depth of about 8 m, i.e. it lives in extreme shallow water, where it feeds almost exclusively on algae. Its intensive grazing ac-

Tripneustes gratilla
Preacher's Cap Urchin

Distribution: Indo-Pacific; common from the Red Sea, East- and South Africa to Hawaii.
Size: Diameter of corona to 12 cm, rarely to 15 cm.
Reproduction: Spawns in autumn or winter, depending upon distribution area. There are a few reports about aquarium spawning, however, apparently only males were seen to release spermatozoa.
Food: Mainly herbivorous with a preference for large, fleshy algae. However, will also feed on planktonic and small benthic invertebrates that are caught and killed by the venomous pedicellaria of this species. Will accept most substitute foods, as well as frozen- and dry food. Should be given supplementary feeding with algae.
Description: One of the most common sea urchins throughout the Indo-Pacific, with a remarkable spectrum of variants, from brownish or orange through bluish or violet to virtually white or black colouration. The spinous areas on the corona are separated from each other by "naked" areas that have a velvet appearance and which are co-

Photos
Specimens of *Tripneustes gratilla* at the Fiji Islands (top and centre; centre photo: S. W. Michael) in the Gulf of Aqaba (bottom left; photo: Dr. I. Illich) and at El Quseir, Red Sea (bottom right; photo: Dr. I. Illich).

vered by venomous pedicellaria, often with contrasting colouration. Preacher's Cap Urchins live principally in the shallow water of lagoons or bays, often among sea grass, at depths from 1 to 8 m, but can also be found as deep as 70 m. It protects and disguises itself with sea grass pieces, shell fragments or other particulate matter, picked up and retained *in-situ* by the suction feet. The pedicellarian toxin is a mixture of proteins and polypeptides, which lowers blood pressure and causes red blood cells to burst, triggers heart attacks and catalyses expelled histamines (tissue hormone). A sting can cause severe pain in humans, but it is generally not dangerous.

Aquarium requirements: Rather hardy and quite suitable sea urchin. Some aquarists have reported that it attacks corals, especially soft corals (e.g. *Xenia*). Yet, we believe that this is an exception rather than the rule, and could possibly be due to a lack of suitable food. However, it cannot totally be trusted in terms of compatibility with other small invertebrates. Will also feed on red, calcareous algae.

GA: +; **SE:** +; **CO:** +/-

Photos

Specimens of *Tripneustes gratilla* in Coral Bay, Western Australia (top; photo: L. Newman and A. Flowers) at Abrolhos Island, Western Australia (centre; photo: C. Bryce) and in the Red Sea (bottom; photos: R. Brons).

Echinometra mathaei - seen here is a "colony" in the shallow water of Suva Reef, Fiji - grazes substrate clean with a surprising speed. During this activity the urchin digs out cavities where it then hides during the day.

tivities can severely affect the substrate, so that carbonate rocks are actually excavated by this little "gnawer" (for details on its significance in the large-scale bio-erosion of coral reefs, refer to page 331). Consequently, keeping this sea urchin in small aquariums is not always an unequivocal success, since it contributes to a rather sterile appearance of the aquarium decoration. However, other invertebrates are generally left in peace (for further details refer to the Fact Sheet on page 360).

The two Caribbean *Echinometra* species, i.e. *E. lucunter* and *E. viridus*, are also distinct algae eaters. Especially, *E. lucunter* (see Fact Sheet on page 359) can readily be considered a "worthy equal" to *E. mathaei* when it comes to rock erosion. OGDEN (1977) has calculated that it erodes up to 7 kg substrate per square metre, off the Bermuda Islands, but this, of course, also depends on the specimen density. For instance, studies at Barbados have revealed that a single specimen on the average "gnaws away" 24 g of calcareous rocks per year. That may not sound like much, but it clearly explains the traces left behind by this species in a small aquarium.

Beyond that, *Echinometra* species are hardy, easy-to-keep aquarium animals, and the least problematic as companion animals for other invertebrates in the same aquarium. Feeding is done primarily with live algae, alternately with fresh or frozen sea algae, possibly also with lettuce or other leafy green vegetables. It is unlikely that these sea urchins require an animal protein component in the diet for their well being.

Genus *Echinostrephus* also contains some outstanding "gnawers"; *E. molaris* and *E. aciculatus* are the representatives in the Indo-Pacific region. *E. molaris* (see Fact Sheet on page 361) has the largest distribution: from the Red Sea and East Africa to the islands of the Central Pacific. *E. aciculatus* has a more eastern distribution toward Hawaii, whereby the western boundary is northern Australia.

The Borer Sea Urchin, *Echinostrephus aciculatus* is extremely philopatric, and it is characterized by excavating deep cavities or channels in to calcareous rocks where it then remains. These cavities are so deep that the sea urchin can hide in it completely, even though they are only a few millimetres wider than the diameter of the sea urchin. This then affords excellent protection against predators. Normally, the long, black spines are extended outward so far that they can readily be seen. This permits the sea urchin to capture organic particles

Hiding in its cavity and with the spines pointing to the outside, *Echinometra mathaei* - seen here at Shark Bay, Western Australia - is well protected against enemies. Photo: C. Bryce

drifting by. In addition, it feeds on organic material growing on the walls of its cave or what accumulates in it. In some places of the reef these small sea urchins (adults have a diameter of only 4 to 5 cm) can occur in dense aggregations, where individuals are virtually side-by-side over the entire reef. Off the Maldives we find *E. molaris* in very shallow water and down to a depth of up to 30 m.

We do not know if the aquarium trade is importing *Echinostrephus* species; the most suitable sea urchins for coral reef aquariums.

Genus *Colobocentrotus*, with three Indo-Pacific species, deserves to be mentioned, especially because their extraordinary appearance that sets them distinctly apart from all other sea urchins. The entire aboral side is covered by extremely short, globe-shaped spines, which are so dense that they form almost a compact mosaic cover over the entire aboral side. Along the lower edge there is a row of large, flat spines that extend closely over the substrate and thus form a lateral, fringed margin. The corona is low and has a flat underside. The entire shape of the animals points to a life in the surf zone, where it is exposed to enormous physical forces because of continuous wave impact. We find similarly appropriate adaptation in snails of order Docoglossa (see this Volume from page 47 onward).

The most widely distributed species is *Colobocentrus atratus*. It is dark blue – lilac or almost black, and lives in most areas of the Indo-Pacific, from the Red Sea and East Africa to Hawaii. Strangely enough, it does not occur in large sections of the western Pacific. There it is replaced by *C. mertensii* and *C. pedifer*. These urchins are possibly being imported, although we have not seen any of them in the aquarium trade. All *Colobocentrotus* species should be very suitable for special wave action- or inter-tidal aquariums. As far as is known they live exclusively on algae, but it cannot be excluded that they also take in some animal food at the same time.

In the Maldives we discovered several of these *Echinostrephus molaris* hiding in their excavated holes.

Family Echinometridae also contains the interesting genus *Heterocentrotus*, with two Indo-Pacific species *H. mammillatus* and *H. trigonarius*. With their thick, pencil-like spines they are somewhat reminiscent of slate-pencil urchins (family Cidaridae), but they are distinguished from the latter in that their spines have a smooth surface and never carry epiphytic growth. Both species occur from the Red Sea and East Africa, through most part of the Indo-Pacific to the central Pacific. *H. mammillatus* is also found around Hawaii.

Heterocentrotus mammillatus (see Fact Sheet on page 362) is beyond doubt the more attractive of the two species; it is also seen more frequently. It looks a lot like *H. trigonarius* but distinguished from this species by (among other features) the fact that it has more slender and evenly-declined primary spines (in contrast to the rather abruptly appearing spines in *H. mammilatus*), Normally, *H. mammilatus* has red-brown (rarely crimson-red) primary spines, sometimes with white cross bands. The secondary spines have usually a distinctly contrasting colouration, e.g. white or dark red. The primary and secondary spines in *H. trigonarius* always have the same colouration, normally olive green or yellow brown, but never white.

With a total diameter of 8 to 10 cm, *Heterocentrotus* species are not overly large. But they are heavy and robust and move rather forcibly through the aquarium (i.e. they turn over rocks and displace coral colonies), much to the annoyance of aquarists. Apart from that, they seem to be excellent aquarium inhabitants that never touch any other food other than algae and possibly dry food in tablet form.

Sea urchins of genus *Colobocentrotus* (seen here is *C. atratus* off Hawaii) are distinguished from other urchins by their extraordinarily short, dome-shaped aboral spines and a row of flattened, elongated spines along the edge of the aboral side. With such a shape these sea urchins are well adapted to a life in a rocky habitat of the inter-tidal zone, where they are exposed to strong wave action. Most of them are uniformly black, however, at particular light angles different colour shades become visible. Photo: Dr. B. Carlson

Family Parasaleniidae

This monotypic family with its sole genus *Parasalenia* is rather insignificant. Nevertheless, these distinctly elongated and rather small sea urchins are of interest to us because they are sufficiently decorative to be regularly available in the aquarium trade. Apart from that, they are food specialists, with coral polyps as the principal component in their diet.

An attractive and rather characteristic species from the Indo-Pacific is the red-brown *Parasalenia poehli*, which has distinctly light coloured, possibly even white, cross bands on its spines (see photo on page 363). Its distribution extends from the Red Sea to the central Pacific. This is a very small sea urchin with a corona diameter of less than 3 cm. In respect to compatibility with corals, this species unfortunately can also not be trusted.

Parasalenia gratiosa is the most frequent species in the Indo-Pacific. Its colouration consists of a very attractive dark brown with distinctly white rings around the base of the spines (see Fact Sheet on page 364). These urchins are rather similar to certain colour varieties of the herbivorous *Echinometra mathaei* that occur over the same geographical range. Clearly there is the potential risk of misidentification. If in doubt, one of the clearest distinguishing features is that *P. gratiosa* reaches a corona diameter of not more than 2 cm, while that of *E. mathaei* reaches at least a diameter of 5 cm. In addition, the corona of *P. gratiosa* is more egg-shaped.

***Heterocentrotus* species (here *H.* cf. *trigonarius* in Papua-New Guinea) is reminiscent of slate pencil urchins of family Cidaridae; however, in contrast to the latter these urchins have rather smooth spines. Photo: P. Schupp**

Superorders Gnathostomata and Atelostomata

These two superorders contain the irregular sea urchins. They are interesting animals, yet the aquarium trade rarely imports them. This is, no doubt, due to the fact that their colouration is rather "boring", i.e. animals that are presumed to be of no interest to coral reef aquarists spoilt by magnificently coloured reef organisms of all kinds.

If we deal with these animals in a somewhat summary fashion, it must not be understood that we consider them to be of less interesting aquarium animals. In fact, the explanation is that there is only limited, documented aquarium experience and virtually no

Echinometra lucunter
Atlantic Rock-boring Urchin

Distribution: Atlantic Ocean; from the Ascension- and St. Helena Islands to West Africa, as well as along the east coast of America; and in the Caribbean region; from North Carolina to Brazil.
Size: Diameter of corona to 10 cm, mostly to 5 cm only. Spines 2 to 4 cm long.
Reproduction: Studies in the Caribbean have shown that there is an annual reproductive cycle. Spawning takes in late summer or autumn. Larvae have been reared under laboratory conditions beyond metamorphosis. There are no reports about breeding in aquariums.
Food: Herbivorous; feeds mainly on drifting algae fragments captured from individual bore holes. Most specimens spend most of their life in such boreholes, only a few will occasionally leave this hole at night to search for food.
Description: Reddish brown or brownish black with a purple or green tinge. Spines can be coloured light green, violet or dark reddish-brown. Lives in holes in coral rocks created by continuous scraping with the teeth of the "Aristotle's Lantern". When it lodges its spines against the walls of the borehole it is virtually impossible to release the urchin. It lives in association with commensal fishes and crustaceans. In view of its boring activities it is a significant factor in respect of bio-erosion of the coral substrate. Rather common on, often on exposed hard fore reef habitats, including the surf zone. Normally lives at depths from 0 to 5 m, but can be found as deep as 45 m. Local aggregations of 10 or more specimens per square meter are not unusual.
Aquarium requirements: Imported only occasionally. Of particular interest for tidal aquariums with strong water movement. It is important to provide suitably fitting, pre-bored holes where this sea urchin can hide. Although it is primarily an algae eater, it cannot be totally trusted when kept together with smaller invertebrates.

GA: +/-; **SE:** +; **CO:** +/-

Photo:
Echinometra lucunter at Puerto Rico.

Echinometra mathaei
Reef Flat Boring Urchin

Distribution: Indo-Pacific; from the Western Indian Ocean to Hawaii, southwards at least to the Cook Islands (personal observations).
Size: Diameter of corona to 10 cm, often less. Spines rarely longer than one-half of corona diameter.
Reproduction: Details unknown.
Food: Diet consists nearly of 100 % of algae, but can pick on encrusting invertebrates. Will accept most substitute foods, but growth is minimal without supplementary algae food.
Description: One of the most conspicuous and common sea urchins on reef flats over its geographical range. Lives in the inter-tidal zone and in shallow water at depth from 1 to 8 m. Can occur in large numbers on the reef flat especially on reefs which have a large algae component. Here the animals typically drill holes and small channels in the rocks. The burrows give excellent protection against waves and enemies and it is really difficult to remove a sea urchin from its burrow without destroying the animal. Colouration variable, but normally brownish. Has short, blunt spines that are usually slightly lighter coloured than the corona, with a white ring around their bases. Very efficient algae grazer that simultaneously rasps away large amounts of calcareous rocks. Often bores holes in to rocks for which teeth and spines are used. Lives occasionally in symbiosis with certain crustaceans, such as *Gnathophyllum*- and *Petrolistes* species.
Aquarium requirements: Hardy aquarium inhabitant, provided it lives in an aquarium decorated with live rocks and there is sufficient epiphytic growth of various algae. Will take most algae species, including red calcareous algae. Mistaking this species for the coral eater *Parasalenia gratiosa* would be fatal (see Fact Sheet on page 364).

GA: +; **SE:** +; **CO:** +

Photos:
Specimens of *Echinometra mathaei* in an aquarium (top), at Palau (centre; photo: Dr. I. Illich) and in the Red Sea (bottom; photo: R. Brons).

Echinostrephus molaris
Pacific Rock-boring Urchin

Distribution: Indo-Pacific; from the Red Sea and East Africa to the Central Pacific.
Size: Diameter of corona to 5 cm, but often less. Spines up to 7 cm long.
Reproduction: Details unknown.
Food: Feeds partially on algae that grow in the vicinity of its bore hole and in part on algae and other vegetable fragments that drift passed. It collects the latter with its spines and with the long suction feet on its aboral side.
Description: An efficient borer that drills a narrow hole in to coral rocks where it spends most of its time. Adult specimens are often larger than the opening of their borehole, so that it is impossible to remove them from it. With long, very thin spines that protrude from the opening.
Aquarium requirements: Only very rarely imported, although it makes a very interesting aquarium inhabitant. However, suitable holes pre-bored in to the substrate must be available, so that it can hide immediately. Because of its "sessility" it is very important to provide for adequate supplementary algae food. Does not appear to harm large invertebrates, such as corals.

GA: +/-; **SE:** +/-; **CO:** +

Photo:
Echinostrephus molaris at the Maldives.

Heterocentrotus mammillatus
Pencil Urchin

Distribution: Indo-Pacific; from the Red Sea, East- and South Africa to Hawaii.
Size: Diameter of corona to 8 cm; primary spines up to 12 cm long.
Reproduction: Forms large spawning aggregations.
Food: Feed preferentially on micro-algae. Will accept frozen- as well as dry food, but will not grow without suitable vegetable food supplement.
Description: Highly variable in its colouration. The large primary spines are usually brownish, with or without whitish bands; sometimes the primary spines are also bright red. The short, secondary spines are blunt and form a mosaic on the corona. They can be pigmented white or dark and form a distinct contrast to the primary spines. Lives usually in crevasses and among coral colonies in shallow reef areas, most commonly found at depths from 0 to 8 m. The typical habitat is in areas of strong currents, which tend to dislodge lots of normally sessile organisms. The strong primary spines enable this urchin to become firmly lodged in cracks and crevasses. Certain pedicellaria are venomous, however, a sting is neither particularly painful nor dangerous to humans. The thick primary spines that were formerly used as writing implements on slate panels are nowadays used by the souvenir industry in suspended mobiles or as jewellery.
Aquarium requirements: Only very rarely imported. Should be kept in aquariums decorated with live rocks, together with lush algae growth and strong water movement. Feeds on most algae species, including red calcareous algae. Normally will not harm larger invertebrates; however, because of its substantial size there can be problems with the aquarium decoration. Appears to be sensitive to adverse transport conditions and hard to acclimatize.

GA: +/-; **SE:** -; **CO:** +/-

Photos:
Specimens of *Heterocentrotus mammillatus* at Hawaii (top), at the Maldives (centre) and in the Red Sea (bottom; photo: S. W. Michael).

details about what species have actually been kept. Moreover, species identification is generally difficult, and is based primarily on anatomical details that require a more or less complex dissection of a specimen. In essence, one distinguishes between species with "Aristotle's Lantern" (Gnathostomata) and those without this anatomical feature.

In total we include about 80 species in these two superorders that inhabit coral reefs in the Indo-Pacific. They are distributed over 27 genera in 13 families. In the Caribbean region there are 15 species in twelve genera and five families.

The general characteristic of irregular sea urchins is that they have developed secondarily a more or less pronounced bilateral symmetry. The shape of the

The Indo-Pacific *Parasalenia poehli*, seen above off Western Australia (photo: C. Bryce) and below in Papua-New Guinea (photo: L. Newman and A. Flowers), is a small and beautiful sea urchin. Although its maximum diameter is less than 3 cm, it is not necessarily a suitable coral reef aquarium inhabitant, since it feeds principally on coral polyps.

Parasalenia gratiosa

Distribution: Indo-Pacific, from East Africa to the South Pacific islands.
Size: Diameter of corona to 2 cm; spines up to 3.5 cm long.
Reproduction: Forms large spawning aggregations.
Food: Food specialist that feeds primarily on the polyps of stony corals. We do not have sufficient aquarium husbandry experiences with this species, to be able to tell to what extent it will take substitute foods.
Description: Occurs at depths from 3 to 12 m among corals that serve as food. The corona is blackish or dark reddish-brown. Spines slightly lighter coloured. Each spine with a white ring around its base. Can be mistaken for young *Echinometra mathaei* (see Fact Sheet on page 360), but it can be distinguished from the latter by its longer overall shape due to the oval-shaped corona. Also, it has fewer spines than *E. mathaei*, which are also longer than the corona diameter.

Aquarium requirements: Occasionally imported. As a food specialist requiring on stony coral polyps, it is not suitable for the traditional coral reef aquarium, but needs a special tank.

GA: -; **SE:** 0; **CO:** –

Photo:
Parasalenia gratiosa at Nusakode Island, Indonesia.
Photo: C. Bryce

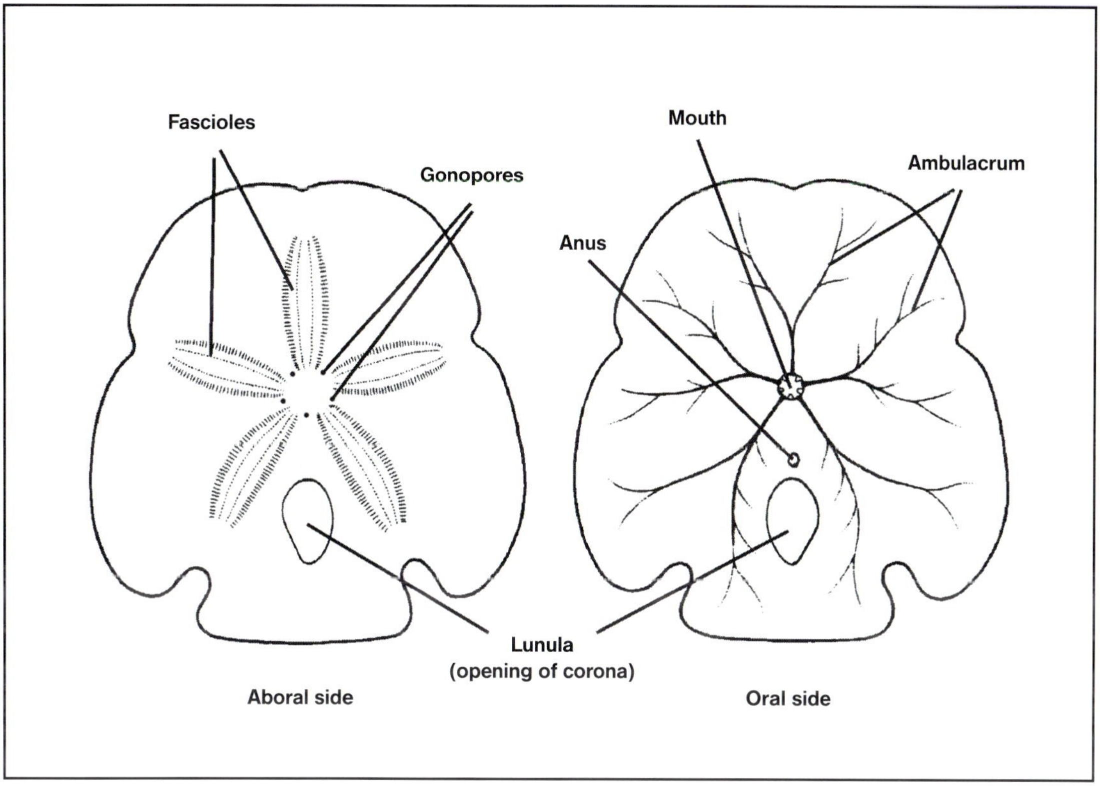

Aboral- and oral side of an irregular sea urchin. **Drawing after GUILLE et al. (1986)**

corona is usually longish oval and more compressed (top to bottom) than in regular sea urchins. There are, however, also species that are completely round. However, in all cases the anal opening is located on the oral side or posterior (on the back of the animal), i.e. never central on the aboral side, as in regular sea urchins.

Nearly all irregular sea urchins are grazers. They spend most of the time on or buried in fine-grained sand- or some other soft bottom, the main reason why they are rarely seen in the trade. The only notable exceptions to the buried life are the small, bean-shaped species of genus *Echinoneus* (Holectypoida: Echinoneidae), e.g. *E. cyclostomus* that has a circum-tropical distribution. This sea urchin reaches a maximal length of 3 cm, and has very short and thin spines. Colouration is variable, but as in many instances, the specimens are relatively uniformly-coloured white-

Dendraster excentricus **– seen here in the Monterey Bay Aquarium – is a typical representative of irregular sea urchins. Because of their flattened shape they are generally referred to as Sand Dollars. This eastern Pacific species lives on sand bottom in protected bays and along the open coast, from the low water mark down to a depth of 40 m.**

Maretia planulata of family Spatangidae, in Sulawesi. Photo: J. Cairns-Michael

Left: *Lovenia elongata* of family Loveniidae at Monte Bellos island, Western Australia. Right: *Brissus latecarinatus* of family Brissidae at Christmas Island. Photos: C. Bryce

yellow or light red-brown. This is not a burying species, but instead remains hidden below rocks that are lying on a sandy bottom.

In general, nearly all irregular sea urchins feed on organic detritus, small live and dead organisms found on or in sandy substrate. Most species are almost omnivorous. A few species have become specialized on vegetable or animal food. In any event, the small species – as detritus feeders – should be of ecological interest to aquarists. Those species that bury in bottom substrate can turn over and aerate a thick bottom layer, which can prove necessary if other burying animals (e.g. sea feathers) are to be kept in the same tank.

In a newly set-up ("sterile") aquarium irregular sea urchins must be able to find sufficient food. Substitute food, such as dry food and frozen plankton can be offered, but that may not always be sufficient. Beyond that the available relevant literature indicates that these urchins are hardy, peaceful and very useful aquarium animals, that need to be able to bury themselves (requiring a bottom layer of at least 10 – 15 cm thickness).

Clypeaster **sp. in the typical habitat (here at Monte Bellos Island, Western Australia) of most irregular sea urchins. Nearly all species feed on dead organisms and scavenge for edible organic material, along sandy or muddy substrate.**
Photo: C. Bryce

Class Holothuroidea - Sea Cucumbers

Sea cucumbers have the same pentamerous symmetry as other echinoderms, but their distinguishing feature is that the polar axis of sea cucumber is substantially lengthened. This gives them a cucumber- or worm-like appearance. Mouth and anus are located on opposing ends of the animal. This body shape forces sea cucumbers to lie on one body side, and not – as we see in most echinoderms – on the oral side. Basically, sea cucumbers have little resemblance to their relative with their distinct, radial symmetry.

Sea cucumbers provide up to 90 % of the biomass in deep-sea habitats, and since the deep-water areas cover about 70 % of the earth surface, it is easy to calculate that sea cucumbers are among the dominating animals of this planet. Moreover "Holothurians feeding on bottom sediments mix and till the surface muds on an enormous scale, producing features more widespread and more visibly evident than those produced by any other animal on earth" (quoted from HEEZEN & HOLLISTER, 1971).

All of the orders in class Holothuroidea contain deep-sea species. Order Elasipodida is particularly fascinating. All of its approximately 90 species are strictly confined to deep sea areas, from the upper Continental Slope down to depths of 10,000 m and more. They represent the most extraordinary creatures known (see drawing on page 368).

The reef flats of the world are literally covered by sea cucumbers, especially by sediment-feeding species (detritus feeders). They crawl about on the substrate in search of all sorts of buried, edible particles. Especially at dusk one can readily see small and large sea cucumbers on the landward sections of reefs and in lagoons. They have left their respective hiding places under rocks and coral colonies behind, and then move along the bottom, sucking up edible material, much like a vacuum cleaner. The two photos on page 368 clearly illustrate how numerous sea cucumbers are on the coral reef.

Sea cucumbers are distinctly more numerous on Indo-Pacific reef than on reefs in the Caribbean, both, overall numerically as well as in the number of species involved. For the shallow water areas of the Indo-Pacific, CLARK & ROWE (1971) list 286 species, while

Sea Cucumbers are common inhabitants of the tropical coral reefs. Sometimes we have to search for them to discover how numerous they really are. At the left is four specimens of the black *Holothuria leucospilota*, a species that is usually common on reef flats and typically hiding below coral boulders. Here on Heron Island the sea cucumbers stretch from their hiding to feed on the sandy bottom. In the right photo (also shot at Heron Island and not showing an arranged scene!) we have turned a small coral boulder close to the beach only to discover that at least seven specimens of sea cucumbers hide here out of sight to the human eye.

HENDLER et al. (1995) report that in the Caribbean region less than 60 species occur at depths accessible by divers.

Most people – possibly with the exception of experienced aquarists – only too often tend to consider sea cucumbers as unattractive, ugly animals. Indeed, many species are uniformly-coloured black, brown, or olive-green or have similarly drab colours. Yet, there are also pink, orange or violet-coloured species, and also some with attractive spotted- or striped patterns. May be presenting sea cucumbers here in some detail will convince more aquarists of the beauty and fascinating behaviours of these animals.

The number of recent species is about 1250 (please see systematic overview on page 369). The size range extends from dwarf species with a length of only a few centimetres up to virtual giants with lengths of at least 2 m and a diameter of 20 to 30 cm.

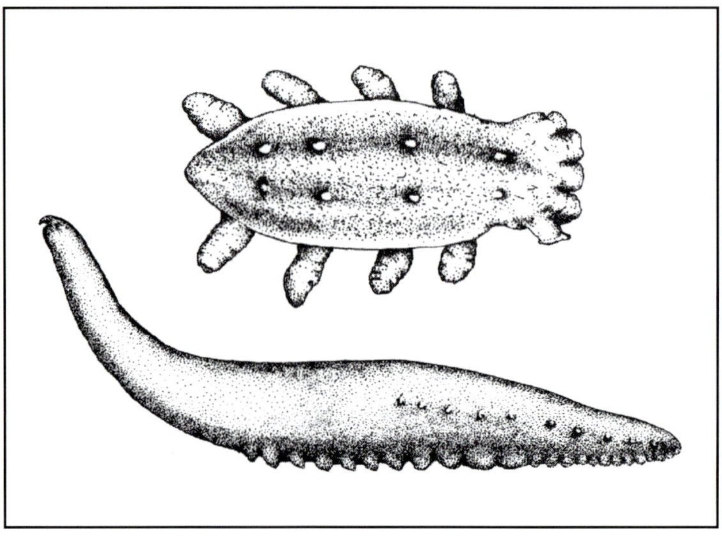

The order Elasipodida contains some spectacular deep-sea cucumbers, that often look like aliens from another planet: *Elpidia glacialis* (above) and *Psychropotes longicauda* (below). Drawing after MARSHALL (1979)

The body of sea cucumbers is divided in to five distinct sections. The ambulacral areas have longitudinal rows of tube feet, while the interambulacral areas are without. The side of sea cucumber in contact with the substratum is referred to as the ventral side, also known as the sole. It is made up of three ambulacral areas and two interambulacral areas. The dorsal side has thus two ambulacral areas and three interambulacral areas.

While it is difficult or even impossible in some sea cucumbers (including those commonly found in aquariums) to determine what is the ventral side and what is the dorsal side, yet in many others this distinction is quite obvious. In the latter, tube feet are either totally absent or are reduced to tubercles or warts on the dorsal side, while the ventral side is fully developed. Species of family Synaptidae provide extreme examples of sea cucumbers without tube feet. In their place are specific, tiny anchor spicules projecting from the skin that feel "sticky" when touched.

Basically, all sea cucumbers

have a leathery or soft body wall, with a huge number of more-or-less microscopically small spicules ("ossicles") embedded as solid skeletal structure. Depending upon the species, these spicules can make up from 3 to 70 % of the dry weight of the body wall. In addition, the spicules reflect a diversity of geometric shapes, but the variation is limited within a particular species. Moreover, spicules are of considerable taxonomic importance. Depending upon their respective appearance they are referred to as buttons, cups, tables, plates, rods, anchors and wheels. A few examples are shown in the drawing on page 370. The epidermis is not ciliated, and is covered by a thin outer layer. Below the epidermis there is a circular muscle layer that covers the single or double bands of longitudinal fibres located in the five ambulacral areas.

The body wall of many sea cucumber species is a delicacy in many far-eastern countries. It is particularly prized in China, possibly because of its significance as an aphrodisiac. Even in Chinese restaurants in Europe, cleaned sea cucumbers are cooked, fried or dried, sometimes smoked, and offered as "Beche de Mer", "Pling Chao", "Hoy Sum" or "Trepang". In Spain (region of Catalonia) sea cucumbers are also traditionally eaten. In restaurants in Barcelona they are offered as a local delicacy under the name of "Espardena" on the menu (DAVIDSON, 1977, 1981).

A tentacular crown, consisting of 10 to 30 very flexible, retractable tentacles, surrounds the mouth of sea cucumbers. They are being used for capturing plankton food particles as well as for searching through the substrate for detritus and microorganisms. Sea cucumbers are suspension feeders as well as detritus feeders.

The tentacles, just like the tube feet, are extensions of the ambulacral system, but better developed and located always around the mouth opening at the anterior end of the animal. The interior of a sea cucumber contains a large body cavity (coelom), which holds the internal organs in an arrangement typical for echinoderm. The ring canal surrounds the pharynx. It is located up front and leads in to five radial canals, which become tentacles at the distal (forward) end, and extend along the sides to form ambulacral areas from where the tube feet project to the outside.

Although the ambulacral system is basically identical to that in other echinoderms, there is, however, one significant difference. In most species the madreporite is no longer connected to the body surface, but instead it is located inside the body cavity. It is suspended just below pharynx, and is connected to the ring canal via a short stone canal. The madreporite has pores and pore canals, so that body (coelomic) fluid enters and leaves the ambulacral system, instead of seawater.

Locomotion in many species is accomplished by the use of tube feets. Other species create wave-like movements by the skin and move rather like a caterpillar. However, most sea cucumbers move only very slowly. They are more sessile animals that tend to remain stationary for prolonged periods of time. It is surprising to

Systematic overview of class Holothuroidea

After HENDLER et al. (1995); GUILLE et al. (1986); CANNON & SILVER (1986, 1994); GEORGE & GEORGE (1979); CLARK & ROWE (1971)

Order Dendrochirotida
- Family Psolidae
  - *Psolus*
- Family Cucumariidae
  - *Aslia, Cucumaria, Duasmodactyla, Havelockia, Hemithyone, Leptopentacta, Ocnus, Orbithyone, Paracucumaria, Pentacta, Plesiocolochirus, Pseudocolochirus, Pseudocnus, Sclerodactyla, Staurothyone, Stolus, Thorsonia, Thyone, Thyonella, Trachythyone*
- Family Sclerodactylidae
  - *Euthyonidiella, Pseudothyone, Sclerodactyla*
- Family Phyllophoridae
  - *Actinocucumis, Afrocucumis, Cladolabes, Globosita, Mensamaria, Neocucumis, Neothyonidium, Ohshimella, Phyllophorus, Phyrella, Selenkiella, Semperiella, Thyonidiella*

Order Dactylochirotida

Order Aspidochirotida
- Family Stichopodidae
  - *Astichopus, Isostichopus, Stichopus, Thelenota*
- Family Holothuriidae
  - *Actinopyga, Bohadschia, Holothuria, Labidodemas*
- Family Synallactidae
  - *Mesothuria*

Order Elasipodida

Order Apodida
- Family Synaptidae
  - *Anapta, Epitomapta, Euapta, Lapidoplax, Leptosynapta, Opheodesoma, Patinapta, Pendekaplectana, Polyplectana, Protankyra, Rynkatorpa, Synapta, Synaptula*
- Family Chirodotidae
  - *Chiridota, Polycheira, Taeniogyrus, Trochodota*

Order Molpadiida
- Family Molpadiidae
  - *Molpadia*
- Family Caudinidae
  - *Acaudina, Paracaudina*

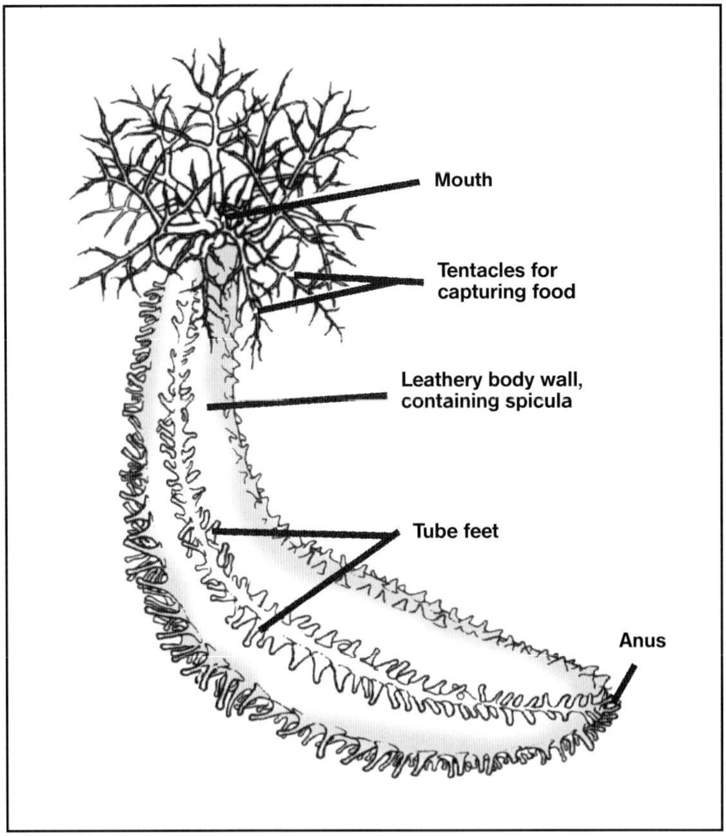

Schematic external structure of a sea cucumber.
Drawing after George & George (1979)

learn that there are sea cucumbers that are able to swim. For instance, family Synaptidae contains certain species that are capable of moving in open water by alternate, U-shaped contractions and extensions of the body.

The dominating detritus feeders on reefs are sea cucumbers of families Holothuridae and Stichopodidae. These have rather short, peltate tentacles (peltatus, Latin = armed with a shield), and they feed by shovelling sediments in to their mouth and digest the micro-fauna contained in it. Suspension feeders, like sea cucumbers of family Cucumariidae, have long, feather-like or bushy tentacles that can be spread out like a fan.

More or less stationary suspension feeders like members of family Cucumariidae, extend their tentacles either in to the open water or they wipes them over the substratum. In both cases food particles adhere to the surface mucus of the tentacles. Eventually the tentacles are placed in to the pharynx and the attached particles are stripped off when the tentacles are pulled out again. The captured food particles are mostly microscopic algae, protozoans, nematodes, ostracods, various larvae as well as organic detritus. Many of those sea cucumbers on coral reef that feed mainly on plankton are not particularly numerous and usually live scattered far apart, which may well be due to prevailing low plankton concentration.

More mobile, benthic species, e.g. those of family Stichopodidae and various deep sea forms, are detritus feeders that graze off the substratum using their tentacles. The worm-shaped species of families Synaptidae and Chirotidae (order Apodida), dig through the substratum and take in the bottom material dug up. In many species the sediment inside the digestive tract is visible through the thin body wall. As detritus feeders they have a preference for organic and inorganic detritus containing microorganisms, protozoans

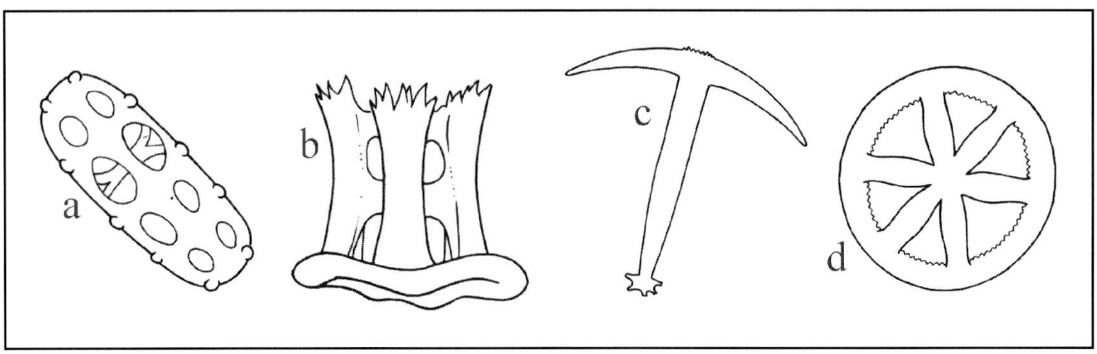

Spicula, embedded as skeletal structures, can have highly variable shapes and play a significant taxonomic role; seen here are: button (a), table (b), anchor (c) and wheel (d). **Drawing based on Cannon & Silver (1986)**

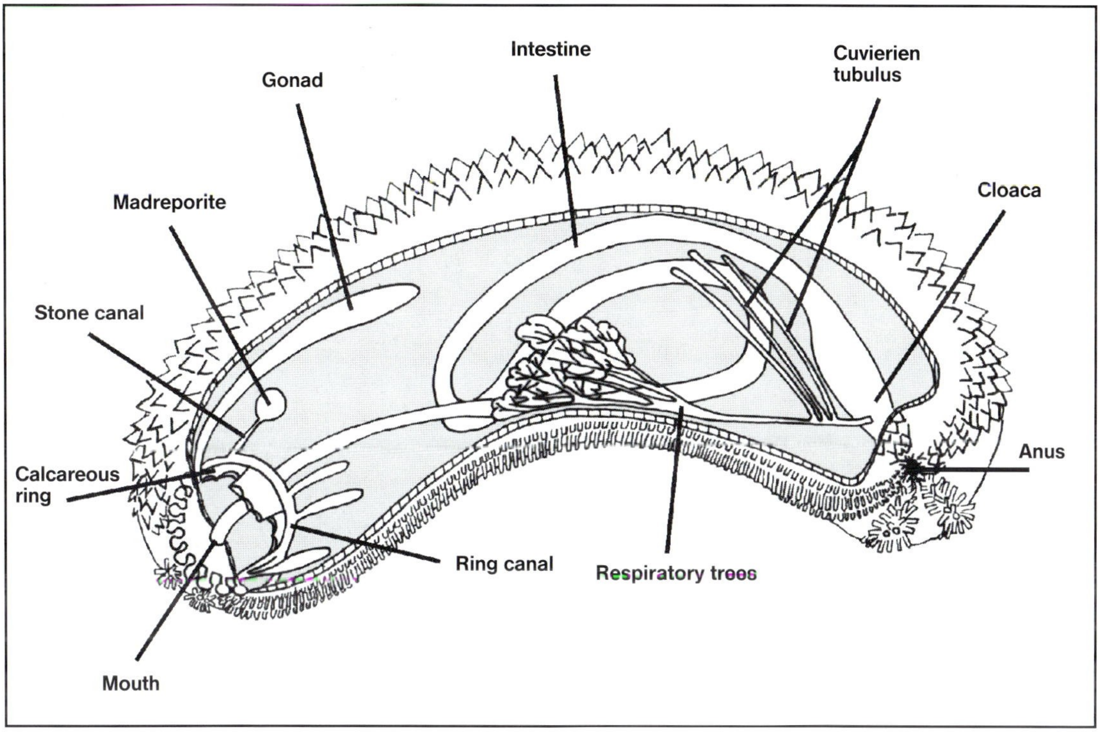

Schematic internal body structure of a sea cucumber. Drawing after CLARK & ROWE (1971)

and the endemic meiofauna. This affords them ample feeding opportunities in reef habitats and so these sea cucumbers occur in large numbers.

Overall, sea cucumbers have conquered various ecological niches and habitats. While some live on soft or hard substratum, others dig through sandy or muddy bottom material or they gnaw their way through calcareous rocks.

The mouth leads directly in to the pharynx, which is surrounded by a calcareous ring at the anterior end. The calcareous ring, which could be viewed as an analogy of the "Aristotle's Lantern", consists normally of five radial and five inter-radial, rectangular plates that arranged like the battens of a wooden barrel. In some sea cucumbers (e.g. members of family Cucumariidae) the pharynx leads in to a slender oesophagus that forms the entrance to a short, muscular stomach. However, in most species the oesophagus is absent so that the pharynx – if the stomach is only poorly developed – leads directly in to the intestine.

The intestine in the coelom is positioned in three loops (along the polar axis of the animal), and is supported by mesenteries. Enzymes break down the food, which is then either transported by amoeboid cells (coelomcytes) to the body or diffused directly through the intestinal wall in to the body fluid. In many sea cucumbers the intestine terminates in a cloaca prior to opening to the out-

It is often difficult for a layman to determine what is the dorsal- and ventral side of a sea cucumber, or what is its front end or back end. Seen here is *Holothuria edulis* in an aquarium.

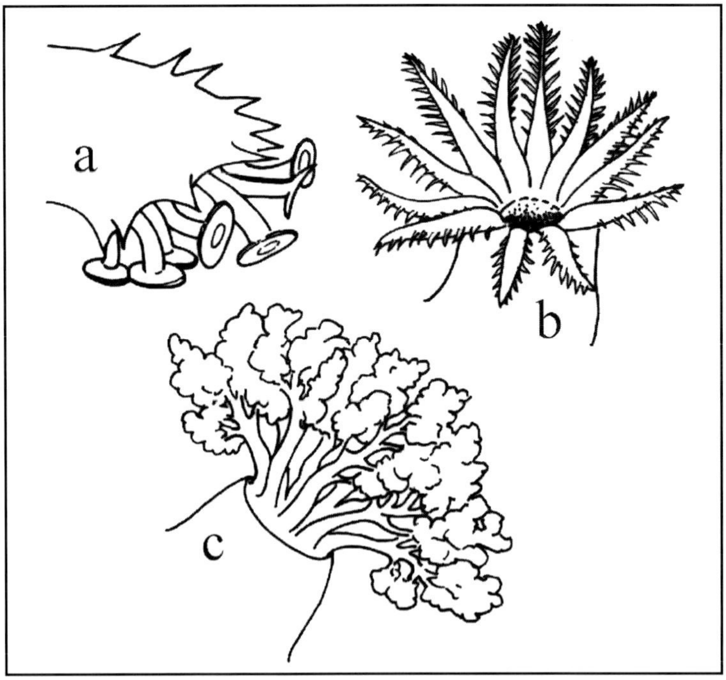

Principal shapes of sea cucumber tentacles: peltate (a), pinnate (b) and bushy (c). **Drawings a and c after CANNON & SILVER (1986), and b after HENDLER et al. (1995)**

side through the anus.

Many detritus feeders produce pellet-shaped, easily recognizable faecal balls, which can measure about 1 cm or larger (from large sea cucumbers). These chain-linked excrements look like sand-coloured sausage chains. All organic material has been digested, while the indigestible components are excreted and because of their distinctly identifiable shape they also clearly reveal the presence of sea cucumbers.

It has been reported, that the 45 cm long Caribbean sea cucumber *Isostichopus badionotus* (family Stichopodidae) fills and empties its digestive tract with sand three-times a day. KAPLAN (1982) estimates that sea cucumbers in a certain area of 4.4 km^2 off Bermuda transport 500 to 1000 tons of sand annually through their bodies. VINE (1986) provides a basic estimate and indicates that sea cucumbers move – on average – about 150 tons of sand per hectare and year. This calculation is based on an estimated population density of 10,000 sea cucumbers per hectare.

Most sea cucumbers are too large and their body walls are too thick in order to be able to facilitate gas exchange through them. Therefore, they have a pair of respiratory trees, of which one each is located to the right and left of the digestive tract inside the coelom. These trees have many branches, and each terminates in a tiny vesicle. The entire respiratory system is slightly reminiscent of a mammalian lung. The ticker ends of the respiratory tress are open towards the cloaca, either individually or through a joint stem.

The respiratory organ can be filled and emptied through the anus, by means of muscular action. During "inhaling" the cloaca expands (dilates) and becomes filled with seawater. At that point the closing muscle of the anus contracts, followed by the cloaca. That action forces the seawater in to the respiratory trees. The thin-walled vesicles at the end of the branches then make it possible that gas diffuses in to the body fluid or in the vascular passages (see further below). Contractions of the vesicles and an opposing reaction of the cloaca force the water out of the body again. "Respiration" takes place only very slowly. For instance, *Holothuria* species require six to ten expansions and contractions of the cloaca in order to fill the respiratory trees. Each of these actions takes about one minute or more. On the other hand, emptying the respiratory trees requires only a single reverse action (BARNES, 1980).

Sea cucumbers, especially those of order Aspidochirotida, have the most advanced blood (hemal) system of all echinoderms. The blood, in terms of its components, is similar to the body fluid. However, the system of dorsal and ventral blood vessels, together with several single-chambered hearts, facilitates an efficient circulation. Peristaltic contractions of the dorsal blood vessels are primarily responsible for proper circulation. Yet, blood circulation takes place only slowly and consists of a type of flow best described as "ebb and flood". The hemal system clearly participates in gas transport, but possibly it also involved in adsorption and nutrient transport; however, its precise total function is still unclear.

Sea cucumbers have an enormous regenerative capability, especially that of the intestine. Expelling the digestive tract together with the respiratory trees is a common response to attacks from predators, something that has been frequently observed in the sea. This phenomenon is referred to as evisceration and occurs in several genera. Under laboratory conditions, evisceration occurred in specimens exposed to unfavourable environmental conditions, such as overcrowding, polluted water or other abnormal conditions. Evisceration taking place in an aquarium can lead to the poisoning of other animals (see below). Following evisceration, the expelled organs are then regenerated.

With a radical loss of body parts, the regenerative abilities vary within the class. In some species, dividing a specimen in to two halves can produce two separate specimens through regeneration, while others can regenerate only from an anterior or posterior part. Invariably, the respective remaining end of the (former) sea cucumber then dies.

The expulsion of sticky tubules from the anus (see photos on page 375) occurs probably only in sea cucumbers, but it is confined to a few species of family Holothuridae. These sea cucumbers have a variable number of white, pink or red tubules, commonly referred to as "Cuvierien Tubules". These are appendages of both respiratory trees. When such a sea cucumber is being irritated or attacked by a predator, it will direct its anus towards the attackers. Then the body wall contracts and the cloaca burst, so that the tubules are thrust out through the anus.

The "Cuvierien Tubules" can be adhesive and trap the intruder in a net of sticky threads and/or can contain a poisonous substance, known as Holothurin. This is a triterpene-glycoside, produced by most sea urchins in smaller or greater quantities. It is a potent poison, which is concentrated

Bohadschia graeffei (seen here at Palau) is a typical detritus feeder that grazes the bottom with its tentacles. Photo: Dr. I. Illich

Pentacta anceps (seen here off Fremantle, Western Australia) is a beautifully coloured, small plankton feeder, that serves as an example here for the suspension feeders among sea cucumbers. Photo: C. Bryce

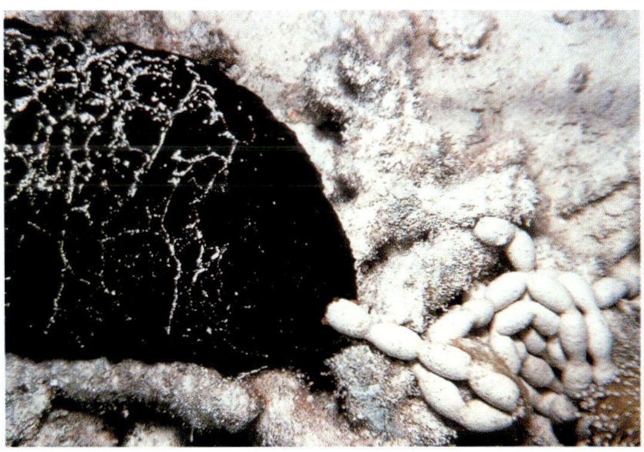

Many suspension feeders produce distinctly shaped and easily recognizable excrements in the shape of pellets. Left: _Actinopyga miliaris_ (seen here at Heron Island, Great Barrier Reef) while depositing its excrements. Right: "Decorative" chains of excrements from an unidentified sea cucumber in Papua-New Guinea.

Photos: L. Newman and A. Flowers

principally in the viscera or in the body wall. It can be released when a sea cucumber is being injured or dies. The natives of the south sea island are utilizing the toxicity by cutting certain sea cucumbers in to small pieces, which are then used to collect fish in tide pools.

Regrettably, many aquarists will have already experienced the toxicity of Holothurin first hand, when fishes suddenly died of acute poisoning. In many such instances appropriate countermeasures were no longer possible. Possible causes for a release of Holothurin could have been a sea cucumber being stung by a sea anemone or by corals, sucked up by a filter intake (and chopped to pieces), or it got burnt when it came in to contact with an unprotected rod heater.

Evisceration can occur when the environmental conditions in an aquarium are changing, becoming poor or unfavourable for a sea cucumber, and it will expel the respiratory trees, the digestive tract and/or the gonads. This too can lead to a poisoning of aquarium co-inhabitants. Poisonings have also been reported in conjunction with sea cucumbers spawning.

Holothurin is a milky, foaming secretion with a soap-like effect (Saponin) that breaks down the surface tension. The toxic effect on fishes is probably directly related to this effect, since it damages their sensitive gill membranes, which blocks oxygen uptake.

There is no reason to believe that the death of a sea cucumber always causes the release of the toxin. We have had sea cucumbers die in an aquarium without any damaging effect on other animals in the same tank. But the fact, that this can happen, should be reason enough to think twice before placing sea cucumbers in an aquarium that accommodates valuable fishes.

Many sea cucumbers are hosts for symbionts. Most of the sea cumbers commonly found on Indo-Pacific reefs are hosts for scaly worms of genus *Gastrolepida*. These worms are distinctly territorial; the reason while there is usually only a single worm per sea cucumber. These worms have a flattened body and adhere to their host by raising their back, so creating a suction effect. When they leave their host, they turn into excellent swimmer, propelling themselves forward with sinusoidal movements. A single worm species can exploit several sea cucumber species, but it takes on only the colouration and markings of its respective host.

A particularly interesting symbiotic relationship exists with the slender pearl fishes of family Carapidae. One pearl fish settles in one of the two respiratory trees of certain sea cucumber species. It leaves its host only at night to search for food. Following its nightly excursion, the fish returns to its host, finding its way through the anus and the cloaca, back to its protective enclosure in the respiratory tree. The majority of Carapidae species does not harm their host. Yet, there are also species that have developed in to true parasites, by feeding on the internal organs of sea cucumbers.

Sea cucumbers have only a single gonad, which is located up front in the coelom, below the central dorsal interambulacral area. Many species have separate sexes, but very few are protandrous hermaphrodites (initially they are males but then turn in to females). Eggs and sperm are normally released in to open water, where the zygotes develop in to free-swimming larvae that undergo various stages of metamorphosis, before they settle on the bottom as juvenile sea cucumbers. One of the developmental stages is known as an auricularia larva, which is rather similar to the bipinnaria larval stage of sea stars.

In a very small number of species – primarily in cold water species – the eggs are collect by the maternal animal with its tentacles, and then incubated directly among the tentacles, or the eggs are transferred to brood sacs on the sole or on the back. Yet, there are also brooders where the eggs are transferred directly from the gonad to the body cavity. Once the young are fully developed they are expelled through the anus. What is not yet known is how fertilization of the eggs takes place in these species.

The size of sea cucumbers is always given as total length, measured from the anterior- to the posterior end. But the size can vary substantially, depending upon whether the animal is contracting or expanding. Consequently, size details in the relevant literature are rather vague and unreliable. The measurements given in this book are to be understood as average values and refer to relaxed specimens.

From the perspective of aquarists there are various reasons for keeping sea cucumbers. Many of the suspension feeders – such as the well known "sea apples" of genus *Pseudocolochirus* – are kept principally because of their attractive appearance and possibly also because of their interesting behaviour. But on the other hand, the detritus-feeding species are the ones most frequently imported. They are not so conspicuously coloured and tend to remain in hiding places most of the time. But for aquarists they are of particular interest, since they create a beneficial biological effect in the aquarium by continuously turning over the substrate and the sediments with their burying activities. On reefs they play an equally important role as that played by algae-eaters. Sea cucumbers are unquestionably the most effective scavengers. Yet, one must never forget that – because of the highly toxic Holothurin – keeping sea cucumbers in an aquarium is always connected with risks. If Holothurin is actually released, there will be an immediate effect on the fishes, and sometimes also on certain invertebrates. Unfortunately, this is usually quickly fatal to fishes. If indeed any life is to be saved, it is absolutely imperative that an injured or dead sea cucumber is removed from the aquarium immediately, followed by a partial water change, an increase in the operating efficiency of the protein skimmer, and the aquarium water be filtered over high-quality, activated carbon.

Apart from their toxicity, sea cucumbers are peaceful and remarkably durable aquarium inhabitants. The detritus feeders do not require any particular care, except for cautious acclimatization. Plankton feeders are fundamentally more complicated in their care. They require frequent feeding with plankton or plankton substitute. Continuous feeding with *Artemia* nauplii has lead to very good results when *Pseudocolochirus* species were kept.

The expulsion of the Cuvierian Tubules is a common defensive action among certain sea cucumber genera. This photo shows the adhesive tubules of *Bodschia argus*, seen here in Indonesia. Photo: C. Bryce

Order Dendrochirotida

This order contains suspension feeders that utilize primarily zooplankton. They catch these organisms by extending their more or less intricately built tentacles in to the open water. A few species live out in the open, but most have their bodies buried in sediments or under rocks. Only the tentacles are then visible.

The order Dendrochirotida encompasses four reef-dwelling families, of which we will discuss only the two largest and most important ones: Cucumariidae and Phyllophoridae.

Plankton-feeding *Pseudocolochirus* species (photo: probably *P. axiologus*) are without doubt the sea cucumbers most frequently available from the aquarium trade.

Family Cucumariidae

The species in this family are small to medium-size sea cucumbers. Characteristically, they have ten bushy, branched tentacles, with the two ventral ones often being somewhat smaller than the other ones. Their geographical distribution extends around the entire globe, where they occur in tropical as well as in subtropical and cold seas, and from shallow inshore water down to great depths. A few species of genus *Cucumaria* are found in the tropical eastern Pacific as well as in the tropical Atlantic at depths down to 4,000 m; however, many species are commonly encountered on coral reefs.

Most species lead a rather cryptic life, hidden under rocks or buried in sand or rubble, where often only the tentacles remain visible. And so it is astonishing that most of the imported species lead a totally different life, i.e. they are openly visible.

The types of sea cucumber of family Cucumariidae most frequently encountered in the aquarium trade are the so-called "sea apples", which are the most frequently imported sea cucumbers overall. These 15 to 20 cm long animals are suspension feeders and now belong in to genus *Pseudocolochirus*, while they were formerly placed in *Paracucumaria*.

In terms of the systematics involved, there is much confusion within genus *Pseudocolochirus*. Consequently, it is difficult to say whether the imported sea cucumbers represent different species, or whether they are all part of the same – yet highly variable – species. All sea apples that we have seen to date had intensive colours, mainly red, yellow and blue. This then fulfils the criteria for species *Pseudocolochirus axiologus*, *P. violaceus* and *P. tricolor* (also refer to Fact Sheet on page 378). Sea apples occur in various areas of the Indo-Pacific, but the imports come mainly from South-East Asia.

The name "sea apple" comes

Close-up photo of the front end of *Pentacta anceps* in an aquarium. In sea cucumbers of family Cucumariidae, the ten tenctacles around the mouth opening form a bushy crown and serve to collect plankton.

Photo: S. W. Michael

from the rather round shape of these sea cucumbers when they are totally contracted. But even when they are fully expanded they are still less elongated than other sea cucumbers, but the relationship is then more obvious. Sea apples have five rows of tube feet, that are more or less equally developed on the ventral- as well as the dorsal side, where they are always well visible, mainly because of the contrasting colours: in red sea apples, the tube feet are frequently yellow and in blue sea apples they are red.

The tentacles around the mouthy opening form a bushy crown, and they often – but not always – display contrasting colours. With the aid of their tentacles, sea apples capture plankton from the current passing by. Afterwards the tentacles are inserted – one-by-one – in to the mouth, where they are virtually "licked off".

When a sea apple is introduced in to an aquarium it will - on its own - select the most suitable "anchor position" for capturing plankton. Normally, that means it will seek out an open location where it is well visible. For such plankton feeder to thrive under aquarium conditions it must receive a regular (preferably daily basis or even a continuous) supply of live or freshly frozen plankton. Ideally suited are brine shrimp (*Artemia*), *Bosminia* and copepods. If sea apples do not receive sufficient food they will soon start to shrivel up and die.

One needs to be aware that a sea apple normally attaches itself extremely firmly to a substrate. In fact, it is almost impossible to remove an established sea apple from its chosen anchor site, without damaging its tube feet. Therefore, it is absolutely imperative to leave such a sea apple alone.

Moreover, it is important for aquarists to be aware that the voluminous sea apples belong to those sea cucumbers that have generated most of the reports about poisoning accidents in aquariums that have resulted in fish mortalities. Even in large tanks, the release of Holothurin can have fatal consequences. However, it should also be noted that such poisonings are more likely to be the exception rather than the rule, if sea apples are maintained properly. Besides, most invertebrates appear to be less sensitive to the poison than fishes, and beyond doubt, sea apples are not more poisonous than other sea cucumbers. The large number of reported poisonings is probably due to the fact that the magnificently coloured sea apples are more popular than other sea cucumbers, because many inexperienced aquarists buy them as their first aquarium inhabitants.

Apart from large *Pseudocolochirus* species there are still two smaller species of family Cucumariidae that are commonly imported: the mainly pink-coloured *Pentacta anceps* (max. size 8 cm), and the intensely yellow *Colochirus robusta* (max. size 6 cm). Although the colours of these two species are rather different, they look rather like miniature versions of their large relatives. Both species (see Fact Sheets on pages 379 and 380) are also plankton feeders and surprisingly durable aquarium animals. However, frequent feeding with plankton or plankton substitute is also very important for these sea cucumbers. Due to their small size, the risk of causing a serious poisoning accident seems to be reduced; at least we have not seen any reports about it.

Family Cucumariidae also contains many other species, especially in the Indo-Pacific. CLARK & ROWE (1971) list 69 shallow water species in 18 genera for the Indo-Pacific. HENDLER et al. (1995) mention three genera with a total of six species for the Caribbean region. As mentioned above, most sea cucumbers lead a cryptic life, and so they are rarely imported, if ever at all. Moreover, they are less conspicuously coloured, often only with a sprinkling of brownish dots and spots.

Neothyonidium magnum **(seen here in Papua-New Guinea) is an interesting and attractive representative of family Phyllophoridae. This view from above on to the (partially) buried specimen reveals the two rings of tentacles: the inner tentacles act as cleaning devices, the long outer ones are used for capturing food.** **Photo: P. Schupp**

Family Phyllophoridae

This family also contains small to medium-size sea cucumbers that have much in common with those in family Cucumariidae, i.e. anatomically as well as in their mode of life. But sea cucumbers in family Phyllophoridae have more than ten tentacles that are often arranged in two or three rings around the mouth opening. Those tentacles in the outer ring are longer than the ones on the one or two inner rings, and serve to collect food organisms, while those on the inner ring(s) are used mainly for cleaning purposes. Another typical feature of family Phyllophoridae is the U-or J-shaped body structure. The body is often swollen in the middle, and tapers off towards both ends.

Most of the Phyllophoridae species also hide under rocks or remain buried in soft substrate, so that only the tentacular crown is visible. In total, 30 species in 15 genera are known to occur in the Indo-Pacific, and four genera with a total of six species in the Caribbean region; however, so far we have not yet encountered any of these species in the aquarium trade. Although there are occasional reports in the aquarium literature about *Thyonidiella oceana*, but these are probably based on incorrect identification of *Colochirus robustus* (family Cucumariidae).

Yet, this is not to say that Phyllophoridae species would not be of interest to the aquarium hobby. After all, there are some very attractive species, especially those of genus *Neothyonidium*, (e.g. *N. magnum* from the Indo-Pacific, with a max. size of 30 cm). In an aquarium this species requires a sufficiently deep bottom substrate where it can readily bury. This is something that could lead to problems in an aquarium. Feeding must be done the same way as described for sea apples.

Order Aspidochirotida

This order contains some of the largest and most conspicuous sea cucumbers seen on coral reefs. They have a distinct cucumber-, cylinder- or sausage-shaped body structure and a very thick body wall. As typical detritus feeders they ingest sand and digest all usable organic components in it.

The two families Stichopodidae and Holothuriidae are of interest to us since both include species that are well suited and very durable for aquarium maintenance. The sea cucumbers imported by the aquarium trade usually originate in South-East Asia, but occasionally Caribbean species are also imported. Here we will discuss only those species that are important from an aquarist's perspective. Those sea cucumbers commonly traded are not only made up of various species, but also involve different genera. It must be noted, however, that unequivocal species identification is not possible without a meticulous comparison to the scientific description of the respective species. Taxonomic work with sea cucumber is – as already mentioned – a difficult matter, and it is based on the shape and number of spicula in the body wall, shape and number of tube feet, etc. Histological studies and microscopic examinations are absolutely essential for this type of work.

It can be categorically stated, that these detritus-feeding sea

Pseudocolochirus spp.
Sea Apples

Distribution: Central Pacific; from Sri Lanka throughout the Indo-Australian Archipelago to Northern Australia and Japan.
Size: Maximum length 20 cm.
Reproduction: Details unknown. Spawning in aquarium has already occurred a few times, unfortunately with fatal consequence for the fishes that had eaten the obviously poisonous gametes.
Food: Plankton feeder. Requires frequent feedings, preferably daily, with *Artemia* nauplii or live or frozen plankton or commercially available plankton substitute of the highest quality. A drop wise or almost "continuous" feeding is most successful.
Description: Incredibly beautiful sea cucumbers with a very characteristic, apple-like shape, which makes it virtually impossible to mistake them for members of any other genus. The spectrum of bright colours is rather wide and can sometimes – but necessarily not – be species specific. Tube feet and tentacles can be rather contrasting in colouration. Sea apples live normally at depths from 10 to 30 m on reef flats in exposed, rocky habitats, but sometimes they can also be found in shallow water. The taxonomy within this genus is unclear. In the literature the taxa *Pseudocolochirus axiologus*, *P. tricolor* and *P. violaceus* have been used uncritically and indiscriminately. *P. area* from South Viet Nam has been described as the fourth species. According to CLARK & ROWE (1971) it is possible that only *P. axiologus* and *P. violaceus* are valid taxa.
Aquarium requirements: These are probably the most frequently imported sea cucumbers. They are rather hardy provided they are properly fed. Sea apples require an aquarium with good water circulation so that the food is distributed throughout the entire tank. If food supply is sparse, the sea apple will rarely open its tentacle crown and gradually fade away becoming smaller and smaller. An aquarist who keeps sea apples together with fishes must be aware that the sea cucumbers represent a potentially lethal danger for the fishes (see page 374). However, sea apples are totally compatible with other invertebrates.

GA: +; **SE:** +; **CO:** +/-

Photos:
Pseudocolochirus cf. *axiologus* in an aquarium (top; photo: S. W. Michael), as well as *P.* cf. *violaceus* in an aquarium (bottom; photo: S. W. Michael).

Pentacta anceps
Red Sea Cucumber

Distribution: Eastern Indo-Pacific; from the Indo-Australian Archipelago to the South Pacific islands.
Size: Maximum length 8 cm, however, usually only 5 to 6 cm.
Reproduction: Details unknown.
Food: Plankton feeder. Requires frequent feedings, preferably daily, with *Artemia* nauplii or live or frozen plankton or commercially available plankton substitute of the highest quality.
Description: Small, pink-coloured sea cucumber with mainly yellow tentacles and normally also with yellow markings over the body. With a robust, square and very hard body. Lives in shallow, exposed, rocky reef habitats from the inter-tidal zone down to a depth of about 25 m; sometimes also on soft substrate.
Aquarium requirements: Regularly imported sea cucumber that makes a good aquarium inhabitant, which can last for many years when given optimum conditions. Requires an aquarium with good water circulation to distribute the food throughout the entire tank. Ideally suited as companion species for other invertebrates. Due to the toxicity of these sea cucumbers, they pose a risk when kept together with fishes.

GA: +; **SE:** +; **CO:** +/-

Photo:
Pentacta anceps in an aquarium. Photo: S. W. Michael

Colochirus robustus
Yellow Sea Cucumber

Distribution: Indo-Australian Archipelago.
Size: Maximum length 6 cm; however, usually only 4 cm.
Reproduction: Details unknown.
Food: Plankton feeder. Requires frequent feedings, preferably daily, with *Artemia* nauplii or live or frozen plankton or commercially available plankton substitute of the highest quality.
Description: Small sea cucumber brightly yellow coloured, sometimes also with greenish areas on the body. Has considerable morphological resemblance to *Pentacta anceps* (see Fact Sheet on page 379), but easily distinguished from the latter by its totally different colouration. Sometimes found in large aggregations along reef edges and – walls in areas with strong currents, from the inter-tidal zone down to a depth of about 25 m. Often miss-identified in the aquarium literature, as *Colochirus quadrangularis*, *Thyodonidella oceana* and *Cucumaria miniata*. *Colochirus luteus* has recently been designated a synonym of *C. robustus*.
Aquarium requirements: Rather frequently imported, primarily from Indonesia. It is a hardy aquarium inhabitant. Requires an aquarium with strong water circulation and frequent feeding with plankton or plankton substitute. Ideally suited as companion species for other invertebrates, but because of its toxicity not without risk for fishes in the same aquarium.

GA: +; **SE:** +; **CO:** +/-

Photos:
Colochirus robustus in an aquarium (top; photo: S. W. Michael) and an aggregation at Padar Island, Indonesia (bottom; photo: C. Bryce).

Stichopus chloronotus **on the Great Barrier Reef: Left – one specimen (below) together with** ***Holothuria atra*** **(above). Right – close-up photo of the body wall showing the elongated papillae with their orange tips.**

cucumbers do remarkably well under aquarium conditions. They should be kept in a healthy and well-conditioned aquarium, set up with live rocks, where there is lots of detritus (detritus develops naturally in an aquarium and can hardly be avoided). Such an aquarium will contain an abundance of microscopically small organisms living in the detritus, which are extremely important in the diet of these sea cucumbers.

Family Stichopodidae

The sea cucumbers in this family are medium-large to large. They can be recognized by an approximately square body cross-section and projecting papillae distributed over the body. Their overall appearance is cylindrical rather than cucumber-like. The sole of these sea cucumbers is rather flattened. They live on sandy or rubble-covered bottom substrate, where they crawl about.

Body shape and –colouration can vary substantially within a particular species, so that identification is a complicated matter. Yet, on the other hand, this family also contains rather typical, easy-to-identify species, and many of these can be observed under water by divers.

In the Indo-Pacific there are the genera *Stichopus*, with about 10 reef-dwelling species, and *Thelenota* with three species, while the Caribbean region is inhabited only by the two species *Astichopus multifidus* and *Isostichopus badionotus*. In addition, this family contains some species that occur worldwide in subtropical, temperate and cold seas, such as *Stichopus regalis* in the Mediterranean Sea and North Atlantic and the intensely red coloured *Parastichopus tremulus* (max. size about 50 cm), that is frequently found along the Norwegian coast and up to Russia.

Stichopus species are medium-size to large sea cucumbers, with lengths from 10 to more than 50 cm. Of those sea cucumbers living on Indo-Pacific reefs, the most frequently imported species are *S. chloronotus* and *S. horrens*.

Stichopus chloronotus attains a length of up to 30 cm. Its body is distinctly square in cross-section, with a uniform colour of dark green or almost black, and covered by numerous longish papillae. The tips of these papillae are orange or brownish-red. Its geographical range in the Indo-Paci-

Stichopus horrens **in Papua-New Guinea (left; Photo: L. Newman and A. Flowers) and at Guam (right; photo: P. Schupp)**

Stichopus variegatus **photographed at Indonesia.**

Stichopus **sp. at Palau.** **Photo: Dr. I. Illich**

fic extends from East Africa to Hawaii, and it is common in shallow water from 0 down to 15 m depth, where it lives on rock- or sand bottom. Juvenile specimens prefer the shallower areas while adults tend to migrate in to deeper water.

The "Horrible" Sea Cucumber, *Stichopus horrens*, is not as terrible as suggested by its name. However, it can indeed be a "terrifying" experience trying to correctly identify this species; especially, so since its body colouration shows enormous intra-specific variability. Yet, most specimens are bronze-coloured with numerous conical papillae that are surrounded at their basis by a series of concentric greenish rings. However, the basic body colouration can also be whitish, greyish or yellowish, and the papillae can be grey or bluish. This species can reach a length of about 40 cm, and it is one of the largest sea cucumbers that can be recommended for aquarium maintenance. It is widely distributed throughout the Indo-Pacific, from the Maldives to Hawaii. During the day it hides under rocks or in crevasses, but at night it moves over the sandy or rocky bottom of the reef flat in search of food. The large snail *Tonna perdix* (see page 81) is the most severe predator of *Stichopus horrens*.

At list of *Stichopus* species that could possibly be imported for the marine aquarium hobby should include the very common but rather large *S. variegatus*, *S. pseudohorrens*, with very long, needle-like papillae, and the rather fascinating *S. vastus*. The latter occurs around New Guinea, Micronesia and Palau. It is an extraordinarily strange sea cucumber, which resembles an accumulation (pile!) of excreta. But apart from that it is rather attractive with its intense yellow and a dark brown stripe, and so it would not be surprising if it appeared periodically in the aquarium trade. All three species can grow to a length of 50 cm, with a possible thickness in excess of 15 cm, so

***Thelenota ananas* on the Great Barrier Reef.**

that these sea cucumbers are really only suitable for large aquariums.

The other genus common throughout the reefs of the Indo-Pacific is *Thelenota*. Here the best-known and probably the most-photographed species is the Pineapple Sea Cucumber, *T. ananas*. This is a virtual giant that can reach a length of 70 cm and a thickness of 15 cm. It is then, a most impressive sight to watch the animal as it slowely wanders across the sandy bottom in 2 to 30 meters depth. The body colour is typically reddish brown, marked with a stellar pattern of large, conical papillae. Often the papillae are of the same colour as that of the body, but they can also be bright orange-coloured. We do not know whether the Pineapple Sea Cucumber is regularly imported for the aquarium trade. But then, we cannot really recommend this species for aquarium maintenance anyway, except for very large display tanks.

The Giant Sea Cucumber, *Thelenota anax* is even more gigantic. Normally, it reaches a length of 80 cm, sometimes even up to 1 m. Apart from its size it is not nearly as spectacular as the Pine Apple Sea Cucumber. The body is coloured light brown, covered by a pattern of darker spots and with a warty skin texture. It is an Indio-Pacific species that occurs from Madagascar to the Marshall Islands, where it lives on a sandy bottom at depths from 12 to 30 m. This is another sea cucumber that

***Thenenota anax* on the Great Barrier Reef.**

Isostichopus badionotus:

Colour variety at Saba, Lesser Antilles. Photo: Dr. I. Illich

Colour variety at Puerto Rico.

Another colour variety at Puerto Rico.

cannot really be recommended for aquarium maintenance, except for very specific purposes.

The third species of genus *Thelenota* is the Red-striped Sea Cucumber, *T. rubralineata*, first described in 1992 by MASSIN & LANE. It has a cream-coloured body, covered by numerous cherry-red stripes. This rather spectacular sea cucumber is known from Indonesia, New Guinea, the Philippines and the Solomon Islands. With a total length of at least 60 cm it is also too large for the standard marine aquarium. This species is only rarely seen in the wild since it tends to inhabit primarily the outer edges of barrier reefs, at depths below 30 m.

The only Caribbean species of family Stichopodidae that is relatively often found in the aquarium trade if the Three-lined Sea Cucumber, *Isostichopus badionotus*. Its popular name refers to the three rows of tube feet on its back, which are often of a darker colour, giving it an appearance as if sprinkled with chocolate. But there are also specimens with a dark body and light-coloured warts, as well as some that have uniform colour pigmentation. When found on reefs, *I. badionotus* occurs mainly on muddy or sandy bottom or in sea grass beds, from the low water line down to a depth of 65 m; sometimes it also lives in rocky habitats. Juvenile specimens tend to hide under rubble or crushed coral. Its geographical distribution ranges from the Bermuda Islands and South Carolina, southward to Brazil.

All stichopodid sea cucumbers feed on detritus and all kinds of organic particles, including microscopically small invertebrates. Under aquarium conditions they accept substitute food items and, if kept under the proper conditions, they can be hardy and durable aquarium inhabitants.

Family Holothuriidae

The size of sea cucumbers in this family varies substantially. There are a few rather small ones

Actinopyga species have five calcified papillae in the anal opening (anal teeth) as a characteristic feature, which are clearly visible in this *A. mauritania* specimen (left; photo: C. Bryce). Right: *A. lecanora* (seen here at Heron Island, Great Barrier Reef) can reach a maximum length of 40 cm (photo: L. Newman and A. Flowers).

Holothuria atra on the Great Barrier Reef.

Left: *Holothuria fuscocinera* at the Fiji Islands (photo: S. W. Michael). Right: *Holothuria fiscorubra* at the Maldives.

***Holothuria nobilis* at the Maldives.**

(e.g. *Holothuria parvula*) that reach a length of barely 7 cm, as well as some incredibly large species (e.g. *H. thomasi*), that are often 2 m long. By the way, these two examples of extremes come from the Caribbean region; however, the majority of tropical species in this family live in the Indo-Pacific.

The leading genera are *Actinopyga*, *Bohadschia*, *Holothuria* and *Labidodemas*, whereby *Holothuria* is unquestionably the largest group also of interest to the aquarium hobby. All of them have a distinct cucumber- or worm-like shape, which is firm and rather thick in *Actinopyga*- and *Bohadschia* species, and softer and more leathery in *Labidodemas*- and *Holothuria* species. All of them have Cuvierien Tubules, but not all species will actually expel them when endangered or being disturbed.

Although the majority of species prefers to hide under rocks or remain buried in sand or rubble, there are also some that can commonly be seen on reefs out in the open. This applies particularly to *Actinopyga*- and *Bohadschia* species, while *Holothuria*- and *Labidodemas* species remain in hiding during the day, and emerge only at dusk or at nightfall, in search of food.

From the viewpoint of aquarists, family Holothuridae is the most important sea cucumber family (apart from family Cucumaria), not because of any ecological significance, but simply because of their colouration. A number of holothuriid cucumbers are more or less regularly imported, in spite of fears among many aquarists for the lifes of their fishes because of the possibility of sudden Holothurin poisoning. As mentioned earlier, such fears are indeed justified, although we have never had any problems in that respect. We believe that this is principally a problem caused by stress. Admittedly, for the dead fish, the causes are no longer of any consequence, and so special caution is advisable.

On coral reefs it is not uncommon, to find specimens that are 50 cm long or even longer, but the majority of imported species is small and this will remain so. There can be considerable variability in colouration among the species discussed here, however, the colour patterns are characteristic for each species that the identification is rather simple. Most of the imported species originate from the Indo-Pacific, where 80 species are known to occur. The majority of them (about 65 species) belong to genus *Holothuria*. This is a genus with considerable diversity, so that it has been subdivided in to 14 subgenera. Thirteen *Holothuria* species and one *Actinopyga* species occur in the Caribbean region. They are imported only infrequently, but a few species appear periodically in the aquarium

Chirodota rigida – seen here at One Tree Island (photo: L. Newman and A. Flowers) – is a typical representative of family Chiridotidae, which are characterized by the presence of peltato-digitate (shield-shaped, with finger-like outgrowths) tentacles (see drawing below, after CLARK & ROWE, 1991).

trade. HENDLER et al. (1995) is an excellent reference publication for their identification. For keying out the Indo-Pacific species, appropriate help can be solicited from the publications by CANNON & SILVER (1986) and GUILLE et al. (1986). A few species of particular interest to the aquarium hobby are discussed in detail in Fact Sheets on pages 388 - 401.

Order Apodida

As is already indicated by its scientific name, the sea cucumbers in this order do not posses tube (A, Greek = without, and pous, genetive podos, Greek = foot); colloquially they are also referred to as "footless sea cucumbers". Basically, they have a far more slender, worm-like shape, with thin and delicate body walls. Respiratory trees are also absent; they are not required since gas exchange is facilitated across the thin skin. Food is taken in either with the help of tentacles that are wiped over the bottom, or by ingesting sand and other sediments, in the same manner as described for order Aspidochirotida.

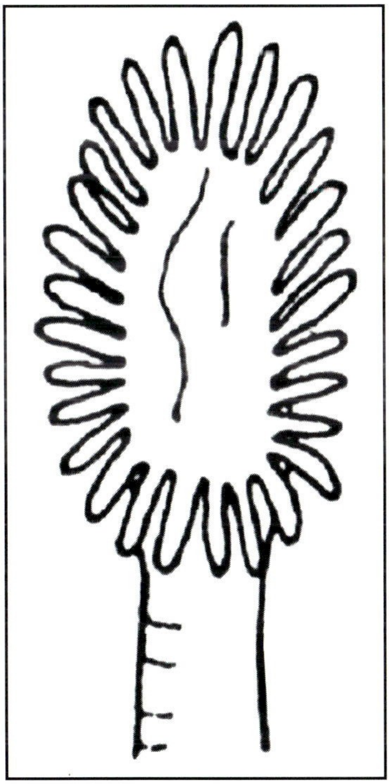

This order contains two tropical families, that is, Synaptidae and Chiridotidae. In terms of their biology and aquarium requirements, both families are rather similar, so that we are only going to discuss family Synaptidae in detail. Both families can be distinguished from each other on the basis of the shape of spicula embedded in the skin. In family Synaptidae these shapes are primarily anchors, but never wheels. Exactly the reverse is true for family Chiridotida. Beyond that, the tentacles in family Chiridotidae

Actinopyga mauritania
Mauritanian Sea Cucumber

Distribution: Common throughout the Indo-Pacific, from the Red Sea to Hawaii.
Size: Maximum length 30 cm.
Reproduction: Details unknown.
Food: Detritus feeder that utilises principally organic particles and microscopic animals of the meiofauna, but probably also feeds on diatoms and other microscopic algae. For this sea cucumber to thrive under aquarium conditions it requires a similar diet, although some substitute foods is accepted.
Description: Sprinkled colour pattern on the dorsal area, consisting of light chestnut brown and white spots; sole normally whitish. Densely covered with tube feet. Has 25 or more tentacles. This species – like all *Actinopyga* species – also has five calcified papillae in the anus (anal teeth). Lives in shallow, exposed reef habitats down to a depth of about 20 m.

Aquarium requirements: Rarely ever imported, however, it makes a good aquarium inhabitant provided the correct food is available. Any aquarist also keeping fish in the same tank needs to be aware that this sea cucumber represents a potential death threat for his aquarium (see page 374). However, this species can be unequivocally recommended as co-inhabitant with other invertebrates.

GA: +; **SE:** +; **CO:** +/-

Photo:
Actinopyga mauritania at the Cocos Islands, Indian Ocean.
Photo: C. Bryce

Bohadschia argus
Eye-spot Sea Cucumber

Distribution: Common throughout the Indo-Pacific, from Madagascar and the Seychelles to the South Pacific.
Size: Maximum length 60 cm.
Reproduction: Details unknown.
Food: Feeds primarily on organic detritus and microscopic organisms of the meiofauna. Requires similar food if it is to do well under aquarium conditions, although some substitute foods is accepted.
Description: Immediately identifiable by the characteristic colour pattern of white-edged, orange eyespots against a grey or grey-brown body colour. The sole is normally light yellow-brown. Rather frequent in some areas. Lives among coral rubble and algae, normally always exposed, at depths from 2 to 40 m. In contrast to many other holothuriids, this species does not bury extensively in substrate.
Aquarium requirements: Rarely imported, moreover, it is too large for the average aquarium; however, will do rather well in large aquariums with extensive substrate areas. Aquarists keeping also fish in the same tank must be aware that this type of sea cucumber can be potentially lethal for fish (see page 374). Can only be recommended unreservedly as a companion animal for other invertebrates.

GA: +/-; **SE:** +; **CO:** +/-

Photos:
Two colour varieties of *Bohadschia argus* at Palau (top; photo: Dr. I. Illich) and in Indonesia (bottom; photo: S. W. Michael).

Bohadschia graeffei
Dotted Sea Cucumber

Distribution: Western Indo-Pacific; from the Red Sea throughout the Indo-Australian Archipelago, eastward to the South Pacific islands
Size: Although several authors listed a maximum size of 50 cm, most specimens do not exceed 30 to 35 cm.
Reproduction: Details unknown.
Food: Detritus feeder, utilising primarily organic particles and microscopic organisms of the meiofauna, probably including diatoms and other microscopic algae. Like most detritus feeders, it feeds by burying through sediments, but has also been seen frequently "cleaning off" sediment deposited on fleshy algae and corals. Requires a similar diet under aquarium conditions if it is to thrive, although some substitute foods is accepted.
Description: Lives on sand, rubble or hard substrate at depths from 5 to 30 m, where it is normally diurnal. Base colouration in adults variable, from light green to light brown with black spots and short tubercles with white tips. Tentacles always black with white crowns. Juveniles with surprisingly variable colour patterns that imitate the appearance of poisonous sea slugs of genus *Phyllidia*.
Aquarium requirements: Rarely ever imported and too large for the average aquarium. However, this sea cucumber does well in large aquariums with extensive bottom substrate areas with sufficient detritus deposits. Aquarists also keeping fish in the same tank must be aware that this type of sea cucumber can be potentially lethal for fish (see page 374). Can only be recommended unreservedly as a companion animal for other invertebrates.

GA: +; **SE:** +; **CO:** +/-

Photos:
Adult specimens of *Bohadschia graeffei* at the Maldives (top) and at Kimberley, Western Australia (centre; photo: C. Bryce) and a juvenile specimen, that imitates a sea slug of genus *Phyllidia* (bottom; photo: L. Newman and A. Flowers).

Bohadschia marmorata
Marbled Sea Cucumber

Distribution: Indo-Pacific; from the Red Sea and East Africa, eastward to the South Pacific Islands.
Size: Maximum size 40 cm.
Reproduction: Details unknown.
Food: Detritus feeder, utilising primarily organic particles and microscopic organisms of the meiofauna. Requires a similar diet under aquarium conditions if it is to thrive, although some substitute foods is accepted.
Description: Lives on sand bottom or among rubble, at depths from 4 to 20 m. Colouration variable mixture of brown and/or yellow, normally with small brown spots on the back, but often also with large, dark angular patches.
Aquarium requirements: Rarely ever imported; an interesting aquarium inhabitant for large tanks. Aquarists also keeping fish in the same tank must be aware that this type of sea cucumber can be potentially lethal for fish (see page 374). Can only be recommended unreservedly as a companion animal for other invertebrates.

GA: +; **SE:** +; **CO:** +/-

Photo:
Bohadschia marmorata at Guam. Photo: P. Schupp

Holothuria atra
Black Sea Cucumber

Distribution: Indo-Pacific; common from the Red Sea and East Africa, eastward to Hawaii.
Size: Maximum length 60 cm.
Reproduction: Apart from sexual reproduction, asexual reproduction by division is also known to occur. In case of a division in to two halves, both parts will regenerate the respective missing half.
Food: Detritus feeder, utilising primarily organic particles and microscopic organisms of the meiofauna. Requires a similar diet under aquarium conditions if it is to thrive, although some substitute foods are accepted.
Description: Uniformly black, although its appearance is often different since its skin is usually covered by sand, revealing only small patches of black. Tentacles also black. Occurs rather frequent and is very conspicuous since it is often seen exposed on sandy reef areas and along the coast, at depth from 0 to 30 m. Sometimes in large aggregations. Belongs to subgenus *Halodeima*, together with the two Indo-Pacific species *Holothuria edulis* (see Fact Sheet on page 393) and *H. pulla*, as well as the three Caribbean species *H. floridana*, *H. grisea* and *H. mexicana)*.
Aquarium requirements: Occasionally imported, but too large for the average aquarium; however, will thrive in large tanks with extensive bottom areas and sufficient detritus deposits. Aquarists also keeping fish in the same tank, must be aware that this type of sea cucumber can be potentially lethal for fish (see page 374). Can only be recommended unreservedly as a companion animal for other invertebrates.

GA: +/-; **SE:** +; **CO:** +/-

Photo:
Holothuria atra at the Cocos Islands, Indian Ocean.
Photo: C. Bryce

Holothuria edulis
Edible Sea Cucumber

Distribution: Indo-Pacific; from the Red Sea and East Africa eastward to Hawaii.
Size: Maximum length 30 cm.
Reproduction: Details unknown.
Food: Detritus feeder that takes principally organic particles and microscopic organisms from the meiofauna. Requires similar diet if it is to do well under aquarium condition, although will take some substitute foods.
Description: As indicated by the scientific and popular species name, this is an edible sea cucumber; however, the name is somewhat misleading since there are other edible sea cucumbers, which are eaten more often and, indeed, taste better. Because of its pink-coloured sole together with the dark red or often black spots on the dorsal side, it is easy to identify. But there are also specimens without any spots. Frequent inhabitant of rocky and sandy habitats, at depths from the tidal zone to 45 m. Can occur out in the open or in hiding in rubble areas. Belongs to subgenus *Halodeima*, together with two Indo-Pacific species *Holothuria atra* (see Fact Sheet on page 392) and *H. pulla*, as well as the three Caribbean species *H. floridana*, *H. grisea* and *H. mexicana*.
Aquarium requirements: Frequently imported and an excellent aquarium inhabitant if the correct food is available. Aquarists also keeping fish in the same tank must be aware that this type of sea cucumber can be potentially lethal for fish. Can only be recommended unreservedly as a companion animal for other invertebrates.

GA: +; **SE:** +; **CO:** +/-

Photo:
Holothuria edulis at One Tree Island, Great Barrier Reef.
Photo: L. Newman and A. Flowers

Holothuria hilla
Sausage Sea Cucumber

Distribution: Indo-Pacific; common from the Red Sea and East Africa eastward to Hawaii.
Size: Maximum length 30 cm.
Reproduction: Details about sexual reproduction unknown. Asexual reproduction by division has occurred under aquarium conditions.
Food: Detritus feeder that ingest bottom substrate. Feeds primarily on organic particles and microscopic organisms of the meiofauna. Requires similar diet under aquarium conditions if it is to thrive, although substitute is accepted.
Description: Slender, sausage-like (hilla, Latin = sausage) sea cucumber, that is easily recognised by its light- to dark brown colouration and the conical papillae scattered over the body. Lives normally among rocks and rubble on reef flats at depth from 1 m to 30 m. Belongs to subgenus *Thymioscycia*, together with *Holothuria impatience* (see Fact Sheet on page 395).

Aquarium requirements: Rather frequently imported and an excellent aquarium inhabitant, provided the correct food is available. Should be kept in a well-established aquarium with sufficient detritus deposits Aquarists also keeping fish in the same tank, must be aware that this type of sea cucumber can be potentially lethal for fish (see page 374). Can only be recommended unreservedly as a companion animal for other invertebrates.

GA: +; **SE:** +; **CO:** +/-

Photo:
Holothuria hilla at Dampier Archipelago, Western Australia.
Photo: C. Bryce

Holothuria impatiens
Impatient Sea Cucumber

Distribution: Cosmopolitan species, which occurs in the Indo-Pacific as well as in the Eastern Pacific, the Caribbean region, the Atlantic and in the Mediterranean Sea.
Size: Maximum length 25 cm, however, most specimens are only 15 cm long.
Reproduction: Details of sexual reproduction unknown. Asexual reproduction by division has been observed under aquarium conditions.
Food: Detritus feeder that ingests bottom substrate, from which it absorbs primarily organic particles and microscopic organisms of the meiofauna. Requires similar food in the aquarium if it is to thrive, although substitute foods are accepted.
Description: Although the term "impatient" (impatiens, Latin = impatient and related meanings) related to a gentle animal such as a sea cucumber is difficult to explain, the name is - nevertheless - quite appropriate. Beyond any doubt, this is the most active of all *Holothuria* species. When removed from its hiding place, it will seek new cover very hurriedly. Until the animal is successful it will continuously twist and turn. The body is sprinkled in grey, pink, brown and/or purple and has variable dark and light cross bands. Lives usually among rocks and rubble on the reef flat, at depths from 0 to 30 m. Belongs to the subgenus *Thymiosycia*, together with *Holothuria hilla* (see Fact Sheet on page 394) and a few other Indo-Pacific as well as Caribbean species.
Aquarium requirements: Rather frequently imported and one of the best detritus-feeding sea cucumbers. Its size too makes it a better aquarium inhabitant than most of the other sea cucumbers. It is extremely adept in crawling in to the smallest of caves and crevasses in live rocks where it removes any accumulated detritus. However, at the same time it is also a "master escapist" by crawling over the surface simmers and right in to the filter sump. Should be kept in a well-established aquarium with sufficient detritus deposits. Aquarists also keeping fish in the same tank must be aware that this type of sea cucumber can be potentially lethal for fish (see page 374). Can only be recommended unreservedly as a companion animal for other invertebrates.

GA: +; **SE:** +; **CO:** +/-

Photos:
Specimens of *Holothuria impatiens* in an aquarium (top) and in Indonesia (middles).

Holothuria leucospilota
White-spotted Sea Cucumber

Distribution: Indo-Pacific; common from the Red Sea and East Africa to Hawaii.
Size: Maximum length 60 cm.
Reproduction: Details unknown.
Food: Detritus feeder that ingests bottom substrate, from which it absorbs primarily organic particles and microscopic organisms of the meiofauna. Requires similar food in the aquarium if it is to thrive, although substitute foods are accepted.
Description: Body colouration normally jet black, with scattered small, white spots (leucos, Greek = white and spilos, Greek = spot); occasionally there can also be dark brown specimens. Soft papillae cover the rather soft and pliable body. Occurs frequently on many Indo-Pacific reefs. Lives in sandy and rubble-strewn habitats on reef flats, at depths from 0 to 10 m. Characteristically, the hind end is hidden under rocks or corals, while the fore end searches along the substrate for food as shown in the photo on page 368. Belongs to subgenus *Mertensiothuria*, together with the Indo-Pacific species *Holothuria fiscocinerea*, *H. papillifera* and *H. pervicax* (see Fact Sheet on page 397).
Aquarium requirements: Occasionally imported, but too large for the average aquarium. Does well in large aquariums with sizeable sand areas and sufficient detritus deposits. Aquarists also keeping fish in the same tank must be aware that this type of sea cucumber can be potentially lethal for fish (see page 374), but can be recommended unreservedly as a companion animal for other invertebrates.

GA: +/-; **SE:** +; **CO:** +/-

Photos:
Holothuria leucospilota at One Tree Island, Great Barrier Reef. The specimen on the bottom photo has expelled its Cuvierian Tubules. Photos: L. Newman and A. Flowers

Holothuria pervicax
Tireless Sea Cucumber

Distribution: Indo-Pacific; common from the Red Sea and East Africa to Hawaii.
Size: Maximum length 35 cm.
Reproduction: Details unknown.
Food: Detritus feeder that tirelessly (= pervicax, Latin) ingests bottom substrate. Feeds primarily on organic detritus and microscopic organisms of the meiofauna. Requires similar food if it is do thrive in an aquarium, although it will accept substitute foods.
Description: Dorsal area brown, sprinkled with lighter and dark shades. Large, distinctly visible papillae. Sole grey with dark spots of tube feet. Lives in sand and rubble habitats on reef flats, at depth from 0 to 10 m. Belongs to subgenus *Mertensiothuria*, together with the Indo-Pacific species *Holothuria fuscocinerea*, *H. leucospilota* (see Fact Sheet on page 396) and *H. papillifera*.

Aquarium requirements: Rather frequently imported and belongs to the most efficient detritus-feeding sea cucumbers. Has lived for many years in our aquariums. Does well in large aquariums with sizeable sand areas and sufficient detritus deposits. Aquarists also keeping fish in the same tank must be aware that this type of sea cucumber can be potentially lethal for fish (see page 374), but can be recommended unreservedly as a companion animal for other invertebrates.

GA: +; **SE:** +; **CO:** +/-

Photo:
Holothuria pervicax in an aquarium.

are peltato-digital (see drawing on page 372), a type never found in family Synaptodae, where they are always feather-like. Moreover, family Chiridotidae with its 12 Indo-Pacific species and Caribbean species, has been studied considerably less than family Synaptidae.

Family Synaptidae

The worm cucumbers – as representatives of family Synaptidae are commonly known – represent a rather interesting animal group. In terms of their length they vary from the miniature forms of genus *Synaptula*, with lengths of only a few centimetres, to virtual giants, such as *Synapta maculata*, with a length of 2.5 m when fully expanded. All species are slender and snake-shaped, and remarkably active. They move with the aid of peristaltic (body) contractions, supported by the adhesiveness of their skin (caused by tiny, protruding anchor spicula).

Identification of most species is very difficult, since morphology and colouration within the family are fairly similar. All previous taxonomic work is based on the shape of spicula, which is virtually impossible to do on live specimens.

Indo-Pacific *Synapta* species, especially *S. maculata*, are often available from the aquarium trade. Species from other genera are sometimes also imported, but only on an irregular basis. Most aquarists consider worm (sea) cucumbers as being rather unattractive and tend to reject them. We do not subscribe to the viewpoint that they are unattractive, but for other reason we have to concede, that for many aquarists these sea cucumbers should not be the "first choice". On one hand, most imported species will get too large for the average home aquarium, although they are only slender and thin. Moreover, it is also important to keep in mind that these cucumbers – because of their fragility and delicateness – are very difficult to transport, so that many specimens offered for sale are damaged and are therefore difficult to acclimate.

Yet, we are of the opinion that it can indeed be a meaningful task for engaged and experienced aquarists to keep worm sea cucumbers. For that it is essential that they be provided with many hiding places and a thick bottom layer of sand. It is absolutely imperative that the aquarium functions flawlessly and that it contains nutrient-rich sediments. In Fact Sheets on pages 399 - 401 we present three species from three different genera, which could be highly interesting aquarium inhabitants.

Most species in this family feed on edible particles from the sediment, such as organic detritus; however, genus *Synapta* contains a few food specialists. Most of them live as commensals on sponges or on other large invertebrates (e.g. Gorgonians), and then feed on sediments found on their host or of metabolic waste products given off by the host. In addition, there are *Synaptula* species that live as epizoites on large

Sea cucumbers of genus *Synapta* (seen here is *Synapta* cf. *maculata*) are fascinating aquarium animals. However, because of their size, transport sensitivity and difficult acclimatization they can really only be recommended for very experienced aquarists.

***Synaptula* sp. "KA6-ECH-06", seen here at Palau.** **Photo: Dr. I. Illich**

Euapta godeffroyi
Godeffroy's Sea Cucumber

Distribution: Indo-Pacific; from the Red Sea and East Africa to Hawaii. Eastern Pacific; from the Gulf of California to Panama.
Size: Maximum length 150 cm.
Reproduction: Details unknown.
Food: Detritus feeder that ingests bottom substrate. Feeds primarily on organic particles and microorganisms of the meiofauna. Requires similar food if it is to do well in an aquarium, although substitute food will be accepted.
Description: Sprinkled creamy white, with grey, green or brown longitudinal stripes. The double yellow longitudinal stripe is a typical feature of this species. Lives on sand bottom or in sea grass beds in shallow water areas, down to a depth of 15 m. Hides among rubble on reef flats or along reef slopes.
Aquarium requirements: Rarely ever imported; of possible interest for large aquariums. This extremely soft and fragile sea cucumber is very sensitive to adverse transport and handling conditions and readily breaks apart if improperly handled. Acclimatisation must take place very gently and carefully. Does also need great care with respect to feeding. Can be recommended only for experienced aquarists. Aquarists also keeping fish in the same tank must be aware that this type of sea cucumber can be potentially lethal for fish (see page 374). However, it can be recommended unreservedly as a companion animal for other invertebrates.

GA: +/-; **SE:** -; **CO:** +/-

Photos:
Euadapta godeffroyi in the Red Sea (top; photo: E. Svensen), in Indonesia (centre; photo: R. Hansen) and at Northwest Cape, Western Australia (bottom; photo: C. Bryce).

Opheodesoma* cf. *australiensis
Australian Worm Sea Cucumber

Distribution: Indo-Australian Archipelago and Western Pacific; from the Great Barrier Reef and the Philippines eastward to Tahiti.
Size: Maximum length 100 cm.
Reproduction: Details unknown.
Food: Detritus feeder that ingests bottom substrate. Feeds primarily on organic particles and microorganisms of the meiofauna. Requires similar food if this sea cucumber is to do well in an aquarium, although substitute food will be accepted.
Description: Sprinkled orange-brown to pink. Lives in sandy habitats and sea grass beds, at depth between 15 and 30 m.
Aquarium requirements: We are not aware that this species has ever been imported, but it could have been possible. It stands to reason that it is just as sensitive to adverse transport conditions as *Synapta*-species. Therefore, it is imperative that it be given a very gradual and careful acclimatisation. Does also need great care with respect to feeding. Can be recommended only for experienced aquarists. Aquarists also keeping fish in the same tank must be aware that this type of sea cucumber can be potentially lethal for fish (see page 374). However, it can be recommended unreservedly as a companion animal for other invertebrates.

GA: +/-; **SE:** –; **CO:** +/-

Photo:
Opheodesoma cf. *australiensis* at Stradbroke Island, Queensland. Photo: L. Newman and A. Flowers

Synapta maculata
Spotted Worm Sea Cucumber

Distribution: Indo-Pacific; from the Red Sea and East Africa to the Society Islands.
Size: Maximum length 250 cm.
Reproduction: Details unknown.
Food: Detritus feeder that ingests bottom substrate. Feeds primarily on organic particles and microorganisms of the meiofauna. Requires similar food if this sea cucumber is to do well in an aquarium, although substitute food will be accepted.
Description: Colouration variable, however, always sprinkled, brown-yellow or bronze-coloured to brown with black markings. Extremely soft and spongy body. Lives in the shallow water of lagoons, beach- and shoreline fringing reefs, as well as in sea grass beds at depths from the tidal zone to 25 m.
Aquarium requirements: Rather frequently imported; because of its size of interest for large aquariums only. This extremely soft and fragile sea cucumber is very sensitive to adverse transport conditions and can break apart when handled roughly. Acclimation must be done very slowly and with great deliberation. Does also need great care with respect to feeding. Can be recommended only for experienced aquarists. Aquarists also keeping fish in the same tank must be aware that this type of sea cucumber can be potentially lethal for fish (see page 374). However, it can be recommended unreservedly as a companion animal for other invertebrates.

GA: +/-; **SE:** –; **CO:** +/-

Photos:
Synapta maculata in its natural habitat on the Great Barrier Reef (top) and close-up photograph of the anterior section of a specimen at Guam (bottom; photo: P. Schupp).

fleshy algae and on calcareous algae, as well as on sea grass. There they feed primarily on epiphytic microscopic algae and diatoms. Although we have not yet seen any *Synaptula* species in the aquarium trade, it is conceivable that they are introduced inadvertently in to our aquarium as "stowaways" on sponges or other hosts. Providing the host can be maintained adequately in the aquarium (which, in many cases, can turn out to be quite difficult), it seems that the aquarium care of *Synaptula* species may well be easier than that of its larger relatives.

Species identification is virtually impossible within this genus if the host cannot be identified. In addition, tissue sections from dead specimens also need to be examined. That is also the reason why we can only show photos here of species that have not been clearly identified.

Synaptula sp. "KA-ECH-07" at the Maldives.

Synaptula sp. "KA-ECH-08" in Indonesia.

Synaptula sp. "KA-ECH-09" at Palau.

Photos: P. Schupp

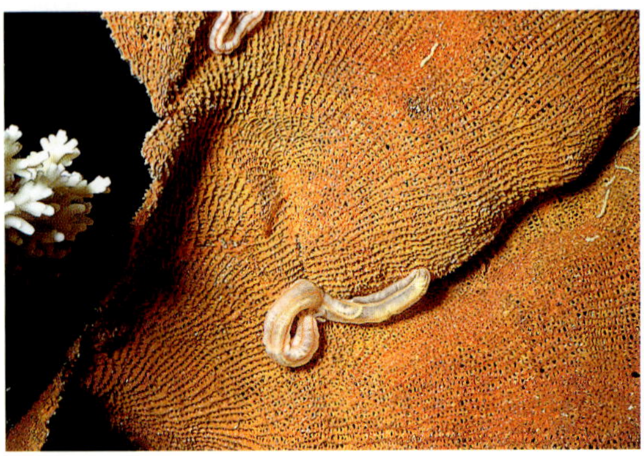

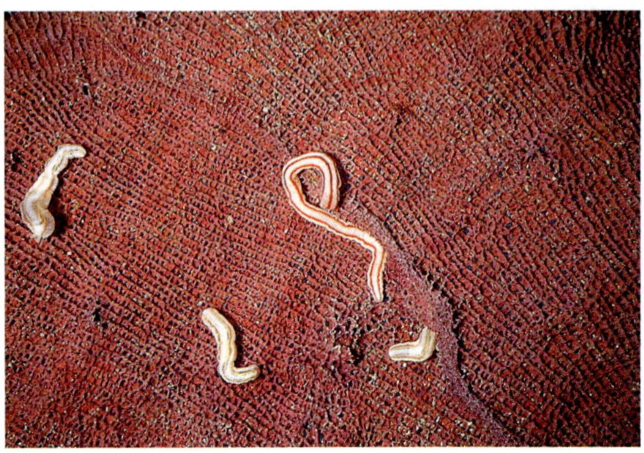

Synaptula sp. "KA-ECH-10" on the Great Barrier Reef.

Photos: E. Svensen

Beautiful echinoderms; the yellow feather star *Comanthina schlegelii* photographed in Indonesia and the Pencil Urchin *Heterocentrotus mammillatus* photographed in Hawaii.

Chapter 4:

Tunicates and Lancelets

Tunicates and lancelets belong to the phylum Chordata, which is subdivided in to three subphyla:

- Tunicata (Urochordata) – tunicates (sea squirts – ascidians; salps; appendicularians – larvaceans)
- Cephalochordata – lancelets
- Vertebrata – hagfishes, lampreys and jawed vertebrates

On the following pages we take a closer look at the Tunicata and refer briefly to the lancelets.

Tunicata and Cephalocordata – in contrast to the Vertebrata – lack backbones. They do, however, have chordate characters that link them to the vertebrate animals: they have gill slits and a pharyngeal groove that fixes iodine (like the vertebrate thyroid gland); and sometimes (often only in the larvae) they have a post-anal tail containing a notochord (like the embryonic backbone of a vertebrate) and a hollow dorsal nerve chord (RUPPERT & BARNES, 1994; HERTLER, 1996). They are important animals that cast light on the evolution of vertebrates from invertebrates. The generally accepted view is that tunicates and lancelets – together with the acorn worms (phylum Hemichordata, see volume 3, page 159) – are cousins of the vertebrates, sharing common ancestors back in geological time.

Tunicates belong to a remarkable and varied animal group that lives in diverse habitats. With few exceptions, tunicates are filter feeders. Two of the exceptions are shown at the left and right.
Left: *Megalodicopia hians*, photographed here at Monterey Bay Aquarium (Monterey, California, USA), a solitary animal living in deep water in all oceans. It uses the large branchial hood around the mouth to trap prey organisms swimming- or drifting past. A predatory habit has evolved to replace filter feeding in deep water where there is little or no organic material in suspension.
Right: The colonial species *Didemnum molle*, photographed here at Kuredu Island, Maldives, which lives in shallow water in a vast area of the Indo-West Pacific and derives some of its nutrient from photosynthesis of its symbiotic algae. It may also filter particles from the water passing through the small branchial sacs. Individual zooids in the colony are very small (1.5 mm).

Subphylum Tunicata

This subphylum contains three classes:

- Ascidiacea – sea squirts
- Thaliacea – salps, doliolids
- Appendicularia (Larvacea) – tailed tunicates

All Tunicata are – with few exceptions – true filter feeders. Ecologically they play an important role since they get their food (organic particles and microorganisms) from the large volumes of water they filter. From an aquarist's perspective, the sessile ascidians are the most interesting class. Salps, dolidids and the tailed tunicates are most spectacular and prolific organisms; but are planktonic and of little interest for the aquarium hobby.

Many members of the tunicate class Ascidiacea (referred to as ascidians) are present and abundant on coral reefs. We found large numbers and diverse communities on reef walls at Bunaken, North Sulawesi where they grew entangled with other sessile invertebrates. Apart from orange-coloured, ahermatypic stony corals of genus *Tubastraea*, this photo shows young *Didemnum molle*, tiny colonies of bright yellow *Trididemnum* sp., a few zooids of *Clavelina robusta* and magnificently blue *Rhopalaea.* In addition, this small (only a few square cms) section supports numerous hydrozoans, brittle stars and algae.

Class Ascidiacea – Ascidians (Sea Squirts, including Sea Tulips, Sea Daisies)

This class contains two orders: Enterogona and Pleurogona. One peculiarity of ascidians in the order Enterogona is their ability to accumulate the element Vanadium. This metallic Vanadium, which has different biological values in the suborders Aplousobranchia and Phlebobranchia, is bound to an organic molecule. While a Vanadium content of 100 ppm dry weight is considered to be normal, the concentrations of Oxo-Vanadium is variable among many of the common species. Furthermore, HAWKINS et al. (1983) found that many species did not accumulate Vanadium concentrations that exceeded the normal level. The ability to store Vanadium is assumed to be a primitive characteristic in the evolution of this animal group (WEBB, 1939; KOTT, 1969). Ascidians in the order Pleurogona do not store Vanadium.

Ascidians are often difficult to

distinguish from each other morphologically. However, fortunately there are a few species that are relatively easy to identify and that can be recognised by marine aquarists. Even at the level of higher taxa (families, suborders, orders) differences are not always obvious, because there are colonial as well as solitary forms in all the higher level taxa. In our opinion, the publication by KOTT in MATHER & BENNETT (1993) currently offers the best popular presentation and general discussion of individual families. The most comprehensive systematic overviews as well as thorough scientific descriptions of genera and species are contained in KOTT (1980, 1981c, 1982, 1985, 1990, 1992 and 2001).

Swimming along a steep reef slope, one quickly discovers numerous ascidians firmly attached to reef walls, in caves and crevices, and growing over rubble and other organisms. They are common under projecting rock or reef structures, under ledges or at the base of corals. Some are upright, like little jugs or pitchers, or there are large, lumpy masses, stalked heads or sessile cushions, paper-thin encrusting sheets or upright fan-shaped lamellae. Some are solitary animals but others (especially in coral reefs) are colonies that can contain hundreds or even thousands of individuals. A single individual in a colony is referred to as a zooid. Many species are transparent, others are brilliantly coloured red, orange, yellow, purple, even green and many solitary species have drab leathery coats with weeds and other organising growing on them or sand entangled in hair-like outgrowths from the surface.

It is unfortunate that ascidians – as a rule – do not do well in an aquarium, since the group includes some of the most beautiful and colourful as well as interesting organisms found anywhere, including coral reefs. Small colonies are sometimes introduced to the reef aquarium on live rocks.

Ascidians are either solitary animals, such as this *Halocynthia papillosa* (top; photo: T. Luther) or colonial with individual zooids embedded in common test like this *Botrylloides leachi* (bottom; photo: L. Newman and A. Flowers).

they do not grow on undersurfaces. Sedimentation may affect colonies under aquarium conditions (Kott, 1989) and it is essential to provide strong, water movement if there is to be any chance of keeping them alive.

Ascidians are highly organised with organs for food intake, digestion and reproduction. Most conspicuous is the large perforated pharynx (branchial sac), its perforations lined with tiny hair-like cilia, which beat to create the current of water drawn in through the mouth, then through the pharyngeal perforations to an atrium and eventually out by an excurrent atrial aperture. They are capable of pumping anywhere from 24 to 540 ml of water per gram wet weight per hour (Kott, 1989). Consequently, an ascidian weighing 10 g can filter about 60 litres of water per day, which amounts of 20,000 litres per year! The food is filtered from the water by a thin net of mucus that is formed by the endostyle (a ciliated groove in the ventral mid-line), and has micropores that capture particles as small as 0.0005 mm in diameter. It is moved up over the pharyngeal wall to the dorsal mid-line where, with the food it has trapped, it is rolled into a rod which passes back into a short oesophagus and subsequently into the stomach, intestine and rectum. Finally, faeces pellets are expelled through the anus into the atrium (or atrial cavity) and to the exterior by the atrial (discharge or excurrent) opening together with the spent water, which passed through the pharyngeal perforations of the branchial wall. In most ascidians the gonoducts also discharge into the atrial cavity. In many colonial species the excurrent streams from the separate atrial apertures of each zooid open into one or more common cloacal chambers in the test before being expelled to the exterior through a common cloacal aperture (see drawings on page 409). The arrangement of the zooids around the common cloacal chamber is a useful taxonomic character.

Ascidians are always fixed to a substrate. Usually they are attached by their posterior or ventral surface, but often they grow on stalks that elevate them above the substrate, thus providing optimum conditions for gathering food, giving them access to sediment-free waters and currents bearing organic particles. An example of the species diversity found on coral reefs is shown in this assemblage at the Abrolhos (a complex of islands and reefs off Western Australia). The red colonies are *Pseudodistoma australe*, the blue colony is *Sigillina cyanea* and the round colony growing in the back is *Polycitor circes* while the translucent orange lobes are probably part of colonial species, *Pseudodistoma aureum*. Photo: C. Bryce

General Biology

Morphology and Feeding

The name of the subphylum Tunicata and its popular name, tunicates, refer to the outer "coat" or "test" (= tunic) that is conspicuous in all sea squirt species. This tunic provides an envelope that surrounds the body or a matrix in which zooids are embedded and in which embryos are incubated. It is formed by the blood vessels that project into it from the outer cell layer (ectoderm) of the animal. However, although it contains some test cells, it is largely acellular, and consists of protein, reinforced by cellulose-like fibres (tunicin). In some families aragonite spicules add strength to the tunic and support the colony or individual. Sand and other particles for added protection and strength, and plant cells in symbiotic relationships, are often embedded in or adhere to the test. It produces the adhesive for attachment to the substrate, or it grows out into stalks or root-like structures that fix the animal in the sediments; and it responds to environmental pressures that affect the growth and form of the colony or individual, forming stalks, lobes or flaps that maximise its interactions with the environment and provide protection against predators as well as against damages from waves and currents.

Ascidians are permanently attached to the substrate and are not able to move about unless they are tethered by a narrow stalk that allows them to move with currents. Most species, especially cushion-shaped or sheet-like colonies, but also larger solitary individuals, can be smothered by sediments if their apertures are directed upwards and their surface are not cleared by currents and wave action or if

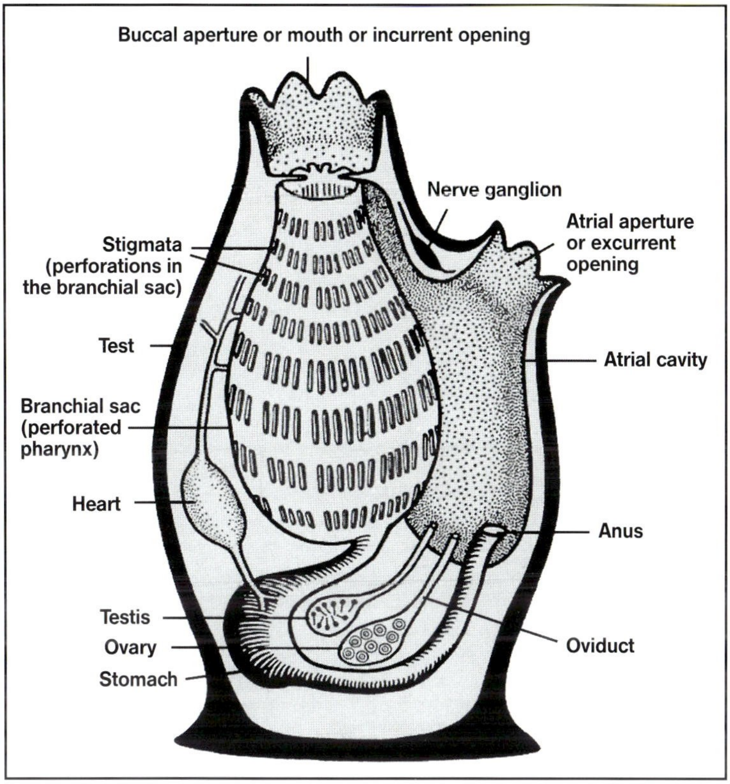

Diagram of the anatomy of an aplousobranch ascidian.
Drawing after Buchsbaum (1976)

A phlebobranch ascidian. Details in this photo correspond to the ascidian anatomy shown in the diagram on the left except that the gut is bent up into the body wall to the left of the branchial sac instead of being behind it. Photo: R. Brons

Since ascidians are unable to move and thus totally dependent on the food supply available at their fixed location, it is important that food uptake is efficient. The available amount of food is dependent on how many particles are present in the water at any given time and the capacity of ascidians to take advantage of external currents to maximise their feeding currents (created by the cilia lining the branchial sac). As food has been removed from the water ejected from the organism, and as it also contains gametes, it is desirable that it does not become mixed with the incurrent water. This is achieved by separation of the incurrent and excurrent streams of water. The excurrent opening usually is smaller than the incurrent (buccal) opening, so that the out-flowing water is ejected under pressure and projected far away from the animal; or the apertures are directed away from one another; or they are well separated either at opposite ends of the body or with one or the other on a long siphon. Generally, these morphological adaptations are a response to the environment and do not have any phylogenetic significance.

In life the pressure that is built up inside both solitary and colonial ascidians keeps them upright and inflated. After collection and preservation their appearance is very different: solitary individuals

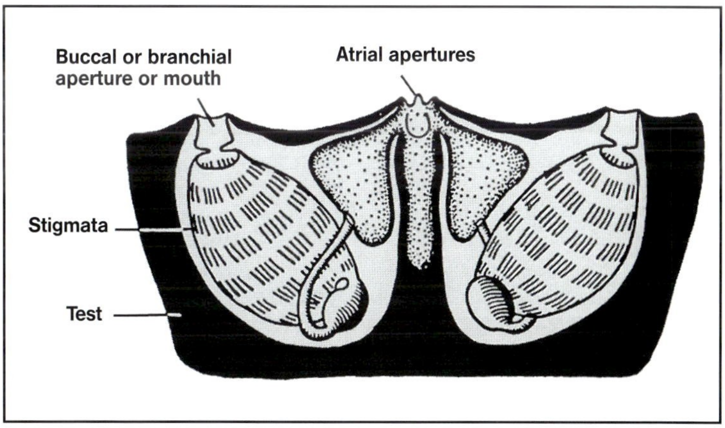

Two zooids of an ascidian colony. Drawing after Buchsbaum (1976)

Schematic outline of the evolution of a common cloacal chamber in colonial ascidians

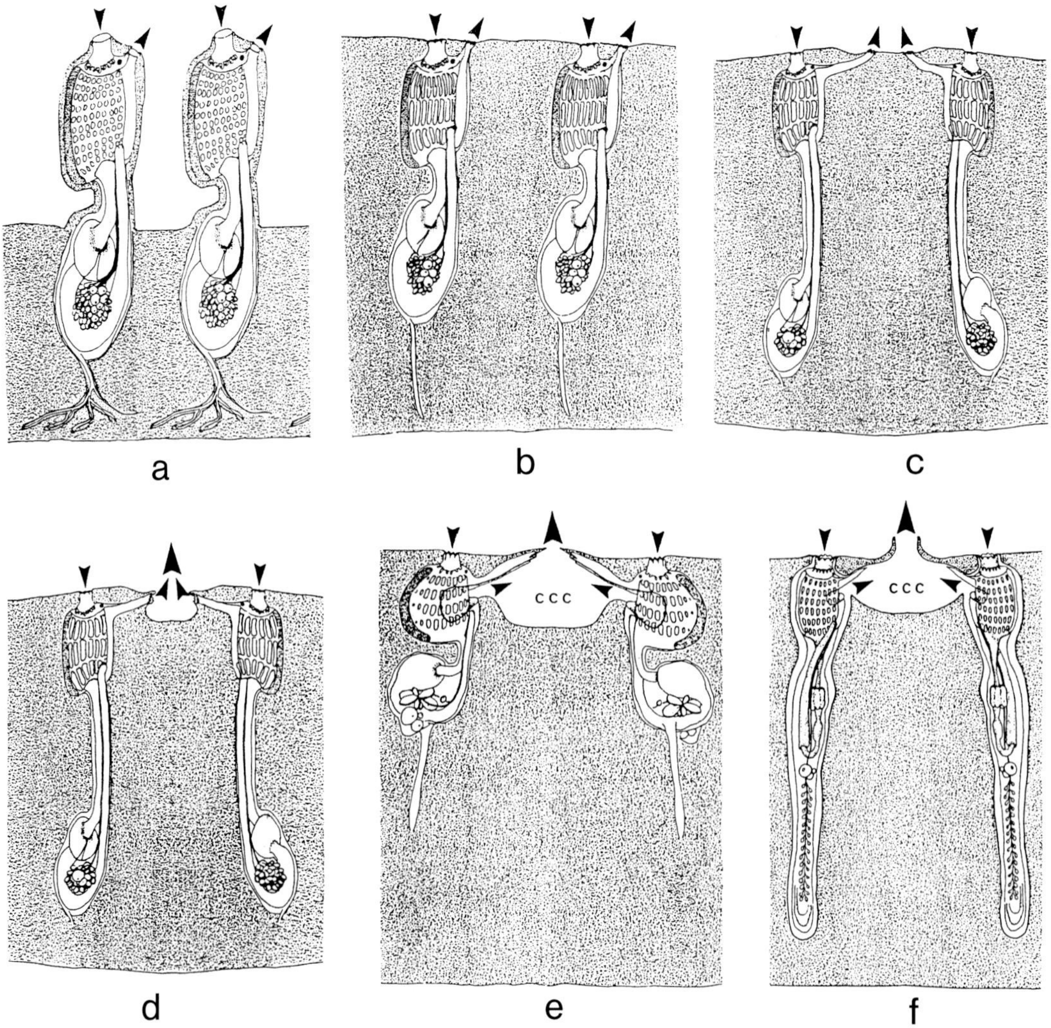

a - partially embedded separately opening zooids *Diazona*, *Clavelina* and *Pycnoclavella*); b – completely embedded separately opening zooids (*Clavelina* spp., *Nephtheis fasicularis*, *Sigillina* spp., *Polydistoma*, *Stomozoa*, *Polycitor* spp., *Eudistoma* spp.); c – separately opening zooids arranged in a circle with centrally located, individual atrial openings forming rudimentary systems (*Eudistoma* spp., *Polycitorella* spp., *Cystodytes* spp.); d – zooids with atrial apertures on siphons opening into either rudimentary (*Eudistoma* spp.) or extensive (*Hypodistoma* spp., *Exostoma*) cloacal chambers; e and f – zooids with well-developed common cloacal systems, and atrial apertures usually sessile often exposing the branchial sac directly to the common cloacal cavity; e – zooids with gonads in intestinal loop (*Distaplia*, *Sycozoa*, Didemnidae); f – zooids with gonads behind the gut loop in a posterior abdomen (Polyclinidae). Symbols: ccc – common cloacal chamber; arrows – direction of water flow.

Drawing from Kott (1990)

and zooids contract, apertures are withdrawn, colonies become flaccid and collapse, and colours oxidise or are lost.

Replication

Like many other invertebrates, ascidians replicate to form colonies and sometimes, like plants, the colony itself subdivides. Replication (or cloning) occurs by division of the zooid and regeneration of whole zooids from the subdivided parts; or it occurs by budding from blood vessels in the test; or by budding from the body wall of the zooid. Occasionally budding can be observed in an aquarium, as for instance in *Ecteinascidia nexa*, a species that is commonly introduced to the reef aquarium by live rocks (see Fact Sheet on page 444). An early form of replication occurs in some more primitive taxa when the damaged anterior part of their body (thorax) can be regenerated on a conserved abdomen as in *Rhopalaea* and *Ciona* species (see Kott, 1981b). Replication is the way the colony grows. In solitary species the individual and all its organs increase in size, but in colonies replication interrupts the growth of individuals and the large numbers of zooids produced compensates for their relatively small size.

Reproduction

At some stage all ascidians reproduce sexually. Individuals are hermaphrodite, containing testis (male reproductive glands), as well as ovaries (female reproductive glands). However these do not usually mature at the same time, so that cross fertilisation does not occur. Gonoducts open into the atrial cavity and their products are released to the exterior with the excurrent water. In most solitary species fertilization takes place outside the body of the parent. In all colonial species sperm, released in the usual way, finds its way into the atrial cavity, or even into the oviduct, where it fertilises eggs, which develop into embryos and subsequently to larvae before being released from the colony. Larvae have a slight resemblance to frog tadpoles – they have an oval trunk containing the developing ascidian, sense organs and adhesive apparatus to attach it to the substrate where it will spend the rest of its life. If it is a colonial species, buds may form early in the larval trunk. The narrow tail with its muscles and rod of notochord cells is attached to the posterior end of the trunk. The tail, with its notochord and other structures is absorbed into the larvae trunk on metamorphosis, when it changes into a sessile ascidian (see photo series on page 412). There are different larval forms and the structure of the larva is an important taxonomic character at family, genus and species levels.

One of the unsolved questions about colonial ascidians is: how does the sperm get into the atrial cavity to fertilise the eggs of these viviparous species without being caught on the mucous filter or without being expelled with the excurrent water being ejected from the atrial cavity?

Symbionts

Symbiotic relationships between zooxanthellae and cnidarians on coral reefs are well known, for it is the zooxanthellae that enable many corals to be autotrophic; and recent studies suggest that it could be the symbiotic algae (and not the corals), which have a decisive influence on the structure of a coral reef (Marshall, 1996). Please refer to Volume 1 and 2 in this book series for more information on zooxanthellae. We have also discussed the symbiotic relationships between blue-green algae and sponges (see Volume 3 of this series).

Some tropical ascidians in the family Didemnidae have a symbiotic relationship with algae. The discovery of the symbiosis was announced in a paper in the journal Nature in 1975 by Newcomb & Pugh who recognised the symbiotic organisms as undescribed procaryotic blue-green algae. These organisms were found to contain chlorophyll-a as well as -b, and their thylacoid membranes (organelles of cells containing chlorophyll) suggested an affinity with green plant (Chlorophyta) chloroplasts. Lewin (1977) established a separate division of the algae (Prochlorophyta) and a new genus *Prochloron*, for these cells, although he later changed his view about Prochlorophyta (Lewin, 1987). RNA sequencing has shown that they have a close affinity with blue-green algae and today it generally is agreed that these algae are, indeed, blue-green algae. *Prochloron* spp. has never been found outside the host ascidians, and may exist only as symbionts with ascidians. Attempts to cultivate it in aquaria or in algal cultures have been unsuccessful.

The didemnid symbioses with blue-green and other algae sometimes are obligatory, the ascidian species never being found without their symbionts. Other symbioses are non-obligatory some individuals of the species not having any symbionts, the host apparently existing without the symbionts and although it provides a habitat for the algal cells it does not gain any obvious advantage from their presence. In these non-obligate associations the plant cells are invariably on the surface of the ascidian colony rather than being in it; and they can be readily brushed off.

Although the *Prochloron* didemnid symbioses have attracted most attention, eucaryotes, and other blue-green algae including *Synechocystis* spp. living in the test and cloacal cavity have obligatory symbioses with several species in the Didemnidae. A few ascidians live in association with only a single algal species; however, in other cases two or three different algae have been found on the same ascidian species (Parry & Kott, 1988). While obligatory, symbiotic relationships have been demonstrated to be present only in family Didemnidae, commensal associations have been found in other families as well (Kott et al., 1984).

Sequence of events from fertilised egg via larval stage to fully developed ascidian

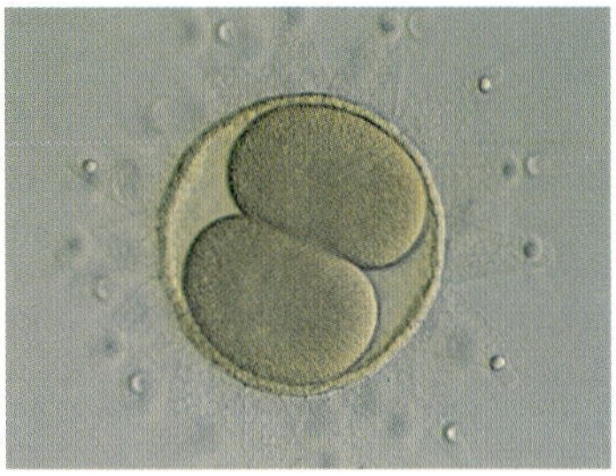
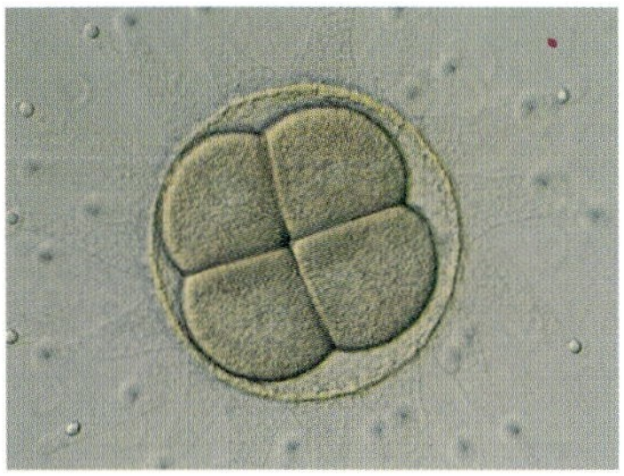
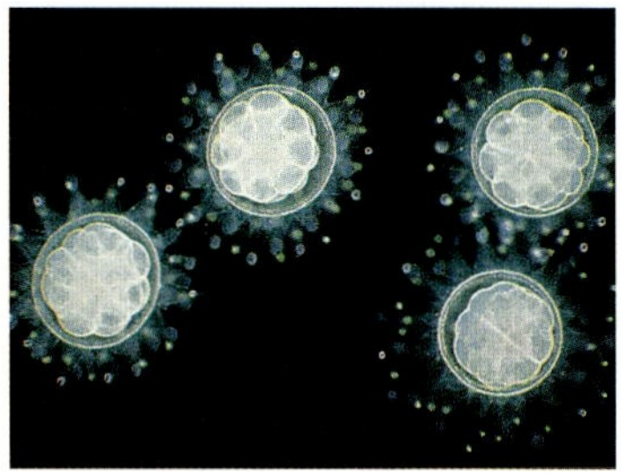

Ascidians are either viviparous, i.e. fertilisation and embryonic development takes place inside the parent or parental colony tailed larvae being liberated; or they are oviparous, i.e. fertilisation and embryonic development occur outside the parent or parental colony. However in both cases, after fertilisation has taken place, the egg cell initially divides in to two (left), then four cells (centre), until a larva starts to develop (right).

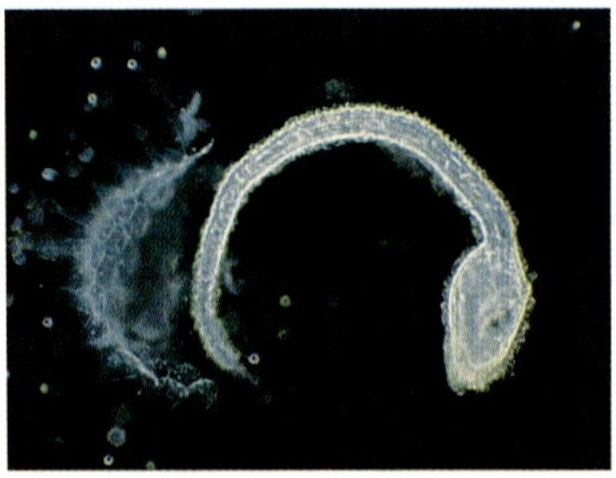
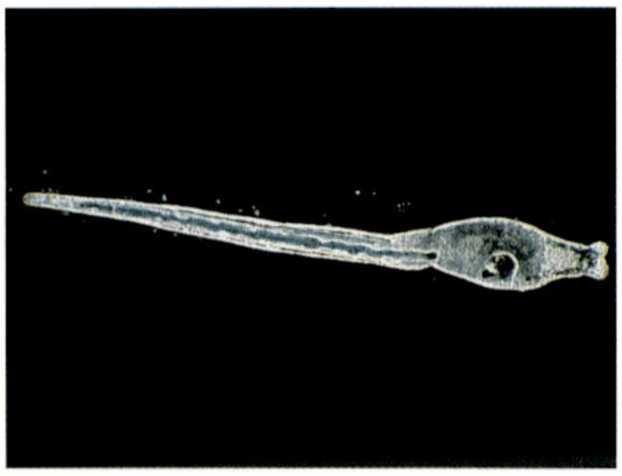
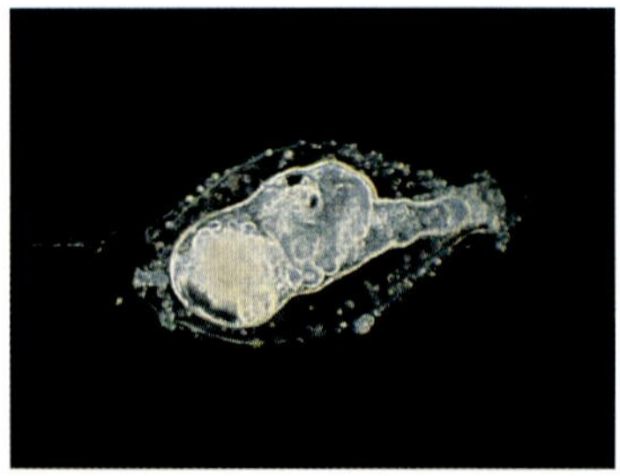

Subsequently the larva escapes from the egg follicle (left). A fully developed larva has a nerve cord, a notochord in the tail and developing oozooid and anterior adhesive organs in the larval trunk (centre). Under normal conditions the larva settles on some substrate within a few days after hatching. The tail and notochord start to regress, sometimes immediately but at most about two days after settlement (right).

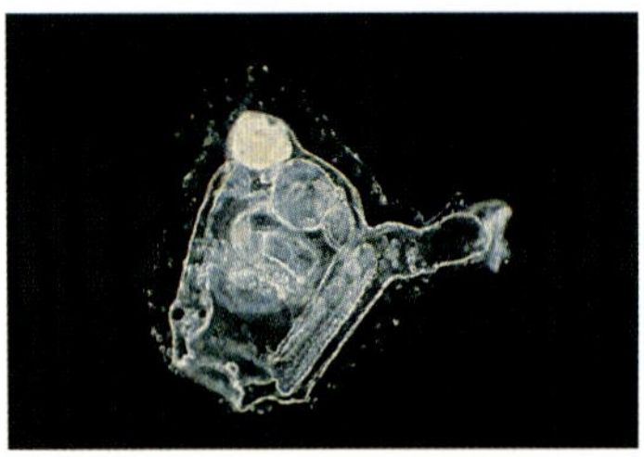
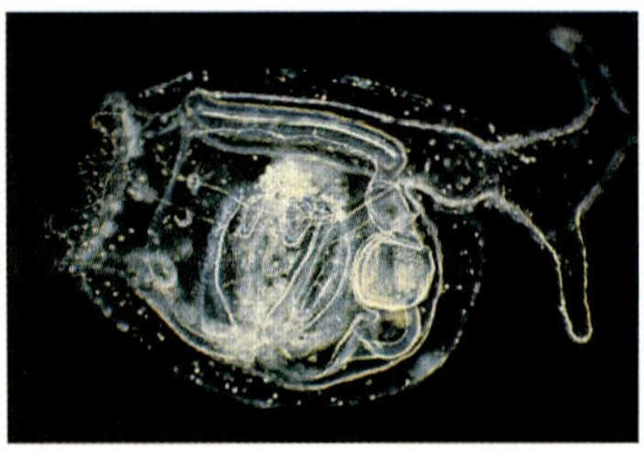

The tail and notochord have almost totally regressed and final development of the internal organs is taking place (left). Five days later most of the adult organs are present in the juvenile (centre). 8 Photos: R. Brons

A juvenile grows in to an adult *Ciona* sp. (right). Photo: E. Svensen

It is a result of their photosynthetic pigments that algal symbionts of didemnids (including *Prochloron* spp. with chlorophyll - a and - b) absorb light as the energy source to produce organic compounds from carbon dioxide and water (GRIFFITH & THINH, 1983; PARRY, 1985). Energy-rich carbon dioxide compounds, which are produced in this process, are transferred as food to the host ascidians.

Ascidians in symbiosis with algae live in the shallows, even in non-cryptic locations on the reef flat where there is plenty of light. Ascidians with symbiotic algae have been shown to adjust the amount of light falling on the algae by varying the pigment content (usually carotenoids), the concentration of calcareous spicules in the surface test, the

Cross-section through *Didemnum molle*, showing *Prochloron* algal cells in the common cloacal chamber; magnification: 24X.

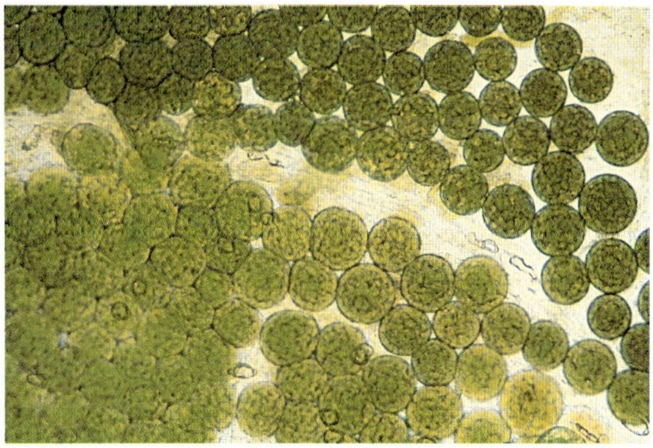

Prochloron algae cells from *Didemnum molle*; magnification: 200X.

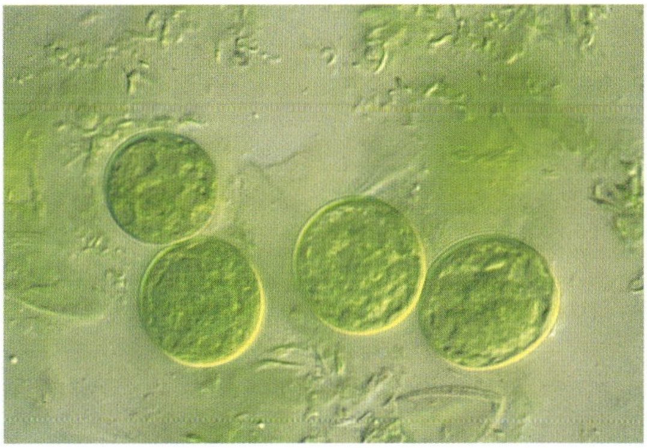

Prochloron algae cells from *Didemnum molle* photographed under Normanskl-Interference Contrast; magnification: 640X.

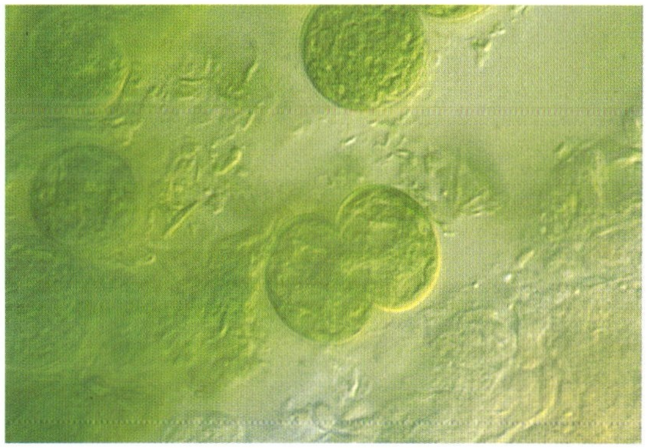

Dividing *Prochloron* algae cells from *Didemnum molle*, photographed under Normanski-Interference Contrast, magnification: 640X. Four photos: G. Cox

shape of the colony and/or the direction of growth relative to the light (phototaxis). Thus the ascidian can regulate algal photosynthesis and the flow of nutrients transferred to it from the symbionts.

The algae, too, adapt to different light conditions by changing the ratio of light-absorbing pigments. It has been found that the adaptations of algae as well as of ascidians are dependent upon light quality and quantity (PARRY, 1987). Moreover, similar adaptations have been observed in ascidians that live symbiotically with red Cyanophyta (*Synechocystis* spp.) – often found together with *Prochloron* (PARRY & KOTT, 1988). *Synechosystis* are coloured red because the algae contain red photosynthetic pigments called phycoerythrins (a group of phycobilin-proteins).

Those algae that live in the common cloaca of ascidians live in an environment in which the metabolic waste products of the ascidians are being discharged. These waste products involve ammonia, which normally does not occur in concentrations higher than 0.2 mM in the nutrient-deficient environment of coral reefs. However, in the cloacal chamber these substances occur in significantly higher concentrations (they commonly compose 95 % of the waste products). Consequently, in an ecosystem that is nutrient-deficient, the algae are being virtually "flooded" with nutrients, which they use for growth.

Thus, algae use the waste products from host ascidians, engage in photosynthesis and so produce food that they use as well as food that is transferred to the host animals. Each represents an adaptation to life in nutrient-deficient surrounding in the form of a unique symbiotic relationship similar to the symbiotic relationships for cnidarian animals and their zooxanthellae (as described in Volume 1 and 2 of this series).

It is important to point out that algae do not live in the tissue or in the actual body of ascidians. They are in the common cloacal chamber of a colony and/or in the test or on the surface of the colony. Therefore, they are not endosymbionts like the intra-cellular zooxanthellae of cnidarians.

The adaptations of didemnids to support obligate symbioses with small algal cells include the

Various *Prochloron* algae cells from different ascidian species.
Scanning electron microscopy: G. Cox

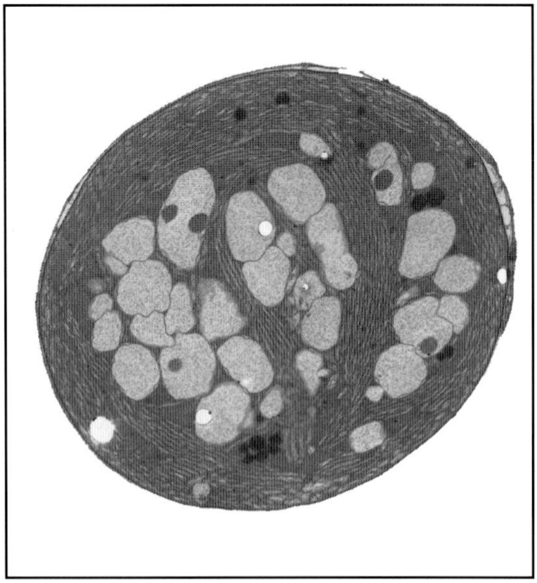

Prochloron **sp. from *Lissoclinum punctatum*; magnification: 4590X.**

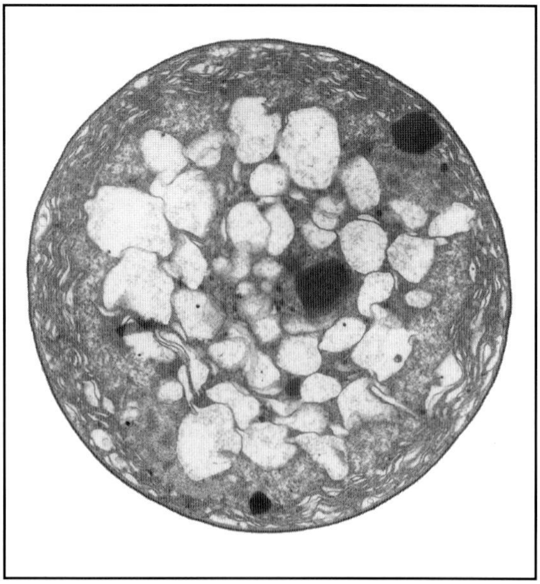

Prochloron **sp. from *Didemnum etiolum*; magnification: 7450X.**

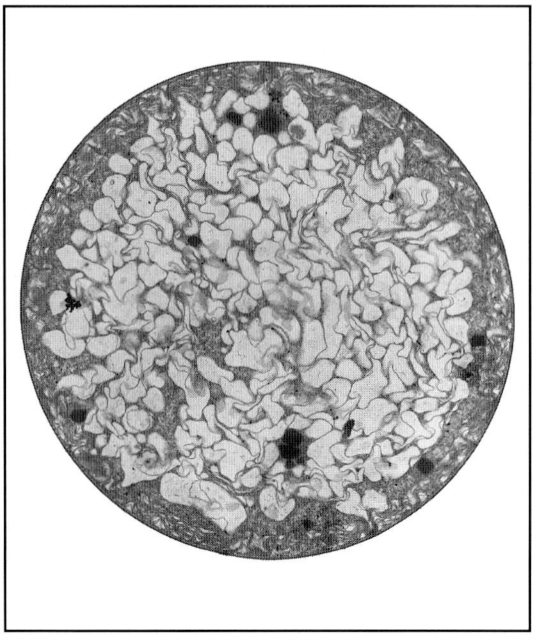

Prochloron **sp. living on the surface of *Aplidium* sp.; magnification: 4740X.**

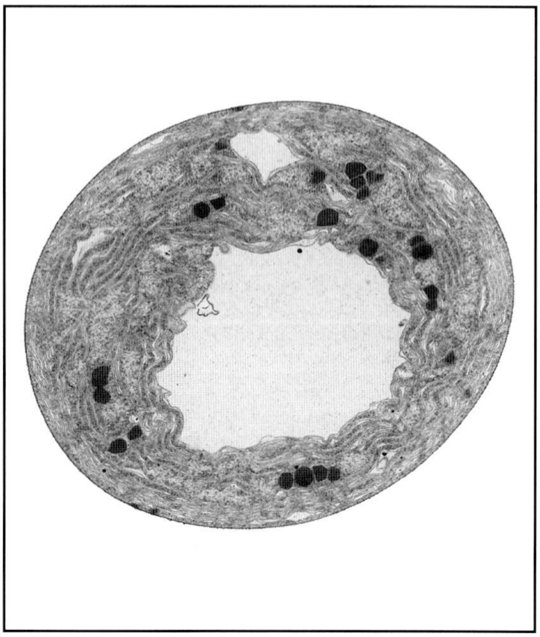

Prochloron **sp. from the cloacal chamber of *Diplosoma simile*; magnification: 6250X.**

adaptation of larvae to carry the symbionts to a new generation colony. Such adaptation involves properties of the larval test that cause the algal cells to adhere to it, either around the posterior end of the trunk or even over the whole trunk (leaving bare places over the sensory and the adhesive organs). In the genus *Diplosoma* a special organ develops at the base of the larval tail called a "rastrum". It is an outgrowth of the trunk ectoderm and is covered by thick larval test, which is separated into hairs that form a sort of brush or "algal rake". As soon as the larva is released into the common cloaca from the test, the rastrum "rakes up" the algae that line the insides of the cavities. When the larva, which has only a brief pelagic stage, leaves the colony through the cloacal opening, it already has the algae entangled in the rastral hairs and these algal cells are thus passed on to the new colony (see drawings and photo on page 416).

Best known among ascidians that enter in to a symbiotic relationship with algae, is *Didemnum molle* (see page 413 and Fact Sheet on pages 434). Most species with symbiotic algae occur in the Indo-Pacific, but such relationship is also known to occur in *Tridemnum solidum* from the Caribbean Sea, and *Tridemnum della vallei* from the Gulf of California (KOTT, 1982). The most important Indo-Pacific species of family Didemnidae that enter in to symbiosis with algae are tabulated in an overview at page 418.

Carnivors

At the other end of the scale of feeding specialisation from tropical symbiosis are the remarkable carnivorous species from deep continental slopes and abyssal ocean depths (1000 m or more) that actively capture their food. In these species, such as *Megalodicopia hians* (see photo at page 404), and the bizarre Hexacrobylidae (see KOTT, 1989), the anterior body muscles have been developed into hoods or arms that

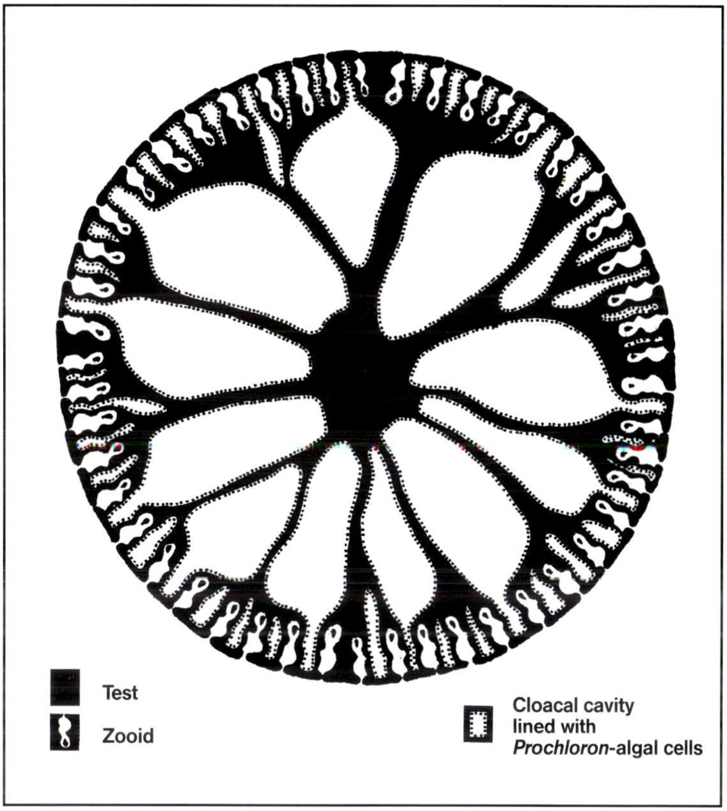

Schematic cross-section through a *Didemnum molle* colony.
Drawing from KOTT (1982)

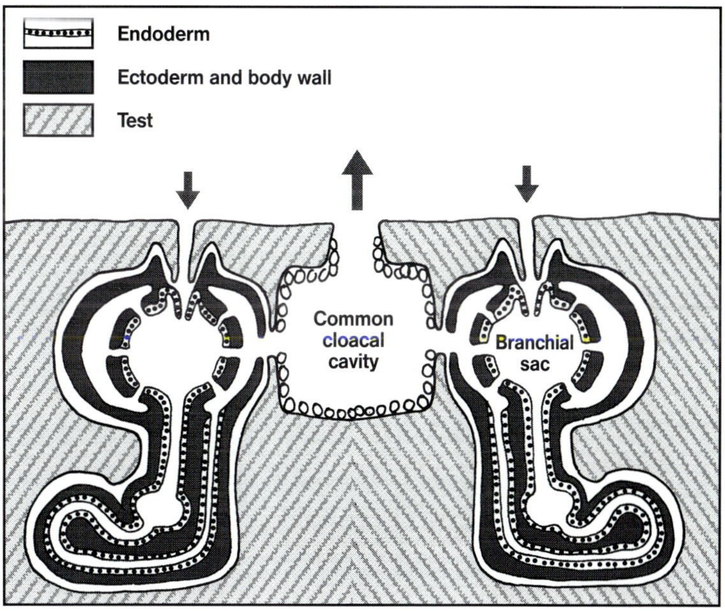

Schematic cross-section through two *Didemnum* zooids.
Drawing as per details provided by KOTT

catch prey from the relatively sterile waters of deep ocean trenches, and force it into the mouth. Their filtering equipment is lost and their gut has become specialised for the digestion of prey.

Ascidians in a Coral Reef Aquarium

It is not very easy to offer reliable advice on how to keep ascidians in a coral reef aquarium. Although many of these animals are relatively small, they belong to a complex and interesting group of animals, often with very appealing colour patterns, which then makes them rather attractive to many aquarists. Generally speaking, ascidians must be considered as difficult aquarium organisms. The best chances of survival appear to be in aquaria that offer more food than usually is available i.e. in an aquarium equipped with biological filtration, but without – or with only limited use of - a protein skimmer. In fact, in just such an aquarium we have had our best success with this animal group (see NILSEN, 1999, 2000 for a description of the system).

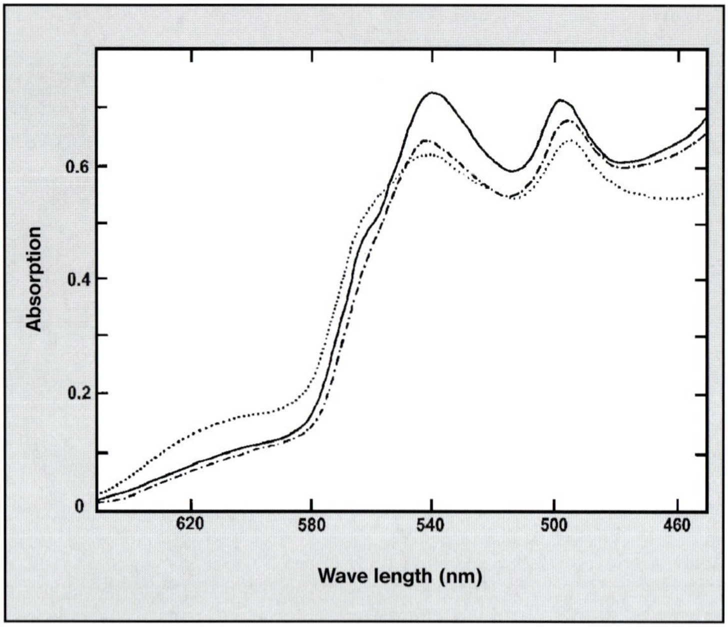

Absorption of Phycoreythrins by Cyanophyta symbionts of *Lissoclinum punctatum* (- - - - -), *Didemnum etiolum* (——) and *Didemnum* sp. (......).

Diagram after PARRY & KOTT (1988)

Nevertheless, we do not believe that all difficulties in the care and maintenance of ascidians are solely due to nutritional problems. These animals are capable of filtering tiny organic particles from the water. Although protein-skimming makes water nutrient-deficient, an abundant quantity of particles always remains, even in a heavy skimmed reef tank. The amount and pattern of water movement might play an important part in the cultivation of ascidians. Also light could be important as

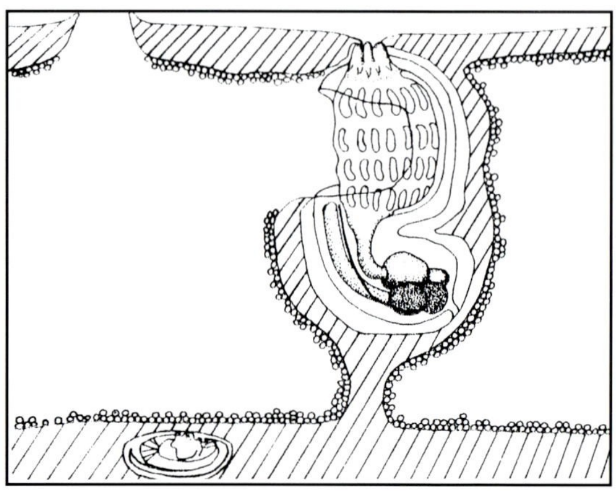

To the left: *Diplosoma simile* is coloured deep blue and often lives between branches of *Acropora* stony corals or other branching stony corals and under undercuts (see Fact Sheet on page 438 for further details on this species). This photo from the Maldives is probably the first record of the species from this location and probably the westernmost record so far of *D. simile.* To the right: Schematic longitudinal section through part of a *Diplosoma simile* colony that shows how symbiotic algal cells line the common cloacal cavity.

Drawing from KOTT (1982)

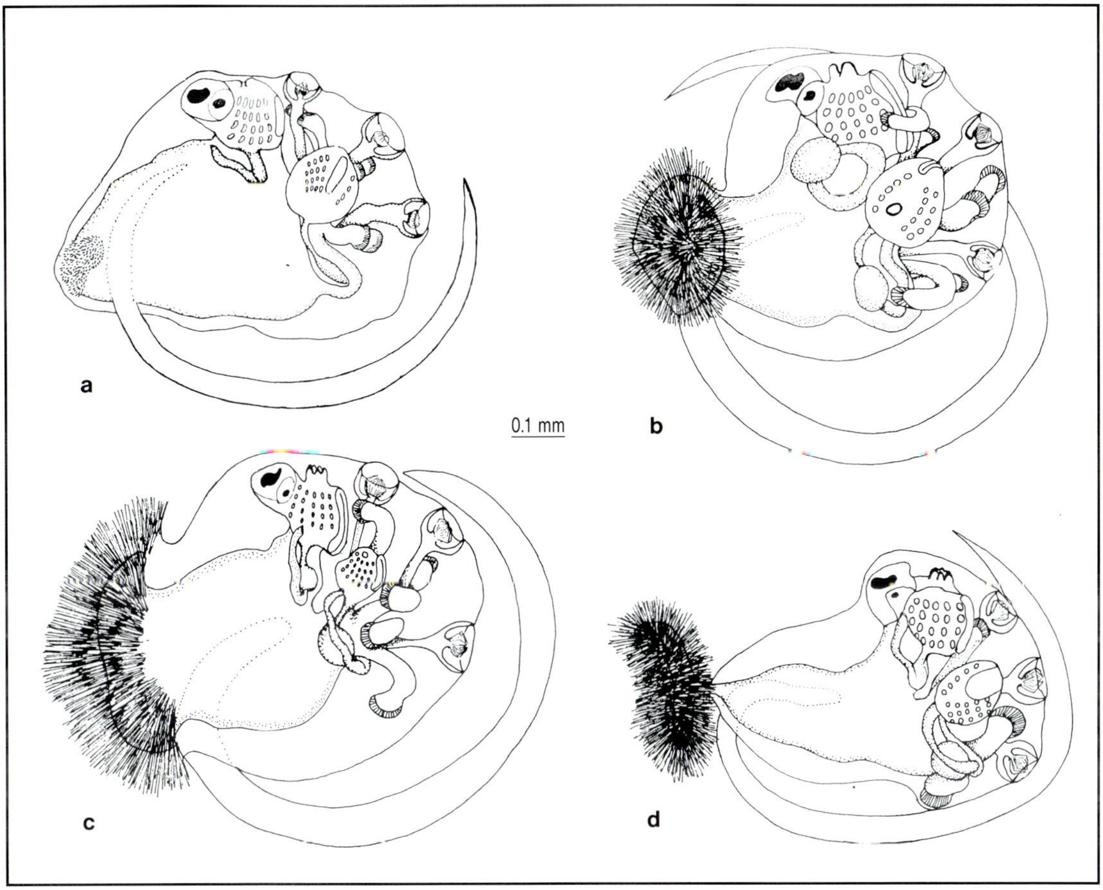

Growth of the rastrum (algae rake) in a *Diplosoma simile* larva: a ectoderm cells on the larval trunk above the base of the tail swell; b-d the larval haemocoel and ectodermal wall projects out and eventually becomes a T-shaped expansion from the trunk, the top of the T lying horizontally. This expansion is covered with a thick coat of larval test differentiated into hair-like strands forming a brush. **Drawing from KOTT (1980)**

many ascidians are restricted to shady microhabitats and are found in caves, undercuts and on the underside of coral boulders. The heavy illumination of a coral reef aquarium might not encourage their settlement, which is often in a shaded place.

If the amount of UV light is high, it could be a factor that would favour species containing photosynthesizing symbionts that use light as a source of energy to produce nutriment for the ascidian. However, the factors that favour the host ascidians are more significant, for without the ascidian host the habitat for the symbionts is not available; and so far these symbioses have not been cultivated in aquaria.

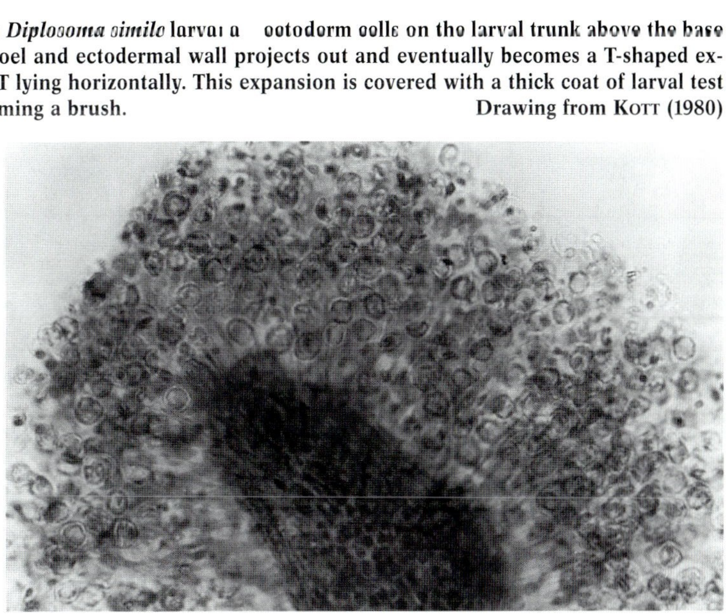

Photomicrograph of a section of the outer margin of a rastrum. Symbiotic algal cells have become trapped among the test hairs. **Photo: P. Kott**

Algae symbionts in ascidians of family Didemnidae (Compiled after KOTT, 1982 and PARRY & KOTT, 1988; related species have been grouped together)		
Host Ascidian species	**Algae symbiont**	**Symbiont diameter (μm)**
Diplosoma simile	*Prochloron*	10-15
D. virens	*Prochloron*	7-15
D. multipapillata	?	8-10
Lissoclinum bistratum	*Prochloron*	10-15
L. voeltzkowi	*Prochloron*	10-15
L. patella	*Prochloron*	10-15
L. punctatum	*Prochloron*	15-20
Echinoclinum triangulum	*Prochloron*	13-15
Didemnun molle	*Prochloron*	15-26
D. viride	Cyanophyta	8-10
D. etiolum	*Prochloron* + Cyanophyta	8-10
Trididemnum cyclops	*Prochloron*	8-19
T. paracyclops	*Prochloron*	10-15
T. solidium	Cyanophyta	8-24
T. della vallei	*Prochloron*	-
T. paraclinides	?	8-15
T. clinides	Cyanophyta	6-8
T. tegulim	Cyanophyta	-
T. nubilum	Cyanophyta	6-12
T. solidum	Cyanophyta	-
T. strigosum	?	4-8
T. miniatum	*Prochloron*	7-10

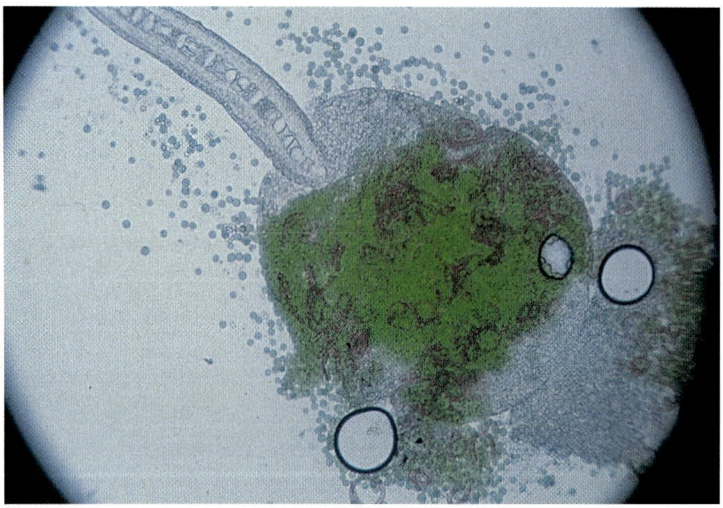

Not all species that live symbiotically with algae develop a larval rastrum. Some gather algal cells around their trunk (in the larval test). This *Trididemnum miniatum* larva has blue-green *Prochloron* and *Oscillatoria spongelia* algal cells (normally a symbiont of sponges) in the larval test.

Photo: G. Cox

We do not know the reasons for our lack of success in cultivating ascidians in aquaria. It could be food, light, water movement or other specific factors or combination of factors. We observed repeatedly that small ascidian colonies would grow in an aquarium for several months and even reproduce, but then suddenly, and without any visible cause, disappear. The only way to find a solution is for aquarists to set up small tank specifically to study ascidians and other organisms attached to living rocks. Then, by varying factors like light, water movement and food supply, the specific requirements of ascidians can be determined. As with all demanding aquarium animals it is advisable to maintain detailed records to find the optimum conditions required for the animals to thrive in an aquarium. Moreover, if the experiences are to be summarized at some later date it is useful to have such records on hand.

Order Enterogona

This order contains two suborders:

- Aplousobranchia
- Phlebobranchia

Suborder Aplousobranchia

Aplousobranchia are tunicates in which the gut and gonads are posterior to the pharynx, dividing the body into a thorax and abdomen and sometimes a posterior abdomen as well. A few species are solitary, but in most, the zooids replicate to form colonies. In the replicating process, the zooids of most families divide horizontally across the abdomen and/or the posterior abdomen and new zooids are generated from endodermal cells of the embryonic epicardial sacs. In the family Clavelinidae (see page 423) new zooids develop from the mesodermal blood vessels in the colony stalk. The capacity to bud has had a profound affect on the evolution of the taxa in this suborder and is considered plesiomorphic (see KOTT 1982 for information on replication in the suborder).

The process of replication interrupts the growth of individuals and, as it evolves and becomes more efficient, more but progressively smaller zooids are produced. With this overall reduction in the size of zooids there is a reduction in the size and simplification of their organs – the pharynx is smaller, stigmata are fewer and internal longitudinal branchial vessels are lost, the gut loop sometimes is relatively short, and gonads are reduced in size. Nevertheless, because there are many small zooids in a colony, their combined productivity (i.e. the productivity of the colony) probably is equivalent to that of a large solitary individual with its relatively large pharynx, gut and gonads. However, there is a striking difference between most solitary and colonial species in all suborders: the eggs of colonial species are large and yolky, are fertilised and develop inside the colony and ovaries are small relative to the testes. The eggs of solitary forms are small with little yolk, are liberated fertilised and develop externally and the ovaries are relatively large.

In Aplousobranchia, the families Cionidae and Diazonidae resemble the primarily solitary Phlebobranchia and Stolidobranchia (Pleurogona) in certain primitive features. They all have large zooids and body organs including a large pharynx with internal longitudinal vessels; and they have small externally fertilised eggs. However, like all aplousobranchs, they have endodermal epicardial sacs, which are involved in organ regeneration in the evolving aplousobranch replicative process; while in Stolidobranchia and Phlebobranchia, the

Suborder Aplousobranchia

Family Cionidae
Family Diazonidae
 Rhopalaea, Pseudodiazona, Diazona**
Family Clavelinidae
 Clavelina, Nephtheis
Family Pycnoclavellidae
 *Pycnoclavella, Euclavella**
Family Holozoidae*
Family Stomozoidae*
Family Polycitoridae
 *Eudistoma, Polycitor, Eucoelium, Cystodytes**
 Brevicollus, Exostoma*, Archidistoma**
Family Euherdmaniidae*
Family Protopolyclinidae*
Family Pseudodistomidae*
Family Placentelidae*
Family Ritterellidae*
Family Polyclinidae
 Polyclinum, Synocium, Aplidium
Family Didemnidae
 Didemnum, Lissoclinum, Diplosoma, Tridemnum
 Leptoclinides, Atriolum, Polysyncraton**

*not discussed in detail here

A beautiful tunicate growing from a live rock imported to a reef tank in Germany. Photo: D. Knop

epicardium has an excretory function and replication is effected by the ectoderm.

Most aplousobranch species live in the shallow waters of continental shelves, and only very few are known to occur at great depths. The systematic overview given at page 419 lists families and genera of the suborder.

Family Cionidae

Cionidae is the only family of the suborder Aplousobranchia in which species are solitary and ***Ciona*** its only known genus. In the few known species of *Ciona* the gut forms a horizontal loop posterior to the thorax and the large branchial sac has internal longitudinal vessels and numerous stigmata. Yellow spots between the branchial lobes in these and other ascidians may be light sensitive. Although a replicative process does not occur, individuals regenerate new thoraces if the body is severed across the epicardium (which is the regenerative tissue).

The best-known species in the genus is *Ciona intestinalis*. It is cosmopolitan and occurs in coastal waters of northern Europe, and in subtropical and tropical regions of the world it is found in harbours. In Australia it once was common but there are few recent reports of it outside Port Phillip Bay. It is common in New Zealand (where it fouls the substrate on which mussels are cultivated) and it is plentiful in Hong Kong harbour. The species is very common along the coast of Norway and is found as far north as Spitsbergen on Svaldbard. It is, however, missing from Iceland (MOEN & SVENSEN, 2000).

Ciona intestinalis often forms large aggregations, hanging head-down from ships' hulls, ropes and other man-made, submarine structures. Individuals are cylindrical, up to 15 cm long, and occasionally up to 27 cm but they have strong longitudinal muscles and are very contractile. The test is soft, translucent and yellowish. Little is known about its reef

Large and typical aggregation of *Ciona intestinalis* photographed off the coast of Norway. Photo: E. Svensen

aquarium requirements, but it is hardy, especially when kept in a special cold-water aquarium. See fact sheet on page 421 for detailed information on this common ascidian

Ciona savignyi is closely related to *C. intestinalis*. It is distributed around Japan and throughout the northern Indo-Pacific. There are also other species described but their status is still uncertain (KOTT, 1990).

Family Diazonidae

As mentioned above, the family Diazonidae has certain primitive characters in common with the family Cionidae, such as inner longitudinal branchial vessels in relatively large branchial sacs, numerous externally fertilized eggs and a gelatinous translucent test. However the family has colonial as well as some solitary species and the former are the only known externally fertilised colonial species in the Ascidiacea. As in most Aplousobranchia, replication is by horizontal division across the abdomen and the epicardium is involved in organ regeneration. The gut forms a long vertical loop (which distinguishes solitary species from the Cionidae).

Genus ***Rhopalaea*** occurs in the central Indo-Pacific, the Mediterranean and the Caribbean. The abdomen is separated from the thorax by a narrow oesophageal neck, and is tightly contained in firm abdominal test (KOTT, 1990). The abdominal section of the zooid is deeply embedded in the substrate and zooids are difficult to remove. Replication does not occur although, if severed, the epicardial epithelium has the capacity to regenerate a new thorax on the abdomen as in the Cionidae.

Rhopalaea crassa is a colourful and most attractive tunicate and is well known and most photographed in the tropical western Pacific. It is solitary but sometimes is found in small groups. Thoraces, projecting from the substrate, are to 6 cm high and about 2 cm in diameter. They are either opaque and a deep violet or blue colour or pale and almost transparent. Some of the oldest specimens are a conspicuous opaque yellow colour which could be due to blood cells containing vanadium with an organic compound that oxidises to olive green when the specimen is removed from the substratum (KOTT, 1990).

In Indonesia we found *R. crassa* most commonly at 2 to 3 m along steep, nearly vertical reef walls, where it seems to grow best in the semi-shade of other organisms. So far we have not yet seen *R. crassa* in the aquarium trade, and are unable to provide relevant aquarium husbandry details. Based on the success with cold-water aquarium keeping of the closely related *Ciona intestinalis*, it might be possible to cultivate the species if plenty of nutrients are supplied.

The other six known species in *Rhopalaea* are of very limited interest to aquarists and are not known to occur in the trade. *Rhopalaea tenius* has been described from Torres Straits (between Cape York Peninsula of northern Australia and Papua-New Guinea) and from deep water of the Java Sea. The purple or purplish brown *Rhopalaea abdominalis*

Ciona intestinalis

Distribution: Cosmopolitan; found in the Arctic and the eastern Atlantic, Svalbard (but not in Iceland), along the coasts of Norway and the rest of western Europe, the Mediterranean, the Black Sea, Cap Verde, widely in the Indo-West Pacific, Japan, New Zealand, California and parts of North and South America. Found from the tidal zone down to several thousand meters depth but generally only in confined places e.g. harbour and fjords. A related species, *Ciona savignyi*, is recorded from Japan (HOSHINO & NISHIKAWA, 1985).

Size and growth form: This solitary species grows fast, newly settled larvae reaching 40 mm in length in a couple of months (MOEN & SVENSEN, 2000). The body is cylindrical, sometimes up to 15 cm long, usually attached by its posterior end to under surfaces such as ships hulls, wharfs and jetty installations, ropes etc. from which it hangs upside down in large, crowded populations. Its presence on ships' hulls could be the explanation for its wide geographic range and its occurrence in harbours especially in larger ports. The species is reported to tolerate brackish water down to 11 ‰ salinity (MOEN & SVENSEN, 2000).

Reproduction: Oviparous, fertilisation is external and the embryos develop in the open ocean as part of the plankton.

Food: Little is known of the feeding requirements, although the organisms have for many decades been kept in laboratory tanks as experimental animals and feeding requirement are probably not very specialised. Populations flourish in waters which there is high organic pollution (e.g. Hong Kong Harbour).

Description: The test is soft and translucent. The siphons are of variable length, the 8-lobed branchial siphon terminal and the 6-lobed atrial siphon antero-dorsal, and they sometimes have a yellow lining. These animals are highly contractile and have 6 long, parallel muscle bands down each side of the body wall, 4 extending from the branchial siphon to the posterior end of the body and 2 from the atrial siphon to the posterior end. Also a pair of ventral muscles is along each side of the endostyle. The large, flat branchial sac has numerous internal longitudinal vessels with spoon-shaped papilla (KOTT, 1990).

Aquarium requirements: This species is rarely introduced to the aquarium trade, although it appears to be relatively hardy. It has been kept in cold-water tanks for more than 4 years, which is four times longer than its natural life span (MOEN & SVENSEN, 2000). Could be tried in the reef aquarium also, but little is known about the keeping of this beautiful and primitive ascidian in the reef aquarium.

GA: 0; **SE:** 0; **CO:** 0

Photo:
Ciona intestinalis from Norwegian waters, close up of a single animal with yellow lining around the edges of the siphons. Photo: E. Svensen

Variations in colour of the Blue Ascidian, *Rhopalaea crassa*

Photo from Indonesia, with *Polycarpa aurata, Didemnum molle* and other *Didemnum* sp. Photo (left): S. W. Michael

Left: Close-up of specimen from Indonesia (photo: C. Bryce). Right: Old, fully grown specimens from central Indo-Pacific with typical yellow rings around the apertures (photo: D. Knop).

Transparent specimens from Madang, Papua-New Guinea. Photos: L. Newman and A. Flowers

occurs in the Caribbean and *Rhopalaea neapolitana* is well documented in the Mediteranean.

Family Clavelinidae

In this family both branchial and atrial apertures are smooth without lobes and although replication always occurs, the process is different from other aplousobranch families. In clavelinids zooids are developed in the terminal ampullae of test vessels, the mesoderm being the regenerative tissue. Some species are solitary, others are colonies with zooids either separate and connected only by basal stolons or stalks, or partially or completely embedded in common test. Many zooids barely reach a height of 1 cm, however a few larger species have individual zooids up to 10 cm high. Many species are attractive, often delicate colours, frequently in different shades of blue. There are two genera: *Clavelina* with a range of species, many being cosmopolitan, and *Nephtheis* with only one species from the Indo-Pacific.

Genus ***Clavelina*** is well known also from European waters, e.g. *Clavelina lepadiformis* is common along the coastline of southern and western Norway where it forms large colonies in shallow water. The genus contains 22 described species from Australia and the western Indo-Pacific (Kott, 1990).

A very attractive species is *Clavelina oliva*, which occurs around the Philippines and along the coastline of northern Australia, from Queensland to Western Australia. The individual zooids are club-shaped and separate from one another, having a short stalk that is branched at its base. The zooids reach a height of 2.5 to 5 cm and usually are transparent with a thin yellow or white ring around the buccal and atrial opening. Occasionally the colours are brighter and the zooids are almost transparent. This is particularly common around Darwin in northern Australia.

Clavelina moluccensis forms large, irregular colonies, with zooids branching from a common stalk or partially embedded. They are light blue with a nearly transparent base and they always have three blue spots in a transverse row between buccal and atrial openings. This is possibly the most common species in the Indo-Pacific but so far has not been reported in the aquarium trade.

Clavelina australis is another attractive species, with zooids that have a sturdy (up to 3 cm long) stalk. The zooids have a transparent test with a blue patch in the mid-line between the apertures and streaks of blue dorsal and ventral to them. This species has been described from Houtman's Abrolhos (off Western Australia) and is common in subtropical waters around Australia.

Clavelina picta, common in the Caribbean Sea, is nearly transparent, and has an intense red ring around the buccal and atrial opening. This ascidian often grows on the branches of gorgonian corals, where it can form colonies of up to 1,000 zooids.

Clavelina **sp. "KA6-ASC-01" from Indonesia.** **Photo: C. Bryce**

Clavelina **sp. "KA6-ASC-02", possibly** ***C. arafurensis*** **from Kimberley, Western Australia.** **Photo: C. Bryce**

Clavelina sp., "KA6-ASC-03" from Madang, Papua-New Guinea.
Photo: L. Newman and A. Flowers

Clavelina moluccensis, seen here on the Great Barrier Reef, forms large branching colonies. It is normally coloured some shade of blue with 3 dark blue spots in a transverse row between the siphons.
Photos: L. Newman and A. Flowers

Clavelina puertosecensis, from the same geographic region, is uniformly sky-blue with a faint, white ring around the openings. According to HUMANN (1994) it is not as common as the previously mentioned species.

The genus ***Nephtheis*** is monotypic with *N. fascicularis* as the only species. The green or blue zooids of this species are conspicuous embedded in longitudinal rows along a conical head of variable length on a fleshy cylindrical stalk. We have treated the species in a Fact Sheet on page 426.

Family Pycnoclavellidae

The species of this family, like Clavelinidae, have zooids with smooth-rimmed apertures opening separately to the surface, the body divided into thorax and abdomen and completely or partially embedded in common test, or almost entirely free of it and connected by basal membranes or stolons. Zooids usually are colourful, though all but the thorax is often covered with sand. Two genera are known, *Pycnoclavella* and the monotypic *Euclavella*. The family is distinguished from Clavelinidae by its long, thread-like zooids with the abdomen many times longer than the thorax, small gonads, fertilisation at the base of the oviduct and embryos developing as they pass up the long abdomen, and larvae with 2 or 3 long, tubular adhesive organs invaginated into the anterior half of the larval trunk (KOTT 1990).

The species of ***Pycnoclavella*** are usually cryptic, and normally not found intertidally. They occur in tropical as well as temperate waters. Six of the 10 known species are from Australian waters. Others are known from the English Channel, California, tropical western Africa and New Zealand.

The small *Pycnoclavella diminuta* is the most common species in this family, widely distributed in the central Indo-Pacific and often noticed and photographed by

Clavelina robusta

Distribution: Eastern Indian Ocean and western Pacific; off western and northern Australia, the Great Barrier Reef, Indonesia, the Philippines, Palau and in tropical Japan.
Size and growth form: Forms large colonies with large zooids, 2 to 4 cm long and 0.5 cm in diameter.
Reproduction: Viviparous; large eggs fertilised at the top of the oviduct where the embryos develop to larvae.
Food: Filter feeder.
Description: The colonies branch, having common basal stalks (of firm test) which subdivide two or three times so that the anterior ends containing thoraces are separated from one another, each with its own covering of soft translucent test. The test is black or dark blue. A greenish-yellow ring surrounds each siphon.
Aquarium requirements: Occasionally available in the aquarium trade, but we have found it difficult to keep in aquaria. Should be kept in a special aquarium with strong water movement and water that has more suspended organic material than is normal for a coral reef aquarium. High illumination intensity must be avoided. Suitable only for experienced aquarists.

GA: +/-; **SE:** -; **CO:** +

Photos:
Top: *Clavelina robusta* colony in an aquarium.
Bottom: Colony on the Great Barrier Reef.

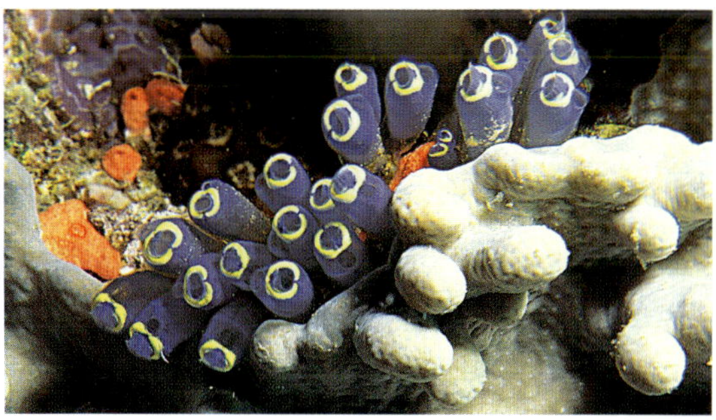

Nephtheis fascicularis
Lollipop Coral

Distribution: Indo-West Pacific; known from western, northern and north-eastern Australia, New Caledonia, Ponape Island, the Philippines and the Sulu Sea, Caroline Islands and Indonesia.
Size and growth form: Colonies of zooids grow as conical or cylindrical heads at the end of firm, thick stalks. The head can grow to a length of 16 cm. The stalk can be up to 20 cm long, but usually it is not longer than 2 to 5 cm, and about 3.5 cm in diameter. The stalk contains an unique two-dimensional network of blood vessels. The head consists of an outer layer of zooids, with their abdomina projecting inwards towards the vascular network of the central axis. New buds develop from the top of this network around the base of the head. Individual zooids are connected to the blood vessel system of the central axis by a posterior vascular stolon. Individual zooids are each about 1 cm long (KOTT, 1990).
Reproduction: Fertilization is internal. Larvae are large, with three adhesive organs, and they develop in brood pouches at the posterior end of the thorax.
Food: Filter feeder.
Description: The individual zooids lie in the colony with their ventral side toward the outside and the base. The branchial aperture is turned ventrally and opens to the surface below the atrial aperture. Both apertures are smooth. Zooids are divided into thorax and abdomen, with the thorax longer than the abdomen. There are 8 branchial tentacles, about 12-14 rows of stigmata and the gut loop is short (KOTT 1990). Colour is variable: often opaque blue, sometimes dark green and yellow and sometimes translucent.
Aquarium requirements: *Nephtheis fascicularis* has been kept in the refugium of a reef aquarium, but it lived for a few months only. During this time the colony shed the zooid-bearing heads regularly, but regenerated them in 3 days (see photos to the left). According to SCHIEMER (pers. comm.) the species must be regarded as difficult, but not impossible to keep. Sufficient water movement and suspended organic food are factors that affect the maintenance of this delicate animal. It should be kept in a special aquarium without grazing animals.

GA: +/-; **SE:** -; **CO:** +

Photos:
Nephtheis development of new zooid-bearing heads in a colony of *N. fascicularis* kept in the refugium of Greg Schiemer's reef aquarium. The photo series spans over a three days period and each photo is taken with approximately one day interval. Photos: G. Schiemer

Two colour varieties of *Pycnoclavella diminuta*. The yellow form (left) was photographed at Bunaken, North Sulawesi, and the purple form (right) on the Great Barrier Reef.

SCUBA divers. Usually it is found in small caves or under ledges, at 5 to 20 m. The zooids are crowded on a common basal mass of test, each with an upper 5 mm wide thoracic region and a 1 to 2 cm long stalk comprising the abdomen. From above, the thin stalks are barely visible, being obscured by the crowded, expanded thoraces. The colour of *Pycnoclavella diminuta* is variable, mostly blue with yellow-brown rings around the apertures, but others are yellow, white, and even orange and blackish with white spots. The wide range in colour does not reflect differences in geographical distribution or morphology (KOTT, 1990).

Pycnoclavella detorta, a common species throughout the Indo-Pacific, is transparent with iridescent green around the atrial opening and a yellow endostyle clearly visible through the test. The species forms large clumps of separate zooids up to 3 cm long, arising from common basal test.

Pycnoclavella aurantia, *P. tabella*, *P. arenosa* and *P. elongata* are other species from Australian waters.

Euclavella is known only from southeast Australia and New Zealand.

Family Polycitoridae

This family is one of the most numerous on coral reefs and indeed in all tropical seas. The species are colonial with zooids that usually are embedded completely in common test. Like Pycnoclavellidae, zooids are divided into a short thorax and a long abdomen that contains the gut and the gonads, and sometimes fertilisation occurs at the base of the abdomen. However, Polycitoridae have 6-lobed (rather than smooth) apertures and lack the tubular larval adhesive organs of Pycnoclavellidae. The zooids withdraw into the base of the colony when disturbed. The family contains 7 genera with *Eudistoma* and *Polycitor* being the two largest.

Eudistoma is the most species rich genus in this family. All species have small zooids with a long oesophageal neck, and gonads and stomach at the posterior end of the long abdomen. In most species the zooids are arranged in circles with the atrial apertures in the centre, usually opening separately, although sometimes they open into a depression, or rudimentary cloacal cavity. The genus is widely distributed throughout all tropical seas. Around Australia alone, there are at least 27 species. Another 11 species recorded from the Indo-West Pacific have not yet been recorded from Australian waters (KOTT, 1990).

Eudistoma gilboviride forms

Pycnoclavella detorta **at One Tree Island, Great Barrier Reef.**
Photo: L. Newman and A. Flowers

Eudistoma sp. "KA6-ASC-04" at One Tree Island, Great Barrier Reef.
Photo: L. Newman and A. Flowers

Eudistoma gilboviride on the Great Barrier Reef.

Eudistoma laysani growing among juvenile colonies of *Didemnum molle* on the Great Barrier Reef. Photo: E. Svensen

colonies of crowded lobes up to 2 cm high with a flat upper surface area of up to 1.5 cm diameter narrowing toward the common, basal test. The colour is usually green to yellow-green or shaded in yellow and green.

In contrast to other species of this genus, *Eudistoma laysani* usually occurs in inter-tidal zones in very shallow water. We have, however, observed the species in moderately deep water on the Great Barrier Reef (see photo on this page at bottom), and MILLAR (1975) reports the species from 22 metres off the Philippines. *E. laysani* is widely distributed in the tropical Indo-West Pacific. The colonies consist of a basal mass of common test supporting many lobes, each with a sometimes branched cylindrical stalk and a slightly expanded head, its upper surface being flat, rounded or conical (KOTT, 1990). The entire colony is from 0.3 to 1.0 cm high. The test is very delicate and nearly transparent. Fully-grown colonies can contain up to 100 zooids.

Genus ***Polycitor*** has quite large zooids with an oesophagal neck that is at least twice as long as the thorax. Colonies are stalked or sessile and cushion-shaped, upright or spherical, ranging in diameter from 1 to 15 cm. Ten species are known from Australian waters, but others occur in all oceans of the world. Some of them, such as *P. circes* and *P. annulus* are indigenous to tropical seas around Australia, while *P. translucidus* and *P. giganteus* are known from other parts of the western Pacific Ocean as well.

Cystodytes and ***Eucoelium*** are genera of this family that are unusual in having saucer-like and/or stellate or burr-like calcareous spicules in the test which sometimes form capsules around the zooids. The genus ***Eucoelium*** is sometimes referred to as *Polycitorella*, a junior synonym of *Eucoelium*. We have observed *Eucoelium* species in several locations in the tropical Indo-West Pacific; for instance, at the Maldives, where pink-coloured colonies (resembling "tennis balls") grow

***Eucoelium* sp. "KA6-ASWC-05" (left; photo: J. Cairns-Michael) and *Eucoelium* sp. "KA6-ASC-06" (right) at the Maldives.**

along the reef edge in strong sunlight.

We have found small colonies of Polycitoridae attached to live rocks, but unfortunately could not identify them, and they did not appear to survive in coral reef aquaria.

Family Polyclinidae

The colonies of species in this family have small, rather slender zooids arranged in circles, or rows around or along common cloacal cavities or canals in the common test. The branchial apertures open to the surface and have regularly lobed rims. The atrial apertures are not regularly lobed and open internally into a common cloacal chamber in the test. Polyclinid colonies often are large, well camouflaged and the gelatinous test is relatively firm and often colourful. In many species sand grains embedded in the test strengthen it. Sometimes, it appears as if sand particles are sorted by size, i.e. the largest at the base of the colony, and the smallest located at the top (KOTT, 1992). The zooids are small and thread-like with a narrow thorax, a short abdomen and long posterior abdomen containing the gonads. The common cloacal systems, internal atrial apertures and gonads in posterior abdo-

***Synoicum castellatum* at Heron Island, Great Barrier Reef.** **Photo: L. Newman and A. Flowers**

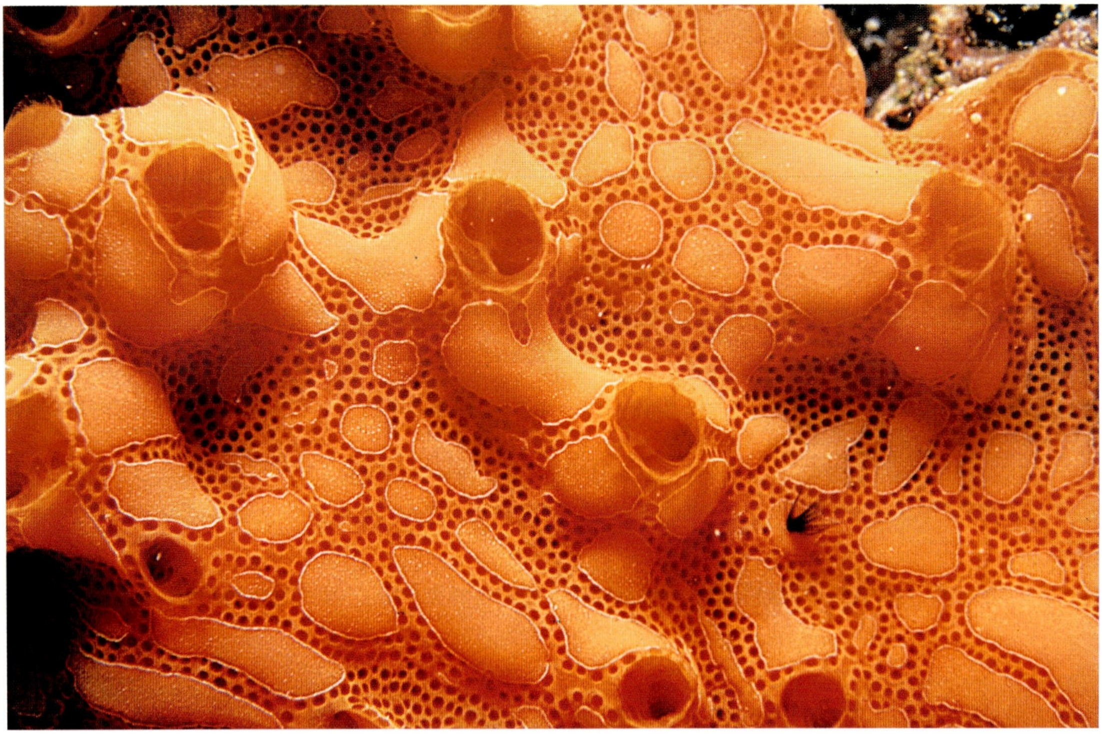

***Aplidium tabascum* at Heron Island, Great Barrier Reef.** **Photo: L. Newman and A. Flowers**

mina distinguish the family from Polycitoridae.

Six genera are in the family, the most speciose being *Polyclinum*, *Synoicum* and *Aplidium*. ***Aplidium*** is well represented on coral reefs and is also common in polar, temperate and tropical waters of all oceans. With more than 200 nominal species, this genus has been described comprehensively, and represents one of the most species-diverse ascidian genera. Yet, the overall taxonomy of *Aplidium* is difficult, species are not readily identified and the actual number of species may be overestimated. A key to Australian species is included in KOTT (1992).

Aplidium tabascum has a characteristic and stable appearance. It is also among the most common of the tropical Australian *Aplidium* species. The colonies are orange-red, robust sheets, up to 7 cm or more in diameter. The zooids are arranged in rows along each side of branching canals that surround irregular stands of solid zooid-free test. They sometimes have green algae of the genus *Prochloron* on the surface

Left: *Synoicum* sp. "KA6-ASC-07" at Madang, Papua-New Guinea (photo: L. Newman and A. Flowers). Right: *Aplidium lenticulum* (orange) and *Lissoclinum* sp. "KA6-ASC-08" (family Didemnidae) on the Great Barrier Reef.

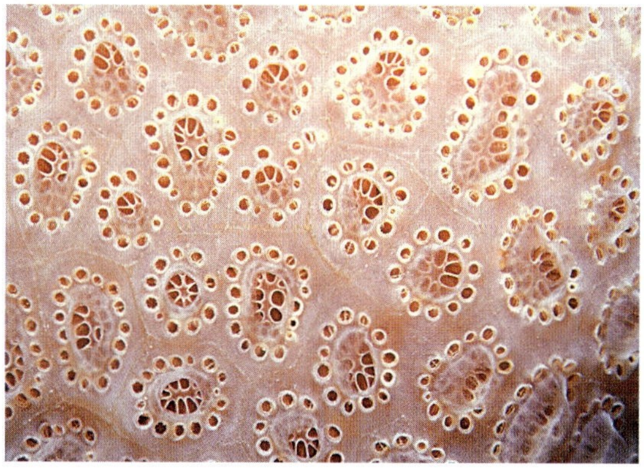

Aplidium longithorax at Heron Island, Great Barrier Reef. Left: entire colony. Right: Clearly visible in this close-up photo of part of the colony are the buccal openings arranged in circles. The atrial openings of the zooids are in a cavity located in the middle of each circle. Photos: L. Newman and A. Flowers

around the common cloacal apertures (KOTT, 1993).

Another species from the Great Barrier Reef, *Aplidium lenticulum* is also recorded from western and south-eastern Australia. It forms robust sheet-like colonies up to 2 cm thick and more than 20 cm wide. The surface is marked by depressions of the thin layer of surface test over the common cloacal canals that surround rounded ridges of zooid-free test. Branchial apertures of individual zooids open along each side of the depressions (KOTT, 1992) and the atrial apertures open into the common cloacal canals. In this species sand particles are in the base of the colony, while the upper section is free of embedded particles. Colour is variable, ranging from black, greenish-black, red, blue to orange.

The species that we have selected here are only two examples of this diverse genus. We have not yet seen these beautiful ascidians in aquaria. Although they may occasionally occur on live rocks, we are not able to provide information on their husbandry and include species of the family Polyclinidae in the group of aquarium animals that are difficult to keep.

Family Didemnidae

Didemnidae has more species on coral reefs than any other family of ascidians on coral reefs, and is especially numerous in Australian waters. Some autotrophic tropical species are especially interesting from an aquarist's point of view.

The family contains 8 genera with a total of about 300 described species world wide. Replication is prolific and colonies contain very numerous, small (never more than 3 or 4 mm) simplified zooids crowded into complex cloacal systems and always completely enclosed in the test. Zooids are divided into a thorax, consisting of a large pharynx perforated by 3 or 4 rows of stigmata, and a compact abdomen containing a short gut loop and small gonads (KOTT, 2001). Muscles are confined to the thorax. Most spe-

Didemnum sp. "KA6-ASC-09" on the Ribbon Reef, Great Barrier Reef.

Didemnum cuculliferum, photographed in Fiji, forms small colonies that can be pink-coloured (top) or magnificently yellow (below).
Photos: J. Cairns-Michael

Didemnum psammatodes, seen here in an aquarium owned by Kjell Nagy, Flekkefjord, Norway, is occasionally found attached to live rocks imported to the reef aquarium. Normally this species occurs in quiet, often polluted shallow water of coastal reefs.

cies have small (usually less than 0.1 mm) calcium carbonate (aragonite) spicules embedded in the test, often strengthening it. Because of the small zooids with contractile thoraces and colonies, which collapse when removed from the water, species are difficult to identify, and although the spicules represent stable and reliable characters, they are seldom unique and similar types occur in more than one species.

About 25 species live in association with blue-green algae including the genus *Prochloron* (see page 411-415) and benefit from the alga's photosynthetic activities. Organisms that are either totally or partially autotrophic may be easier to keep in an aquarium than those that are heterotrophic and if this is so, autotrophic species of the family Didemnidae may be good aquarium inhabitants. The *Prochloron* algae – in contrast to zooxanthellae that have a free-living development stage – have never been found outside the host animal and the host provides the alga with its environmental requirements. Therefore, if the ascidian can survive in the aquarium, its algal symbionts will too – and in return they will provide nutriment to the host – a true symbiosis. Nevertheless species of the family Didemnidae are very rarely reported in the aquarium trade. There are a few reports of didemnids occasionally being imported to the reef aquaria, and some have been observed to grow well in captivity, but these animals are largely unknown with respect to aquarium care.

Normally tunicates grow in shaded areas under rocks or overhangs or in deeper water where the light intensity is reduced compared to that closer to the surface. Some of the photosynthesising species of the family Didemnidae are, however, able to live in relatively shallow water on reef flats, in tidal pools, and even intertidally, in strong sunlight. The majority of didemnid species lack symbionts, are hetrotrophic filter feeders, and occupy a wide range

Various *Didemnum* species at the Maldives
Photos (except bottom right): J. Cairns-Michael

of habitats under rocks and in crevices, in deeper waters on reef slopes and in sea floor habitats on continental shelves. The family becomes less diverse toward the poles and is not known in abyssal waters.

Didemnum is the most specious genus in the family. KOTT (2001) lists 69 species from Australia (including the tropical western Pacific) and there are probably another 30 undescribed species from these waters. Only about 50 species are known from other parts of the world. The genus has small zooids (about 1 mm long) and a male duct coiled around an undivided testis.

Didemnum molle is by far the best known of the tropical members of the genus. Anyone swimming on a coral reef simply cannot help but notice these attractive and extraordinary colonial animals resembling small vases or upright barrels growing in non-cryptic habitats on coral skeletons and other hard substrates. The colonies contain hundreds of zooids, each with its own buccal aperture on the outer surface, while their excurrent

Didemnum molle
Green Reef Sea Squirt

Distribution: Indo-West Pacific: from East Africa via the Maldives, Indonesia, western- and northern Australia and the Great Barrier Reef to the Philippines and Okinawa (Japan). The habitat of this ascidian extends from shallow water areas along reef flats (exposed to bright sunlight) down to depths of about 40 m and rarely to 70 m (SLUITER, 1909).
Size and growth form: Colonies usually to 4 cm high and 2 to 4 cm, but occasionally to 10 cm diameter (KOTT, 1980a); zooids minute, to 1.5 mm long.
Reproduction: Viviparous. The larval trunk is 0.9 mm long and the tail about twice that length. *Prochloron* algae adhere to the test around the posterior end of the larval trunk and are thus transported to the new generation colony (see photo on page 417). Breeds throughout the year. Colonies divide into two across the common cloacal aperture as they grow.
Food: The species has symbiotic *Procholoron* algal lining a common cloacal chamber around the centre of the colony (see photos on page 414 and drawing on page 416). To some extent at least, nutriment comes from the symbiotic algae, which produce carbohydrate in the presence of sunlight and nutrients. In addition, the zooids filter particulate matter fom the incoming water.
Description: (Based on KOTT, 2001). The colonies are dome-, barrel- or irregular vase-shapes, each with a large terminal common cloacal aperture. They are firmly attached to the substrate (dead corals or rocks) by means of fine strands of spicule-filled test. The surface of the colony is smooth and has a thin layer of small spicules (0.005 - 0.015 mm). The distribution of spicules and brown pigment and the green symbionts affect the colour of the colonies, which range from grey or whitish to reddish brown, occasionally purple or even black. The green symbiotic algae are clearly visible through the large common cloacal opening. Juvenile colonies are very small, flat and predominantly greenish in colour (KOTT, 1980a). Colonies of this species can move and have been seen climbing up aquarium walls as well as up to the tip of coral substrates (KOTT, pers. comm.).

Photos:
Top: Colony at Madang, Papua-New Guinea, with strongly concentrated pigment resulting in the brown colour.
Photo: S. W. Michael
Centre: Close-up photograph of colony surface area. Buccal openings of zooids clearly visible, as well as the terminal opening of the common cloacal chamber.
Bottom: Colonies at 15 m depth in the Maldives.

continuation *Didemnum molle*

Aquarium requirements: Hardly anything is known about aquarium care. Colonies are fragile and hard to transport (requiring special packing). In general the species must be regarded as difficult to keep and care for in captivity, although symbiotic algae provide some of its nutriment nothing is known of the other parameters needed to maintain a healthy colony in which the symbionts can live. Nevertheless *D. molle* is the greatest opportunist among ascidians, as is evident by its large geographic range and its ability to accommodate diverse habitats. Reports on experiences with this species in captivity would be much appreciated!

GA: 0; **SE:** 0; **CO:** 0

Top: Small colonies of *Didemnum molle* together with hydroids and sponges in North Sulawesi.
Centre: One large and several small colonies in North Sulawesi.
Bottom right: Tiny colonies in Fiji. Photo: J. Cairns-Michael
Bottom left: Colonies at One Tree Island, Great Barrier Reef.
Photo: L. Newman and A. Flowers

Lissoclinum patella **on the Great Barrier Reef.**

Didemnum **(?) sp. "KA6-ASC-10" at Papua-New Guinea.**

Photo: S. W. Michael

(atrial) apertures open into an internal common cloacal cavity with a terminal common cloacal opening to the exterior (please refer to Fact Sheet on pages 434/435 for further details on this species).

Didemnum cuculliferum is another species common throughout the central Indo-West Pacific, having been reported from Indonesia, the Philippines, Palau, Eniwetok, Western Australia, the Great Barrier Reef and Fiji (KOTT, 1981c, 2001). The colonies are variable – from small round or oval cushions less than 1 cm in extent up to sheet-like colonies. Usually a pointed papilla protrudes from the surface of the colony above the ventral lobe of each branchial aperture giving the colony a spiky, echinated appearance. Colour is usually lilac-red or pink, occasionally yellow. In this and other species the small colonies form a sort of mosaic produced by subdivision (lobulation) of larger colonies. *Prochloron* sometimes is present on the surface as it is on many other species. The identity of the type material of *D. cuculliferum* is at present subject to change and a revision is expected.

Didemnum psammatodes is occasionally introduced to reef aquaria attached to live rocks. It occurs over wide areas in the Indo-West Pacific, from East Africa to Fiji (KOTT, 1981c). Colonies are an inconspicuous grey or cream, or yellow-brown, depending on the colour of the sediments ingested with the food. These sediments are incorporated into oval faecal pellets, which are expelled from the rectum of each zooid into the test (rather than into the common cloacal cavity – which is their usual route to the exterior). These faecal pellets remain in the test and probably have a skeletal role, helping to support the large irregular colonies with irregular, twig-like branches up to 8 cm high that are formed in deeper waters. In shallower water the species usually forms large irregular mats over hard subtrates. The species seems to thrive in slightly polluted habitats where the current is weak and the water carries significant loads of suspended, fine-grained material. We have found *D. psammatodes* growing on live rocks, yet we have been unable to keep this species alive in an aquarium for any significant length of time. After a while it disappears.

A well known species around the Caribbean is *Didemnum conchyliatum*. The species is one with a layer of crowded spicules in the surface that looks like a veil (HUMANN, 1994). Usually these colonies are yellow. Other species of *Didemnum* are shown in photos on page 433.

Lissoclinum is another genus rich in species, most showing a wide geographical range in the tropics where it is most numerous: KOTT (2001) lists 24 species from Australia alone. The branchial sac is large, but delicate, with 4 rows of long stigmata. The genus is further characterised by having a straight testis duct (vas deferens) and large open common cloacal cavities. Many species have cushion-, plate-, and sheet-like colonies, often brittle owing to the calcareous spicules crowded in the test. The rather inconspicuous, translucent, turquoise-coloured *Lissoclinum patella* has obligate *Prochloron* symbionts and is unusual in being a bulky gelatinous species growing to more than a metre in diameter and up to 5 cm or more, thick and often hemispherical, with surface depressions between high, gelatinous, smoothly rounded elevations. We have encountered this species on the Great Barrier Reef in semi-shaded sites, although KOTT (pers. comm.) reports that it is also found in very shallow water on reef flats where it is exposed to full sunlight. Sometimes it grows over live corals. *Lissoclinum patella* occurs around Indonesia and Borneo, as well as along the coasts of western, northern and eastern Australia. Up to now, we are not aware that *L. patella* or any other species of *Lissoclinum* has been kept in a coral reef aquarium, but the genus might very well be introduced on live rocks.

Different colour variations of *Diplosoma simile* put together on the reef flat for comparison. Please refer to Fact Sheet on page 438 for further details on this beautiful tunicate. **Photo: J. S. Ryland**

Diplosoma is a less diverse genus. KOTT (2001) records six species from Australia, with four being tropical Indo-West Pacific species, one cosmopolitan, and only one an indigenous species from southern Australia. Like *Lissoclinum*, the genus has large thoraces, each with a wide open atrial aperture exposing much of the branchial sac directly to the common cloacal cavity, and its zooids have a straight vas deferens. Its most conspicuous feature is its lack of spicules.

Diplosoma listerianum is a cosmopolitan species found in all temperate and tropical oceans, apparently common in, but not exclusive to, slightly polluted locations e.g. in harbours.

Diplosoma simile and *D. virens* are two of the most common of

Colonies of *Trididemnum miniatum*, seen here on the Great Barrier Reef. This species, is reminiscent of young *Didemnum molle*, but the colonies never exceed a size of 1 cm.

Diplosoma simile

Distribution: Indo-West Pacific: the Maldives, the Philippines, New Caledonia, Indonesia, Singapore, Western Australia, The Great Barrier Reef, New South Wales, and Japan.
Size and growth form: (Based on KOTT, 2001). The colonies form robust encrusting sheets, about 2 mm thick, and often extensive. They adhere closely to, and the basal test thickens to fill irregularities in, the substratum. The thickness of the colony is therefore uneven, although the surface is smooth, almost slippery, and even. Individual zooids are only about 1 mm long. This species is common at the reef edge (in the surf zone), where it encrusts under surfaces of reef boulders and can be seen encroaching from undercuts onto thc upper surface of the reef through gaps in consolidated reef. It also is found between the branches of stony corals and is sometimes in pools exposed to direct sunlight during low tides. The replication rate is very high and the two-dimensional growth rate of this species is fast. It appears to bind together a lot of the loose rubble on the reef flat.
Reproduction: Internally fertilised larvae are large, and develop in the parent colony. They have a T-shaped rastrum ("algal rake") above the base of the tail. It is covered with filaments of larval test that enmesh the symbiotic algae and carry them on to the new colony.
Food: The species also benefits from nutriment produced by the photosynthesis of its symbiotic algae (*Prochloron* sp.). It is probable that it also filters particulate matter from the sea water passing through the small pharynx.
Description: (Based on KOTT, 2001). One of the most common of the didemniid-*Prochloron* obligate symbioses in the tropical Pacific. The colonies are translucent and almost any shade of green to dark blackish-blue, or green with blue iridescent surface. The colours vary with the incident light, and may be affected by the colour of the *Prochloron*-algae that line the extensive common cloacal cavity. Normally the surface layer of the test with its embedded zooids, the common cloacal cavity and the thick basal test each occupy one third of the thickness of the colony. The small zooids have well developed branchial siphons. The atrial aperture is large exposing much of the branchial sac directly to the cloacal cavity. Four rows of 6 stigmata are found in the branchial sac.
Aquarium requirements: Nothing is known about keeping this species in aquaria, but as it is distributed in the areas where collecting for aquarium purposes is done and as it commonly occupies holes and crevices in coral boulders and is entangled in branches of coral, it is likely that it will show up in the reef aquaria sporadically. Careful observations of this interesting species should be recorded and we would be very happy to receive reports on any experience with it. Also refer to notes for *Didemnum molle* at page 435.

GA: 0; **SE:** 0; **CO:** 0

Photos:
Left: Typical appearance of blue-green *Diplosoma simile* growing among rocks and rubble on a coral reef.
Right: Close up of colonies growing among coral rubble.
Photos: J. S. Ryland

the didemnid-*Prochloron* obligate symbioses in the Indo-West Pacific. *Diplosoma simile* often is found on the reef crest in the surf zone, where it forms translucent or green iridescent, gelatinous, encrusting sheets, often deep among coral branches or in small crevices and undercuts, and often growing up through holes in consolidated surfaces (see photo on page 437 as well as Fact Sheet on page 438). The larvae of all *Diplosoma* species in obligate symbioses have a T-shaped rastrum ("algal rake") above the base of the tail. It is covered with filaments of larval test that enmesh the symbiotic algae and carry them on to the new colony. *Diplosoma virens* resembles *D. simile* but instead of growing in large sheets it subdivides (lobulates) to form small circular or oval cushions.

Genus ***Trididemnum***, which probably contains about 60 species world wide (34 in Australian and Indo-West Pacific waters), is distinguished from other genera in the Didemnidae by its three (instead of four) rows of stigmata (Kott, 2001). The most attractive ascidian in this genus is probably the small green jewel-like *Trididemnum miniatum*, known from the Philippines and the Great Barrier Reef. Its colonies are small cushions, to 1 cm diameter, resembling miniature *Didemnum molle*. Their white or lime-green colour depends on the concentration of spicules and algal cells, both embedded in the test. The species has both *Prochloron* and red filamentous blue-green algal symbionts.

Trididemnum solidum is widely distributed throughout the Caribbean Sea, and around Florida and the Bahamas. Its test is thick, and colonies are generally blue-green or green-white, depending on the concentration and distribution of the chlorophyll-containing blue green symbiotic algae embedded in the test. This species thrives in areas where there is good stony coral growth, and sometimes it even grows over stony corals (Humann, 1994).

Leptoclinides dubius **on the Great Barrier Reef.** **Photo: E. Svenson**

The genus ***Leptoclinides*** contains large, robust colonies that have large zooids (2 to 4 mm). The atrial apertures are large and posteriorly oriented on long siphons. Species of this genus have been found to accumulate large amounts of vanadium, an element found in small amounts in specimens of *Atriolum robustum*, but not among other genera of the Didemnidae (Kott, 2001). Although a few species in this genus sometimes have symbiotic algae on the surface, obligate symbioses are not known. There are at least 58 species of *Leptoclinides* from the Indo-West Pacific and Australasia (including New Zealand), but few species are known from the northern hemisphere. The genus has not been recorded from the eastern Pacific or the eastern Atlantic but three species are known from the western Atlantic.

The attractive species *Leptoclinides dubius* forms large, encrusting sheets with the surface raised into low domes or long curved ridges. The zooids are numerous and about 3 mm long. *Leptoclinides dubius* is known from the central Indo-Pacific and Australia (Kott, 2001).

Leptoclinides **sp. "KA6-ASC-11", seen here at the Maldives.**
Photo: J. Cairns-Michael

Ascidia sp. "KA6-ASC-12" and juvenile *Didemnum molle*, at Fiji.
Photo: J. Cairns-Michael

Suborder Phlebobranchia

The species in this suborder mostly are large, solitary ascidians, usually with a gelatinous and translucent test. Sand grains are only occasionally present in the test. The branchial sac is large and occupies the length of the body, the gut being folded up into the body wall outside the atrial cavity where the gonads also are embedded. Fertilization (except in the colonial family Perophoridae) is external and larvae are primitive, with simple, unstalked, adhesive organs arranged in a triangle. Replication to form colonies occurs in families Perophoridae and Plurellidae only.

The colour in these tunicates usually results from pigmented blood cells present in a vascular network in the body wall, branchial sac and test vessels. Members of Phlebobranchia generally contain Vanadium as yellow-coloured Vanadium chromogen in vanadocytes (morula cells) and compartment cells. Red and orange pigments cells are also present in many species (KOTT, 1985).

The suborder Phlebobranchia contains many tropical species, some of which are rather big and easily detected on coral reefs. There are, however, also many smaller species that are normally found on the underside of boulders and reef rocks. Among these are some species that are commonly introduced to the reef aquaria on live rocks. The families Ascidiidae, Plurellidae and Perophoridae are commonly found in the shallow waters of the tropics, while Agnesiidae and Corellidae are better represented in abyssal depths or in northern cold seas than in shallow tropical waters. The systematic overview below to the left lists the further division of this suborder into families and genera.

Suborder Phlebobranchia

Family Ascidiidae
Ascidia, Phallusia, Psammascidia, Ascidiella**
Family Hypobythiidae*
Family Agnesiidae*
Family Corellidae*
Family Plurellidae*
Family Perophoridae
Perophora, Ecteinascidia

* not dealt with here

Family Ascidiidae

In this family, the genera *Ascidia* and *Phallusia* are discussed below. The species of both are solitary. The test is gelatinous and often – especially in genus *Phallusia* – robust, firm and rigid, but it is hard, brittle or tough only rarely.

Ascidia species are often attached, to the underside of living rocks, along most of the left side of the body. In most species the test is firm, naked, gelatinous and translucent, but in a few species (especially where it adheres to the substrate) it can be thin and papery. Transverse or branching bands of muscles are well developed forming an irregular mesh in the right body wall and anterior to the gut on the left. About 30 species are known from the central Indo-West Pacific, including 19 from Australia (KOTT, 1985). Many species are known from cold and temperate waters, *Ascidia mentula* and *Ascidia virginea*, being common species along European coasts.

Ascidia archaia often accommodates symbiotic amphipods in its cloacal chamber (KOTT, 1993). The test is nearly always transparent, and acts as a lens through

Ascidia archaia **is very often found on the underside of large rocks and boulders (left), and is often introduced in to aquaria attached to living rocks or to the bases of coral heads (right). The test is transparent. Sometimes symbiotic amphipods (small crustaceans) live inside this ascidian.**

which these crustaceans can be seen. We have found *A. archaia* repeatedly on living rocks, but we have not been able to keep it alive in an aquarium for any significant length of time. The species occurs widely in the Indo-Pacific, from Indonesia and the Philippines northward to Japan, southward to Australia and eastward to the Marshall islands and Fiji. Individual specimens grow to 5 cm long.

Genus ***Phallusia*** is difficult to separate from *Ascidia* being distinguished only by the presence of the accessory openings of the neutral gland into the atrial cavity and by its rather more robust and larger individuals and firmer test. The species *Phallusia julinea*, *P. arabica* and *P. millari* are all widely distributed throughout the Indo-West Pacific, but in general the genus is not diverse. *Phallusia arabica* has the largest geographic range, being known from the Red Sea, the Gulf of Suez, Sri Lanka, the Philippines and Australia. It is cylindrical, sometimes slightly flattened and up to 20 cm long. The test is nearly transparent milky white.

Phallusia julinea occurs from Madagascar eastward to Palau and New Caledonia. Over its geographic range it lives in very shallow water, but it has also been found down to a depth of 30 m (Kott, 1985). *Phallusia julinea* is variable in shape, in many instances long and club-shaped, or

Phallusia arabica **at One Tree Island, Great Barrier Reef.**
Photo: L. Newman and A. Flowers

Phallusia julinea **at Heron Island, Great Barrier Reef.**
Photo: L. Newman and A. Flowers

Phallusia obesa at Esperance, Western Australia. Photo: C. Bryce

The growth of ascidians underneath rocks and boulders can be rather diverse, as seen here at One Tree Island, Great Barrier Reef. The yellow colony in the middle of the photo is a *Botrylloides* species, the yellow zooids along the upper edge of the photo are those of *Ecteinascidia nexa*, while the red-white colony at the left photo margin is probably an *Aplidium* species. Photo: L. Newman and A. Flowers

curved and slightly irregular and often laterally flattened. Posteriorly the body is rounded. On its dorsal side – far from the buccal opening – is the atrial opening. The test has an unique colour pattern, being creamy yellow or brown with strong chromatophores and groups of brown pigment cells.

Phallusia obesa is a robust ascidian that may reach 16 cm in length. It is longer than wide and tapers to a terminal branchial aperture, on a short siphon, often curved to the right and posteriorly (KOTT, 1985). The atrial opening always protrudes from the anterior third of the dorsal surface. The surface of the test may be raised in rounded longitudinal ridges, separated by deep crevices, or these ridges are divided into rounded protrusions with central tubercles. Colour is variable, and can be cream, brown, blue-grey or often dark grey to black. Pigment spots are sometimes present. *Phallusia obesa* is known to occur from Western Australia and South Australia and along the southeastern Australian coastline to Moreton Bay in southeastern Queensland. Except for its lighter colour, it closely resembles the glossy black *Phallusia nigra* of the Caribbean.

Ecteinascidia diaphanis at Nusakode Island, Maldives. Photo: C. Bryce

Family Perophoridae

In this colonial family, the zooids are minute, often bubble-like, round and connected to each other by basal stolons. The borders of the atrial aperture are divided into a variable (up to 14) number of lo-

Ecteinascidia bandaensis at the Maldives.

bes. The test often contains small pigment spots over the anterior end of each zooid, around the apertures. The branchial sac is flat, without folds, but with internal longitudinal vessels and up to 25 rows of stigmata. New zooids are formed from ectodermal and mesenchymal cells in the stolon. Species are generally viviparous, new individuals developing in a brood chamber in the posterior part of the atrial cavity. Larvae are large and have a well-developed yolk mass. Two genera are in the family: *Perophora* and *Ecteinascidia*.

The species of genus ***Ecteinascidia*** have larger, more colourful zooids than *Perophora* and its species occur only in tropical seas. The branchial sac has many stigmata, more than 20 rows with more than 20 perforations per row. The gut loop is curved with the rectum extending anteriorly and the testis has very numerous

Ascidians can be among the first organisms to re-colonise a destroyed reef, such as Osprey Reef in the Coral Sea, seen here six months after being devastated by a cyclone in January 1991. Left: The rocks are covered by juvenile *Didemnum molle* and crowded zooids of colonies of *Perophora modificata*. Right: Close-up photo of a *P. modificata* colony.

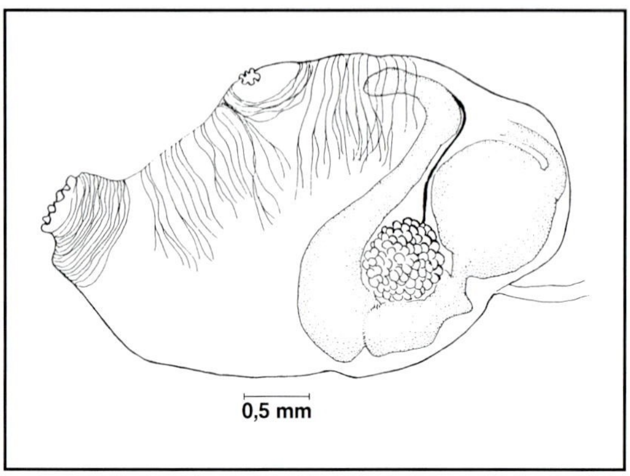

Ecteinascidia nexa

Distribution: Indo-West Pacific: Sri Lanka, Indonesia, Fiji, tropical Japan (Tokahara Island), Great Barrier Reef and Lord Howe Island (Kott, 1985). Typically found on the underside of coral boulders and often exposed to surf and wave action. Never recorded deeper than 3 metres.
Size and growth form: Colonies form extensive mats with small, crowded zooids, usually less than 6 mm long. Colonies grow by budding from the vascular stolons that join the zooids and attach them to the substrate.
Reproduction: Viviparous. Gonads consisting of a circle of berry-like male follicles and a small ovary, are in the intestinal loop. Embryos develop in a brood pouch on the right side of the body in the atrial cavity.
Food: As with other ascidians the animal depends for food on the minute organisms and organic particles suspended in the water passing through the pharynx.
Description: (Based on Kott, 1985). Zooids are transparent and pale yellow to olive-yellow, sometimes pale iridescent blue with a yellow intestine. Pale and transparent zooids have clearly visible yellow rings around the edge of their apertures. Zooids lie on their ventral surfaces or the anterior end is bent into an upright position. The atrial opening is in the middle of the dorsal surface. A stalk from the posterior end of the ventral surface is of variable length, often rather thick and is attached to the basal stolon. The body wall is thin and delicate, but muscular. The intestinal loop occupies more than half of the posterior part of the left side of the body.
Aquarium requirements: Although *Ecteinascidia nexa* is frequently found attached to living rocks, it is often overlooked because of its small size. It sometimes does rather well in a coral reef aquarium, but apparently dislikes water that has been extensively foam-fractionated. We had great success with this species in one of our experimental aquariums (Nilsen, 1999, 2000) where the tiny tunicate established small colonies in crevices and holes in the live rocks that made up the aquarium's decorations. In this aquarium the amount of protein skimming was limited and the amount of organic material present was somewhat higher than in many other reef tanks. In general *E. nexa* belongs to the more durable acidians.

GA: +/-; **SE:** +/-; **CO:** +

Photos:
Top: *Ecteinascidia nexa* attached to a piece of live rock from the author's first reef aquarium kept as early as in 1979.
Centre: *Ecteinascidia nexa* at the base of a stony coral head.
Bottom: Zooid with ripe gonads, intestinal loop and muscles (drawing from Kott, 1985).

follicles. At least 12 species have been recorded from the Indo-West Pacific (KOTT, 1985).

Ecteinascidia nexa has repeatedly been introduced into aquaria on live rocks. This is one of the few ascidians that appear to thrive and even breed when given optimum conditions in a coral reef aquarium (see Fact Sheet on page 444).

Ecteinascidia diaphanis has zooids that are up to 1 cm tall and 0.5 mm wide, upright and cylindrically shaped and taper to a stalk posteriorly. Each zooid has a thin, glassy transparent test. As a rule, the zooids have a colourful ring around the buccal and atrial openings.

Ecteinascidia bandaensis is a small species originally described from Indonesia. We have also found it in the Maldives, but the species has not yet been reported from Australia (KOTT, 1985).

Ecteinascidia turbinata is a species of this genus in the Caribbean Sea, and around Florida and the Bahamas. The zooids are orange-coloured and grow in small groups, from a few individuals to several hundred. HUMANN (1994) refers to this species as the Mangrove Sea Squirt, because it is commonly found on the roots of mangroves and in areas with marine grasses; however, it also occasionally occurs among corals.

The genus ***Perophora*** has small zooids which normally do not grow taller than 5 mm, never have more than 8 rows of stigmata, with the gut forming a simple horizontal loop and the testis with 1 to 20 lobes. The genus is most frequently found in temperate waters, but there are also at least 12 tropical species, 6 of which have been described from Australia (KOTT, 1985).

We have encountered large populations of *Perophora modificata* on Osprey Reef in the Coral Sea, in July 1991, about 6 months after the reef was severely damaged by a tropical cyclone. Most corals had disappeared, there was a predominance of various algal species, and this ascidian had become established in many areas. According to KOTT (1985) it has so far been recorded only from the Philippines, the Coral Sea and the Great Barrier Reef. Intensely yellow zooids about 1 om high grow up from a ctrongly interwoven stolon mat.

The tiny *Perophora multiclathrata*, which is very common and widely distributed in tropical seas, is usually found under rocks, living among other sessile organisms. It is quite possible that this species is occasionally introduced in to aquaria, together with live rocks. This photo was taken in the Maldives and shows only an 8.5 cm² section.

Perophora multiclathrata is widely distributed in the Indo-West Pacific as well as the tropical West Atlantic. The small, 2 to 3 mm tall zooids are located on a short stem and appear almost square. This species is so common that is it conceivable that it occasionally will be discovered growing in a reef aquarium.

Order Pleurogona

The order Pleurogona contains the suborder Stolidobranchia, with a total of 6 families (see systematic overview on page 446). The order is represented in tropical, temperate and cold waters and includes a number of species found on the coral reefs. In contrast to the members of the Order Enterogona where the atrial cavity develops from a pair of dorsal invaginations on each side of the midline, members of Pleurogona have the atrial cavity developing from a single median dorsal invagination. The pharyngeal wall usually is folded along its length, which gives a greater filtering area than the flat pharyngeal wall of members of the order Enterogona. Gonads vary in numbers, but always are present on both sides of the body. The test usually is tough, often opaque and leathery and epibionts often are present on its surface. There is no abdomen and heart, gut and gonads are located in the parietal body wall alongside the large branchial sac but separated from it by the peribranchial (atrial) cavity. Most species in Pleurogona are large and solitary, but members of subfamilies Polyzoinae (page 449) and Botryllinae (page 454) replicate to form colonies of small and often simplified zooids. Generally, solitary members of Pleurogona are oviparous,

Order Pleurogona

Family Styelidae

Subfamily Styelinae

Polycarpa, Monandrocarpa, Bathyoncus, Pelonaia*, Styela*, Cnemidocarpa*, Asterocarpa*,Hemistyela*, Minostyela*, Dicarpa*, Dextrocarpa*, Azygocarpa*, Dendrodoa*, Podostyela**

Subfamily Polyzoinae

Polyandrocarpa, Eusynstyela, Oculiinaria, Cyndrocarpa*, Syncarpa*, Seriocarpa*, Kukenthalia*, Polyzoa*, Symplegma*, Botryllocarpa*, Berrilia*, Metandrocarpa*, Stolonica*, Amphicarpa*, Protostyela, Theodorella*, Arnbackia*, Oligocarpa*, Distomus*, Alloecarpa*, Chorizocarpa**

Subfamily Botryllinae

Botrylloides, Botryllus

Family Hexacrobylidae*

Family Pyuridae

Microcosmus, Herdmania, Pyura, Eupera, Heterostigma*, Bolteniopsis*, Cratostigmata*, Bathypera*, Pyureela*, Hartmeyeria*, Boltenia*, Ctenyura*, Halocynthia*, Ctenicella*, Paraculeolus*, Bathypyurea*, Culeolus*,*

Family Mogulidae*

* not discussed in detail here

being externally fertilized with the small, primitive larvae (Kott, 1985). However, in a few species of *Polycarpa* and *Molgula* the eggs are retained in the atrial cavity where they are fertilised and the embryos are incubated, being released as swimming larvae. In all colonial pleurogonid species, eggs are fertilised and embryos incubated in the colony (as they are in colonial species of all suborders).

Family Styelidae

These ascidians have up to 4 branchial bolds per side, simple unbranched branchial tentacles a simple plain-edged mid-dorsal lamina in the branchial sac and they lack liver diverticulae of the gut. Ascidians from the three subfamilies of the family Styelidae are imported either relatively frequently or only sporadically to the aquarium trade.

The photos on the left and right show the colour variation in the Golden Ascidian, *Polycarpa aurata*. This species occurs in the Indo-West Pacific, from Sri Lanka eastward to the Philippines and to Palau, as well as the Great Barrier Reef and throughout the Coral Sea. It normally is encountered along steep drop-offs at depths from a few metres down to 30 m. Individuals are up to 10 cm long, robust and the test is white or cream-coloured with purple or yellow markings. The edges of the rather short siphons are golden yellow. The test is tough, thick and smooth on the outside (description after Kott, 1985). So far nothing is known about the aquarium care of this species, but its aquarium maintenance appears to be difficult.

Left: *Polycarpa nigricans.* Right: A large specimen of *P. pigmenta* (brown with blue siphon linings) is seen here overgrown by a few young *Didemnum molle,* and often is found with *P. aurata.* Both photos were taken on the Great Barrier Reef.

The species of subfamily **Styelinae** are solitary with many large species with a stout test and body wall with strong muscles. This subfamily contains 14 genera distinguished mainly by their sex organs (KOTT, 1985). Three of these genera ***Polycarpa***, ***Styela*** and ***Cnemidocarpa*** are commonly encountered. Generally there is some degeneration of the larval light sensitive ocellus.

The Stalked Ascidian, *Polycarpa clavata* is widely distributed in subtropical and tropical waters around Australia. The individuals are up to 8 cm long and 3 cm wide and the supporting stalk is often twice as long as this.

Genus ***Polycarpa*** is represented by a relatively few common species on coral reefs, but appears also to be well adapted to open sea-floor habitats, where the necessity for light sensitive larvae to seek shaded settlement sites is removed; and where larval vulnerability to dispersal is reduced by reduction of the free-swimming larval life: short oviducts liberate eggs into the atrial cavity rather than directly to the exterior through the atrial aperture (Kott, 1985). Adaptations to a life on the sea-floor are a thin test with incorporation of sand grains to increase its hardness, test hairs to which sand also adheres and which fix the organism in sea-floor sediments, and numerous gonads that produce large numbers of gametes (Kott, 1985). Thirty-nine species of *Polycarpa* are known to occur throughout the tropical Indo-West Pacific. *Polycarpa aurata* is the most conspicuous of these (see Fact Sheet at page 450). Species also occur in cold seas, in shallow as well as in deeper waters, such as *Polycarpa pomaria* found

A crowded population of *Polycarpa procera* at One Tree Island, Great Barrier Reef.

Photo: L. Newman and A. Flowers

from the Arctic to the Mediterranean and *Polycarpa rustica* commonly occurring along the coast of the British Isles. See also KOTT (1985) for a more detailed account of this genus.

While *P. aurata* is often attractively coloured in white, lilac or yellow, several other *Polycarpa* species are brown or inconspicuously grey, often encrusted with many other organisms, so that only the siphons are visible. Occasionally, solitary *Polycarpa* species are totally covered epibionts, including colonial ascidians.

Polycarpa nigricans usually grows in small aggregations or in compact, irregular colonies where the individual zooids are sometimes separated from one another or occasions apparently embedded in tough test. Individuals are about 4 cm long and 3 cm wide. The test is grey to black, very hard and covered by a thin layer of sand. White bands ring the siphons. *P. nigricans* is known from Mauritius, Indonesia, the Philippines, Western Australia and the Great Barrier Reef. It always grows at sites with strong currents, down to a depth of up to 20 m.

Polycarpa pigmentata is one of the most common ascidians in tropical seas. It has a wide range in the Indo-West Pacific and occurs eastward to the Solomon Islands and Fiji. Large populations can be found in shallow water. Individuals are laterally flattened, upright and more or less pear-shaped (KOTT, 1985). The tough and leathery test, up to 1 cm thick in some places, with a rough and uneven surface often covered with epibionts, is brown with deep to pale blue siphonal linings. The atrial aperture is on the dorsal surface slightly posterior to the terminal branchial aperture. Both apertures are located on very short siphons giving them a wart-like appearance when closed. *P. pigmentata* is a large ascidian, up to 12 cm or more long.

Polycarpa clavata is a conspicuous species closely related to *P. aurata* but largely restricted to the coast of Australia and the Coral Sea. It occurs on sandy as well as on rocky bottoms from shallow waters down to depths of about 40 meters. It has a distinctive thick, firm, spectacularly coloured test with rounded longitudinal ridges and a unique thick fleshy, stalk that is joined to the anterior end of the head, rather than the posterior end. The animal sways with the current that flows past it, presenting its mouth to the on-coming water, so reinforcing the ciliary feeding current drawn in through the month. Individuals are up to 8 cm long and 3 cm wide and the supporting stalk is often twice that length. The brilliant yellow or orange colour is caused by pigments mostly in the surface layer of test, which is richly supplied with terminal ampullae and blood vessels (KOTT, 1985). We have observed *P. clavata* on several occasions while diving on the Great Barrier Reef and in the Coral Sea, but have no reports of this magnificent species being kept in captivity.

Polycarpa procera is known from western, southern and eastern Australia, as well as from Sri Lanka, Indonesia and Japan. The species has an oval, elongate body to 4 cm long, and 3 cm diameter. It usually has an orange colour. The test is thin, hard and brittle and can contain sand particles embedded in the tissue. Aggregates of individuals appear to be connectedly hairs and root-like projections of the test, which are longer around the ventral, posterior and/or postero-dorsal aspects of the body.

In the Caribbean Sea this genus is represented by *Polycarpa spongiabilis*. Although it has a wide range on corals reefs in the Caribbean region, it is not very frequently encountered (HUMANN, 1994). As in other species of this genus, algae and other organisms usually grow on the test obscuring the individuals.

Genus ***Monandrocarpa*** contains small, solitary species that rarely grow longer than 1 cm, and are found on sandy sediments. So far only three species have been described (KOTT, 1985). *Monandrocarpa incubita* is known from the western Pacific, however, it does not occur around Australia. *Monandrocarpa tritonis* from South Africa, closely resembles *M. incubita*. *Monandrocarpa plana* occurs down to 457 m in waters around the Philippines and also has been found at 160 meters depth at Cronulla, New South Wales, Australia. We discovered this particular ascidian in the coral reef aquarium of Karl Ole Dahl in Bergen (Norway), where it thrived very well and became established everywhere throughout the tank. This probably indicates that the species occurs in shallow waters of the central Indo-West Pacific since, without doubt, it was introduced to the aquarium with live rocks imported from this area. For further details see Fact Sheet on page 451.

In subfamily **Polyzoinae** the species replicate to form colonies, although zooids always have their own separate buccal and atrial openings to the exterior, and there is no common cloacal chamber. Individual zooids are connected either by a basal stolon or are embedded in a common test. The zooids are almost always less than 1 cm long. This subfamily contains 20 known gonora (KOTT, 1985). Eight are well known in the tropical Indo-West Pacific but only 3 are discussed below.

The genus *Polyandrocarpa* differs from *Polycarpa* only in its colonial status and the (related) small zooids. The genus contains at least 12 species, but is possibly polyphyletic as some of the species have closer relationships with species of *Cnemidocarpa* than with species of *Polycarpa* (KOTT, 1985). Many species are known only from Australian locations, often from tropical and subtropical locations although *Polyandrocarpa sagamiensis* is widely distributed in the Indo-West Pacific. *Polyandrocarpa tumida* inhabits reefs of the Caribbean where it is frequently found in areas of dead coral and ship wrecks (HUMANN, 1994).

The genus *Eusynstyela* resembles *Monandrocarpa*, differing

Polycarpa aurata
Golden Ascidian

Distribution: Indo-West Pacific: Sri Lanka in the west, the Coral Sea, Great Barrier Reef, the Philippines to the Palau Is in the east. Normally found on vertical drop-offs in a few meters depth down to 30 metres. Occupies sea-floor habitats and seems to tolerate moderate to rather strong sunlight, but never found in extreme light-exposed areas.
Size and growth form: Individuals are large and solid solitary ascidians to 10 cm long. Usually growing alone or in small groups.
Reproduction: Externally fertilised, the embryos developing to larvae in the open sea as part of the plankton. The large gonads are deeply embedded in the body wall.
Food: It is most likely that these solitary ascidians, with their large apertures firmly turned away from one another, and their long branchial folds providing a vast filtering area and generating a considerable current through the pharyngeal wall, have a filtering capacity of considerable magnitude to satisfy the nutritional requirements of these large robust organisms.
Description: (Based on KOTT, 1985). The colour is striking, being opaque white or cream with purple and yellow markings. The siphonal lining is bright yellow. The body is slightly flattened laterally and strongly curved along the longitudinal axis with deep longitudinal creases between the long, curved ridges. The body narrows toward the anterior branchial aperture while the posterior end is wide and rounded. Siphons are short but conspicuous. The test is tough, thick and smooth and lacks epiphytic growth. The branchial tentacles are widely spaced, often fewer than 10 in larger specimens (see photo). The branchial folds are narrow but flat, with 17-25 internal longitudinal vessels crowded on them and 3-5 more widely spaced between the folds. Up to 12 stigmata per mesh. The gut forms a fairly narrow, deeply curved loop around the wide posterior end of the body.
Aquarium requirements: Nothing is known about the aquarium requirements for this species, but we regard it as difficult to keep. Sufficient water movement and suspended organic food are factors that must be available in order to succeed. It should be kept in a special aquarium without grazing animals.

GA: 0; **SE:** 0; **CO:** 0

Photos:
Left: Two specimens of *Polycarpa aurata* growing on the branches of a stony coral from the genus *Hydnophora*.
Right: A specimen of *Polycarpa aurata* seen from above showing clearly the branchial tentacles.

Monandrocarpa plana

Distribution: West Pacific: the Philippines, Mariana Islands, and the coast of New South Wales, south-eastern Australia at depths from 160 to 475 m (KOTT, 1986).
Size and growth form: Solitary, however, it usually is in small aggregations in substrate depressions, caves and rocky crevasses. Adult specimens are oval, slightly flattened laterally and about 1.5 cm long including the sandy covering.
Reproduction: Probably fertilised externally. Up to five gonads are in a row on each side of the body. Specimens examined by us had only a single testis follicle associated with each small sac like ovary.
Food: The small branchial sac with 4 folds suggest that the species has a significant filtering capacity in relation to its size, although the sessile apertures indicate that the feeding currents derive from water in the immediate vicinity of the organisms.
Description: (Based on KOTT, 1985). Individuals are coloured red or orange-red. The apertures are sessile on the upper surface, about one-third to half of the body-length apart. Long test hairs around the mid-line of the body have sand grains adhering to them as well as to the remainder of the test. The body wall is thin, but has delicate muscles, except around the apertures where they are stronger and more conspicious. There are about 40 crowded tentacles. There are 4 branchial folds and about 5 oval stigmata per mesh. The gut forms a short and compact loop in the posterior third of the body. The intestine is wide but short. Gonads are arranged in a single row along each side of the body.
Aquarium requirements: It is a bit unexpected that this species, which is only collected from relatively deep water, shows up in a home reef aquarium far away from its recorded geographic range. This may indicate that the distribution of the species is much wider than recorded so far or (a more likely explanation) is that it has been imported on live rocks. According to our observations *Monandrocarpa plana* is one of the first ascidians known to have become properly established in a coral reef aquarium, various scattered colonies have thrived for several years. It seems to settle only in shady areas. The 400-litre aquarium had a very strong current (11,000 litres/hour).
GA: +; **SE:** +; **CO:** +

Photos:
Top: *Monandrocarpa plana* in a coral reef aquarium.
Centre: Close-up photo of a single specimen.
Bottom: Schematic body structure (drawing from KOTT, 1985).

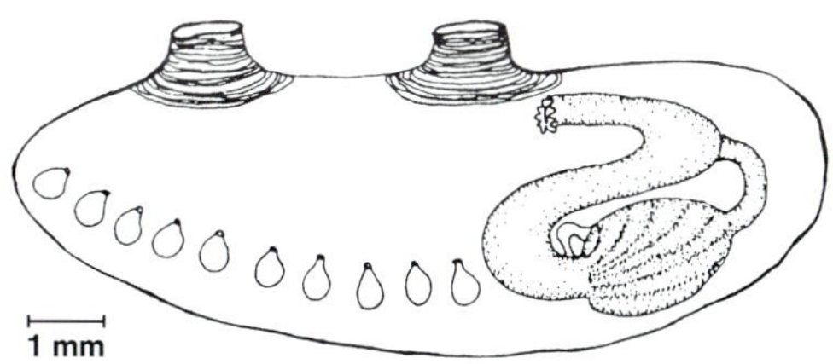

Pink-coloured colony of a *Eusynstyela* species with *Polycarpa aurata* and numerous other organisms, attached to a reef wall at a depth of 10 m, at Bunaken, North Sulawesi.

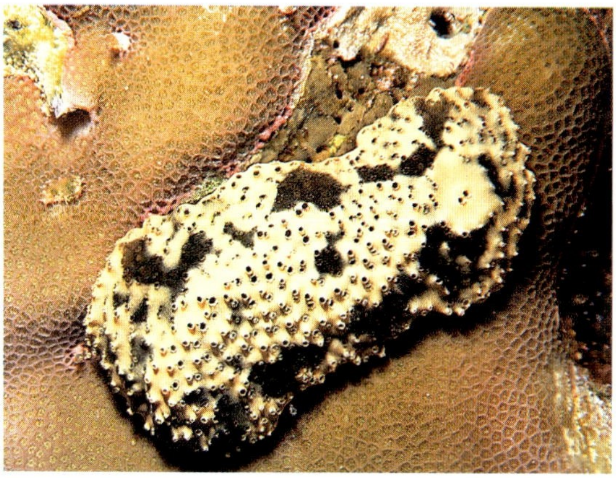

Two *Polyandrocarpa* species: left: one in the Maldives (photo: J. Cairns-Michael) and on the right one in Fiji (photo: S. W. Michael).

mainly in the form of its colonies. It contains few species, all with small zooids usually embedded in common test. *Eusynstyela latericius* occurs frequently throughout the Indo-West Pacific, where it often grows as an epibiont on other ascidians.

The four species of the genus ***Symplegma*** all lack branchial folds and have four internal longitudinal branchial vessels and a single hermaphrodite gonad on each side of the body. *Symplegma stuhlmanni* has been recorded from Sri Lanka, Thailand, China, Palau, Fiji and along the coast of Australia. We often found this species on the underside of coral boulders on the reef flat of Heron Island (Great Barrier Reef). The zooids, completely embedded in the transparent test, are bright or pastel-coloured purple

Eusynstyela latericius at Wisari Reef, Great Barrier Reef. Photo: L. Newman and A. Flowers

Two species of *Symplegma* growing on the underside of coral boulders on the reef flat, Heron Island, Great Barrier Reef. Left: *S. stuhlmanni* and right *S. reptans.*

or yellow, orange-red or pale or greenish lemon. *Symplegma reptans*, which is recorded from the Great Barrier Reef (Heron Island and the adjacent Wistari Reef), Hong Kong and Japan, forms thin colonies that stretch over the substratum. The colony is pink with large, triangular white patches arranged in circles. The patterns indicate the circular arrangement of the zooids with their atrial apertures in the circle centres. The white patches, clearly visible in the photo above, are the prebranchial areas around the branchial apertures (KOTT, 1985).

In the Caribbean the genus is represented by the beautiful *Symplegma viride* which, according to HUMANN (1994), inhabits reefs and reef walls and often encrusts areas of dead coral and the surface of large sponges.

Species of the colonial subfamily **Botryllinae** have small zooids completely embedded in common test and forming systems around common cloacal cavities. The smooth circular branchial apertures open to the exterior, and usually the large atrial apertures open directly to the common cloacal cavities in the test. Eggs are fertilised internally in brood pouches in the body wall and species are viviparous. Larvae are large with a large yolk mass and well developed larval organs (KOTT, 1985). Some species appear sporadically in coral reef aquariums (e.g. see BIRKHOLZ, 1983). There are two genera, *Botryllus*, where the brood pouches protrude into the atrial cavity, and *Botrylloides* where they protrude outside the body wall. Only in *Botryllus primigenus*, known from the Western Pacific, has each zooid its own atrial opening leading to the exterior. The species in this subfamily have a highly variable colour pattern. Consequently, it is virtually impossible to identify respective species on the basis of colour. For instance, 48 different colour patterns have been found in the rather common *Botryllus schlosseri* (KOTT, 1985). The reason for this is the changing content of different blood pigments, which not only vary from one colony to the next, but their concentration also increases with advancing age (SABADIN, 1979; SABADIN & GRAZIANI, 1967). Similarly, the co-

Two colour varieties of *Botryllus schlosseri.* **Photos: E. Svensen**

Botrylloides leachi has diverse colour patterns: this brown colony is from Madang, Papua-New Guinea (left; photo: L. Newman and A. Flowers). The close-up photo of _Botrylloides magnicoecus_ from Ulladulla, New South Wales (Australia) shows a magnificently yellow colony (right; photo: K. Kates).

lony shape also changes with age. The species of this subfamily are numerous in shallow water and in the inter-tidal zone of coral reefs, and along tropical coastlines.

Botryllus is a relatively small genus with six known species from the central Indo-West Pacific (KOTT, 1985), and it also occurs in the Caribbean region (HUMANN, 1994). The zooids are often arranged along each side of canals, or in circles around a central common cloacal canal like rosettes spread over some underlay.

Botryllus schlosseri is the best known species with the widest

A group of large _Herdmania momus_ totally overgrown by _Botrylloides leachi._ Photo: K. Kates

geographic range. It has been reported mostly from subtropical seas, such as the Mediterranean Sea and the Black Sea, and it even occurs frequently in our own coastal waters where it is distributed along the entire coast of Norway as far north as Svalbard. However, there have also been several observations in tropical latitudes, such as from Wistari Reef, Great Barrier Reef, and from Western Australia, Hong Kong and Japan. Attempts to keep this species in a coral reef aquarium could indeed be worthwhile, even with specimens collected in temperate waters. The colonies vary in shape; they can be leaf-like, flat or stalked or they may develop fleshy lobes. Their maximum length is 5 cm, with a width of 3 cm. The zooids usually are arranged in circular or elongate systems. Some of the common cloacal openings are long although mostly they are large and round. The colour of *B. schlosseri* is highly variable, and can be brown, yellow, orange or grey. The test and zooids often have different colours giving the colony a mixed and beautiful colour pattern.

Botryllus tuberatus has been found sporadically around Indonesia, Western Australia, Palau, the Gilbert Islands and Japan, however, the type material came originally from Santa Barbara, southern California (but has never been found there again). *B. tuberatus* grows in colonies with a maximum number of 12 zooids, forming a thin, crust-like layer over the substrate.

The most frequently occurring species of the genus ***Botrylloides*** is *Botrylloides leachi*, a cosmopolitan species (see Fact Sheet on page 457). The rather attractive *Botrylloides magnicoecus* is known from West- and South Africa, the Indian Ocean, Indonesia, the western and eastern coast of Australia and the Great Barrier Reef. This is one of the most impressive-looking ascidians known. As with other species of this genus, colour varies from white and/or yellow to light-

We found this ascidian (family Pyuridae), possibly a *Pyura*- or *Microcosmus* species, among encrusting anemones in an aquarium at Zoo Welke, Lünen. The "hairy fringes", typical for this family, are clearly visible.

and/or dark lilac. The zooids are crowded in straight or curved double rows, along common cloacal canals that converge to a few (sometimes terminal) large common cloacal openings. Occasionally, colonies found in Australian waters are stalked (Kott, 1985). Two other species from the genus are *Botrylloides perspicuum* and *B. violaceum*.

Family Pyuridae

The invariably large, solitary species of the 17 genera belonging to this family have well developed gastric folds and/or diverticulae, branched branchial tentacles and more than 4, (usually 6-10), branchial folds on each side of the body. These folds give the organism a large filtering area. The surface layer of the test that lines the siphons is produced into a variety of hooks and spines (siphonal armature) that may discourage intruders from crawling into them, or stop them being sucked in. The test is usually very tough and leathery and often the superficial cuticle is thickened into scales. Replication is not known to occur in Pyuridae, and all known species are solitary. Eggs are fertilised externally.

The majority of species and genera are known from subtropical seas, yet some species of the genera *Pyura* and *Microcosmus* are relatively common in tropical latitudes. ***Pyura*** is a very large genus with about 55 species (Kott, 1985). It is interesting to note that one species, *Pyura stolonifera*, was kept in an aquarium in Brisbane (Australia), where it was fed on cornflour for many years (Kott, pers. comm.). This clearly indicates that this species can indeed be kept in an aquarium when there is an adequate food supply.

Herdmania contains about 12 species with *H. momus* being one of the most common ones. It is the species usually found in coral reef communities in the Indo-West Pacific and occasionally is imported to the aquarium trade, usually attached to the under surfaces of live rocks or at the base of corals. It is an attractive ascidian, with red and white colour markings (for further details refer to Fact Sheet on page 458).

Family Molgulidae

Species of this family, have branchial folds, branched tentacles and well developed gastric folds (like Pyuridae). It is distinguished from the solidobranch families by the large kidney vesicle on the right side of

Botrylloides leachi

Distribution: North-eastern Atlantic north to Norway, along the coast of Norway as far north as western Finnmark, Mediterranean and Black Sea, Red Sea, tropical Indo-West Pacific, western and northern Australia, Great Barrier Reef and subtropical waters along the coast of south-eastern Australia.
Size and growth form: Forms thin, encrusting sheet-like to irregularly shaped colonies sometimes with flat lobes projecting from the surface. The zooids are up to 2 mm long and are arranged along each side of curved and branched common cloacal canals with large randomly spaced common cloacal apertures.
Reproduction: Ovaries and testis are located on each side of the body. Fertilization is internal. Embryos are incubated in the body wall in a brood pouch, which protrudes from the outside of the zooid.
Food: As most ascidians, this species is a filter feeder.
Description: (Based on KOTT, 1985). Colour of colonies is variable. Some specific colour morphs can be recognised, but there does not appear to be any limit to the range of colours in colonies of *B. leachi*. Each small zooid has a wide atrial aperture, which exposes a large section of the branchial sac directly to the cloacal cavity. There are 4 large branchial tentacles and 4 smaller ones, alternating with rudimentary tentacles. The 10 or 11 rows of stigmata have about 16 stigmata per row.
Aquarium requirements: Small colonies are sometimes found attached to living rocks or to the base of coral heads. The specimen, described by BIRKHOLZ (1983b), probably was *Botrylloides leachi*. We have observed substantial growth of colonies kept in sales tanks of various aquarium shops. However, we do not know of any aquarium where *B. leachi* has become well established for a long time and we regard it as difficult to keep. Water movement and sufficient suspended organic food probably are important but there may be other parameters, essential to the aquarium keeping of this beautiful, colonial tunicate.

GA: +/-; **SE:** +/-; **CO:** +

Photos:
Top: Dark colour variety of *Botrylloides leachi* in a sales tank of Kölle Zoo, Stuttgart.
Centre: *B. leachi* at the Maldives.
Bottom: Red and white coloured colonies growing on the underside of a boulder at Heron Island reef flat.

Herdmania momus

Distribution: Sometimes thought to contain a single species, *Herdmania momus*, the genus is now known to contain at least 12 species, mostly from tropical locations although species are known from southern Australia and Japan. *Herdmania momus*, the species most often present on coral reefs in the Indo-West Pacific is recorded from: the Red Sea, Indian Ocean, Indonesia, Philippines, north-eastern Australia, Great Barrier Reef and eastward to Fiji, Melanesia and the Mariana Islands. Not found in the eastern Pacific.

Photos:
Top: *Herdmania momus* attached to live rock.
Bottom: *Herdmania* sp., seen here at Rottnest Island (Western Australia) totally overgrown by other ascidians, mainly the yellow *Pycnoclavella diminuta*. The species is distinguished from *H. momus* by having only one vas deferens opening. Photo: S. W. Michael

Size and growth form: Solitary, to 8cm.
Reproduction: Oviparous.
Food: Both apertures are large, and in their normal (feeding) position are open, on short cylindrical siphons turned away from each other. Undoubtedly large quantities of particulate organic material and micro-organisms are sieved from the vast amounts of water pumped through these large prolific individuals. However there is no specific information or the amount or type of food they require (pers. comm.).
Description: (According to Kott, pers. comm.) The test is smooth, peach coloured and translucent the body almost spherical, and large and conspicuous. The species has a large branchial sac forming 8 to 12 wide longitudinal folds per side. The gut loop and gonads can be seen through the semi-opaque body wall forming a large mass across the lower half of the left side of the body behind the muscles which are only in the anterior half of the body wall on the left. Long (to 1.5 mm) needle-like calcium-carbonate spicules, barbed along the their length, are embedded in the body wall. Care needs to be taken when handling these organisms to avoid these spicules impregnating the skin. A fan-shaped hood of membrane from the body wall shields the oviducal opening. The testis follicles undulate along the lop of the tightly undulating ovary and open by many short vasa deferentia.
Aquarium requirements: Often small individuals are encountered on living rocks or at the bases of coral heads. However, it is difficult to keep alive for prolonged periods and must be kept in a special aquarium with medium light intensity, strong current flow and water with sufficient organic particles in suspension to satisfy its food requirements.

GA: +/-; **SE:** +/-; **CO:** +

the body and the spirally curved pharyngeal perforations. Individuals are solitary, although they often are found in aggregates, adhering to the sand attached to and embedded in their thin often brittle test.

It is a family well represented in open, sea floor habitats but is only rarely encountered in coral reefs (KOTT, 1985) and has not be found in aquaria. The family contains 12 known genera with *Molgula* being the most common one with about 17 known species. They all have a branchial sac that is gathered into 6-8 folds. Most of the species are adapted to a life on the open floor where they are usually found lying on or rooted in the sand.

Typical appearance of *Hermania momus*. Photo from Great Kepple Island, Australia. Photo: N. Coleman

Classes Thaliacea (Salps and Doliolids) and Appendicularia (Larvaceans)

The two remaining classes of tunicates, Thaliacea and Appendicularia (sometimes known as Larvacea) are free-swimming and planktonic. Although some species are numerous in the waters around coral reefs, they are animals of the open oceans, and are not appropriate for the home reef aquarium. Therefore they are summarised only briefly in this volume.

Salps and Doliolids (**Class Thaliacea**) are part of the "jelly" (gelatinous) plankton. They are suspended and move in the water masses of the ocean currents; and so have vast geographic ranges. Like ascidians, they are filter feeders, the cilia lining the pharyngeal perforations moving water through the body from the buccal aperture at one end, through the pharynx and the atrial cavity and out through atrial aperture at the posterior end of the body. As well as providing the food strained from it by the mucous net over the wall of the pharynx this current moves the animal forward by jet propulsion.

Salps and Doliolids have a complex life cycles often involving more than one stage between the larvae and the adult. Usually at least one of the stages is a colonial sexually reproducing phase alternating with a solitary asexual replicative phase (as in certain hydroids: see Volume 2, page 36). Colonies can be very large as indicated in the figure on page 460. Some *Pyrosoma* species form oval colonies large enough for a diver to swim into.

This animal group does not have very many species and only six genera are known. Species of the colonial genus *Pyrosoma* are brilliantly luminescent. Two luminescent organs, one on each side of the pharynx, probably contain bioluminescent bacteria that produce intense light when required to do so by their host (BONE, 1998). *Pegea*, *Salpa* and *Thalia*

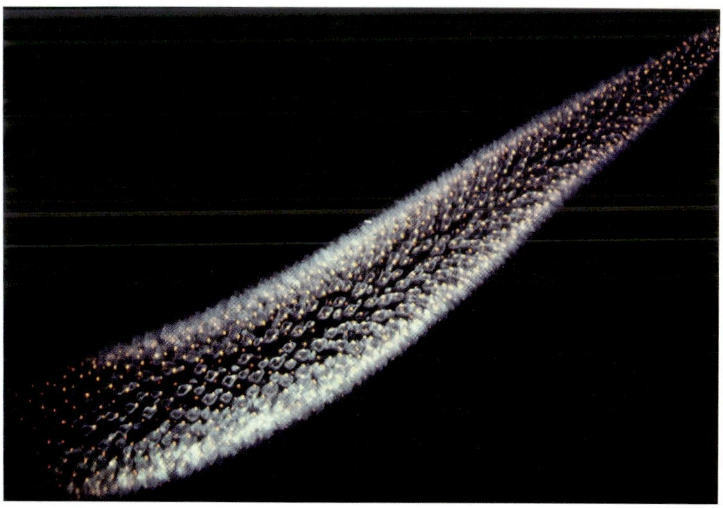

***Pyrostremma agassizi* forms colonies that can be up to 30 cm long, while the individual zooids are only 5 to 6 mm long. Photo: R. van Soest**

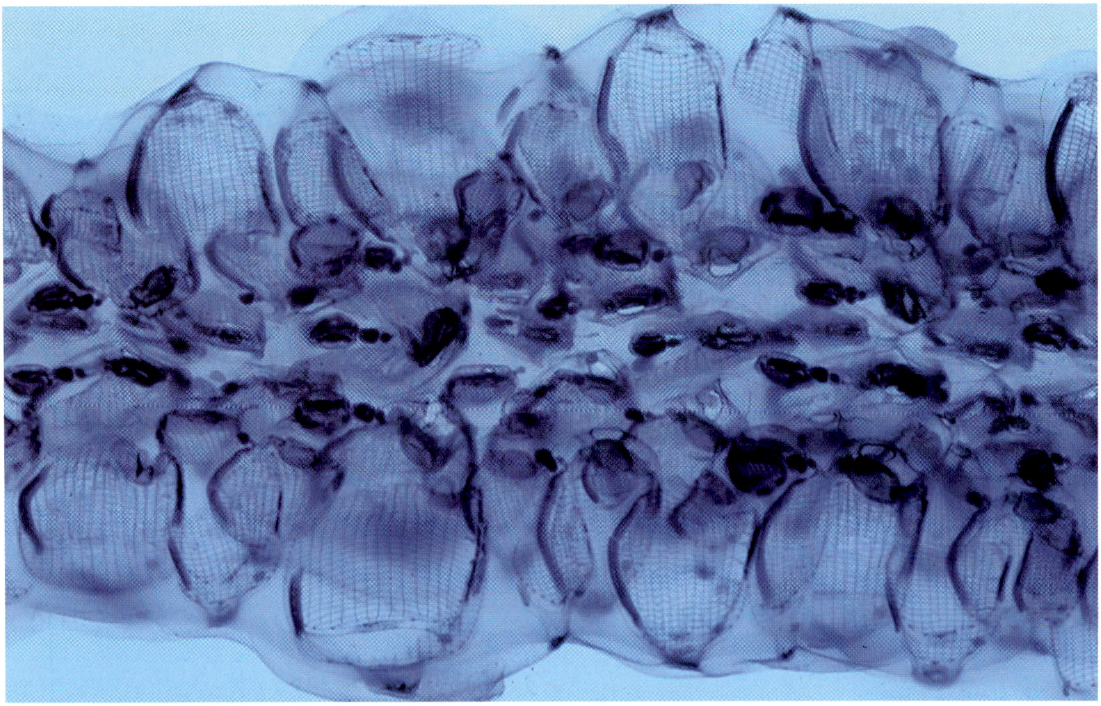

Salps sometimes form large band-or oval-shaped colonies, however, the zooids are only very small. This is a close-up photo of some of the zooids of *Pyrosoma atlanticum*. Photo: R. van Soest

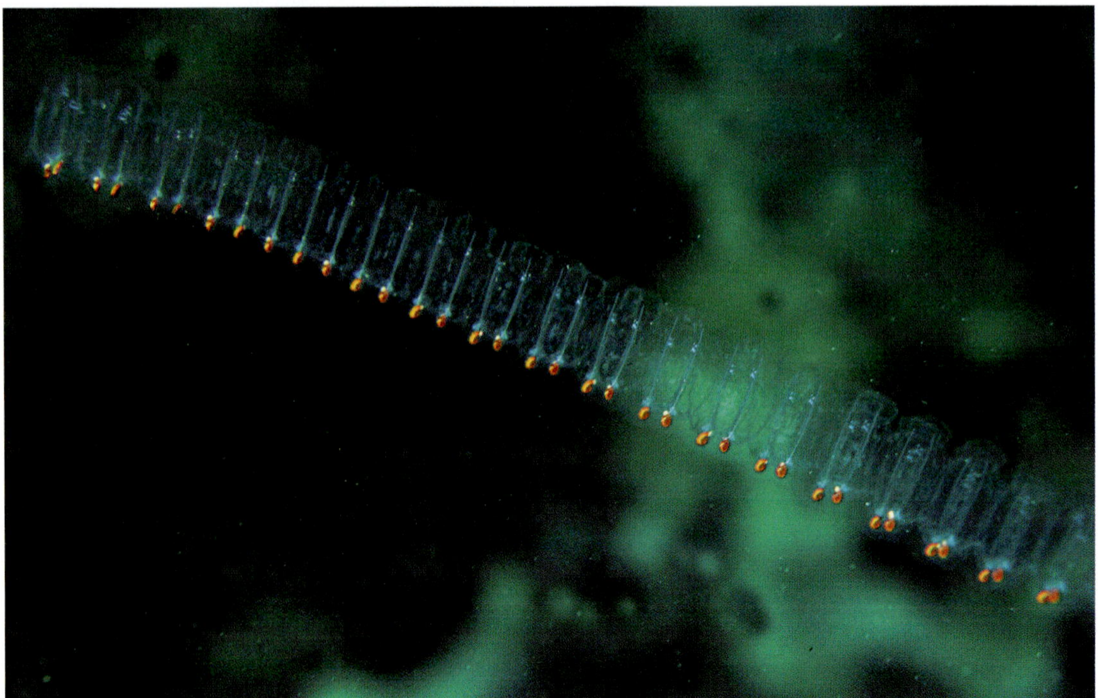

Chain of salps photographed at Phi-Phi island, Thailand. The chain contains several sexual individual salps (*Pegea* sp.) that are produced by budding from the solitary asexual form shown.

are three other well known genera of salps (for further details see van SOEST 1972, 1973a, 1973b, 1974a, 1974b, 1975a, 1975b; BONE, 1998).

Class Appendicularia contains about 70 species, all small, transparent tailed organisms, possibly the most complex of all tunicates (RUPPERT & BARNES, 1994; FENAUX, 1998). Larvaceans are found, often in very high numbers in the surface marine plankton all over the world. A population density of 25,620 individuals of the species *Oikopleura dioica* per cubic meter of water was reported from British Columbia (RUPPERT & BARNES, 1994). These tailed tunicates feed on very small phytoplankton (nannoplankton) that is not being utilised by many other plankton feeders. Food particles are filtered out of the water that is driven by the animal's tail through a sieve in the mucous "house" that envelopes the trunk or encloses the entire body. One such animal can filter daily 250,000 phytoplankton cells from 300 ml of water. Like ascidian larvae, the tail contains notochord-like cells, although in Appendicularia there is no larva, and the tail is retained through life. As in Thaliacea, certain species are luminescent, the light being produced by bodies included in the mucous house of the organism in a process that is not completely understood (GALT & FLOOD, 1998).

Pegea confoederata. **Photo: R. van Soest**

Band-shaped salp colony (aggregate, sexual phase) in the open sea off eastern Australia. **Photo: M. Cufer**

Below: Tailed tunicates, seen here are two different species of the genus *Oikopleura*, left probably *O. cornutogastra* and right probably *O. fusiformis*. **Photos: R. Brons**

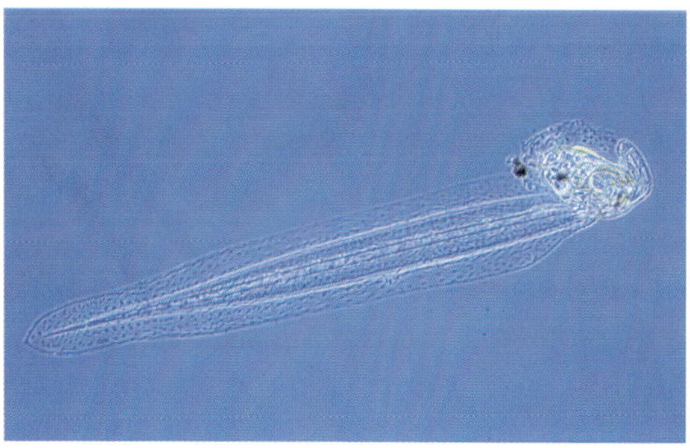

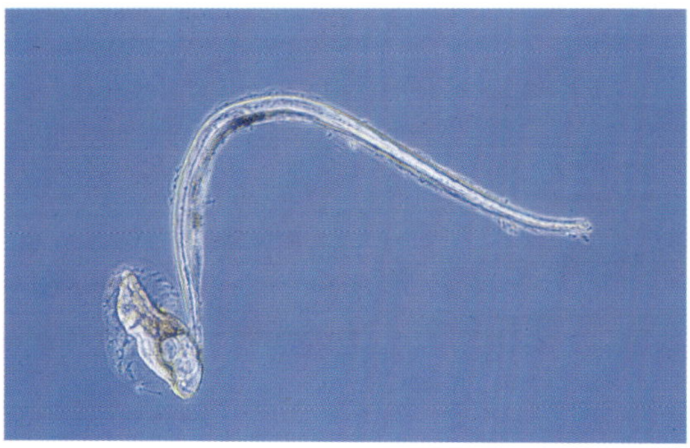

A Lancelet (*Branchiostoma* sp.) is resting on the bottom at 8 metres depth close to the harbour of Stavanger, Norway. The white dots are the gonads and the beige notochord is also visible. Photo: E. Svensen

Subphylum Cephalochordata - Lancelets

This subphylum represents only a very small part of phylum Chordata, only two genera of lancelets with less than 20 species being known. These genera occur in temperate (including Norwegian waters), as well as in subtropical and tropical seas. They live buried in the bottom sediments, often with their posterior end pointing upward.

Lancelets have neither a skull nor jaws; and brain, eyes and other sensory organs are absent from the head. They differ from invertebrates in having a hollow dorsal nerve chord, a perforated pharynx, a rod of notochord cells, well developed myomeres and a simple vascular system. Species of the genus *Branchiostoma* have a lip equipped with tentacular cirri, instead of jaws. These cirri form a strainer in front of the mouth opening, so that large particles are excluded and only suitable food organisms can pass through into the mouth. The water current also flows through numerous gill clefts that absorb oxygen. An endostyle, as found in tunicates (ascidians, thaliaceans and appendicularians) produces

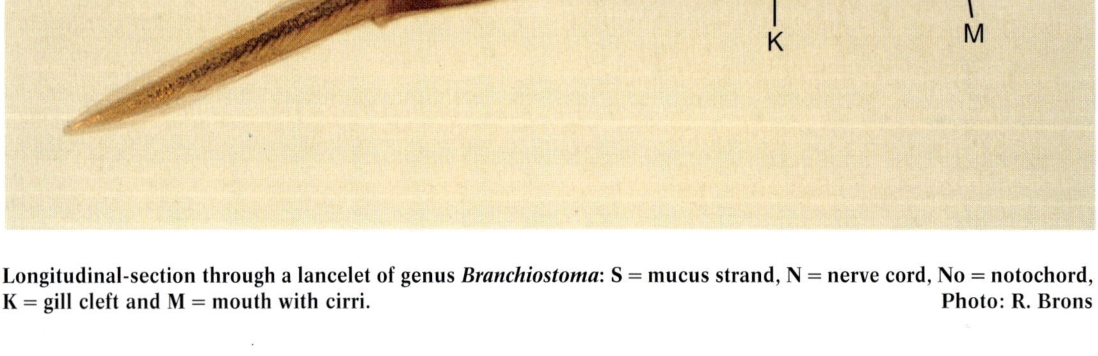

Longitudinal-section through a lancelet of genus *Branchiostoma*: S = mucus strand, N = nerve cord, No = notochord, K = gill cleft and M = mouth with cirri. Photo: R. Brons

mucus that enmeshes the food, which is moved into the oesophagus and subsequently to the remainder of the gut for digestion.

Right: Cross-section through a lancelet of genus *Branchiostoma*:
No = notochord
N = nerve cord
Ep = epidermis
My = myomere (a muscle segment)
K = gill bars separated by gill slits
En = endostyle
Go = gonads
L = liver
S = mucus strand
Photo: R. Brons

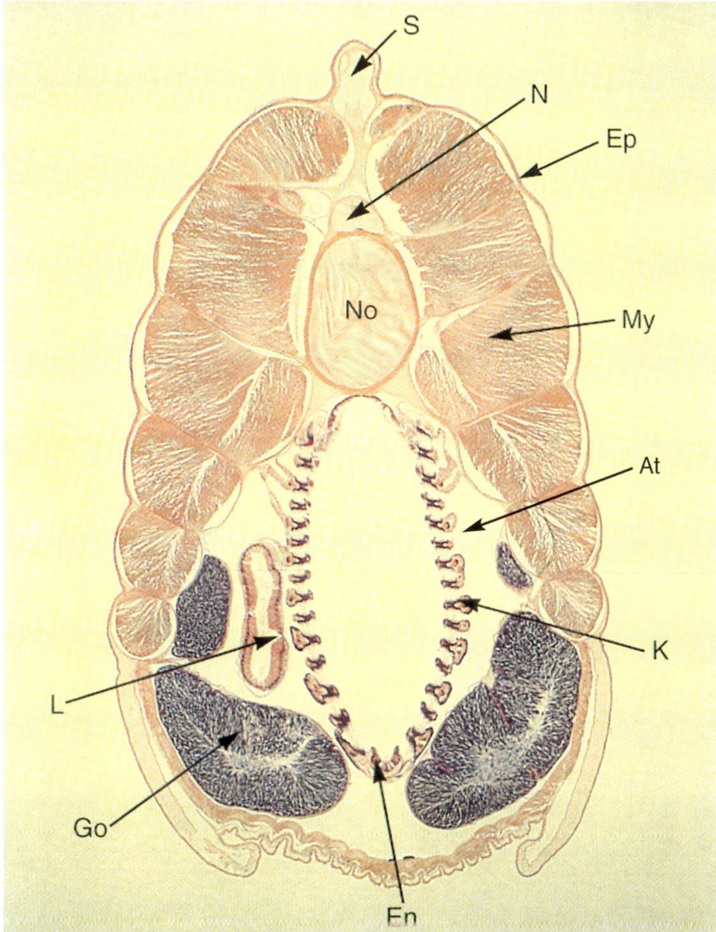

Finally

Finally we have reached the end of the invertebrates. From the single-celled organisms and sponges we have covered them all and ended where the vertebrates begin. We do hope that you as readers have enjoyed the journey and have picked up some useful information along the road.

The picture to the right, which was taken in Indonesia, can very well stand as a representative illustration to most of the invertebrate chapters. If you look closely you will among the variety of algae discover tiny foraminiferans, a lot of sponges, tiny hydrozoans, bryozoans, a colourful nudibranch and colonial tunicates. The area photographed is no bigger than 50 square centimetres. As visitors to the tropical corals reefs we are stunned by their beauty, magnificent colours and enormous diversity. The organisms found on the reefs are unique and it is our responsibility as enthusiasts and nature lovers to treat them with great care and to protect the natural habitat.

Sponges, sea squirts, and algae growing on a root of a mangrove in a lake with brackish water on Kataban Island, Kalimantan, Borneo. **Foto: Dr. M. Bergbauer**

List of References

ABBOTT, R. T. (1960): The genus Strombus in the Indo-Pacific. *Indo-Pacific Mollusca*. **1** (2): 33 144, Pls. 11-117.

ABBOTT, R. T. (1961): The genus Lambis in the Indo-Pacific. *Indo-Pacific Mollusca*. **1** (3): 147-174, Pls. 118-134.

ABBOTT, R. T. (1972): *Kingdom of the Seashells*. Rutledge Books, New York, USA.

ABBOTT, R. T. (1991): *Seashells of South East Asia*. Tynron Press, Thornhill, Scotland.

ABBOTT, R. T. & S. P. DANCE (1982): *Compendium of Seashells*. E. P. Dutton, New York, USA.

ABBOTT, R. T. & P. A. MORRIS (1995): *A Field Guide to Shells - Atlantic and Gulf Coasts and the West Indies*. 4th ed. The Peterson Field Guide Series. Houghton Mifflin Company, Boston, New York, USA.

ALLDREDGE, A. L. (1978): Appendicularians. *Scientific American* **235**: 95-102.

ALLDREDGE, A. L. (1972): Abandoned larvacean houses: A unique food source in the pelagic environment. *Science* **177**: 885-887.

ALLEN, G. R. (1978): *Butterfly and Angelfishes of the World*, Vol. 2. Mergus Verlag, Melle, Germany.

ALLEN, G. R. & R. SREENE (1994): *Indo-Pacific Coral Reef Field Guide*. Tropical Reef Research, Singapore.

ANDREWS, J. (1994): *A Field Guide to the Shells of the Florida Coast*. Gulf Publishing Company, Huston, Texas, USA.

ANSARI, Z. A. & B. S. INGOLE (1983): Meiofauna from some sandy beaches of Andaman Islands. *Indian Jour. Mar. Sci.* **12**: 245-246.

ARNOLD, J. M. & B. A. CARLSON (1986): Living Nautilus Embryos: Preliminary Observations. *Science* **232**: 73-76.

AYLING, A. M. & A. L. AYLING (1987): *Ningaloo Marine Park: preliminary fish density and habitat survey with information on coral damage due to Drupella cornus grazing*. Report prepared for the Department of Conservation and Land Management, Western Australia.

AYLING, A. M. & A. L. AYLING (1992): Preliminary information on the effects of Drupella spp. Grazing on the Great Barrier Reef, in Drupella cornus: a symposium, ed. S. Turner, *Western Australia Department of Conservation and Land Management, CALM Occasional Paper* No. 3/92.

BAENSCH, H. A & H. DEBELIUS (1992): *Meerwasser Atlas*, Vol. 1. Mergus Verlag, Melle, Germany.

BAIRD, A. (1999): A large aggregation of Drupella rugosa Following the Mass Bleaching of Corals on the Great Barrier Reef. *Reef Research* **9** (2).

BANDEL & KADOLSKY (1982): *The Veliger* **25**: 1-42.

BARNES, R. D. (1980): *Invertebrate Zoology*, 4th. ed., Saunders College, Philadelphia, USA.

BEESLEY, P., G. J. B. ROSS & A. WELLS (eds.) (1996): *Mollusca. Fauna of Australia*. Vol. 5. Australian Biological Resource Study, Canberra, ACT, Australia.

BERGER, W. H. (1978): Deep-sea carbonate: Pteropod distribution and aragonite compensation depth. *Deep-Sea Research* **25**: 447-452.

BERRILL, N. J. (1950): The Tunicata. *Ray Soc. Publ.* (133): 1-354.

BERTSCH, H. & S. JOHNSON (1981): *Hawaiian Nudibranches*. Oriental Publishing Company, Honolulu, Hawaii, USA.

BEU, A. G. & W. O. CERNOHORSKY (1986): Taxonomy of gastropoda of the families Ranellidae (= Cymatiidae) and Bursidae, Pt. I. *N. Z. Jour. Zool.* **13**: 241-266.

BETZER, P. H., R. H. BYRNE, J. G. ACKER, C. S. LEWIS, R. R. JOLLEY & R. A. FEELY (1984): The oceanic carbonate system: A reassessment of biogenic controls. *Science* **226**: 1074-1077.

BIELER, R. (1990): Haszprunars «Clado-Evolutionary» Classification of the Gastropoda - a Critique. *Malacologia* **31** (2):371-380.

BIELER, R. (1992): Gastropod Phylogeny and Systematics. *Ann. Rev. Ecol. Sys.* **23**: 311-338.

BIRKHOLZ, J. (1983a): Parasitische Schnecke Spurilla neapolitana. *das Aquarium*. **17** (7): 375-376.

BIRKHOLZ, J. (1983b): Wir stellen vor: Botryllus sp. *das Aquarium* **17** (10): 540-541.

BLACK, R. & M. S. JOHNSON (1994): Growth rates in outbrake populations of the corallivorous gastropod Drupella cornus (Röding 1798) at Ningaloo Reef, Western Australia. *Coral Reefs* **13** (3): 145-150.

BLACKBURN, M. (1956): Sonic scattering layers of heteropods. *Nature* (177): 374-375.

BOMLEY, R. G. (1975): Comparative analysis of fossil and recent echinoid bioerosion. *Paleontology* **18** (4): 725-739.

BONE, Q. (1998): *The Biology of Pelagic Tunicates*. Oxford University Press, Oxford, UK.

BOSCH, T., S. P. DANCE, R. G. MOOLENBECK & P. G. OLIVER (1995): *Seashells of Eastern Arabia*. Motivate Publishing, Dubai, Abu Dhabi and London.

BRALEY, R. D. (ed.). (1984): The giant clam: a hatchery and nursery culture manual. *ACAIR Monograph No. 15*: 144 pp.

BRUSCA & BRUSCA (1990): *Invertebrates*. Sinauer Associates, Sunderland, Massachusetts, USA.

BRYAB, P. J., W. Y. YOSHIDA, J. B. MCCLINTOCK & B. J. BAKER (1995): Ecological role for pteroenone, an antifeedant from the conspicuous Antarctic pteropod Clione antarctica (Gymnosomata: Gastropoda). *Marine Biology* (122): 271-277.

BUCHSBAUM, R. (1976): *Animals without Backbones*. The University of Chicago Press, Chicago, USA.

CALUMPONG, H. P. (ed.) (1992): The giant clam: an ocean culture manual. *ACAIR Monograph No. 16*.

CANNON, L. R. G. & H. SILVER (1986): *Sea Cucumbers of Northern Australia*. Queensland Museum, South Brisbane, Australia.

CANNON, L. R. G. & H. SILVER (1994): *North Australian Sea Cucumbers*. The World Biodiversity Database CD-ROM. Springer Electronic Media, Heidelberg, Germany.

CARLSON, B. A. (1979): Chamberes Nautilus - a new challenge for aquarists. *FAMA* **2** (8): 48-67.

CARLSON, B. A. (1983): Hatching and Rearing the Flash-Back Cuttlefish, Sepia latimanus. *AAZPA Annual proceedings* (1983): 217-219.

CARLSON, B. A. (1987): Collection and Aquarium Maintenance of Nautilus. In: SAUNDERS, W. B. & N. H. LANDMAN (eds.): *The biology and Paleobiology of a living fossil*: 563-583. Plenum Publishing Corporation.

CARLSON, B. A. & L. R. TAYLOT (1981): Holacanthus griffisi - A New Species of Angelfish from the Central Pacific Ocean. *FAMA* **4** (5): 8-11.

CARLSON, B. A., J. N. MCKIBBEN & M. V. DEGRUY (1984): Telemetric Inverstigation of Vertical Migration of Nautilus belauensis in Palau. *Pacific Science* **38** (3): 183-188.

CARLSON, B. A., M. L. AWAI & J. M. ARNOLD (1992): Hatching and Early growth of Nautilus belauensis and Implications on the Distribution of the Genus Nautilus. *Proceedings of the Seventh International Coral reef Symposium, Guam*, Vol. 1: 587-592.

CARROL, D. J. & S. C. KEMPF (1990a): Laboratory Culture of the Aeolid Nudibranch Berghia verrucicornis (Mollusca, Ophistobranchia): Some Aspects of Its Development and Life History. *Biol. Bull.* **179**: 243-253.

CARROL, D. J. & S. C. KEMPF (1990b): On Aiptasia eating Nudibranch. *Biol. Bull.* **179**: 243-253.

CARROL, D. J. & S. C. KEMPF (1994): Changes occur in the general nervous system of the nudibranch Berghia verrucicornis (Mollusca, Ophistobranchia) during development. *Biol. Bull.* **186** (2): 243-253.

CAYCEDO, I. E. (1979): Observaciones de los equinodermos en las Islas del Rosario. *Ann. Inst. Invest. Mar. de Punta de Betin* **11**: 39-47.

CERNOHORSKY, W. O. (1967a): *Marine Shells of the Pacific - 1.* Pacific Publications, Sydney, Australia.

CERNOHORSKY, W. O. (1967b): *Marine Shells of the Pacific - 2.* Pacific Publications, Sydney, Australia.

CERNOHORSKY, W. O. (1967c): *Marine Shells of the Pacific - 3.* Pacific Publications, Sydney, Australia.

CHEN, C. (1964): Pteropod ooze from the Bermuda Pedestal. *Science* **144** (3614): 60-62.

CHESTER, C. (1993): Comparative feeding biology of Acteocina canaliculata (Say, 1826) and Haminoea solitaria (Say, 1822) (Ophistobranchia: Cephalaspidea). *American Malacol. Bull.* **10** (1): 93-101.

CHOI, D. R. (1984): Ecological succession of reef cavity dwellers (ceolobites) in coral rubble. *Bull. Mar. Sci.* **35** (1): 72-79.

CLARK, K. B. (1994): Ascoglossan (= Sacoglossan) molluscs in the Florida Keys: rare marine invertebrates at special risk. *Bull. Mar. Sci.* **54** (3): 900-916.

CLARK, S. & A. J. EDWARDS (1995): Coral transplantation as an aid to reef rehabilitation: evaluation of a case study in the Maledive Islands. *Coral Reefs* **14** (4): 201-213.

CLARK, A. M. & F. W. E. ROWE (1971): *Monograph of Shallow-Water Indo-West Pacific Echinoderms.* British Museum (Natural History), London, UK.

CLARK, A. M. & M. E. OWNEY (1992): *Starfishes of the Atlantic.* Chapmann and Hall, London, UK.

COBB, W. D. (1939): The part of Allah. *Nat. Hist. N. Y.* **44** (4): 197-202.

COLEMAN, N. (1987): *Australia Sea Life south of 30 °S.* Doubleday Australia Pty Limited, Sydney, Australia.

COLEMAN, N. (1989): *Nudibranches of the South Pacific.* Nevil Coleman's Sea Australia Resource Centre, Springwood, QLD, Australia.

COLEMAN, N. (1991): *Encyclopedia of Marine Animals.* Blandford, London, UK.

COLEMAN, N. (1994): *Sea Stars of Australia and their Relatives.* Neville Coleman's Underwater Geographic Pty. Ltd., Springwood, Australia.

COLIN, P. L. (1978): *Caribbean reef invertebrates and plants.* TFH Publications, Inc., Neptune City, N. J., USA.

COLIN, P. L. & C. ARNESON (1995): *Tropical Pacific Invertebrates.* Coral Reef Press, California, USA.

COPLAND, J. W. & J. S. LUCAS. (1988): Giant clams in Asia and the Pacific. *ACIAR Monograph No. 9*: 274 pp.

CRAME, J. A. (1986): Late Pleistocene molluscan assemblages from the coral reefs of Kenya coast. *Coral Reefs* **4** (3): 183-196.

CRAWFORD, C. & W. NASH. (1986): Giant Clams. *Oceanus* **29** (2): 60-61.

CUMMING, R. L. (1988): Pyramidellid parasites of giant clam marinculture systems. In: COPELAND, J. W. & J. S. LUCAS (eds.): *Giant Clams in Asia and the Pacific.* ACIAR Monograph No. 9.

DALZELL, P., P. COPRUZ, R. GANADAN & D. PAULY (1987): Estimation of the maximum sustainable yield and maximum economic rent from the Philippine small pelagic fisheries. *Bureau of fisheries and Aquatic resources, Tech. Rep.* **10**: 1-23.

DANCE, P. (1974): *The Encyclopedia of Shells.* Blandford Press Ltd., London, UK.

DANCE, P. (1977): *Das grosse Buch der Meeresmuscheln.* Eugen Ulmer Verlag, Stuttgart, Germany.

DAVIDSON, A. (1977): *Seafood of South-East Asia.* Federal Publications, Singapore.

DAVIDSON, A. (1981): *Mediterranean Seafood* 2nd ed. Penguin Cookery Library, Penguin Books Ltd, Harmondsworth, Middlesex, UK.

DEBELIUS, H. (1998): *Nudibranchs and Sea Snails, Indo-Pacific Field Guide.* IKAN-Unterwasserarchiv, Frankfurt, Germany.

DELBEEK, J. C. & J. SRUNG (1994): *The Reef Aquarium* Vol. 1. Ricordea Publishing, Miami, USA.

DENTON, G. R. W. & L. WINSOR (1986): Giant Clams as Pollution Indicators. *Oceanus* **29** (2): 63.

EARLE, J. L. & R. L. PYLE (1997): Hoplolatilus pohlei, a new species of sand tilefish (Perciformes: Malacanthidae) from the deep reefs of the D'Entrecasteaux Islands, Papua New Guinea. *Copeia* 1997 (2): 383-387.

ELLIS, R. (1996): *Deep Atlantic.* Alfred A. Knopf, N.Y., USA

ELLIS, R. (1998): *The search for the Giant Squid.* The Lyons Press, N. Y., USA.

ERHARDT, H. & H. MOOSLEITNER (1995): *Meerwasser Atlas* Band 2 and 3. Mergus Verlag, Melle, Germany.

FENAUX, R. (1998): Anatomy and functional morphology of the Appendicularia,pp.25-34. In: BONE, Q. (ed.): *The Biology of Pelagic Tunicates.* Oxford University Press, Oxford, UK.

FISHER, C. R., W. K. FITT & R. K. TRENCH (1985): Photosyntesis and respiration in Tridacna gigas as a function of irradiance and size. *Biol. Bull.* **169**: 230-245.

FITT, W. K., C. R. FISHER & R. K. TRENCH (1984): Larval biology of tridacnid clams. *Aquaculture* **39**: 189-195.

FITT, W. K., C. R. FISHER & R. K. TRENCH (1986): Contribution of the symbiotic dinoflagellate Symbiodinium microadriaticum to the nutrition, growth and survival of larval and juvinile tridacnid clams. *Aquaculture* **55**: 5-22.

FITT, W. K. & R. K. TRENCH (1981): Spawning, development, and acquisition of zooxanthellae by Tridacna squamosa (Mollusca, Bivalvia). *Biol. Bull.* **161**: 213-235.

FOREY, P. L. (ed.) (1981): *The evolving Biosfere. Chance, Change and Challenge.* BMNH and Cambridge University Press, London, U.K.

FOSSÅ, S. A. (1995): Die Zebra-Schnecke: Ein gestreifter Algenfresser von den Bahamas. *das Aquarium* **29** (3): 25-27.

FOSSÅ, S. A. & A. J. NILSEN (1996): *The Modern Coral Reef Aquarium,* Vol. 1. Birgit Schmettkamp Verlag, Bornheim, Germany.

FOSSÅ, S. A. & A. J. NILSEN (1998): *The Modern Coral Reef Aquarium,* Vol. 2. Birgit Schmettkamp Verlag, Bornheim, Germany.

FOSSÅ, S. A. & A. J. NILSEN (2000): *The Modern Coral Reef Aquarium.* Vol. 3. Birgit Schmettkamp Verlag, Bornheim, Germany.

FOSSÅ, S. A. & A. J. NILSEN (2001): *Korallenriff-Aquarium,* Vol. 1, 4th. ed., Birgit Schmettkamp Verlag, Bornheim, Germany.

FRANKBONER, P. J. (1971): Intracellular digestion of symbiotic zooxanthellae by host amoebocytes in giant clams (Bivalvia: Tridacnidae), with a note on the nutritinoal role of the hypertrophied siphonal epidermis. *Biol. Bull.* **141**: 222-234.

FREEMAN, W. H. & B. BRACEGIRDLE (1971): *An Atlas of Invertebrate Structure.* Heineman Educational Books, London, UK.

GAGE, J. D., M. PEARSON, A. M. CLARK, G. L. G. PATERSON & P. A. TYLER (1983): Echinoderms of the Rockall Trough and adjacent areas. I: Crinoidea, Asteroidea and Ophiuroidea. *Bull. Brit. Mus. (Nat. Hist.) Zool.* **45**: 263-308.

GALT, C. P. & P. R. FLOOD (1998): Bioluiminescence in the Appendicularia. In: BONE, Q. (ed.): *The Biology of Pelagic Tunicates*: 215-230. Oxford University Press, Oxford, UK.

GARRARD, T. A. (1972): *Journ. Malacolog. Soc. Australia* **2** (3): 267-338.

GARRARD, T. A. (1974): *Journ. Malacolog. Soc. Australia* **3** (1): 15-17.

GARRARD, T. A. (1978): *Journ. Malacolog. Soc. Australia* **5** (3-4): 195-200.

GEORGE, D. & J. GEORGE (1979): *Marine life.* Harrap, London, UK.

GILL, A. C., R. L. PYLE & J. L. EARLE (1996): Pseudochromis ephippiatus, new species of dottyback from southeastern Papua New Guinea (Teleostei: Perciformes: Pseudochromidae). *Revue fr. Aquariol.* **23** (3-4): 97-100.

GILMER, R. W. (1974): Some aspects of feeding in thecosomatous pteropod molluscs. *Jour. Exp. Mar. Biol. Ecol.* **15**: 127-144.

GILMER, R. W. & G. R. HARBISON (1986): Morphology and field behaviour of pteropod molluscs: Feeding methods in the families Cavoliniidae, Limacinidae and Peraclididae (Gastropoda: Thecosomata). *Marine Biology* **91**: 47-57.

GOMEZ, E. D., P. M. ALINO, H. T. YAP & W. Y. LICUANAN (1994): A review of the status of Philippine reefs. *Mar. Poll. Bull.* **29**: 62-68.

GOREAU, T. F., N. I. GOREAU & C. M. YONG (1973): On the utilization of photosyntetic products from zooxanthellae and dissolved amino acid in Tridacna maxima cf. elongata (Mollusca: Bivalvia). *Jour. Zool. (Lond.)* **169**: 417-454.

GOSLINER, T. M. (1995): The genus Thuridilla (Ophistobranchia: Elysiidae) from the tropical Indo-Pacific, with revision of the phylogeny and systematics of the Elysiidae. *Proc. Calif. Acad. Sci.* **49** (1): 1-54.

GOSLINER, T. M., D. W. BEHRENS & G. C. WILLIAMS (1996): *Coral Reef Animals of the Indo-Pacific.* Sea Challengers, Monterey, Ca., USA.

GRIFFITHS, D. J & L.-V. THINH (1983): Transfer of photosynthetically fixed carbon between thr procaryot green alga Prochloron and its ascidian host. *Aust. J. Mar. Freshw. Res.* **34**: 431-440.

GROBER, M. S. (1989): Starlight on the reef. *Natural History* **10**: 72-80.

GROSSKOPF, J. (1994): *Das Korallenriff im Wohnzimmer* 4th ed.. Dähne Verlag, Ettlingen, Germany.

GUILLE, A., P. LABOUTE & J.-L. MENOU (1986): *Guide des étoiles de mer, oursins et autres échinodermes du lagon de Nouvelle-Calédonie.* Éditions de l'Orstom, Paris, France.

HAMMER, W. M., L. P. MADIN, A. L. ALLDREDGE, R. W. GILMER & P. P. HAMMER (1975): Underwater observations of gelatinous zooplankton: Sampling problems, feeding biology and behaviour. *Limnology and Oceanography* **20**: 907-917.

HARBISON, G. R. & R. W. GILMER (1992): Observations on the swimming and buoyancy of Cymbulia peroni (Gastropoda: Thecosomata) made from a submersible. *Jour. Mar. Biol. Assoc. UK* **72**: 435-446.

HARRIOT, V. J. & D. A. FISK (1988): Coral transplantation as reef management option. *Proc. 6th Internat. Coral Reef Symp., Australia*: 375-379.

HASZPRUNAR, G. (1985): The Heterobranchia - a new concept in the phylogeny of the higher Gastropoda. *Zeitschr. Phylog. Syst. Evol.* **23** (1): 15-37.

HASZPRUNAR, G. (1988): A preliminary phylogenetic analysis of the streptoneurous gastropods. *Malacological Review*, Supplement 4: 7-16.

HAWKINS, C. J., P. KOTT, D. L. PARRY & J. H. SWINEHART (1983.): Vanadium content and oxidation state related to Ascidian phylogeny. *Comp. Biochem. Physiol.* **76B** (3):555-558.

HAY, M. E., J. R. PAWLIK, J. EMMET DUFFY & W. FENICAL (1989): Seaweed-herbivore-predator interactions: host-plant specialization reduces predation on small herbivores. *Oecologia* **81**: 418-427.

HEEGER, T., M. CASHMAN & F. SOTTO (1999): Coral farming as alternative livelihood, for sustainable natural resource management and coral reef rehabilitation. *Proc. Ocean. Internat. Pacific Rim, Singapore*: 171-185.

HEEGER, T., F. SOTTO, J. GATUS & C. LARON (2000): *Community-based coral farming for reef rehabilitation, biodiversity conservation and as a livelihood option for fisher folk.* Proc. of Worhshop on Responsible Aquaculture Development in Southeast Asia (ADSEA), Seafdec, Philippines.

HEEZEN, B. C. & C. D. HOLLISTER (1971): *The Face of the Deep.* Oxford Univ. Press, Oxford, UK.

HENDLER, G. (1979): Reproductive periodicity of ophiuroids (Echinodermata: Ophiuroidea) on the Atlantic and Pacific coasts of Panama. In: S. E. STANCYK (ed.): *Reproductive Ecology of Marine Invertebrates*: 145-156. Univ. of South Carolina Press,USA.

HENDLER, G., J. E. MILLER, D. L. PAWSON & P. M. KIER (1995): *Sea Stars, Sea Urchins, and Allies.* Smithsonian Institution Press, Washington, USA.

HERTLER, C. (1996): Die Tunicaten und der Ursprung der Wrbeltiere. Eine historisch-metodische Fallstudie. *Natur und Museum* **126** (2): 55-68.

HESLINGA, G.A. & T. C. WATSON (1985): Recent advances in giant clam marine-culture. *Proc. Int. Congr. Coral Reefs* **5**: 531-537.

HESLINGA, G. A. & W. K. FITT (1987): The Domestication of Reef-dwelling Clams. *BioScience* **37** (5): 332-339.

HESLINGA, G. A., F. E. PERRON & O. OORAK (1984): Mass culture of giant clams (f. Tridacnidae) in Palau. *Aquaculture* **39**: 197-215.

HESLINGA, G. A., T. WATSON & T. ISAMU (1990): *Giant Clam Farming.* Pacific Fisheries Development Foundation (NMFS/NOAA), Honolulu, Hawaii, USA.

HICKMANN, C. S. (1988): Archaeogastropod evolution, phylogeny and systematics: a re-evaluation. *Malacological Review*, Supplement 4: 17-34.

HINTON, A. (1972): *Shells of New-Guinea and the Central Indo-Pacific.* Robert Brown and Associates, Port Moresby, Australia.

HINTON, A. (1978): *Guide to Australian Shells.* Robert Brown and Associates, Port Moresby, Australia.

HOFF, F. H. & T. W. SNELL (1989): *Plankton culture manual.* 2nd ed. Florida Aqua Farms, Florida, USA.

HOSHINO, Z. & T. NISHIKAWA (1985): Taxonomic studies of Ciona intestinalis (L.) and its allies. *Publs. Seto mar. biol. Lab.* **30** (1-3): 69-79.

HOUBRICK, J. R. (1987): *Smithsonian Contributions to Zoology* **445**: 1-57.

HOUBRICK, J. R. (1988): *Malacological Review*, Supplement 4: 88-128.

HOUBRICK, J. R. & V. FRETTER (1969): Some aspects of the functional anatomy and biology of Cymantium and Bursa. *Proc. Malacol. Soc. London* **38**: 415-429.

HUMANN, P. (1994): *Reef Creatures* 4th ed. New World Publications, Inc., Florida, USA.

ILIFFE, T. M. & J. S. PEARSE (1982): Annual and lunar reproductive rhytms of the sea urchin Diadema antillarum Philippi in Bermuda. *Internat. Jour. Invertebr. Repr.* **5**: 139-148.

ITZKOWITZ, M. & T. KOCH (1991): Relationships between damselfish egg loss and brittlestars. *Bull. Mar. Sci.* **48**: 164-166.

JACKSON, J. (1977): Competition on marine hard substrata: the adaptive significance of solitary and colonial strategies. *The American Naturalist*, Vol. III, No. 980.

JAMIESON, B. J. M. & L. J. NEWMAN (1989): The phylogenetic position of the heteropod Atlanta gaudichaudi Souleyet, 1852 (Gastropoda, Mollusca), a spermatological investigation. *Zoological Scripta* **18**: 269-278.

JAPLAN, E. H. (1982): *A Field Guide to Coral Reefs, Caribbean and Florida.* Houghton Mifflin Company, Boston, USA.

JAUBERT, J. (1989): An integrated nitrifying-denitrifying biological system capable of purifying sea water in a closed circuit aquarium. *Bulletin de l'Institut oceanographique, Monaco*, Spécial No. 5: 101-106.

JENSEN, K. (1981): Observations on feeding methods in some Florida Ascoglossans. *Journ. Moll. Stud.* **47**: 190-199.

JENSEN, K. (1982): Anatomy of some Indo-Pacific Elysidae (Ophistobranchia: Sacoglossa (= Ascoglossa)), with a discussion of the generic division and phylogeny. *Journ. Moll. Stud.* **58**: 257-296.

JENSEN, K. (1983): Factors affecting feeding selectivity in herbivorous Ascoglossa (Mollusca, Ophistobranchia). *Jour. Exp. Mar. Biol. Ecol.* **66**: 135-148.

JENSEN, K. (1989): Learning as a factor in diet selection by Elysia viridis (Montagu) (Ophistobranchia). *Jour. Moll. Stud.* **55**: 79-89.

JENSEN, K. (1992): Anatomy of some Indo-Pacific Elysidae with discussion of the Generic Division and Phylogeny. *Proc. Malacol. Soc. London* **58**: 257-296.

JENSEN, K. (1993a): Morphological adaptations and plasticity of radular teeth of the Sacoglossa (= Ascoglossa) (Mollusca: Ophistobranchia) in relation to their food plants. *Biol. Jour. Linnean Soc.* **48**: 135-155.

JENSEN, K. (1993b): Sacoglossa (Mollusca: Ophistobranchia) - specialist Herbivorous and partial Predators: Intergrating Ecological, physiological and Morphological data. *Proc. first Intern. Conf. Mar. Biol. of Hongkong and the South China Sea, Hong Kong (1990)*: 437-458.

JENSEN, K. (1993c): Evolution of Buccal Apparatus and Diet Radiation in the Sacoglossa (Ophistobranchia). *Bull. Malacol.* **29** (5-8): 142-172.

JENSEN, K. (1994): Behavioural adaptations and diet specificity of sacoglossan ophistonbranchs. *Ethology, Ecology and Evolution* **6**: 87-101.

JOHNSON, C. W. (1930): The giant clam (Tridacna gigas). *The Nautilus* **43** (4): 141.

JOHNSON, M. S. & R. L. CUMMING (1995): Genetic distinctness of three widespread and morphologically variable species of Drupella (Gastropoda, Muricidae). *Coral Reefs* **14** (2): 71-78.

KABAT (1990): Predator ecology of naticid gastropods with a review of shell boring predation. *Malacologia* **32** (1): 155-193.

KAPLAN, E. H. (1982): *A Field Guide to Coral Reefs, Caribbean and Florida.* Houghton Mifflin Company, Boston, USA.

KARLSON, R. H. (1983): Disturbance and monopolization of a spatial resource by Zoanthus sociatus (Coelenterata: Anthozoa). *Bull. Mar. Sci.* **33**: 118-131.

KASTL, J. (1993): Leicht zu vermehrende und nützliche Restevertilger im Korallenrlff-Aquarium. *das Aquarium* **27** (9): 29-30.

KAWAGUTI, S. (1950): Observations on the heart shell, Corculum cardissa (L.) and its associated zooxanthellae. *Pacific Science* **4** (1): 43-49.

KEMPF, S. C. (1991): A «Primitive» Symbiosis between the Aeolid Nudibranch Berghia verrucicornis (A. Costa, 1867) and a Zooxanthella. *Jour. Moll. Stud.* **57**: 75-85.

KEMPF, S. C. & M. BRITTSAN (1996): Berghia verrucicornis, a Nudibranch Predator of the Aquarium «Weed» Anemone Aiptasia. *Proc. Americ. Zoo and Aquarium Assoc. Ann. Conf.*.

KERSTITCH, A. (1989): *Sea of cortez Marine Invertebrates - A Guide for the Pacific Coast, Mexico to Equador.* Sea Challengers, Monterey, Ca., USA.

KLUMPP, D. W. & J. S. LUCAS (1994): Nutritional ecology of the giant clams Tridacna tevoroa and T. derasa from Tonga: influence of light on filter-feeding and photosynthesis. *Mar. Ecol. Prog. Ser.* **107**: 147-156.

KNOP, D. (1996): *Giant Clams: A Comprehensive Guide to the Identification and Care of Tridacnid Clams.* Dähne Verlag, Ettlingen, Germany.

KNOWLTON, N., J. C. LANG, M. C. ROONEY & P. CLIFFORD (1981): Evidence for delayed mortality in hurricane-damaged Jamaican staghorn corals. *Nature* **294**: 251-252.

KOBLUK, D. (1988): *Cryptic faunas in reefs: Ecology and geologic importance.* The Society of Economic paleontologists and Mineralogists. PALAIOS, 1988, V. 3: 379-390.

KOTT, P. (1969): Antarctic Ascidiacea IV. A monographic account of the known species based on specimens collected under U.S. Government auspecies (1947 to 1963). *Antartic Research Series* **13** : i-xv, 1-239.

KOTT, P. (1972): The ascidians of South Australia I. Spencer Gulf, St. Vincent Gulf and Encounter Bay. *Trans. R. Soc. S. Austr.* **96** (1): 1-52.

KOTT, P. (1980): Algal-Bearing Didemnis Ascidians in the Indo-West Pacific. *Mem. Queensland Mus.* **20** (1): 1-47, pls. 1-4.

KOTT, P. (1981a): Didemnid-algal symbiosis: Algal transfer to a new Host Generation. *Proceeding of the Fourth International Coral reef Symposium, Manila, 1981*, Vol. 2: 721-723.

KOTT, P. (1981b): Replication in the Ascideacea: an adaptive strategy in the Coral Reef environment. *Proceeding of the Fourth International Coral reef Symposium, Manila, 1981* Vol. 2: 725-733.

KOTT, P. (1981c): The Ascidians of the Reef Flats of Fiji. *Proc. Linnean Soc. N.S.W.* **105** (3): 147-212.

KOTT, P. (1982): Didemnid-Algal Symbiosis: Host Species in the Western Pacific with notes on the Symbiosis. *Micronesica* **18** (1): 95-127.

KOTT, P. (1983): Related speciews of Trididemnum in Symbiosis with Cyanophyta. *Proc. Linnean Soc. N.S.W.* **107** (4):515-520.

KOTT, P. (1985): The Australian Ascidiacea. Part 1, Phlebobranchia and Stolidobranchia. *Mem. Qld. Mus.* **23** :1-440.

KOTT, P. (1988): Co-Symbiosis in the Ascidiacea. *Bull. Mar. Sci.* **42** (1): 149-153.

KOTT, P. (1989): Form and function in the Ascidacea. *Bull. Mar. Sci.* **45** (2): 253-276.

KOTT, P. (1990): The Australian Ascidiacea. Part 2, Aplousobranchia (1). *Mem. Qld. Mus.* **29**: 1-298.

KOTT, P. (1992): The Australian Ascidiacea. Part 3, Aplousobranchia (2). *Mem. Qld. Mus.* **32**: 375-655.

KOTT, P. (2001). The Australian Ascidiacea. Part 4, Aplousobranchia (3), Didemniidae. *Mem. Qld. Mus.* **47** (1): 1-410.

KOTT, P., D. L. PARRY & G. C. COX (1984): Procaryotic Symbionts with a range of Ascidian Hosts. *Bull. Mar. Sci.* **34** (2): 308-312.

KRÜGER, F. J. (1997): Brutpflege bei Seeigeln. *Natur und Museum* **127** (4): 202-211.

KRÜGER, F. J. (1998): Bioerosion durch reguläre Echiniden. *Natur und Museum* **128** (3): 65-73.

LALLI, C. M. & R. W. GILMER (1989): *Pelagic Snails. The Biology of Holoplanktonic Gastropod Molluscs.* Stanford University Press. Stanford, USA.

LAMPRELL, K. (1987): *Spiny oyster shells of the world. Spondylus.* Robert Brown, Bathurst, NSW, Australia.

LAMPRELL & WHITEHEAD (1992): *Bivalves of Australia* Vol. 1. Crawford House Press; Bathurst, NSW, Australia.

LANDMAN, N. H., J. KIRK COCHRAN, D. M. RYE, K. TANABE & J. M. ARNOLD (1994): Early life history of evidence from isotopic analyses of aquarium-reared specimens. *Paleobiology* **20** (1): 40-51.

LANGE, J. & R. KAISER (1991): *Niedere Tiere tropischer und kalter Meere.* Ulmer Verlag, Stuttgart, Germany.

LASERON, C. F. (1959): The Family Pyramidellidae (Mollusca) from Northern Australia. *Austr. Jour. Mar. Freshw. Res.* **10**: 177-267.

LAXTON, J. H. (1971): Feeding in some Australian Cymantiidae (Gastropoda: prosobranchia). *Zool. Jour. Linnean Soc.* **50**: 1-9.

LEMCHE, H. (1959): Molluscan phylogeny in the light of Neophilina. *Proc. Int. 15th Int. Cong. Zool. London.*

LEMCHE, H. & K. G. WINGSTRAND (1959): The anatomy of Neophilina galatheae. *Galathea Rep.* **3**: 9-71.

LEWIN, R. A. (1986): The Phylogeny of Prochloron. *Giorn. Bot. Ital.* **120**: 1-14.

LEWIN, R. A. (1977): Prochloron, type genus of the Prochlorophyta. *Phycologia* **16**: 217.

LEWIS, A. D. & E. LEDUA (1988): A possible new species of Tridacna (Tridacnidae: Mollusca) from Fiji. In: COPLAN, J. W. & J. S. LUCAS (eds.): *Giant Clams in Asia and the Pacific.* ACIAR Monograph No. 9.

LINDAL, U. (1998): Low-tech rehabilitation of degraded coral reefs through transplantation of staghorn corals. *Ambio* **27**: 645-650.

LINDNER, G. (1976): *Snegler og muslinger fra alle hav.* J. W. Cappelens Forlag, Oslo, Norway.

LING, S.-W. (1977): *Aquaculture in Southeast Asia - A Historical Overview.* Washington Sea Grant Publication. University of Washington Press, Seattle, USA.

LUCAS, J. S., E. LEDUA & R. D. BALEY (1990): A new species of giant clam (Tridacnidae) from Fiji and Tonga. *ACIAR Working Paper No. 00.*

MADDOCKS, R. R. (1987): An ostracode commensal of an ophiuroid and other new species of Pontocypria (Pontocopida: Cypridacea). *Jour. Crust. Biol.* **7**: 727-737.

MADSEN, F. J. (1955): A note on the sea star genus Acanthaster. *Vidensk. Medd. dansk naturh. Foren* **117**: 179-192.

MARSH, L. & S. SLACK-SMITH (1986): *Sea Stingers.* Western Australian Museum, Perth, Australia.

MARSHALL, J. F. (1996): Calcification in Hermatypic and Ahermatypic Corals. *Science* **271**: 637-639.

MARSHALL, N. B. (1979): *Developments in Deep-Sea Biology.* Blandford Press, Poole, Dorset, UK.

MASSIN, C. L. & D. J. W. LANE (1991): Description of a new species of sea cucumber (Stichopodidae, Holothuroidea, Echinodermata) from the eastern Indo-Malayan Archipelago: Thelenota rubralineata n. sp. *Micronesia* **24** (1): 57-64.

MATHER, P. & I. BENNETT (1993): *A Coral Reef Handbook*, 3. ed.. Surrey Beatty and Sons Pty. Ltd.,Chipping Norton, NSW, Australia.

McCLANAHAN, T. R. (1990): Kenyan coral reef associated gastropod assemblages: distribution and diversity patterns. *Coral Reefs* **9** (2): 63-74.

McCLINTOCK, J. B. & J. JANSSEN (1990): Pteropod abduction as a chemical defence in a pelagic Antarctic amphipod. *Nature* **346**: 462-464, 2 tabs, 2 figs.

McPHERSON, B. F. (1968): Contributions to the biology of the sea urchin Eucidarus tribuloides (Lamarck). *Bull. Mar. Sci.* **18**: 400-443.

MEBS, D. (1989): *Gifte im Riff.* Wissenschaftliche Verlaggesellschaft mbH, Stuttgart, Germany.

MEBS, D. (1992): *Gifttiere.* Wissenschaftliche Verlaggesellschaft mbH, Stuttgart, Germany.

MEINKOTH, N. A. (1994): *National Audubon Society Field Guide to North American Seashore Creatures*, 7th printing. Alfred A. Knopf, Inc., N. Y., USA.

MERGNER, H. (1979): Quantitative ökologische Analyse eines Rifflaguneareals bei Aqaba, (Golf von Aqaba, Rotes Meer). *Helgol. Wiss. Meeresunter.* **32**: 476-507.

MILADENOV, P. V., R. H. EMSON, L. V. COLPITTS & I. C. WILKIE (1983): Asexual reproduction in thc West-Indian brittle star Ophiocomella ophiactoides (H. L. Clark) (Echinodermata: Ophiuroidea). *Jour. Exp. Mar. Biol. Ecol.* **72**: 1-23.

MILLAR, R. H. (1963): Australia Ascidian in the British Museum (Natural History). *Proc. Zool. Soc. London* **141** (4): 689-746.

MILLAR, R. H. (1975): Ascidians from the Indo-West Pacific region in the Zoological Museum, Copenhagen (Tunicata, Ascidiacea). *Steenstrupia* **3** (20): 205-336.

MILLOTT, N. (1956): The covering reaction of sea urchins. 1. A preliminary account of covering in the tropical echinoid Lytechinus variegatus (Lamarck), and its relation to light. *Jour. Exp. Biol.* **33**: 508-523.

MIRONOV, A. N. (1971): Soft Sea-urchins of the family Echinothuriidae collected by the RV Vityaz and Akademik Kurchatov in the Pacific and Indian Oceans. *Fauna of the Kurile-Kamchatka Trench. Tr. Inst. Okeanol.* **92**: 317-325.

MOE, M. Jr. (1992): *The Marine Handbook - Beginner to Breeder*, 2nd ed. Green Turtle Publication, Florida, USA.

MOE, M. Jr. (1997): *Breeding the Orchid Dottyback*. Green Turtle Publication, Florida, USA.

MOEN, F. E. & E. SVENSEN (2000): *Dyreliv i Havet*, 2nd ed. Kom Forlag, Kristiansund, Norway.

MOFFAT, D., M. N. NGOILE, O. LINDEN & J. FRANCIS (1998): The reality of the stomach: Coastal management at the local level in Eastern Africa. *Ambio* **27**: 590-598.

MOORE, R. C. (ed.) (1960): *Treatise on Invertebrate Palentology.* Part 1. Mollusca 1. University of Kansas Press and Geological Society of America Inc., USA.

MOOSLEITNER, H. (1997): Shallow water Asteroida (Echinodermata) from the coral reefs of the Maldive Islands. *Revue fr. Aquariol.* **24** (1-2): 3-14.

MORIN, J. G. (1988): Piscivorous behavior and activity patterns in the tropical ophiuroid Ophiarachna incrassata (Ophiuroidea: Ophiodermatidae). In: BURKE, R. D., P. V. MLADENOV, P. LAMBERT & R. L. PARSLEY (eds.): *Echinoderm Biology*: 401-407. Balkema, Rotterdam, The Netherlands.

MORTENSEN, T. (1938): Contributions to the study of the development and larval forms of Echinoderms, IV. *K. danske Vidensk. Selsk. Skr. (naturv.-math.) (9)* **7** (3): 1-59.

MORTON, J. E. (1958): Observations on the gymnosomatous pteropod Clione limacina (Phipps). *Jour. Mar. Biol. Assoc. U. K.* **37**: 287-297.

MOUGIN, J. (1979): Reproduction en aquarium d'une Ophiure tropicale (Ophiarachna incrassata Lam.) *Rev. fr. Aquariol.* **6** (1): 15-16.

NESIS, K. N. (1987): *Cephalopods of the World.* TFH Publications, Inc., Neptune City, N. J., USA.

NEWCOMB, E. H. & T. D. PUGH (1975): Blue-green algae associated with ascidians of the Great Barrier Reef. *Nature* **253**: 533-534.

NEWMAN, L. J. (1987): Swarms of sea-butterflies. *Hawaiian Shell News* **35** (12): 3.

NEWMAN, L. J. (1990): The taxonomy, distribution and biology of Atlanta gaudichaudi Souleyet (Gastropoda, Heteropoda) from the waters of the Australian Great Barrier Reef. *Amer. Malacol. Bull.* **8**: 85-94.

NEWMAN, L. J. (1998a): Superfamily Carinioidea. In: BEESLEY, P., G. J. B. ROSS & A. WELLS (eds.): *Mollusca. Fauna of Australia.* Vol. 5. Part B: 804-808. Australian Biological Resource Study, Canberra, ACT, Australia.

NEWMAN, L. J. (1998b): Order Thecosomata. In: BEESLEY, P., G. J. B. ROSS & A. WELLS (eds.): *Mollusca. Fauna of Australia.* Vol. 5. Part B: 980-985. Australian Biological Resource Study, Canberra, ACT, Australia.

NEWMAN, L. J. (1998c): Order Gymnosomata. In: BEESLEY, P., G. J. B. ROSS & A. WELLS (eds.): *Mollusca Fauna of Australia.* Vol. 5. Part B: 985-989. Australian Biological Resource Study, Canberra, ACT, Australia.

NEWMAN, L. J. & J. G. GREENWOOD (1988): A new gymnosomatous pteropod, Pneumodermopsis spoeli nov. sp, (Gastropoda, Opisthobranchia) from the Great Barrier Reef, Australia. *The Veliger* **31** (1/2): 75-79.

NILSEN, A. J. (1999): A small experimental Aquarium. *Marine Fish and Reef* **1** (1):102-120.

NILSEN, A. J. (2000): Ein Experimentier-Aquarium, parts 1-4. *das Aquarium* 1: **34** (3): 48-55, 2: (4): 44-50, 3: (5): 50-56, 4: (6): 52-57.

OGDEN, J. C. (1977): Carbonate-sediment production by parrot fish and sea urchins on Caribbean reefs. *Studies in Geology* **4**: 281-288.

OTT, M. (1989): Ein seltener Aquariengast. Der Gemeine Krake, Octopus vulgaris. *das Aquarium* **23** (240): 363-366.

OVERATH, H. (1981): Verbesserte Aufzuchtmöglichkeiten von Sepia officinalis. *das Aquarium* **15** (148): 537-540.

PARRY, D. L. (1985): Nitrogen assimilisation in the symbiotic marine algae Prochloron spp. *Marine Biology* **87** (2): 19-22.

PARRY, D. L. (1987): *Selected Chemistry of the Ascidian.* Ph. D. thesis: University of Queensland, Australia.

PARRY, D. L. & P. KOTT (1988): Co-Symbionts in the Ascidiacea. *Bull. Mar. Sci.* **41** (1): 149-153.

PETEUCH, E. J. & D. M. SARGENT (1986): *Atlas of the Living Oliv Shells of the World.* Coastal Education and Research Foundation, Inc., Fort Lauderdale, Florida, USA.

PONDER, W. & A. WAREN (1988): Prosobranch Phylogeny. *Malacol. Rev.* Supplement 4.

POWELL, A. B. W. (1964): The family Turridae in the Indo-Pacific. Part 1. The subfamily Turrinae. *Indo-Pacific Mollusca* **1** (5): 227-345.

POWELL, A. B. W. (1967): The family Turridae in the Indo-Pacific. Part 1. The subfamily Turrinae concluded. *Ibid* **1** (7): 409-431.

POWELL, A. B. W. (1969): The family Turridae in the Indo-Pacific. Part 1. The subfamily Turriculinae. *Ibid* **2** (10): 207-415.

PRASHAD, B. (1932): The Lamellibranchia of the Siboga Expedition, Systematic Part, II, Pelecypoda (exclusive of the Pectinidae). *Siboga-Expeditie Vol. xxxiv, Livre LIIIc.*

PYLE, R. L. (1991a): Ein Schatz aus der Zwielichtzone. *Unterwasserfotografie* (3):114-117.

PYLE, R. L. (1991b): Rare and Unusual Marines: The Hawaiian deep anthias Holanthias fuscipinnis (Jenkins). *FAMA* **14** (12):74-78.

PYLE, R. L. (1992): The peppermint angelfish: Centropyge boylei, n. sp. Pyle and Randall. *FAMA* **15** (7): 16-18.

PYLE, R. L. (1994) Rare and unusual marines: The Narc Angelfish Centropyge narcosis Pyle and Randall. *FAMA* **7** (4): 8-19.

PYLE, R. L. (1996a): How much coral reef biodiversity are we missing? *Global biodiversity* **6** (1): 3-7.

PYLE, R. L. (1996b): The Twilight Zone. *Natural History* **105** (11): 59-62.

PYLE, R. L. (2000): Assessing Undiscovered Fish Biodiversity on Deep Coral Reefs Using Advanced Self-Contained Diving Technology. *Mar. Techn. Soc. Jour.* **34** (4): 82-91.

PYLE, R. L. & J. E. RANDALL (1992): A new species of Centropyge (Perciformes: Pomacanthidae) from the Cook Islands, with a redescription of C. boylei. *Revue fr. Aquariol.* **19** (4): 115-124.

RANDALL, J. E. & R. L. PYLE (2001a): Three new species of labrid fishes of the genus Cirrhilabrus from islands of tropical Pacific. *Aqua* **4** (3): 89-98.

RANDALL, J. E. & R. L. PYLE (2001b): Four new serranid fishes of the anthiine genus Pseudanthias from the South Pacific. *Raffles Bull. Zool.* **49** (1): 19-34.

REISINGER, E. & S. KELBETZ (1964): Feinbau und Entlandungsmechanismus der Rhabditen. *Z. Wiss. Mikroskopie und mikr. Technik* **65** (8): 472-508.

RICHTER, G. & R. R. SEAPY (1999): Heteropoda, pp. 621-647. In: BOLTOVSKOY(ed.): *South Atlantic Zooplankton*, Vol 1. Bacyuys Publishers: Leiden, The Netherlands.

ROSEWATER, J. (1965): The family Tridacnidae in the Indo-Pacific. *Indo Pacific Mollusca* **1** (6): 347-395.

ROSEWATER, J. (1982): A new species of Hippopus (Bivalvia: Tridacnidae). *The Nautilus* **96** (1): 3-6.

ROTHANS, T. C. & A. C. MILLER (1991): A link between biologically imported particulate organic nutrients and the detritus food web in reef communities. *Mar. Biol. (Berlin)* **110**: 145-150.

RUDMAN, W. B. (1973): On some species of Chelidonura (Ophistobranchia: Aglajidae) from Zanzibar and Fiji. *Zool. Jour. Linnean Soc.* **52**: 201-215.

RUDMAN, W. B. (1974): A comparison of Chelidonura, Navanax and Aglaja with other genera of Aglajidae (Ophistobranchia: Gastropoda). *Zool. Jour Linnean Soc.* **54**: 185-212.

RUMRILL, S. S. (1982): *Contrasting reproductive patterns among ophiuroids (Echinodermata) from Southern Monterey Bay, USA.* M.S. Thesis, Univ. of California, Santa Cruz, Ca. USA

RUPPERT, E. E & R. D. BARNES (1994): *Invertebrate Zoology*, 6th ed. Saunders College Publishing, N. Y., USA.

RUPPERT, E. & R. FOX (1988): *Seashore Animals of the Southeast. A Guide to Common Shallow-water Invertebrates of the Southeastern Atlantic Coast.* University of South Carolina Press, Columbia SC, USA.

RUSSO, R. (1994): *Hawaiian Reefs.* Wavecrest Publications, San Leandro, Ca, USA.

RYLAARSDAM, K. W. (1983): Life histories and abundance patterns of colonial corals on Jamaican reefs. *Mar. Ecol. Progr. Ser.* **13**: 249-260.

SABBADIN, A. (1979): Ascidian colonial structure and genetics. In: LARWOOD, G. L. & R. B. ROSEN (eds.): *Biology and Systematics of Colonial Organisms*: 433-444. Academic Press, London, U.K.

SABBADIN, A. & G. GRAZIANI (1967): New data on the inheritance of pigments and pigmentation patterns in the colonial ascidian Botryllus schlosseri. *Riv. Biol.* **60**: 559-598.

SAUNDERS, W. B. (1981a): A new species of Nautilus from Palau. *The Veliger* **24** (1): 1-7.

SAUNDERS, W. B. (1981b): The species of living Nautilus and their distribution. *The Veliger* **24** (1): 8-18.

SAUNDERS, W. B. (1987): The species of Nautilus. In: SAUNDERS, W. B. & N. H. LANDMAN (eds.): *Nautilus. The Biology and Paleobiology of a Living Fossil.* Plenum Press, N.Y., USA.

SCHEIBLING, R. E. (1982): Feeding habits of Oreaster reticulatus (Echinodermata: Asteroidea). *Bull. Mar. Sci.* **32**: 504-510.

SEAPY, R. R. (1974): Distribution and abundance of the epipelagic heteropod mollusk Carinaria japonica in the waters off southern California. *Marine Biology* **24**: 243-250.

SEAPY, R. R. (1996): Atlantid heteropods as living substrata for eggs of Halobates sericcus (Heteroptera: Gerridae). *Pacific Science* **50** (1): 84 -92.

SEAPY, R. R. & C. SKOGLUND (2001): First records of atlantid heteropod mollusks from the Golfo de California. *The Festivus* **34**: 33-44.

SEAPY, R. R. & R. E. YOUNG (1986): Concealment in epipelagic pterotracheid heteropods (Gastropoda) and cranchiid squids (Cephalopoda).*J. Zool. London* (A) **210**: 137-147.

SEMB-JOHANSSON, A. (ed.) (1988): *Verdens Dyr, Band 13, Virvelløse Dyr II.* Cappelens Forlag, Norway.

SHEPPARD, A. L. S. (1984): The Molluscan Fauna of Chagos (Indian Ocean) and an Analysis of its Broad Distribution Patterns. *Coral Reefs* **3** (1): 43-50.

SHOKITA, S. (ed.) (1991): *Aquaculture in Tropical Areas.* Midori Shobo Co., Ltd., Tokyo, Japan.

SIRENKO, D. I. & O. A. SCARLATO (1991): Tridacna rosewateri sp.n. A new species of giant clam from Indian Ocean. *La Conchiglia* **261**: 4-9.

SLUITER, C. P. (1909): Die Tunicata der Siboga Expedition Pt. 2. Die merosomen Ascidien. *Siboga Exped.* **56B**: 1-112.

SMITH, D. C. & A. E. DOUGLAS (1987): *The Biology of Symbiosis.* Edward Arnold, London, UK.

SMITH, S. M. & D. HEPPEL (1991): *Checklist of British Marien Mollusca.* National Museums of Scotland Information, Series No. 11.

SOEST, R. W. M. VAN (1972): Latitude variation in Atlantic Salpa fusiformis Cuvier, 1804 (Tunicata, Thaliacea). *Beaufortia* **20** (262): 59-68.

SOEST, R. W. M. VAN (1973a): The genus Thalia Blumenbach, 1798 (Tunicata, Thaliacea) with description of two new species.*Beaufortia* **20** (271): 193-212.

SOEST, R. W. M. VAN (1973b): A new species in the genus Salpa Forskål, 1775, (Tunicata, Thaliacea). *Beaufortia* **21** (272): 9-15.

SOEST, R. W. M. VAN (1974a): A revision of the genera Salpa Forskål, 1775, Pegea Savigny, 1816, and Ritteriella, Metcalf, 1919, (Tunicata, Thaliacea). *Beaufortia* **22** (293): 153-191.

SOEST, R. W. M. VAN (1974b): Juvenile colonies of the genus Pyrostremma Garstang, (1929 (Tunicata, Thaliacea). *Bull. Zool. Mus. Univ. Amsterdam* **4** (4): 23-33.

SOEST, R. W. M. VAN (1975a): Observations on taxonomy and distribution of some salps (Tunicata, Thaliacea) with description of three new species. *Beaufortia* **23** (302): 105-130.

SOEST, R. W. M. VAN (1975b): Zoogeography and speciation in the Salpidae (Tunicata, Thaliacea). *Beaufortia* **23** (307): 181-215.

SOEST, R. W. M. VAN (1979): Revised classification of the order Pyrosomatida (Tunicata, Thaliacea), with description of a new genus. *Steenstrupia* **5** (11): 197-217.

SPIES, G. (1983): Ein verspieltes „Ungeheuer" Der Gemeine Krake ist zutraulich und intelligent. *das Aquarium* **17** (168): 315-318.

SPOEL, S. VAN DER (1967): *Euthecosomata, a Group with Remarkable Developmental Stages (Gastropoda, Pteropoda).* J. Noorduijn en Zoon. N.V., Gorinchem, The Netherlands.

SPOEL, S. VAN DER (1976): *Pseudothecosomata, Gymnosomata and Heteropoda (Gastropoda).* Bohn, Scheltema, and Holkema, Utrecht, The Netherlands.

SPOEL, S. VAN DER (1992): A note of Pteropod ink. *Beaufortia* **56**: [illegible]-103.

SPOEL, S, VAN DER & R. P. HEYMAN (1983): *A Comparative Atlas of Zooplankton - Biological Patterns in the Oceans.* Springer-Verlag, Berlin, Germany.

SPOEL, S. VAN DER, L. J. NEWMAN & K. ESTEP (1996): *Pelagic Molluscs of the World.* ETI, Amsterdam, The Netherlands. CD-ROM Series 5. 325 pp. + 2000 images.

SPRUNG, J. & C. DELBEEK (1997): *The Reef Aquarium*, Vol. 2. Ricordea Publ., Miami, Florida, USA.

STODDART, J. (1989): Fatal Attraction. *Landscope* **4** (4): 14-20.

TAYLOR, J. D. (1971): Reef Associated Molluscan Assemblanges in the Western Indian Ocean. *Symp. Zool. Soc. London* **28**: 501-534.

TESCH, J. J. (1949): Heteropoda. *Dana Report* **34**: 1-53.

TESCH, J. J. (1950): The Gymnosomata. II. *Dana Report* **36**: 1-55.

TOMEY, W. A. (1988): Kraken. *das Aquarium* **22** (225): 163-167.

TREEK, P. VAN & H. SCHUHMACHER (1997): Initial survey of coral nubbins transplanted by a new coral transplantation technology: options for reef rehabilitation. *Mar. Ecol. Prog. Ser. 150*: 287-292.

TRENCH R. K., D. S. WETHEY & J. W. PORTER (1981): Some observations on the symbiosis with zooxanthellae among Tridacnidae (Mollusca: Bivalvia). *Biol. Bull.* **161**: 180-198.

TURNER, S. J. (1994): The biology and population outbreaks of the corallivorous gastropod Drupella on Indo-Pacific reefs. *Oceanogr. Mar. Biol. Ann. Rev.* **32**: 461-530.

VALENTINE, J. W. (1973): *Evolutionary paleoecology of the Marone Hemisphere.* Prentice-Hall, Inc., N. J., USA.

VAUGHT, K. C. (1989): *A Classification of the Living Mollusca.* American Malacologists Inc., Melbourne, Florida, USA.

VINE, P. (1986): *Red Sea Invertebrates.* Immel Publ., London, UK.

WALBRAN, P. D., R. A. HRNDERSON, J. W. FAITHFUL, H. A. POLACH, R. J. SPARKS, G. WALLACE & D. C. LOWE (1989): Crown of thorns starfish outbreaks on the Freat Barrier Reef: A geological perspective based upon the sediment record. *Coral Reefs* **8** (2): 67-78.

WALLACE, R. A., J. L. KING & G. P. SANDERS (1986): *Biology the Science of Life.* Scott, Forseman and Company, London, U.K.

WALLS, J. G. (1982): *Encyclopedia of Marine Invertebrates.* TFH Publications, Inc., Neptune City, N. J., USA.

WARD, P. & W. B. SAUNDERS (1997): Allonautilus: a new genus of living nautiloid cephalopod and its bearing on phylogeny of the Nautilida. *Jour. Paleontol.* **71**: 1054-1064.

WEBB, D. A. (1939): Observations on the blood of certain ascidians with special references to the biochemistry of vanadium. *Jour. exp. Biol.* **16**: 499-523.

WELLS, F. E. & C. W. BRYCE (1988): *Seashells of Western Australia.* Western Australian Museum, Perth, Australia.

WELLS, F. E. & C. W. BRYCE (1993): *Seaslugs of Western Australia.* Western Australian Museum, Perth, Australia.

WELLS, H. W., M. J. WELLS & I. E. GRAY (1961): Food of the sea-star Astropecten articulatus. *Biol. Bull. (Woods Hole)* **120**: 265-271.

WHITE, A. T. & A. CRUZ-TRINIDAD (1998): *The values of Philippine coastal resources: Why protection and management are critical*. Coastal Resource Management Project, Cebu City, Philippines.

WILKENS, P. (1987): *Niedere Tiere im tropischen Seewasseraquarium*, Band 2, 2nd ed. Engelbert Pfriem Verlag, Wuppertal-Elberfeldt, Germany.

WILKERSON, J. (1997): C-Quest – The Leading Marine Aquarium Fish Hatchery for 1997. *Aquarium Frontiers On-Line*; July (1997). (http://www.aquariumfrontiers.com).

WILKINSON, C. R. (1990): Acanthaster planci. *Coral Reefs* **9** (3): 93-172

WILKINSON, C. R. & J. G. MACINTYRE (eds.) (1992): The Acanthaster debate. Special Issue. *Coral Reefs* **11** (2): 51-122.

WILSON, B. (1993/1994): *Australian Marine Shells*, Vol. 1 & 2. Odyssey Publishing, Kallaro, Western Australia.

YONGE, C. M. (1936): Mode of life, feeding, digestion and symbiosis with zooxanthellae in the Tridacnidae. *Sci. Rep. Great Barrier Reef Exped.* **1** (11): 283-321.

YONGE, C. M. (1975): Giant Clams. *Scientific American* **23**: 96-105.

YONGE, C. M. (1980): Functional morphology and evolution in the Tridacnidae (Mollusca; Bivalvia; Cardiacea). *Rev. Austral. Mus.* **33** (17): 735-777.

ZANN, L. P. (1980): *Living together in the Sea*. TFH Publications, Inc., Neptune City, N. J., USA.

ZANN, L. P. et al. (1987): The Crown of Thorns Starfish. Special Feature. *Austr. Sci. Mag.* 1987 (3): 14-54.

Index

All scientific genera and species names are printed in *italics*. Page numbers printed in **bold** types refer to illustrations and/or major discussions of the topic. The Letter 'S' behind a page number indicates that the page in qustion contains a summary (Fact Sheet) on the taxon in question.